TOPOLOGICAL VECTOR SPACES, DISTRIBUTIONS AND KERNELS

Pure and Applied Mathematics

A Series of Monographs and Textbooks

Edited by

Paul A. Smith and Samuel Eilenberg

Columbia University, New York

Pure and Applied Mathematics

A Series of Monographs and Textbooks

TOPOLOGICAL VECTOR SPACES, DISTRIBUTIONS AND KERNELS

François Treves

PURDUE UNIVERSITY
LAFAYETTE, INDIANA

1967

ACADEMIC PRESS New York • London

A Subsidiary of Harcourt Brace Jovanovich, Publishers

ACADEMIC PRESS, INC.
111 Fifth Avenue, New York, New York 10003

United Kingdom Edition published by
ACADEMIC PRESS, INC. (LONDON) LTD.
Berkeley Square House, London W1X 6BA

LIBRARY OF CONGRESS CATALOG CARD NUMBER: 66-30175

Fourth Printing, 1973

PRINTED IN THE UNITED STATES OF AMERICA

To Gioia and Dario Treves

Preface

This is not a treatise on functional analysis. It is a set of lecture notes aimed at acquainting the graduate student with that section of functional analysis which reaches beyond the boundaries of Hilbert spaces and Banach spaces theory and whose influence is now deeply felt in analysis, particularly in the field of partial differential equations. It is admittedly with an eye on this kind of application that the material has been selected. To the student who works his way through Part I (Topological Vector Spaces—Spaces of Functions) and Part II (Duality —Spaces of Distributions), all the essential information on these subjects has been made available. Part I starts at a very elementary level by recalling the definitions of a vector space and of a topological space; later, the completion of a topological vector space is described in detail. Inevitably, the difficulty of the reading increases, but I have tried to make this increase very gradual, at the risk of irritating some readers with my overexplanations. The student who has gone as far as Chapters 31 and 32 should jump, at this point, to Part III and read the first three or four chapters there—so as to learn something, if he has not done so yet, about tensor products and bilinear mappings. Further progress, through the remainder of Part III (Tensor Products—Kernels), as well as the end of Part II, although not difficult in any sense, presupposes a good assimilation of the exposition that precedes.

For Parts I and II, the prerequisites are a standard course in point-set topology, a decent undergraduate course on the theory of functions of one complex variable, and a standard course on the theory of functions of real variables and Lebesgue integration. Some knowledge of measure theory is assumed here and there. No serious result of linear algebra (such as the reduction to Jordan canonical form) is ever used, but it is clear that the student who has never heard of linear spaces or linear mappings should be deterred from opening this book.

Pedagogical considerations have been given dominant weight, sometimes at the expense of systematic exposition. I have made a special point of breaking the monotony of the text by alternating topics—from functional analysis to analysis proper and back again. In teaching this material, I have found that such tactics were successful. The book concentrates on what, in the author's opinion, are key notions and key results (quotient spaces, transposes, the open mapping theorem, and the theorems of Hahn-Banach, Banach-Steinhaus, and Mackey, to mention only a few examples in functional analysis and not to speak of Part III). Other concepts have seemed to be of less crucial importance, or else easier to reconstruct from the context, and are learned only by use, without any formal definition (this applies, for instance, to product spaces, to linear subspaces, and to induced topologies). Many important concepts are missing. There are various reasons for this. Sometimes these concepts partake of a degree of specialization higher than the surrounding material (e.g., bornological spaces versus barreled spaces, Laplace transform versus Fourier transform). In other instances, they have seemed somewhat alien to the general trend of the book (I think mainly here of ordered topological vector spaces and of extremal points of compact sets). But the most compelling factor, at the root of those omissions, has been the lack of space. Some standard notions are not defined until the very moment they are needed: such is the case of general inductive and projective limits and topological direct sums, whose definitions will not be found before Chapter 49! All this is based on the belief that, once the main strongholds are secured, the conquest of larger territory should not prove difficult. It does however require further reading and recourse to the true treatises on the subject (see General Bibliography at the end of the book).

The advanced theory of Hilbert spaces and Banach spaces, and of their linear operators, constitutes, in a sense, the other wing of functional analysis. Its importance and depth cannot be overestimated. But there is no reason why I should have embarked on a description, even fragmentary, of this theory. Its ramifications toward C^* and von Neumann algebras, its applications to harmonic analysis and to group representation, give it a highly distinctive character that sets it aside. Furthermore, there are excellent books on the subject. Most graduate schools offer a course on spectral theory—which is more than they do for the brand of functional analysis upheld here! This is not to say, of course, that the basic facts about Hilbert and Banach spaces are not to be found in this book. They are duly presented and, because of their great importance, the applications of the general theorems to these spaces are carefully pointed out. The spaces L^p and l^p are looked at from a variety of viewpoints, and much of the "illustrating" material originates with them.

The contents of Parts I and II correspond, more or less, to a one-year course given at Purdue University. Some additions are aimed at making the book self-contained. The subdivision of Part III is more fictitious, as I have never taught a course on topological tensor products. Part III has been added because I firmly believe that analysts should have some familiarity with tensor products, their natural topologies, and their completions.

I wish to thank Mrs. Judy Snider who typed, with great competence, the manuscript of Parts I and II.

FRANÇOIS TREVES

April 1967
Paris

Contents

PART I

Topological Vector Spaces. Spaces of Functions

Part I is devoted to the basic definitions and properties about topological vector spaces, with no detailed reference to duality. We use filters quite systematically, after having introduced them in Chapter 1 (which is devoted to recalling what a topological space is, with emphasis on the filter of neighborhoods of a point). Chapter 2 recalls what a vector space is, without going into any algebraic subtlety, as we shall never need any in what follows. The scalars are always complex numbers; no other field is ever considered (we shall switch to the field of real numbers for a very short moment, when talking about the theorem of Hahn-Banach, in Part II, and switch quickly back to complex numbers). Chapter 3 makes the synthesis, in a sense, combining the topology with the linear structure, under the natural compatibility conditions. In relation to this, I have thought that the students might like to see a meaningful example of a topology on an algebra A which does *not* turn A into a topological vector space. Such an example is provided by the algebra $C[[X]]$ of formal power series, equipped with the topology defined by the powers of the maximal ideal. It is instructive to compare it with the topology of simple convergence of the coefficients, noting that $C[[X]]$ is metrizable and complete for both topologies. Chapter 4 is devoted essentially to quotient topological vector spaces; quotient spaces will be much needed in Part II, when they will be carrying many a different topology, and it is the experience of the author that this is what students find most difficult: to visualize and manipulate correctly the various useful topologies, related to weak duality, on quotients of duals and on duals of quotients, and this is unfortunately inevitable.

Chapter 5 is devoted to the completion of a (Hausdorff) topological vector space; the theorem on completion is proved in all details. Although it is by no means a deep theorem, I have chosen to devote an entire Chapter to it so as to familiarize the reader with Cauchy filters, and also because I have too often been faced with audiences who simply did not know what a complete (nonmetric) uniform space is.

Chapter 6 is devoted to compact subsets of a topological vector space. Chapter 7 introduces locally convex spaces and seminorms. Needless to say we shall not be dealing with any TVS that is not locally convex. In Chapter 8 we look at metrizable TVS, that is to say TVS whose

3

topology can be defined by a metric. We construct a metric when the underlying space is locally convex; we do this out of a sense of duty, because people in a nondistant past liked to think in terms of metrics. Inspection of the reasoning shows quickly that, most of the time, the consideration of a metric is unnecessary, that countable bases of neighborhoods would do. Furthermore, metrics have generally the drawback that the balls are not convex; and outside of local convexity there is little hope for salvation! There is no doubt that the proof of the open mapping theorem would be quite awkward without a metric, but the actual form of the latter is irrelevant (provided that it is translation invariant). The criteria about continuity, compactness, and completeness in metrizable spaces are given, together with the proof of the fact that every complete metrizable space is a Baire space. In Chapter 9 we study rapidly the most primitive examples of metrizable and complete TVS: the finite dimensional Hausdorff spaces! Linear subspaces of finite codimension are considered, as they will be useful later, and the correspondence between closed hyperplanes and kernels of continuous linear functionals is shown. The next three chapters are devoted to examples of locally convex metrizable spaces. These are (in decreasing order of generality): Fréchet spaces, normed and Banach spaces, Hilbert spaces. Fréchet spaces will be the most important topological vector spaces for our purposes. About Banach and Hilbert spaces, the student should expect to find in this book only the most basic and elementary information: for instance, in the chapter devoted to Hilbert spaces we prove the projection theorem and its consequence, the fundamental theorem of Hilbert space theory: the canonical isomorphy between a Hilbert space and its antidual. After this we describe rapidly (finite) Hilbert sums and orthonormal bases. The basic examples of F-spaces, B-spaces, and Hilbert spaces are introduced:

$\mathscr{C}^k(\Omega)$, $L^p(\Omega)$ (Ω, open subset of $\mathbf{R}^n$; $0 \leqslant k \leqslant \infty$, $1 \leqslant p \leqslant \infty$);

$H(\Omega)$, space of holomorphic functions in an open subset Ω of $\mathbf{C}^n$;

$\mathscr{C}^0(K)$, $\mathscr{C}^k(\bar{\Omega})$ (K, compact subset of $\mathbf{R}^n$; Ω, bounded open subset of $\mathbf{R}^n$);

$\mathscr{S}$, space of $\mathscr{C}^\infty$ functions in $\mathbf{R}^n$, rapidly decaying at infinity.

Chapter 13 is devoted to a class of locally convex spaces which are not metrizable (in general) but which are of great importance to us: the spaces LF (strict inductive limits of a sequence of F-spaces); as examples, we present the space of polynomials $\mathbf{C}[X]$ (as inductive limit of the finite dimensional spaces of polynomials of bounded degree) and the spaces of functions with compact support, $\mathscr{C}^k_c(\Omega)$, $L^p_c(\Omega)$. The space $\mathscr{C}^\infty_c(\Omega)$ will be the space of *test functions* on which *distributions* in Ω will be defined: the distributions in Ω are the continuous linear functionals on $\mathscr{C}_c^\infty(\Omega)$, i.e., the elements of the dual of $\mathscr{C}_c^\infty(\Omega)$.

After this string of examples of spaces of functions, we return to the general theory and introduce bounded subsets of topological vector spaces. Bounded sets will be much used in relation with duality. By using Ascoli's theorem we prove that many of the spaces used in the theory of distributions (namely $\mathscr{C}^\infty(\Omega)$, $\mathscr{S}$, $\mathscr{C}_c^\infty(\Omega)$) have the property that closed bounded sets are compact (this property never holds in infinite dimensional normed spaces). In the infinite dimension situation, this property was first encountered by Montel in the study of normal family of holomorphic functions. We prove also Montel's theorem which states that in $H(\Omega)$ any closed bounded set is compact ($H(\Omega)$: space of holomorphic functions in the open set $\Omega \subset \mathbf{C}^n$).

Chapter 15 is devoted to approximation techniques in the standard spaces of functions, $\mathscr{C}^k(\Omega)$, $L^p(\Omega)$, $H(\Omega)$, $\mathscr{C}_c^k(\Omega)$. We study approximation by entire functions, by polynomials, and by $\mathscr{C}^\infty$ functions with compact support (in the cases where this makes sense!). In Chapter 16 we apply some of these approximation results to showing that, given an arbitrary locally finite covering of an open subset Ω of $\mathbf{R}^n$, there is a partition of unity in $\mathscr{C}^\infty(\Omega)$ subordinated to it. The last chapter of Part I, Chapter 17, is devoted to the statement and the proof of the open mapping theorem.

1
Filters. Topological Spaces.
Continuous Mappings

A topological vector space E is, roughly speaking, a set which carries two structures: a structure of topological space; a structure of vector space. Furthermore, some kind of compatibility condition must relate these two structures on E. We begin by recalling briefly what each one of them is, in the absence of any relation between the two.

One defines usually a *topology* on a set E by specifying what the *open* subsets of E are going to be. However, in dealing with topological vector spaces, as we are going to do in this book, it is more convenient to define a topology by specifying what the *neighborhoods* of each point are going to be. It is well known that the two approaches are equivalent: an open set will be a set which, whenever it contains a point, contains a neighborhood of this point; one can also say that an open set is a set which *is* a neighborhood of each one of its points; on the other hand, a neighborhood of a point x of E is simply a set which contains some open set containing x.

In order to define a topology by the system of the neighborhoods of the points, it is convenient to use the notion of *filter*. This is a very primitive notion, and the student should find it easy to become familiar with it, and to learn how to use filters, just as he learned how to use sequences. The notion of filter is perfectly independent of topology. A filter is given on a *set* which need not carry any other structure. Let E be the set. A filter $\mathscr{F}$ is a family of subsets of E, submitted to three conditions:

(F_1) *The empty set $\varnothing$ should not belong to the family $\mathscr{F}$.*

(F_2) *The intersection of any two sets, belonging to the family, also belongs to the family $\mathscr{F}$.*

(F_3) *Any set, which contains a set belonging to $\mathscr{F}$, should also belong to $\mathscr{F}$.*

The simplest example of a filter on a set E is the family of all subsets of E which contain a given subset A, provided the latter is nonempty.

With every *infinite* sequence of points of E is associated a filter. Let x_1, x_2,... be the sequence under consideration. The associated filter is the family of all subsets of E which have the following property:

(AF) *The subset of E contains all elements x_1, x_2,... except possibly a finite number of them.*

A family $\mathscr{B}$ of subsets of E is a *basis* of a filter $\mathscr{F}$ on E if the following two conditions are satisfied:

(BF$_1$) $\mathscr{B} \subset \mathscr{F}$, *i.e., any subset which belongs to $\mathscr{B}$ must belong to $\mathscr{F}$.*

(BF$_2$) *Every subset of E belonging to $\mathscr{F}$ contains some subset of E which belongs to $\mathscr{B}$.*

A familiar example of a basis of filter on the straight line is given by the family of all intervals $(-a, a)$ with $a > 0$: it is a basis of the filter of the neighborhoods of zero in the usual topology on the real line. Another useful example is the following one: let $\mathscr{F}$ be the filter associated with a sequence $S = \{x_1, x_2, ..., x_n, ...\}$. For each $n = 1, 2, ...$, let us set

$$S_n = \{x_n, x_{n+1}, ...\}$$

and view S_n as a subset of E. Then the sequence of subsets $S = S_1 \supset S_2 \supset \cdots S_n \supset \cdots$ is a basis of $\mathscr{F}$.

Let $\mathscr{A}$ be some family of subsets of our set E. We may ask the question: is there a filter $\mathscr{F}$ having $\mathscr{A}$ as a basis (note that a filter can have several different bases)? In view of the filters axioms, (F_1), (F_2), (F_3), that filter $\mathscr{F}$, if it exists, is completely and uniquely determined: it is the family of subsets of E which contains some subset belonging to $\mathscr{A}$. Observe that the latter property defines perfectly well a certain family, which we have called $\mathscr{F}$, of subsets of E. Then our question can be rephrased as follows: is $\mathscr{F}$ a filter? Obviously $\mathscr{F}$ satisfies (F_3); it also satisfies (F_1) if we take care of requiring that no set belonging to $\mathscr{A}$ be the empty set. As for (F_2) it is equivalent, as we see easily, with the following property of $\mathscr{A}$:

(BF) *The intersection of any two sets, belonging to $\mathscr{A}$, contains a set which belongs to $\mathscr{A}$.*

The difference with Condition (F_2) is that the intersection of two subsets which belong to $\mathscr{A}$ is not requested to belong to $\mathscr{A}$, but only to contain some set belonging to $\mathscr{A}$. Thus we may state: *a basis of filter on E is a family of nonempty subsets of E satisfying Condition* (BF). The filter generated by the basis is uniquely determined: by Condition (BF$_2$).

Next step: *comparison of filters*. We want to be able to say: this filter is finer than this other filter. Keep in mind that filters are sets of sets, or rather of subsets. In other words, filters are subsets of the set of subsets of E, usually denoted by $\mathfrak{P}(E)$. As filters are subsets (of some set, in this case $\mathfrak{P}(E)$), there is a natural order relation among them: the inclusion relation. We can write $\mathscr{F} \subset \mathscr{F}'$ if $\mathscr{F}$ and $\mathscr{F}'$ are two filters on the same set E. It means that every subset of E which belongs to $\mathscr{F}$ also belongs to $\mathscr{F}'$ (the converse being in general false). Instead of saying that $\mathscr{F}$ is contained in $\mathscr{F}'$, one usually says that $\mathscr{F}'$ is *finer* than $\mathscr{F}$, or that $\mathscr{F}$ is less fine than $\mathscr{F}'$. Let $\mathscr{F}$ (resp. $\mathscr{F}'$) be the family of all subsets of E which contain a given subset A (resp. A') of E; $\mathscr{F}'$ is finer than $\mathscr{F}$ if and only if $A' \subset A$.

A *topology* on the set E is the assignment, to each point x of E, of a filter $\mathscr{F}(x)$ on E, with the additional requirement that the following two conditions be satisfied:

(N$_1$) *If a set belongs to $\mathscr{F}(x)$, it contains the point x.*

(N$_2$) *If a set U belongs to $\mathscr{F}(x)$, there is another set V belonging also to $\mathscr{F}(x)$ such that, given any point y of V, U belongs to $\mathscr{F}(y)$.*

When these conditions are satisfied we say that we have a topology on E and we call $\mathscr{F}(x)$ the *filter of neighborhoods* of the point x. At first sight Condition (N$_2$) may seem involved. It expresses, however, a very intuitive fact. Roughly speaking, it says that given any point z near x (i.e., z is a generic element of U), if a third point y lies sufficiently near to x (the sufficiently near is made precise by the neighborhood of x, V, of which y is an element), then z lies near to y (i.e., $z \in U \in \mathscr{F}(y)$). In the language of open sets, (N$_2$) becomes evident: since U is a neighborhood of x, U contains an open set containing x; let V be such an open set. Since V is open, and $V \subset U$, U is obviously a neighborhood of each point of V. A basis of the filter $\mathscr{F}(x)$ is called a *basis of neighborhoods* of x. This simple notion will play an important role in the forthcoming definitions.

Once we have the notion of filter of neighborhoods of a point, hence of *neighborhood* of a point (any subset of E belonging to the filter of neighborhoods), we can review quickly the concepts that are used to describe a topology. As we have already said, an *open set* is a set which is a neighborhood of each one of its points. A subset of E is *closed* if its complement is open. The *closure* of a set $A \subset E$ is the smallest closed set containing A. It will be denoted by $\bar{A}$. The following is easy to check: a point belongs to $\bar{A}$ if and only if everyone of its neighborhoods meets A (that is to say, has a nonempty intersection with A). The *interior* of a set is the largest open set contained in it; if A is the set, its interior will be denoted by $\mathring{A}$.

A very important notion is the one of a set A *dense* in another set B; both A and B are subsets of the same topological space E. Then, one says that A is dense in B if the closure $\bar{A}$ of A contains B. In particular, A is said to be dense in E (or everywhere dense) if $\bar{A} = E$. To say that A is dense in B means that, given any neighborhood of any point x of B, $U(x)$, there is a point y of A which belongs to $U(x)$, i.e., $A \cap U(x) \neq 0$. A standard example of a set everywhere dense is the set of rational numbers $\mathbf{Q}$, when regarded as a subset of the real line $\mathbf{R}$ (with the usual topology); note that the complement $\mathbf{R} - \mathbf{Q}$ of $\mathbf{Q}$ is also dense in $\mathbf{R}$. Examples of sets which are dense and open are given by the complement of a straight line in the plane or in space, by the complements of a plane in space, etc. Easy to check are the basic intersection and union properties about open or closed sets: that the intersection of a finite number of open sets is open (this follows immediately from the fact, itself obvious in virtue of Axiom (F_2), that the intersection of a finite number of neighborhoods of a point is again a neighborhood of that point); that the union of any number of open sets, be that number finite or infinite, is open (this follows from the fact that the union of a neighborhood of a point with an arbitrary set is a neighborhood of the same point: Axiom (F_3)). By going to the complements, one concludes that finite unions of closed sets are closed, arbitrary intersections of closed sets are also closed, etc.

Observe that a set E may very well carry several different topologies. When dealing with topological vector spaces, we shall very often encounter this situation of a set, in fact a vector space, carrying several topologies (all compatible with the linear structure, in a sense that is going to be specified soon). For instance, any set may carry the following two topologies (which, in practice, are almost never used):

(1) *the trivial topology:* every point of E has only one neighborhood, the set E itself;

(2) *the discrete topology:* given any point x of E, every subset of E is a neighborhood of x provided that it contains x; in particular, $\{x\}$ is a neighborhood of x, and constitutes in fact a basis of the filter of neighborhoods of x.

We may compare topologies, in analogy with the way we have compared filters. Let $\mathscr{T}$, $\mathscr{T}'$ be two topologies on the same set E. We say that $\mathscr{T}$ *is finer than* $\mathscr{T}'$ if every subset of E which is open for $\mathscr{T}'$ is also open for $\mathscr{T}$, or equivalent, if every subset of E which is a neighborhood of a point for $\mathscr{T}'$ is also a neighborhood of that same point for the topology $\mathscr{T}$. Let $\mathscr{F}(x)$ (resp. $\mathscr{F}'(x)$) be the filter of neighborhoods of an arbitrary point x of E in the topology $\mathscr{T}$ (resp. $\mathscr{T}'$): $\mathscr{T}$ is finer than $\mathscr{T}'$, which we shorten into $\mathscr{T} \geqslant \mathscr{T}'$, if, for every $x \in E$, $\mathscr{F}(x)$ is finer than $\mathscr{F}'(x)$. Given two topologies on the same set, it may very well happen that none is finer than the other. If one is finer than the other, one says sometimes that they are comparable. The discrete topology is finer, on a set E, than any other topology on E; the trivial topology is less fine than all the others. Topologies on a set form thus a partially ordered set, having a maximal and a minimal element, respectively the discrete and the trivial topology.

The notion of a topology has been introduced in order to provide a solid ground for the notions of *convergence* and of *continuity*. Of course, the latter were correctly manipulated (or most of the time, at least) well before anybody thought of topology. We proceed now to give their general definition.

Convergence. This concerns filters: filters are the "objects" which may (or may not) converge. When do we say that a filter $\mathscr{F}$ on a topological space E converges? We should recall that $\mathscr{F}$ is a family of subsets of E. If $\mathscr{F}$ is to converge to a point x of E, it means that elements of $\mathscr{F}$, which, we repeat again, are subsets of E, get "smaller and smaller" about x, and that the points of these subsets get "nearer and nearer" to x. This can be made precise in terms of the neighborhoods of x, which we have at our disposal, since E is a topological space: we must express the fact that, however small a neighborhood of x is, it should contain some subset of E belonging to the filter $\mathscr{F}$ and, consequently, all the elements of $\mathscr{F}$ which are contained in that particular one. But in view of Axiom (F_3), this means that the neighborhood of x under consideration must itself belong to the filter $\mathscr{F}$, since it must contain some element of $\mathscr{F}$. The phrase "however small a neighborhood of x is" has to be made mathematically meaningful: it simply means "whatever is the neighborhood of x." In brief, we see that *the filter $\mathscr{F}$ converges to the point x if every neighborhood of x belongs to $\mathscr{F}$, in other words if $\mathscr{F}$ is finer than the filter of neighborhoods of x, $\mathscr{F}(x)$.* This is what the convergence to a point of a filter means.

We recall how the *convergence of a sequence* to a point is defined. Let $S = \{x_1, x_2, ...\}$ be the sequence. We say that S converges to x if, given an arbitrary neighborhood U of x, there is an integer $n(U)$ such that $n \geqslant n(U)$ implies $x_n \in U$. Let $S = S_1 \supset S_2 \supset \cdots \supset S_n \cdots$ be the subsequences introduced on p. 7: S converges to x if to every $U \in \mathscr{F}(x)$ there is an integer $n(U)$ such that $S_{n(U)} \subset U$. As the subsets S_n of E form a basis of the filter associated with the sequence S, we see immediately that *a sequence S converges to x if and only if the associated filter converges to x.*

Note that a filter may converge to several different points. Suppose, for instance, that E carries the trivial topology (p. 9): then every filter on E converges to every point of E. Note also that a filter may not converge: for instance, if it is the filter associated with some sequence and if this sequence does not converge. Another example is given by a filter on E which is *not* the filter of all subsets of E which contain a given point x—when E carries the discrete topology: in this topology, the only converging filters are the filters of neighborhoods of the points. So much for convergence in general topological spaces.

Continuity. This concerns mappings. In point set topology, a map $f : E \to F$, this is to say a map from a topological space E into another topological space F, is said to be continuous if any one of the following two conditions is satisfied:

(a)   given any point x of E and any neighborhood V of the *image* $f(x) \in F$ of x, the *preimage* of V, that is to say the set

$$f^{-1}(V) = \{x \in E; f(x) \in V\},$$

is a neighborhood of x. In short,

$$\forall x \in E, \; V \in \mathscr{F}(f(x)) \qquad \text{implies} \qquad f^{-1}(V) \in \mathscr{F}(x);$$

(b)   the preimage of any open subset $\mathcal{O}$ of F,

$$f^{-1}(\mathcal{O}) = \{x \in E; f(x) \in \mathcal{O}\},$$

is an open subset of E.

The student will easily check the equivalence of (a) and (b). As for the intuitive meaning of these conditions, we may say the following. If the mapping f is to be continuous at the point x, it should mean that if $x' \in E$ "converges to x," then $f(x')$ should converge to $f(x)$. Note that "$f(x')$ converges to $f(x)$" can be made precise in the following way: given an arbitrary neighborhood of $f(x)$, $f(x')$ should eventually belong to it; and the "eventually" means here: provided that x' is sufficiently near to x. Thus given an arbitrary neighborhood V of $f(x)$, if x' belongs to a sufficiently small neighborhood of x, then $f(x') \in V$. The "sufficiently small" can only be determined by the existence of a certain neighborhood U of x, such that, as soon as $x' \in U$, then $f(x') \in V$. This is exactly Property (a): to every neighborhood V of $f(x)$ there is a neighborhood U of x such that

$$x' \in U \qquad \text{implies} \qquad f(x') \in V.$$

It is immediately seen that, if a sequence $\{x_1, x_2, ...\}$ converges in E to a point x, and if f is a continuous function from E into F, then the sequence $\{f(x_1), f(x_2), ...\}$ converges to $f(x)$ in F. Convergence of filters is also easily related to continuity of mappings. Let

$$f : E \to F$$

be a mapping from a *set E* into a *set F*. Let $\mathscr{F}$ be a filter on E. The *image* $f\mathscr{F}$ of $\mathscr{F}$ under f is defined as being the filter having the basis

$$(f\mathscr{F})_0 = \{f(U) \in F; \ U \in \mathscr{F}\}.$$

Observe that, in general, $(f\mathscr{F})_0$ is not itself a filter; it is always the basis of a filter (the student may check this point as an exercise). Now, *if the filter $\mathscr{F}$ converges to a point x in E and if f is a continuous function, then $f\mathscr{F}$ converges to $f(x)$ in F*. Indeed, the continuity of f implies that $f\mathscr{F}(x)$ is finer than $\mathscr{F}(f(x))$; this is simply a restatement of Property (a) above. If then $\mathscr{F}$ is finer than $\mathscr{F}(x)$ (which means that $\mathscr{F}$ converges to x), $f\mathscr{F}$ is finer than $f\mathscr{F}(x)$ and a fortiori finer than $\mathscr{F}(f(x))$.

We have only considered continuous functions, which is to say functions defined everywhere and continuous everywhere. Of course, one may prefer to talk about functions continuous at a point. This is defined by the condition (where x is the point under consideration):

for every $V \in \mathscr{F}(f(x))$, $f^{-1}(V)$ belongs to $\mathscr{F}(x)$,

or, equivalently,

$$f\mathscr{F}(x) \geqslant \mathscr{F}(f(x)).$$

Let us insist on the fact that *all* the functions or mapping which will be considered in this book are *defined everywhere*.

As a last remark, let us consider the case where F is identical with E *as a set*, but carries a different topology from the one given on E, and where f is the *identity mapping* of E onto F, I. The following two properties are obviously equivalent:

(i) $I : E \to F$ is continuous;

(ii) the topology of E is finer than the topology of F (these two topologies are defined on the same set).

Exercises

1.1. Let X be a topological space, A, B two subsets of X. Prove that if A is open we have

$$A \cap \bar{B} \subset \overline{A \cap B}.$$

Give an example of sets A and B such that A is not open and that the preceding inclusion is not true.

1.2. Prove that the image of a dense set under a map, which is continuous and surjective (i.e., onto), is dense.

1.3. Let X, Y be two topological spaces, $f : X \to Y$ a continuous function, B a subset of Y, and $\bar{B}$ its closure. Do we always have

$$\text{closure of } f^{-1}(B) = f^{-1}(\bar{B})?$$

1.4. Give an example of the following situation: X and Y are topological spaces, $f : X \to Y$ is a continuous mapping, A is a closed subset of X, and $f(A)$ is *not* closed in Y.

1.5. Consider a straight line L in the plane $\mathbf{R}^2$. The filter *of neighborhoods of L in $\mathbf{R}^2$* is the filter formed by the sets which contain an open set containing L. Prove that there is no basis of this filter which is *countable*.

1.6. Let f be a real-valued, continuous function on the closed interval $[0, 1]$. Show that there is a "natural" filter $\mathscr{F}$ on the real line, defined by means of the Riemann sums

$$\sum_{j=0}^{r} (t_j - t_{j+1}) f(\tau_j),$$

where $0 = t_0 < t_1 < \cdots < t_j < \cdots < t_{r+1} = 1$ and $t_j \leqslant \tau_j \leqslant t_{j+1}$ for each j. Show that the filter $\mathscr{F}$ converges to the integral

$$\int_0^1 f(t)\, dt.$$

1.7. Prove that the filter of neighborhoods of the closed unit disk $\{(x, y) \in \mathbf{R}^2; x^2 + y^2 \leqslant 1\}$ in the plane has a countable basis (cf. Exercise 1.5).

1.8. Let X_α ($\alpha \in A$) be a family of topological spaces. Consider their product set

$$X = \prod_{\alpha \in A} X_\alpha .$$

Let us denote by p_α the *projection* mapping on the αth *coordinate axis* X_α :

$$p_\alpha : x = (x_\alpha) \to x_\alpha .$$

The *product topology* on X is defined in the following way: a subset U of X is a neighborhood of one of its points $x = (x_\alpha)$ if, for every α, $p_\alpha(U)$ is a neighborhood of x_α and if, for all α except possibly a *finite* number of them, $p_\alpha(U) = X_\alpha$. Prove that this is the least fine topology on X such that all the mappings p_α be continuous.

1.9. Let $\mathscr{F}$ be a filter on the product space X of Exercise 1.8. Let us denote by $\mathscr{F}_\alpha$ the image of $\mathscr{F}$ under the projection p_α . Show that $\mathscr{F}_\alpha$ is the family of subsets of X_α of the form $p_\alpha(M)$ as M ranges over $\mathscr{F}$, and that $\mathscr{F}$ converges if and only if every $\mathscr{F}_\alpha$ does.

1.10. Let us say that a set A is *predirected* if there is a preorder relation $a \leqslant b$ on A and if, for any pair of elements a, b of A, there is $c \in A$ such that $a \leqslant c$, $b \leqslant c$ (the relation $a \leqslant b$ is a *preorder relation* if it is reflexive, i.e., $a \leqslant a$ for all a, and transitive, i.e., $a \leqslant b$ and $b \leqslant c$ imply $a \leqslant c$; it is an *order relation* if, furthermore, $a \leqslant b$ and $b \leqslant a$ imply $a = b$). Let Φ be the set of all filters on X, Φ' the set of all mappings of predirected sets into X. Prove that there is a canonical mapping of Φ' onto Φ (this mapping is not one-to-one). Under this mapping, the image of a function f on the "predirected" set of positive integers into X is the filter associated with a *sequence* in X.

2

Vector Spaces. Linear Mappings

We recall first what a vector space is. The vector spaces we shall consider will be defined only on one of the two "classical" fields: the field of real numbers, $\mathbf{R}$, or the field of complex numbers, $\mathbf{C}$. As a rule, we shall suppose that the field is $\mathbf{C}$. When we specifically need the field to be $\mathbf{R}$, we shall always say so. In other words, we deal always with *complex* vector spaces. A vector space E over $\mathbf{C}$ is a system of three objects (E, A_v, M_s) consisting of a set E and of two mappings:

$$A_v : E \times E \to E, \qquad (x, y) \rightsquigarrow x + y,$$
$$M_s : \mathbf{C} \times E \to E, \qquad (\lambda, x) \rightsquigarrow \lambda x.$$

Of course, there are conditions to be satisfied by these objects. The mapping A_v, called *vector addition*, must be a commutative group composition law, i.e., it must have the following properties:

(associativity): $\quad (x + y) + z = x + (y + z)$;

(commutativity): $\quad x + y = y + x$;

(existence of a
neutral element): There exists an element, denoted by 0, in E such that $\forall x, x + 0 = x$;

(existence of an
inverse): To every $x \in E$ there is a unique element of E, denoted by $-x$, such that $x + (-x) = 0$.

Of course, we write $x - y$ instead of $x + (-y)$. The mapping M_s is called *scalar multiplication*, or *multiplication by scalars*, and should satisfy the following conditions:

(i) $\lambda(\mu x) = (\lambda \mu) x$;

(ii) $(\lambda + \mu) x = \lambda x + \mu x$;

(iii) $1 \cdot x = x$;

(iv) $0 \cdot x = 0$;

(v) $\lambda(x + y) = \lambda x + \lambda y$.

14

We do not recall the meaning of such notions as linear independence, basis, vector (or linear) subspace, etc. A mapping $f : E \to F$ of a vector space E into another, F, is called *linear* if for all x, $y \in E$, λ, $\mu \in \mathbf{C}$,

$$f(\lambda x + \mu y) = \lambda f(x) + \mu f(y).$$

Let us recall that a linear mapping $f : E \to F$ is *one-to-one* if and only if $f(x) = 0$ implies $x = 0$. Indeed, if f is one-to-one, $f(x) = 0$ must imply $x = 0$. Conversely, suppose that $f(x) = 0$ implies $x = 0$, and let x, $y \in E$ be such that $f(x) = f(y)$. This equation can be written $f(x-y) = 0$, implying then $x - y = 0$.   Q.E.D.

A notion with which the student may not be so familiar is the one of *quotient space*. As it will play a crucial role in the sequel, we shall recall its definition.

Let E be a vector space (over $\mathbf{C}$) and M a linear subspace of E. For two arbitrary elements x and y of E, the property

$$x - y \in M$$

defines an *equivalence relation*: it is reflexive, since $x - x = 0 \in M$ (every linear subspace contains the origin); it is symmetric, since $x - y \in M$ implies $-(x - y) = y - x \in M$ (if a linear subspace contains an element, it contains its inverse); it is transitive, since

$$x - y \in M, \quad y - z \in M \quad \text{implies} \quad x - z = (x - y) + (y - z) \in M$$

(when a linear subspace contains two vectors, it also contains their sum). Then we may define the *quotient set* E/M: it is the set of equivalence classes for the relation $x - y \in M$. There is a "canonical" mapping of E onto E/M: the mapping which, to each $x \in E$, assigns its class *modulo* the relation $x - y \in M$. It helps the intuition to visualize the class of elements equivalent to x *modulo* M, that is to say the y's such that $x - y \in M$, as a *linear subvariety*: indeed, they constitute the set

$$M + x = \{x' + x \mid x' \in M\},$$

which is the translation of M by x. Observe the following, which is easy to check (using the fact that M is a linear subspace):

(2.1) if $x \sim y \bmod M$, and if $\lambda \in \mathbf{C}$, then $\lambda x \sim \lambda y \bmod M$;

(2.2) if $x \sim y \bmod M$, and if $z \in E$, then $x + z \sim y + z \bmod M$.

Thus we define vector addition and scalar multiplication in E/M: if $\phi(x)$ is the class of $x \bmod M$, $\lambda\phi(x) = \phi(\lambda x)$ and $\phi(x) + \phi(y) = \phi(x + y)$.

These definitions are unambiguous by virtue of (2.1) and (2.2); they turn E/M into a vector space, and $\phi : E \to E/M$, the canonical mapping, is then a linear map. It is, of course, *onto*.

Now let E, F be two linear spaces (over **C**), and f a linear map $E \to F$. We define the *image* of f, and denote it by Im f, as the subset of F:

$$\text{Im} f = \{y \in F; \text{ there exists } x \in E \text{ such that } y = f(x)\}.$$

We define the *kernel* of f, and denote it by Ker f, as the subset of E:

$$\text{Ker} f = \{x \in E; f(x) = 0\}.$$

Both Im f and Ker f are linear subspaces (of F and E resp.). We have then the diagram

$$E \xrightarrow{\ \ f\ \ } \text{Im} f \xrightarrow{i} F$$

$$\phi \downarrow \quad \nearrow \tilde{f}$$

$$E/\text{Ker} f$$

where i is the *natural injection* of Im f into F, that is to say the mapping which to each element y of Im f assigns that same element y, regarded as an element of F; ϕ is the canonical map of E onto its quotient, $E/\text{Ker} f$. The mapping $\tilde{f}$ is defined so as to make the diagram commutative, which means that the image of $x \in E$ under f is identical with the image of $\phi(x)$ (i.e., the class of x modulo Ker f) under $\tilde{f}$. The mapping $\tilde{f}$ is well defined by the equation

$$\tilde{f}(\phi(x)) = f(x).$$

Indeed, if $\phi(x) = \phi(y)$, in other words if $x - y \in \text{Ker} f$, then $f(x) = f(y)$. It is an immediate consequence of the linearity of f and of the linear structure of the quotient space $E/\text{Ker} f$ that $\tilde{f}$ is also linear. Now, $\tilde{f}$ is a *one-to-one* linear map of $E/\text{Ker} f$ onto Im f. The onto property is evident from the definition of Im f and of $\tilde{f}$. As for the one-to-one property, observe that, if $\tilde{f}(\phi(x)) = \tilde{f}(\phi(y))$, it means by definition that $f(x) = f(y)$, hence $f(x - y) = 0$ or $x - y \in \text{Ker} f$, which means that $\phi(x) = \phi(y)$. Q.E.D.

Let E be an arbitrary set (not necessarily a vector space) and F a vector space. Let us denote by $\mathscr{F}(E; F)$ the set of all mappings of E into F. It can be equipped with a natural structure of vector space. We must first define the sum of two mappings $f, g : E \to F$. It must be a function of the same kind, and we must therefore say what its value should be at an arbitrary point x of E. Naturally, we take this value to be equal to the sum of the values of the factors, $f(x)$, $g(x)$:

$$(f + g)(x) = f(x) + g(x) \qquad \text{(this is a definition).}$$

Similarly, to define λf, where λ is an arbitrary scalar, we define its value at an arbitrary point x of E. We set (this is again a definition):

$$(\lambda f)(x) = \lambda f(x).$$

When E also is a vector space (over the same scalar field as F, for us the field of complex numbers $\mathbf{C}$), we will be particularly interested in the linear mappings of E into F. They form a linear subspace of $\mathscr{F}(E; F)$, as immediately checked, which we shall denote by $\mathscr{L}(E; F)$.

When $F = \mathbf{C}$, $\mathscr{L}(E; F)$ is denoted by E^* and called the *algebraic dual* of E. When E is a topological vector space (see next chapter) we shall be interested in a "smaller" dual of E, namely the linear subspace of E^* consisting of the linear mappings $E \to \mathbf{C}$ which are continuous; this will be called the *dual* of E and denoted by E'. One should always be careful to distinguish between E^* and E' (except in exceptional cases, e.g., when E is finite dimensional[†]). The elements of E^* are most of the time referred to as the *linear functionals*, or the *linear forms* on E.

If E, F, G are three vector spaces over $\mathbf{C}$, and $u : E \to F$, $v : F \to G$ two linear mappings, it is clear that the *compose* $v \circ u$, defined by

$$(v \circ u)(x) = v(u(x)), \qquad x \in E,$$

is a linear map of E into G. If $G = \mathbf{C}$, v is a linear functional on F, i.e., v is an element x^* of the algebraic dual F^* of F; the compose $x^* \circ u$ is a linear functional on E. We obtain thus a mapping $x^* \rightsquigarrow x^* \circ u$ of F^* into E^* for each given $u \in \mathscr{L}(E; F)$. This mapping is obviously linear. It is called the *algebraic transpose* of u; we shall denote it by u^*. As is readily seen, $u \rightsquigarrow u^*$ is a linear mapping of $\mathscr{L}(E; F)$ into $\mathscr{L}(F^*; E^*)$.

Exercises

2.1. Give an example of a linear space E and of two linear mappings u, v of E into itself with the following properties:

(i) u is injective (i.e., one-to-one) but not surjective (i.e., onto);

(ii) v is surjective but not injective.

[†] Also Hausdorff; see Chapter 9.

2.2. Let E be a vector space, M a linear subspace of E, and u a linear map of M into a vector space F. Prove that there is a linear map $v : E \to F$ which extends u, i.e., such that $u(x) = v(x)$ for all $x \in M$.

2.3. Let E, F be two vector spaces, $u : E \to F$ a linear map, and $u^* : F^* \to E^*$ the algebraic transpose of u. Prove that the following properties are equivalent:
(a) u is surjective;
(b) u^* is injective.

2.4. Let E, F be two vector spaces, and $u : E \to F$ a linear map. Let M (resp. N) be a linear subspace of E (resp. F), and ϕ (resp. ψ) the canonical mapping of E (resp. F) onto E/M (resp. F/N). Prove the equivalence of the following two properties:
(a) $u(M) \subset N$;
(b)   there exists a linear map v such that the following diagram is commutative:

$$
\begin{array}{ccc}
E & \xrightarrow{\ u\ } & F \\
{\scriptstyle \phi}\downarrow & & \downarrow{\scriptstyle \chi} \\
E/M & \xrightarrow{\ v\ } & F/N.
\end{array}
$$

Prove that, if (a) holds, the mapping v above is unique.

2.5. Let M be a linear subspace of E, j the natural injection of M into E, and ϕ the canonical map of E onto E/M. Let us set

$$M^\perp = \{x^* \in E^*; \text{ for all } x \in M, \, x^*(x) = 0\}.$$

(i)   Prove that there is a linear map k such that the following diagram is commutative:

$$
\begin{array}{ccc}
E^* & \xrightarrow{\ j^*\ } & M^* \\
{\scriptstyle \psi^*}\downarrow & \nearrow{\scriptstyle k} & \\
E^*/M^\perp, & &
\end{array}
$$

where ψ^* is the canonical mapping onto the quotient vector space. Moreover, prove that k is an isomorphism onto.

(ii)   Prove that $\phi^* : (E/M)^* \to E^*$ is one-to-one and that the image of ϕ^*, $\phi^*((E/M)^*)$, is equal to $M^\perp$.

2.6. Let (E_α) ($\alpha \in A$) be an arbitrary family of vector spaces over the complex numbers. Consider the product set

$$E = \prod_{\alpha \in A} E_\alpha \, ;$$

it carries a vector space structure where vector addition and scalar multiplication are performed componentwise. The direct sum of the E_α is the linear subspace of E consisting of those elements $x = (x_\alpha)_{\alpha \in A}$ for which all the components x_α are equal to zero with the possible exception of a finite number of them; we shall denote by E_0 this direct sum. Prove that there is a canonical isomorphism between the algebraic dual of the direct sum E_0 and the product of the algebraic duals E_α^* of the E_α.

2.7. Let us keep the notation of Exercise 2.6. Let $\alpha \in A$. We denote by j_α the (linear) mapping of E_α into E defined as follows: if $z \in E_\alpha$, $j_\alpha(z)$ is the element $x = (x_\beta)_{\beta \in A}$ of E such that $x_\beta = 0$ if $\beta \neq \alpha$ and $x_\alpha = z$. It is evident that j_α is one-to-one; if p_α is the coordinate projection $x = (x_\beta) \rightsquigarrow x_\alpha$, we have $p_\alpha \circ j_\alpha = $ identity of E_α. Let $j_\alpha^* : E^* \to E_\alpha^*$

be the algebraic transpose of j_α . We may define the following linear map of E^* into the product $\tilde{E}^* = \prod_{\alpha \in A} E_\alpha^*, j^*$:

$$E^* \ni x^* \rightsquigarrow j^*(x^*) = (j_\alpha^*(x^*))_{\alpha \in A} \in \tilde{E}^*.$$

Prove the equivalence of the following properties:
  (a) the set of indices A is finite;
  (b) j^* is one-to-one;
  (c) j^* is an isomorphism of E^* onto $\tilde{E}^*$.

  2.8. Prove that every vector space E is isomorphic to the direct sum of a family of one-dimensional vector spaces. Then, by making use of the results stated in Exercises 2.6 and 2.7, prove that the following properties are equivalent:
  (a) E is finite dimensional;
  (b) the canonical mapping

$$x \rightsquigarrow (x^* \rightsquigarrow x^*(x))$$

is an isomorphism of E onto the algebraic dual E^{**} of its own algebraic dual E^*.

3

Topological Vector Spaces. Definition

Let E be a vector space over the field of complex numbers $\mathbf{C}$ (in short, a vector space). Let

$$A_v : E \times E \to E, \qquad (x, y) \rightsquigarrow x + y,$$

$$M_s : \mathbf{C} \times E \to E, \qquad (\lambda, x) \rightsquigarrow \lambda x,$$

be the vector addition and the scalar multiplication in E. A topology $\mathscr{T}$ in E is said to be *compatible with the linear structure of E if A_v and M_s are* continuous when we provide E with the topology $\mathscr{T}$, $E \times E$ with the product topology $\mathscr{T} \times \mathscr{T}$, and $\mathbf{C} \times E$ with the product topology $\mathscr{C} \times \mathscr{T}$, where $\mathscr{C}$ is the usual topology in the complex plane $\mathbf{C}$. We recall the meaning of a "product topology." Consider two topological spaces E, F. In order to say what the product topology on $E \times F$ is, it suffices to exhibit a basis of the filter of neighborhoods of each point (x, y) of $E \times F$. Such a basis is provided by the rectangles

$$U \times V = \{(x', y') \in E \times F;\ x' \in U,\ y' \in V\},$$

where U (resp. V) is a neighborhood of x (resp. y) in E (resp. F). That these rectangles form a basis of filter is trivial; they obviously do not form a filter (except in trivial cases), since a set which contains a rectangle does not have to be a rectangle. It remains to check that the filters thus defined, for each pair (x, y), indeed can be taken as filters of neighborhoods of (x, y) in a topology on $E \times F$ (Axiom (N_2), p. 8, has to be verified). We leave this point to the student. The topology $\mathscr{C}$ assigns to each point λ of the complex plane a remarkable basis of neighborhoods, the disks, open or closed, with center at this point (and with positive radius p). When provided with a topology compatible with its linear structure, E becomes a *topological vector space*, which we shall abbreviate into TVS.

Suppose that E is a TVS. Then its topology is "translation invariant," which, roughly speaking, means that, topologically, E looks about any point as it does about any other point. More precisely: *the filter of neighborhoods $\mathscr{F}(x)$ of the point x is the family of sets $V + x$, where V varies over the filter of neighborhoods of the neutral element, $\mathscr{F}(0)$.* Proof of this statement: Let U be an arbitrary neighborhood of x. As the mapping $y \rightsquigarrow y + x$ from E into (as a matter of fact, onto) itself is continuous, which follows immediately from the continuity of the mapping $A_v : (x, y) \rightsquigarrow x + y$, the inverse image of U under this mapping must be a neighborhood of the preimage of x under this mapping; this preimage is obviously the neutral element 0. Let V be the inverse image of U. We have $U = V + x$. Conversely, given an arbitrary neighborhood V of 0, $V + x$ is a neighborhood of x by virtue of symmetry, or by virtue of the continuity of the mapping $y \rightsquigarrow y - x$. Thus: *in order to study the topology of a topological vector space E, it suffices to study the filter of neighborhoods of the origin.*

In practice, one always begins by giving the filter of neighborhoods of the origin, or (more frequently) a basis of this filter. It follows from there that we need some criteria on a filter which would insure that it is the filter of neighborhoods of the origin in a topology compatible with the linear structure of E.

THEOREM 3.1. *A filter $\mathscr{F}$ on a vector space E is the filter of neighborhoods of the origin in a topology compatible with the linear structure of E if and only if it has the following properties*:

(3.1)   *The origin belongs to every subset U belonging to $\mathscr{F}$.*

(3.2)   *To every $U \in \mathscr{F}$ there is $V \in \mathscr{F}$ such that $V + V \subset U$.*

(3.3)   *For every $U \in \mathscr{F}$ and for every $\lambda \in \mathbf{C}$, $\lambda \neq 0$, we have $\lambda U \in \mathscr{F}$.*

(3.4)   *Every $U \in \mathscr{F}$ is absorbing.*

(3.5)   *Every $U \in \mathscr{F}$ contains some $V \in \mathscr{F}$ which is balanced.*

We have used two words, *absorbing* and *balanced*, which have not yet been defined.

Definition 3.1.   *A subset A of a vector space E is said to be absorbing if to every $x \in E$ there is a number $c_x > 0$ such that, for all $\lambda \in \mathbf{C}$, $|\lambda| \leqslant c_x$, we have $\lambda x \in A$.*

In more colorful but less precise language, we may say that A is absorbing if it can be made, by dilation, to swallow any single point of the space.

Definition 3.2.   *A subset A of a vector space E is said to be balanced if for every $x \in A$ and every $\lambda \in \mathbf{C}$, $|\lambda| \leqslant 1$, we have $\lambda x \in A$.*

The only balanced subsets of the complex plane are the open or the closed disks centered at the origin.

Proof of Theorem 3.1.   We begin by proving the necessity of Properties (3.1)–(3.5). The necessity of (3.1) goes without saying.

Necessity of (3.2).   By $V + V$ we mean the set of points $x + y$, where x and y run over V. Let U be an arbitrary neighborhood of the origin. Its preimage under the mapping $(x, y) \rightsquigarrow x + y$ of U must be a neighborhood of 0, therefore must contain a rectangular neighborhood $W \times W'$, where W and W' are neighborhoods of 0 in E. But then it contains a "square," namely $(W \cap W') \times (W \cap W')$. If we take $V = W \cap W'$, this means precisely that $V + V \subset U$.

Necessity of (3.3).   By λU we mean the set of vectors λx, where x varies over U. Because of the continuity of the mapping $(\lambda, x) \rightarrow \lambda x$ from $\mathbf{C} \times E$ into E, if we fix $\lambda \in \mathbf{C}$, $\lambda \neq 0$, the map $x \rightarrow \lambda^{-1} x$ of E into itself must be continuous. The preimage of any neighborhood U of the origin in E must be such a neighborhood; this preimage is obviously λU.

Necessity of (3.4).   Again we use the continuity of the mapping $(\lambda, x) \rightarrow \lambda x$, this time at a point $(0, x)$ where x is an arbitrary point of E. The preimage of a neighborhood U of 0 in E must be a neighborhood of $(0, x)$, since $(0, x)$ is mapped into 0. Hence that preimage must contain a rectangle $N \times W$ where N (resp. W) is a neighborhood of 0 (resp. of x) in $\mathbf{C}$ (resp. in E). By definition of the topology of a TVS, W is of the form $W' + x$, where W' is a neighborhood of 0 in E (see p. 21). On the other hand, N contains a disk of the complex plane, centered at the origin, $D_\rho = \{\lambda \in \mathbf{C}; \ |\lambda| \leqslant \rho\}$, $\rho > 0$. Thus we see that, for all $y \in W' + x$ and all complex numbers λ such that $|\lambda| \leqslant \rho$, we have $\lambda y \in U$. In particular, we may take $y = x$.

Necessity of (3.5).   We duplicate the proof of the necessity of (3.4) but taking this time $x = 0$, hence $W' = W$. We have seen that the preimage of U contains a rectangle $D_\rho \times W$, which means that the set

$$V = \bigcup_{|\lambda| \leqslant \rho} \lambda W$$

is contained in U. This set V is obviously balanced. It is a neighborhood of zero, since each λW, $\lambda \neq 0$, is one (in view of (3.3)).

Sufficiency of Conditions (3.1)–(3.5).   We must first of all show that, if we define the filter of the neighborhoods of an arbitrary point x of E as the image of the given filter $\mathscr{F}$ under the translation $y \rightsquigarrow x + y$, we have indeed a topology on the set E. Once we have proved this, we must show that this topology is compatible with the linear structure of E. Let us call $\mathscr{F}(x)$ the image of $\mathscr{F}$ under $y \leadsto x + y$, that is to say the family of subsets $U + x$, where U varies over $\mathscr{F}$. Since $0 \in U$, x belongs to $U + x$. Thus Axiom (N_1), p. 8, is satisfied. Let $V \in \mathscr{F}$ be such that $V + V \in U$. Take an arbitrary point y of the set $V + x$; then $U + x$ contains $V + (V + x)$, hence $V + y$. But $V + y$ belongs to $\mathscr{F}(y)$, and therefore so does $U + x$. Thus Axiom (N_2), p. 8, is satisfied when we take $U + x$ and $V + x$, respectively, in the place of U and V in the statement of p. 8. We conclude that we have indeed a topology on E. The last two steps consist in proving that the mappings A_v and M_s are continuous. The continuity of A_v follows immediately from (3.2). Indeed, let (x, y) be an arbitrary element of $E \times E$; let W be a neighborhood of its image, $x + y$. We know that $W = U + x + y$, $U \in \mathscr{F}$. Choose $V \in \mathscr{F}$ such that $V + V \subset U$. Then $(V + x) + (V + y) \subset W$, which means that the image of the neighborhood of (x, y),

$$(V + x) \times (V + y),$$

is contained in W. Then the preimage of W contains that same neighborhood of (x, y) and, consequently, is a neighborhood of (x, y).

Last step: continuity of

$$M_s : (\lambda, x) \leadsto \lambda x.$$

Let U' be a neighborhood of $\lambda_0 x_0$; U' is of the form $U + \lambda_0 x_0$, where U is a neighborhood of zero in E. Let us select another neighborhood of 0, W, such that

(1) $W + W + W \subset U$;

(2) W is balanced.

Such a neighborhood of zero, W, exists in view of Properties (3.2) and (3.5). In view of (3.4), W is absorbing. In other words, there is a number $\rho > 0$, which we may as well take $\leqslant 1$, such that

$$\lambda \in \mathbf{C}, \quad |\lambda| \leqslant \rho, \quad \text{implies} \quad \lambda x_0 \in W.$$

Let D_ρ be the disk centered at the origin, in the complex plane, with radius ρ. Suppose first that $\lambda_0 = 0$, which implies $\lambda_0 x_0 = 0$ and $U' = U$. Then we look at the image under M_s of the set $D_\rho \times (W + x_0)$; it is the set

(3.6) $$\{\lambda y + \lambda x_0; |\lambda| \leqslant \rho, y \in W\}.$$

As $|\lambda| \leqslant \rho \leqslant 1$ and as W is balanced, $y \in W$ implies $\lambda y \in W$. As

$|\lambda| \leqslant \rho$, we have also $\lambda x_0 \in W$. We conclude that the set (3.6) is contained in $W + W$, hence in U. Thus the preimage of U contains $D_\rho \times (W + x_0)$, which is a neighborhood of $(0, x_0)$; the preimage of U is a neighborhood of $(0, x_0)$. Let us suppose finally that $\lambda_0 \neq 0$. In this case we look at the image under M_s of the set

$$(3.7) \qquad (D_\sigma + \lambda_0) \times (|\lambda_0|^{-1} W + x_0),$$

where $\sigma = \inf(\rho, |\lambda_0|)$, the smallest of the two numbers ρ, $|\lambda_0|$. The image of (3.7) is the set

$$(3.8) \qquad \{\lambda|\lambda_0|^{-1}y + \lambda x_0 + \lambda_0|\lambda_0|^{-1}y + \lambda_0 x_0;\ |\lambda| \leqslant \sigma, y \in W\}.$$

Since the complex numbers $\lambda|\lambda_0|^{-1}$, $\lambda_0|\lambda_0|^{-1}$ both have absolute value $\leqslant 1$, and since W is balanced, the sum

$$\lambda|\lambda_0|^{-1}y + \lambda_0|\lambda_0|^{-1}y$$

belongs to $W + W$. Since $|\lambda| \leqslant \sigma \leqslant \rho$, $\lambda x_0 \in W$, so that the set (3.8) is contained in

$$W + W + W + \lambda_0 x_0 \subset U + \lambda_0 x_0.$$

In other words, the preimage of $U + \lambda_0 x_0$ contains (3.7) and, therefore, it is a neighborhood of (λ_0, x_0). Q.E.D.

The following property of the filter of neighborhoods of zero in a TVS E is important:

PROPOSITION 3.1. *There is a basis of neighborhoods of zero in a* TVS E *which consists of closed sets.*

Proof. It suffices to show that an arbitrary neighborhood of zero U in E contains a closed neighborhood of 0. Let V be another neighborhood of 0 such that $V - V \subset U$. I contend that $\bar{V} \subset U$. Indeed, let $x \in \bar{V}$, which means that every neighborhood of x, in particular $V + x$, meets V. Thus, there are elements $y, z \in V$ such that $z = x + y$—in other words,

$$x = z - y \in V - V \subset U.$$

COROLLARY. *There is a basis of neighborhoods of 0 in E consisting of closed balanced sets.*

Indeed, every neighborhood U of 0 in E contains a closed neighborhood of 0, V, which in turn (Theorem 3.1) contains a balanced neighborhood of 0, W. Then $\bar{W}$ is closed and balanced; $\bar{W} \subset V \subset U$. Q.E.D.

The student will easily see that, whatever may be the vector space E, the trivial topology (p. 9) is always compatible with the linear structure of E, and the discrete topology (p. 9) never is—unless E consists of a single element (the origin). We proceed now to discuss a less trivial example.

Example. Let us denote by $\mathbf{C}[[X]]$ the ring of *formal power series* in one variable, X, with complex coefficients. Such a formal power series is written

$$u = u(X) = \sum_{n=0}^{\infty} u_n X^n,$$

where the coefficients u_n are complex numbers. It is the same thing as a power series as encountered in the theory of analytic functions, except that one does not care if it converges or not. Essentially, it is a sequence of complex numbers $(u_0, u_1, u_2, ..., u_n, ...)$. Addition and multiplication are immediately defined, by just extending what one does with polynomials or with Taylor expansions of analytic functions about the origin. If

$$v = \sum_{n=0}^{\infty} v_n X^n,$$

we have

$$u + v = \sum_{n=0}^{\infty} (u_n + v_n) X^n,$$

$$uv = \sum_{n,p=0}^{\infty} u_n v_p X^{n+p} = \sum_{n=0}^{\infty} \left(\sum_{p=0}^{n} u_{n-p} v_p \right) X^n.$$

Multiplication by scalars is defined in the obvious way:

$$\lambda u = \sum_{n=0}^{\infty} (\lambda u_n) X^n.$$

Addition and multiplication by scalars turn $\mathbf{C}[[X]]$ into a vector space; multiplication of formal power series turn it then into an *algebra*. There is a unit element in this algebra: the formal power series 1, that is to say the series u having all its coefficients u_n equal to zero if $n \geqslant 1$, and such that $u_0 = 1$. The following fact is not difficult to prove:

For a formal power series u to have an inverse, it is necessary and sufficient that its first coefficient, u_0, be different from zero.

Let us denote by $\mathfrak{M}$ the set of elements which do not have an inverse, that is to say the set of formal power series u such that $u_0 = 0$. The set $\mathfrak{M}$ is an *ideal* of the algebra $\mathbf{C}[[X]]$, which means that

(1) $\mathfrak{M}$ is a vector subspace of $\mathbf{C}[[X]]$;

(2)   for all $u \in \mathfrak{M}$ and all $v \in \mathbf{C}[[X]]$ we have $uv \in \mathfrak{M}$.

Both properties are evident. The student may easily check that $\mathfrak{M}$ is the largest proper (i.e., different from the whole algebra $\mathbf{C}[[X]]$) ideal of $\mathbf{C}[[X]]$. It suffices to observe that any ideal which is not contained in $\mathfrak{M}$ must contain an invertible element, hence must contain the series 1, in view of (2), and hence must be identical with the whole algebra.

For $n > 1$, let us denote by $\mathfrak{M}^n$ the set of formal power series u such that $u_p = 0$ if $p < n$. Any element of $\mathfrak{M}^n$ can be written

$$u(X) = X^{n-1}u_1(X),$$

with $u_1 \in \mathfrak{M}$. As the series X (i.e., all coefficients except the one of X^1 are equal to zero, and the coefficient of X^1 is equal to 1) belongs obviously to $\mathfrak{M}$, we see that every element of $\mathfrak{M}^n$ is the product of n elements of $\mathfrak{M}$, which justifies the notation. It is also easily checked that each $\mathfrak{M}^n$ is an ideal of $\mathbf{C}[[X]]$. The intersection of the $\mathfrak{M}^n$, as $n \to +\infty$, is obviously the zero power series (i.e., the power series having all its coefficients equal to zero). As the sequence of sets

$$\mathfrak{M}^0 = \mathbf{C}[[X]] \supset \mathfrak{M}^1 = \mathfrak{M} \supset \mathfrak{M}^2 \supset \cdots \supset \mathfrak{M}^n \supset \cdots$$

is totally ordered for inclusion, it certainly satisfies Axiom (BF), p. 7, for bases of filters. Let $\mathscr{F}$ be the filter it generates: a set U of formal power series belongs to $\mathscr{F}$ if it contains $\mathfrak{M}^n$ for large enough n. Let u be an arbitrary formal power series, and let $\mathfrak{M}^n + u$ be the set of formal power series $v + u$, where $v \in \mathfrak{M}^n$. Let us denote by $\mathscr{F}(u)$ the filter generated by the basis $\mathfrak{M}^n + u$, $n = 0, 1, 2,\ldots$. Observe that $\mathscr{F}$ satisfies Condition (3.2) in Theorem 3.1. Indeed, each $\mathfrak{M}^n$ being a vector subspace, we have

(3.9) $\mathfrak{M}^n + \mathfrak{M}^n \subset \mathfrak{M}^n.$

This implies immediately that the filters $\mathscr{F}(u)$ are the filters of neighborhoods of the points u in a topology on $\mathbf{C}[[X]]$: if $v \in \mathfrak{M}^n + u$, $\mathfrak{M}^n + u$ contains $\mathfrak{M}^n + v$ and hence is a neighborhood of v. In other words, $\mathfrak{M}^n + u$ is a neighborhood of each one of its points, i.e., $\mathfrak{M}^n + u$ is open. But (3.9) also implies that the preimage of

$$\{\mathfrak{M}^n + u + v; u, v \in \mathbf{C}[[X]]\},$$

under the addition mapping, viewed as a map from the topological space $\mathbf{C}[[X]] \times \mathbf{C}[[X]]$ into the topological space $\mathbf{C}[[X]]$, contains

$$(\mathfrak{M}^n + u) \times (\mathfrak{M}^n + v),$$

and hence is a neighborhood of the pair (u, v). This proves the continuity of addition.

The continuity of multiplication,

$$\mathbf{C}[[X]] \times \mathbf{C}[[X]] \to \mathbf{C}[[X]] : (u, v) \rightsquigarrow uv,$$

follows from the obvious inclusion

$$\mathfrak{M}^p \cdot \mathfrak{M}^q \subset \mathfrak{M}^{p+q},$$

where the left-hand side is the set of products uv where $u \in \mathfrak{M}^p$ and $v \in \mathfrak{M}^q$.

These continuity properties turn $\mathbf{C}[[X]]$ into what is called a *topological ring*.

However *they do not turn $\mathbf{C}[[X]]$ into a topological vector space*. The reason for this fact is that there are neighborhoods of the origin which are not absorbing. Indeed, $\mathfrak{M}^n$ is not absorbing as soon as $n > 0$: for there is no complex number $\lambda \neq 0$ such that $\lambda \cdot 1 \in \mathfrak{M}^n$. Thus the multiplication by scalars $(\lambda, u) \rightsquigarrow \lambda u$, viewed as a map from $\mathbf{C} \times \mathbf{C}[[X]]$ into $\mathbf{C}[[X]]$, is not continuous (although it is continuous if we identify λ with the formal power series u such that $u_0 = \lambda$, $u_p = 0$ for $p > 0$, and if we view the multiplication by scalars as a mapping from $[\mathbf{C}[X]] \times \mathbf{C}[[X]]$ into $\mathbf{C}[[X]]$).

Observe furthermore that the $\mathfrak{M}^n$ are open. They are also linear subspaces. Now the following is easy to check:

PROPOSITION 3.2. *In a TVS E, if a vector subspace M is open, we have $M = E$.*

Indeed, M being open is a neighborhood of each one of its points, in particular of the origin, hence must be absorbing (Property (3.4) of Theorem 3.1). But if $\lambda x \in M$ with $\lambda \neq 0$, then $x = \lambda^{-1}(\lambda x) \in M$.

One sees easily that every ideal $\mathfrak{M}^m$ ($m = 0, 1, 2, \dots$) is closed, so that the basis of neighborhoods of zero $\mathfrak{M}^m$ consists of sets which are both closed and open (cf. Exercise 3.4).

The topology which we have just described is actually used in algebra. Note that every point has a *countable* basis of neighborhoods in that topology.

There is another topology which is used on $\mathbf{C}[[X]]$, and which is compatible with the linear structure of $\mathbf{C}[[X]]$. It is *the topology of simple convergence of the coefficients*. A formal power series $u = \sum_{n=0}^{\infty} u_n X^n$ is said to converge to another formal power series $v = \sum_{n=0}^{\infty} v_n X^n$ if, for each n separately, the complex number u_n converges to v_n. Note that in the first topology described, u did converge to v if the numbers $p = 0, 1,...,$ such that $u_n = v_n$ for $n < p$ converged to infinity. The latter convergence therefore obviously implies the simple convergence of the coefficients. In other words, the topology defined by the ideals $\mathfrak{M}^n$ is finer than the topology of simple convergence of the coefficients which we are now going to define in a precise way. (Obviously a topology $\mathscr{T}$ is finer than another topology, $\mathscr{T}'$, on the same set, if any filter which converges for $\mathscr{T}$ also converges for $\mathscr{T}'$.)

As we shall always do in these chapters, we define a topology on a vector space, compatible with its linear structure, by exhibiting a basis of neighborhoods of zero. In our case, the basis will be the collection of the following sets of formal power series:

$$V_{m,n} = \left\{ u = \sum_{p=0}^{\infty} u_p X^p \in \mathbf{C}[[X]]; \forall\, p \leqslant n,\, |u_p| \leqslant 1/m \right\}.$$

Here m and n are integers, $n = 0, 1,..., m = 1, 2,....$ We leave to the student, as an exercise, the task of checking that the filter $\mathscr{F}$ generated by the basis $\{V_{m,n}\}$ $(m = 1, 2,..., n = 0, 1,...)$ satisfies Conditions (3.1)–(3.5) in Theorem 3.1. That the $V_{m,n}$ indeed form a basis of a filter is an obvious consequence of the fact that

$$V_{m,n} \cap V_{m',n'} \supset V_{\sup(m,m'),\sup(n,n')},$$

where $\sup(a, b)$ means the greatest of the two numbers a, b. Let $\{u^{(\nu)}\}$ $(\nu = 1, 2,...)$ be a sequence of formal power series. It converges to a series u if and only if, to every pair of integers $m \geqslant 1$, $n \geqslant 0$, there is another $\nu(m, n) \geqslant 1$ such that

$$\nu \geqslant \nu(m, n) \qquad \text{implies} \qquad u^{(\nu)} \in V_{m,n} + u.$$

This means, roughly speaking, that $u^{(\nu)}$ converges to u if more and more coefficients of $u^{(\nu)}$ get nearer and nearer to the coefficients with the same index of u. This is precisely what expresses the name: "topology of simple convergence of the coefficients." For this topology, the ideals $\mathfrak{M}^n$ are closed, as is immediately seen (if a formal power series u is a limit of formal power series v such that $v_p = 0$ for $p < n$, in the sense that the coefficients of u are the limits of the corresponding coefficients of the v's, we must have $u_p = 0$ for $p < n$). They are not open in view of

Proposition 3.2. It should also be noted that in the topology of simple convergence of the coefficients, the origin, and therefore each point, has a *countable* basis of neighborhoods. This property was also valid for the first topology we have defined on $\mathbf{C}[[X]]$.

Exercises

3.1. Let X be a set. Let us assign, to each $x \in X$, a topological vector space E_x ; let us denote by $\mathbf{E}$ the disjoint union of the spaces E_x as x varies over X and by $\Gamma(X; \mathbf{E})$ the set of mappings f of X into $\mathbf{E}$ such that, for every $x \in X$, $f(x) \in E_x$. Show that there is a "natural" structure of vector space over $\Gamma(X; \mathbf{E})$. Next, consider a finite subset S of X and, for every $x \in S$, a neighborhood of zero, U_x , in E_x . We may then consider the following subset of $\Gamma(X; \mathbf{E})$:

$$\{f \in \Gamma(X; \mathbf{E}); \text{ for all } x \in S, f(x) \in U_x\}.$$

Show that, when S varies in all possible ways and so do also the neighborhoods of zero U_x in each E_x , the above sets form a basis of a neighborhood of zero in $\Gamma(X; \mathbf{E})$ for a topology compatible with the linear structure of $\Gamma(X; \mathbf{E})$ (called topology of pointwise convergence in X).

3.2. Prove that, in a topological vector space E over the field of complex numbers, a set different from $\varnothing$ and from E cannot be both open and closed.

3.3. Let E be a vector space, $\{E_\alpha\}$ ($\alpha \in A$) a family of topological vector space and, for each α, H_α a set of linear mappings of E into E_α . Prove that there is at least one topology on E, compatible with the linear structure of E, such that, for every α, all the mappings belonging to the sets H_α are continuous. Describe the least fine topology with these properties (it is called *projective limit* of the topologies of the E_α with respect to the sets of mappings H_α).

Suppose that E is the *product vector space* of the E_α and that each one of the sets H_α consists of only one map, the projection p_α on the "coordinate axis" E_α (cf. Exercise 2.6). Prove that the projective limit topology on E and the product topology are identical.

3.4. Let $\mathbf{N}^n$ be the set of n-tuples $p = (p_1, ..., p_n)$ of n integers $p_j \geqslant 0$ $(1 \leqslant j \leqslant n)$. Show that the following vector spaces are naturally isomorphic:

(a)   the space $\mathscr{F}(\mathbf{N}^n; \mathbf{C})$ of complex-valued functions defined in $\mathbf{N}^n$;

(b)   the product space

$$\mathbf{C}^{\mathbf{N}^n} = \prod_{p \in \mathbf{N}^n} \mathbf{C}_p \qquad (\mathbf{C}_p \cong \mathbf{C} \text{ for all } p);$$

(c)   the space of complex sequences depending on n indices $\sigma = (\sigma_p)_{(p \in \mathbf{N}^n)}$;

(d)   the space $\mathbf{C}[[X_1, ..., X_n]]$ of formal power series.

Prove that the "natural" isomorphisms extend to the topologies when (a) carries the topology of pointwise convergence in $\mathbf{N}^n$ (cf. Exercise 3.1), (b) carries the product topology, (c) carries the topology of convergence of each term σ_p , and (d) carries the topology of simple convergence of the coefficients (cf. p. 28). Show that the isomorphism between (a) and (d) does *not* extend to the ring (or multiplicative) structure when $\mathscr{F}(\mathbf{N}^n; \mathbf{C})$ carries the usual multiplication of complex functions and $\mathbf{C}[[X_1, ..., X_n]]$ carries the multiplication of formal power series (see p. 25).

3.5. Let us denote by $\mathscr{F}_c(\mathbf{N}; \mathbf{C})$ the vector space of complex-valued functions in the set $\mathbf{N}$ of integers $\geqslant 0$ which vanish outside a finite subset of $\mathbf{N}$. Let $\{\varepsilon_n\}$ $(n = 0, 1, 2,...)$ be a sequence of numbers > 0 such that $\varepsilon_n > \varepsilon_{n+1} \to 0$. The subsets of $\mathscr{F}_c(\mathbf{N}; \mathbf{C})$,

$$\mathscr{U}(\{\varepsilon_n\}) = \{f; p \geqslant n \Rightarrow |f(p)| \leqslant \varepsilon_n\},$$

form, when the sequence $\{\varepsilon_n\}$ varies in all possible ways, a basis of neighborhoods of zero for a topology compatible with the linear structure of $\mathscr{F}_c(\mathbf{N}; \mathbf{C})$. Prove this statement. Prove also that there is no basis of neighborhoods of zero in this topology which is countable.

3.6. We keep the notation of Exercise 3.5. We denote by $\mathscr{F}_c(n)$ the linear subspace of $\mathscr{F}_c(\mathbf{N}; \mathbf{C})$ consisting of the functions f such that $f(p) = 0$ for all $p > n$. Prove the following assertions: (i) $\mathscr{F}_c(n)$ is a linear space of dimension $n + 1$; (ii) $\mathscr{F}_c(n)$ is closed in the TVS $\mathscr{F}_c(\mathbf{N}; \mathbf{C})$ (equipped with the topology defined in Exercise 3.5); (iii) the topology induced by $\mathscr{F}_c(\mathbf{N}; \mathbf{C})$ on $\mathscr{F}_c(n)$ is identical with the topology carried over from $\mathbf{C}^{n+1}$ by using any isomorphism $\mathscr{F}_c(n) \cong \mathbf{C}^{n+1}$.

3.7. Let E be a vector space, and U a subset of E which is convex (i.e., if $x, y \in U$, $tx + (1 - t)y \in U$ for all $t, 0 \leqslant t \leqslant 1$), balanced, and absorbing. Prove that the sets $(1/n)U$ $(n = 1, 2,...)$ form a basis of neighborhoods of zero in a topology on E which is compatible with the linear structure of E.

Is this always true when we drop the assumption that U is convex?

4

Hausdorff Topological Vector Spaces. Quotient Topological Vector Spaces. Continuous Linear Mappings

Throughout this chapter, we denote by E a TVS over the field of complex numbers.

Hausdorff Topological Vector Spaces

A topological space X is said to be *Hausdorff* if, given any two distinct points x and y of X, there is a neighborhood U of x and a neighborhood V of y which do not intersect, i.e., such that $U \cap V = \emptyset$. A very important property of Hausdorff topological spaces is the so-called *uniqueness of the limit:*

A filter on a Hausdorff topological space X converges to at most one point.

Indeed, suppose that a filter $\mathscr{F}$ on X would converge to two distinct points x and y. Let U (resp. V) be a neighborhood of x (resp. y) such that $U \cap V = \emptyset$. But both U and V must belong to $\mathscr{F}$, which demands that their intersection be nonempty!

In a Hausdorff space, any set consisting of a single point is closed (there are topological spaces with the same property which are not Hausdorff; but such spaces are not TVS, as will be seen). A TVS E is Hausdorff if, given any two distinct points x and y, there is a neighborhood U of x which does not contain y. As a matter of fact, we have the following result:

PROPOSITION 4.1. *A TVS E is Hausdorff if and only if to every point $x \neq 0$ there is a neighborhood U of 0 such that $x \notin U$.*

The necessity of the condition is trivial. Suppose it is satisfied. Let x, y be two distinct points of E, which means that $x - y \neq 0$. Then there is a neighborhood U of 0 such that $x - y \notin U$. Choose a balanced neighborhood of 0, V, such that $V + V \subset U$ (Theorem 3.1, p. 21). Since V is balanced, we have $-V = V$, hence $V - V \subset U$. Suppose that the intersection

(4.1) $$(V + x) \cap (V + y)$$

is nonempty, and let z be one of its points: $z = x + x'$, $z = y + y'$, with $x', y' \in V$. We have

$$x - y = y' - x' \in V - V \in U,$$

which contradicts our choice of U. Thus (4.1) must be empty.

PROPOSITION 4.2. *In a TVS E, the intersection of all neighborhoods of the origin is a vector subspace of E, which is the closure of the set $\{0\}$.*

Let us first prove that the intersection of all the neighborhoods of the origin, which we denote temporarily by N, is a vector subspace of E. Let U be an arbitrary neighborhood of the origin, x, y two elements of N, and α, β two complex numbers, which are not both equal to zero. Let V be a neighborhood of 0 such that $V + V \subset U$; assume furthermore that V is balanced (Definition 3.2, Theorem 3.1, p. 21). As $x, y \in N$, we have

$$x, y \in (\alpha^2 + \beta^2)^{-1} V;$$

hence, $\alpha x + \beta y \in \alpha(\alpha^2 + \beta^2)^{-1} V + \beta(\alpha^2 + \beta^2)^{-1} V \subset V + V \subset U$. This implies that $\alpha x + \beta y \in N$ since U is arbitrary.

Let x belong to N. Then every neighborhood U of 0 contains x, which can also be written

$$0 \in (-U) + x.$$

But $(-U) + x$ is an arbitrary neighborhood of x (since multiplication of vectors by -1 is a homeomorphism, that is to say a bicontinuous one-to-one mapping onto). Thus every neighborhood of x contains the origin, which means that $x \in \overline{\{0\}}$ (see p. 8). Conversely, suppose that an arbitrary neighborhood of x contains 0; such an arbitrary neighborhood can be written $-U + x$, where U is an arbitrary neighborhood of 0; and $0 \in -U + x$ is equivalent to $x \in U$, which means that $x \in N$.

Q.E.D.

COROLLARY. *For a TVS E to be Hausdorff, it is necessary and sufficient that the set $\{0\}$ be closed in E, or that the complement of the origin be open in E.*

Indeed, to say that $\{0\}$ is closed in E is equivalent to saying that $N = \{0\}$ or that no point $x \neq 0$ may belong to all the neighborhoods of 0.

An important consequence of the corollary of Proposition 4.2 is the next result:

PROPOSITION 4.3.   *Let f, g, be two continuous mappings of a topological space X into a Hausdorff TVS E. The set A in which f and g coincide,*

$$A = \{x \in X; f(x) = g(x)\},$$

is closed in X.

Indeed, A is the preimage of the closed set $\{0\} \subset E$ under the continuous mapping $x \leadsto f(x) - g(x)$.

PROPOSITION 4.4.   *Let X, E, f, g be as in Proposition 4.3. If f and g are equal on a dense subset Y of X, they are equal everywhere in X.*

Indeed, $f = g$ on a closed subset of X (Proposition 4.3) containing Y.

Quotient Topological Vector Spaces

Let M be a vector subspace of E, and let us consider the quotient vector space (p. 15) E/M and the canonical map $\phi : E \to E/M$ which assigns to every $x \in E$ its class $\phi(x)$ modulo M. We know that the mapping ϕ is linear. On E we have a topology (since E is a TVS). We may define then, in a canonical way, a topology on E/M which is called the *quotient topology* on E/M. As always, we say what the filter of neighborhoods of the origin in E/M is going to be: it is simply the image under ϕ of the filter of neighborhoods of the origin in E. This is the same as saying the following: *a subset $\dot{U}$ of E/M is a neighborhood of zero for the quotient topology if and only if there is a neighborhood U of zero in E whose image under ϕ is equal to $\dot{U}$*, i.e., $\dot{U} = \phi(U)$. The neighborhoods of zero in E/M are the direct images under ϕ of the neighborhoods of 0 in E.

Note that ϕ transforms neighborhoods of a point into neighborhoods of a point. This is not true in general about continuous functions: the preimage of a neighborhood under a continuous function is a neighborhood, but nothing is said about the image. On the other hand, we do not know *a priori* if ϕ is continuous. But it is easy to see that this is indeed so: let $\dot{U}$ be a neighborhood of the origin in E/M; there is a neighborhood U of zero in E such that $\phi(U) = \dot{U}$, hence $U \subset \phi^{-1}(\dot{U})$, which proves that $\phi^{-1}(\dot{U})$ is a neighborhood of zero in E.

Going to open sets, we see that ϕ transforms open sets into open sets and the preimages of open sets under ϕ are open sets.

It is not true, in general, that the direct images of closed sets, under ϕ, are closed sets.

A familiar counterexample is the following one. Consider in the plane $\mathbf{R}^2$ the hyperbola $\{(x_1, x_2) \in \mathbf{R}^2; \; x_1 x_2 = 1\}$. Take for M one of the coordinate axes. Then E/M can be identified with the other coordinate axis and ϕ with the orthogonal projection on it; all these identifications are also valid for the topologies. The hyperbola above is closed in $\mathbf{R}^2$ but its image under ϕ is the complement of the origin on a straight line, which is open.

The student may easily verify the following point: *the quotient topology on E/M is the finest topology on E/M such that ϕ is continuous.*

From our definition it follows immediately that the quotient topology on E/M is compatible with the linear structure of E/M (see p. 20).

PROPOSITION 4.5.   *Let E be a TVS, and M a vector subspace of E. The two following properties are equivalent:*

(a)   *M is closed;*

(b)   *E/M is Hausdorff.*

In view of the corollary of Proposition 4.2, (b) can be restated as saying that the complement of the origin is open in E/M. But the complement of the origin is exactly the image under ϕ of the complement of M, and ϕ maps open sets into open sets, and is continuous, whence the equivalence of (a) and (b).

COROLLARY.   *The TVS $E/\overline{\{0\}}$ is Hausdorff.*

The TVS $E/\overline{\{0\}}$ is said to be the *Hausdorff topological vector space associated with the* TVS E. When E itself is Hausdorff, $\phi : E \to E/\overline{\{0\}}$ (canonical mapping) is one-to-one onto, since then $\overline{\{0\}} = \{0\}$, and $E/\overline{\{0\}}$ is identified with E.

Continuous Linear Mappings

Let E, F be two TVS, and f a linear map of E into F. We suppose that F is Hausdorff and that f is continuous, in the usual sense (see p. 11 et seq.). Then the kernel (p. 16) of f is closed. Indeed, Ker f is the preimage of the set $\{0\} \subset F$, which is closed when F is Hausdorff. Of course, Ker f might be closed also when F is not Hausdorff (Example 1, $f = 0$; Example 2, f is one-to-one and E is Hausdorff; in this case, Ker $f = \{0\}$ is closed in E).

Let us consider the usual diagram (p. 16):

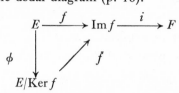

where i is the natural injection, ϕ the canonical map, and $\check{f}$ the unique linear map which makes the diagram commutative.

PROPOSITION 4.6.   *The map f is continuous if and only if the map $\check{f}$ is continuous.*

Suppose f continuous and let Ω be an open subset in F or in Im f (an open set in Im f is the intersection of an open set in F with Im f). The preimage of Ω under $\check{f}$ is equal to the image, under ϕ, of the preimage of Ω under f. By hypothesis, $f^{-1}(\Omega)$ is open, and ϕ transforms open sets into open sets; therefore the preimage of Ω under $\check{f}$ is open.

When both ϕ and $\check{f}$ are continuous, so is $f = \check{f} \circ \phi$.   Q.E.D.

In general, the inverse of $\check{f}$, which is well defined on Im f, since $\check{f}$ is one-to-one, will not be continuous; in other words, $\check{f}$ will not be bicontinuous.

Definition 4.1.   If $\check{f}$ is continuous and if the inverse of $\check{f}$, defined on Im f (this subspace of F being equipped with the topology induced by F), is also continuous, we say that f is a homomorphism. If furthermore f is one-to-one, we say that f is an isomorphism of E into F or onto Im f.

The set of *continuous* linear maps of a TVS E into another TVS F will be denoted by $L(E; F)$. Of course, it is a subset of $\mathscr{L}(E; F)$, the vector space of linear maps, continuous or not, from E into F. It is evident that $L(E; F)$ is a vector subspace of $\mathscr{L}(E; F)$, hence is a vector space, for the natural addition and multiplication by scalars, of functions. When $E = \mathbf{C}$, one denotes usually $L(E; F)$ by E' and calls this vector the *dual* of E (sometimes, the topological dual of E, in order to underline the difference between E' and E^*, the algebraic dual of E; see p. 17). Naturally, E' is a vector subspace of E^*; E' is the vector space of *continuous linear functionals*, or *continuous linear forms*, on F. Elements of E' will usually be denoted by x', y', etc. The vector spaces E' and $L(E; F)$ will play an important role in the forthcoming and will be equipped with various topologies.

We conclude this section with a property of continuous linear mappings which is well known, and reflects the "homogeneity" of the topology in a TVS:

PROPOSITION 4.7.   *Let E, F be two TVS, u a linear map of E into F. The mapping u is continuous if* (and only if!) *u is continuous at the origin.*

Indeed, an arbitrary neighborhood of $u(x)$ $(x \in E)$ in F is of the form $V + u(x)$, where V is a neighborhood of 0 in F. Since u is linear, we have

$$u^{-1}(V + u(x)) \supset u^{-1}(V) + x.$$

If u is continuous at the origin, $u^{-1}(V)$ is a neighborhood of zero in E.

Exercises

4.1.   Consider the topological vector space $\Gamma(X; E)$ defined in Exercise 3.1. Prove that it is Hausdorff if and only if the spaces E_x are Hausdorff for all x.

4.2.   Prove that the product of a family of TVS E_α ($\alpha \in A$) is Hausdorff if and only if every E_α is Hausdorff.

4.3.   Let M be a linear subspace of a TVS E. Another linear subspace, N, of E is called an *algebraic supplementary* of M in E if the mapping $(x, y) \rightsquigarrow x + y$ of $M \times N$ into E is an isomorphism onto E for the vector space structure; N is called a *topological supplementary* of M if $(x, y) \rightsquigarrow x + y$ is an isomorphism of $M \times N$ onto E for the TVS structure. One says then that E is the *topological direct sum* of M and N.

Prove the equivalence of the following two properties:

(a) N is a topological supplementary of M;
(b)   the restriction to N of the canonical mapping of E onto E/M is an isomorphism (for the TVS structure) of N onto E/M.

Prove that M has at least one topological supplementary in E if there is a continuous linear map p of E onto M such that $p \circ p = p$ (then $p(x) = x$ is equivalent with $x \in M$).

4.4.   Let f be a continuous linear map of a TVS E *onto* another one, F. Prove the equivalence of the following properties:

(a) $\operatorname{Ker} f$ has a topological supplementary in E (cf. Exercise 4.3);
(b)   there is a continuous linear map g of F into E such that $f \circ g = $ identity of F.

4.5.   Let E be a TVS, and M a linear subspace of E. For every TVS G, the restriction to M of the continuous linear mappings $f : E \to G$ defines a linear mapping of $L(E; G)$ into $L(M; G)$. Prove that this mapping is *onto* for every TVS G if and only if M has a topological supplementary in E.

5

Cauchy Filters. Complete Subsets. Completion

The definition of a *Cauchy sequence* in a TVS E is simple enough. Let $S = \{x_1, x_2, ...\}$ be the sequence; S is a Cauchy sequence if to every neighborhood U of the origin in E, there is an integer $n(U)$ such that

(5.1) $\qquad\qquad n, m \geqslant n(U) \qquad$ implies $\qquad x_m - x_n \in U$.

This definition obviously agrees with the usual one when the TVS E is the complex plane (it suffices then to take for U a disk of radius $\varepsilon > 0$, centered at the origin). Let us introduce the usual subsequences

$$S_n = \{x_{n+1}, x_{n+2}, ...\}.$$

We see that (5.1) simply means

$$S_{n(U)} - S_{n(U)} \subset U.$$

Observing that the S_n form a basis of the filter associated with the sequence S, this suggests what the definition of a Cauchy filter should be:

Definition 5.1. A filter $\mathscr{F}$ on a subset A of the TVS E is said to be a Cauchy filter if to every neighborhood U of 0 in E there is a subset M of A, belonging to $\mathscr{F}$, such that

$$M - M \subset U.$$

It may help to illustrate this definition by an example in a metric space. Suppose that there is a metric $d(x, y)$ on $E \times E$ defining the topology of E. Choose U such that, for some number $\varepsilon > 0$, the relation $x - y \in U$ means exactly that $d(x, y) < \varepsilon$. If M is any subset of E, the diameter of M is defined as being the supremum of the positive numbers $d(x, y)$ when both x and y vary over M. Now $M - M \subset U$ simply means that the diameter of M is $\leqslant \varepsilon$. Definition 5.1 can then be rephrased

37

as follows in the case of a metric TVS E: a filter $\mathscr{F}$ on $A \subset E$ is a Cauchy filter if it contains subsets of A of arbitrarily small diameter.

Going back to the general case, the following statement is obvious:

PROPOSITION 5.1.   *The filter associated with a Cauchy sequence is a Cauchy filter.*

Also obvious are the following statements:

PROPOSITION 5.2.   (a) *The filter of neighborhoods of a point $x \in E$ is a Cauchy filter.*

(b)   *A filter finer than a Cauchy filter is a Cauchy filter.*

(c)   *Every converging filter is a Cauchy filter.*

(a) follows from the fact that if U is a neighborhood of zero there is another neighborhood of 0, V, such that $V - V \subset U$, hence such that

$$(V + x) - (V + x) \subset U.$$

(b) is evident; (c) follows from (a) and (b), since a filter converges to a point x if it is finer than the filter of neighborhoods of that point.

It is well known that the converse of (c) is false, in other words that not every Cauchy filter converges.

Definition 5.2.   A subset A of E is said to be complete if every Cauchy filter on A converges to a point x of A.

It makes sense to ask if E itself is complete. We also use the term *sequentially complete* for any set $A \subset E$ such that any Cauchy sequence in A converges to a limit in A. Complete always implies sequentially complete, the converse being in general false. We shall encounter an important class of TVS, the so-called *metrizable spaces*, for which the converse is true.

As an exercise, the student may attempt to prove the next two propositions:

PROPOSITION 5.3.   *In a Hausdorff TVS E, any complete subset is closed.*

PROPOSITION 5.4.   *In a complete TVS E, any closed subset is complete.*

We must now describe an abstract procedure which, to an arbitrary *Hausdorff* TVS E, associates—in a canonical way—a complete (and Hausdorff) TVS $\hat{E}$, called its *completion*.

But before doing this, we must establish a certain number of properties of *uniformly continuous* functions, defined in a subset A of a TVS E and valued in a TVS F. Here is the definition of these functions:

Definition 5.3.   *A mapping $f : A \to F$ is said to be uniformly continuous if to every neighborhood of zero, V, in F, there is a neighborhood of zero, U, in E, such that for all pairs of elements x_1, $x_2 \in A$,*

$$x_1 - x_2 \in U \qquad \text{implies} \quad f(x_1) - f(x_2) \in V.$$

The student may compare this definition with the usual one, when $E = F = \mathbf{R}$ or $\mathbf{C}$. Any uniformly continuous map is continuous at every point; the converse is false. The following two statements have easy proofs, left to the reader:

PROPOSITION 5.5.   *Every continuous linear map of a linear subspace A of TVS E into a TVS F is uniformly continuous.*

PROPOSITION 5.6.   *Let f be a uniformly continuous map of $A \subset E$ into F. The image under f of a Cauchy filter on A is a Cauchy filter on F.*

From there follows (as will be shown now) the main extension result about uniformly continuous functions:

THEOREM 5.1.   *Let E, F be two Hausdorff TVS, A a dense subset of E, and f a uniformly continuous mapping of A into F.*

If F is complete, there is a unique continuous mapping $\tilde{f}$ of E into F which extends f, i.e., such that for all $x \in A$,

$$\tilde{f}(x) = f(x).$$

Moreover, $\tilde{f}$ is uniformly continuous, and $\tilde{f}$ is linear if A is a linear subspace and if f is linear.

Proof.   As we have said, it is essentially based on Proposition 5.6. The uniqueness of the extension $\tilde{f}$ follows immediately from Proposition 4.4. We shall therefore prove its existence.

Let x be an arbitrary point of E, $\mathscr{F}(x)$ the filter of its neighborhoods. Each neighborhood of x intersects A, since A is everywhere dense, therefore none of the sets $V \cap A$ is empty, when $V \in \mathscr{F}(x)$, and therefore they form the basis of a filter on A. The filter generated by this basis on A is called the *trace* of $\mathscr{F}(x)$ on A. It is obviously a Cauchy filter; let us denote it by $\mathscr{F}(x) \cap A$. Thus its image in F is a Cauchy filter, because of Proposition 5.6. As F is complete and Hausdorff, this Cauchy filter has a unique limit in F, which we call $\tilde{f}(x)$. If $x \in A$, then $\mathscr{F}(x) \cap A$

is the filter of neighborhoods of x in A for the topology induced by E, and since f is continuous $\tilde{f}(x) = f(x)$.

Let us prove, now, that $\tilde{f}$ is continuous. Let x be a point in E, and V a *closed* neighborhood of $\tilde{f}(x)$. By definition of $\tilde{f}(x)$, V belongs to the image of $\mathscr{F}(x) \cap A$, i.e., there is U, neighborhood of x, such that

$$U \cap A \subset f^{-1}(V).$$

Let y then belong to the closure of $U \cap A$; this means that none of the sets $U' \cap U \cap A$ is empty, when U' varies over $\mathscr{F}(y)$; hence these sets form the basis of a filter on A, which is obviously a Cauchy filter, finer than $\mathscr{F}(y) \cap A$; hence its image is a filter $\mathscr{G}$ on F which converges to $\tilde{f}(y)$. This means that every neighborhood of $\tilde{f}(y)$ contains a set of the form $f(U' \cap U \cap A)$, hence some point belonging to $f(U \cap A) \subset V$. As V is closed, this implies that $\tilde{f}(y)$ belongs to V. Thus,

$$\overline{U \cap A} \subset f^{-1}(V).$$

But now observe that we may take U *open*. Let, then, $y \in \bar{U}$. Every neighborhood of y intersects U; choose an open neighborhood of y, U'. Since $U \cap U'$ is open and A is dense, $U \cap U'$ intersects A, hence U' intersects $U \cap A$, which means that $y \in \overline{U \cap A}$. This proves that

$$\bar{U} = \overline{U \cap A} \subset f^{-1}(V).$$

This implies that the preimage of V under $\tilde{f}$ is a neighborhood of x, which is what we wanted to prove.

We shall leave the proof of the uniform continuity of $\tilde{f}$ as an exercise to the student.

Suppose now that f is linear. Consider the following two mappings:

$$(x, y) \leadsto \tilde{f}(x + y), \qquad (x, y) \leadsto \tilde{f}(x) + \tilde{f}(y),$$

defined on $E \times E$ and valued in F. Since addition is continuous in E and in F, and since f is continuous, these mappings are continuous in $E \times E$. But they coincide on $A \times A$, which is a dense subset of $E \times E$. Hence they coincide everywhere, by Proposition 4.4. This proves that

$$\tilde{f}(x + y) = \tilde{f}(x) + \tilde{f}(y).$$

A similar argument holds for proving that $\tilde{f}(\lambda x) = \lambda \tilde{f}(x)$ for all $x \in E$, $\lambda \in \mathbf{C}$. This proves the last part in the statement of Theorem 5.1.

We proceed now to state and prove the theorem on completion of topological vector spaces:

THEOREM 5.2.   *Let E be a TVS. If E is Hausdorff, there exists a complete Hausdorff TVS Ê and a mapping i of E into Ê with the following properties*:

(a)   *The mapping i is an isomorphism* (for the TVS structure) *of E into Ê.*

(b)   *The image of E under i is dense in Ê.*

(c)   *To every complete Hausdorff TVS F and to every continuous linear map $f : E \to F$, there is a continuous linear map $\hat{f} : \hat{E} \to F$ such that the following diagram is commutative*:

(5.2)
$$\begin{array}{ccc} E & \xrightarrow{\;f\;} & F \\ {\scriptstyle i}\downarrow & \nearrow_{\hat{f}} & \\ \hat{E} & & \end{array}$$

Furthermore:

(I)   *Any other pair $(\hat{E}_1, i_1)$, consisting of a complete Hausdorff TVS $\hat{E}_1$ and of a mapping $i_1 : E \to E_1$ such that Properties (a) and (b) hold with $\hat{E}_1$ substituted for $\hat{E}$ and i_1 substituted for i, is isomorphic to (E, i), which means that there is an isomorphism j of $\hat{E}$ onto $\hat{E}_1$ such that the following diagram is commutative*:

(5.3)
$$\begin{array}{ccc} E & \xrightarrow{\;i_1\;} & \hat{E}_1 \\ {\scriptstyle i}\downarrow & \nearrow_{j} & \\ \hat{E} & & \end{array}$$

(II)   *Given F and f as in Property* (c), *the continuous linear map $\hat{f}$ is unique.*

Proof of Theorem 5.2

We prove the existence of $\hat{E}$ by constructing it. First we construct the *set* that is going to be $\hat{E}$, next we define vector addition and multiplication by scalars on this set, definitions which turn $\hat{E}$ into a vector space, then we define the topology of $\hat{E}$ (this topology is going to be compatible with the linear structure). This gives us the TVS $\hat{E}$. We prove that $\hat{E}$ is Hausdorff. Then we construct the isomorphism i of E into $\hat{E}$ (called the *natural injection* of E into $\hat{E}$): we define it first as a mapping, show that it is linear (this will be quite evident), and show that it is an homeomorphism, i.e., bicontinuous and one-to-one. We prove next that Im i is dense in $\hat{E}$, then that $\hat{E}$ is complete, and finally that Property (c) holds. This will conclude the existence part of the proof. The second and last part will be the uniqueness (up to isomorphisms), that is to say the proofs of Properties (I) and (II).

(1)   *The Set* $\hat{E}$

The set $\hat{E}$ is going to be the quotient of the set of all Cauchy filters on E modulo the equivalence relation:

(R) $\mathscr{F} \underset{R}{\sim} \mathscr{G}$ if to every neighborhood U of 0 in E there is an element A of $\mathscr{F}$ and an element B of $\mathscr{G}$ such that $A - B \subset U$.

If the topology of E is defined by a metric d, we may take as U the neighborhood of 0 such that $x - y \in U$ means exactly $d(x, y) < \varepsilon$. Then $A - B \subset U$ means that A is contained in the open neighborhood of order ε of B and that B is contained in the open neighborhood of order ε of A (the open neighborhood of order ε of a set S is the set of points $x \in E$ such that

$$\inf_{y \in S} d(x, y) < \varepsilon).$$

Whether this indeed defines an equivalence relation has to be checked. If $\mathscr{F}$ is a Cauchy filter, given any neighborhood of 0, U, there is $A \in \mathscr{F}$ such that $A - A \subset U$ (reflexivity of R). The symmetry of R comes from the fact that $A - B \subset U$ implies $B - A \subset -U$, and that $-U$ is a generic neighborhood of 0 in the same right as U. As for the transitivity of R, let V be a neighborhood of 0 such that $V + V \subset U$. Let $\mathscr{F}$, $\mathscr{G}$, $\mathscr{H}$ be three Cauchy filters and suppose that we have

$$\mathscr{F} \underset{R}{\sim} \mathscr{G}, \qquad \mathscr{G} \underset{R}{\sim} \mathscr{H}.$$

Then there exist $A \in \mathscr{F}$, B, $B' \in \mathscr{G}$, and $C \in \mathscr{H}$ such that

$$A - B \subset V, \qquad B' - C \subset V.$$

This immediately implies

$$A - B \cap B' \subset V, \qquad B \cap B' - C \subset V.$$

By adding we obtain

$$A - C \subset (A - B \cap B') + (B \cap B' - C) \subset V + V \subset U.$$

This proves that R is an equivalence relation. Then $\hat{E}$ is the set of equivalence classes modulo R.

(2) *Vector Addition and Multiplication by Scalars in $\hat{E}$*

From now on, elements of $\hat{E}$ will be denoted by $\hat{x}$, $\hat{y}$, etc. As $\hat{x}$ is an equivalence class of Cauchy filters we may talk about its elements or representatives.

If λ is a scalar $\neq 0$, the element $\lambda\hat{x}$ of $\hat{E}$ will be the equivalence class mod R of the filter

$$\lambda\mathscr{F} = \{\lambda A; A \in \mathscr{F}\},$$

where $\mathscr{F}$ is any representative of $\hat{x}$. That this definition does not depend on a specific choice of a representative $\mathscr{F}$ is easy to see. Indeed, if $\mathscr{F}'$ is another representative of $\hat{x}$ and if U is an arbitrary neighborhood of 0, there must exist subsets of E, $A \in \mathscr{F}$, $A' \in \mathscr{F}'$, such that

$$A - A' \subset \lambda^{-1}U, \quad \text{whence} \quad \lambda A - \lambda A' \subset U,$$

which proves that $\lambda\mathscr{F}$ and $\lambda\mathscr{F}'$ are equivalent mod R.

If $\lambda = 0$, we have $\lambda \cdot \hat{x} = \hat{o}$, where $\hat{o}$ is the equivalence class mod R of the filter of neighborhoods of the origin (or, which is the same, of the Cauchy filter consisting of *all* the subsets of E which contain 0).

Let now $\hat{x}$, $\hat{y}$ be two arbitrary elements of E, and $\mathscr{F}$ (resp. $\mathscr{G}$) a representative of $\hat{x}$ (resp. $\hat{y}$). Let us denote by $\mathscr{F} + \mathscr{G}$ the filter generated by the basis of filter

$$(\mathscr{F} + \mathscr{G})_0 = \{A + B; A \in \mathscr{F}, B \in \mathscr{G}\}.$$

That $(\mathscr{F} + \mathscr{G})_0$ is indeed the basis of a filter is easy to check. None of its elements is the empty set, and if A, $A' \in \mathscr{F}$, $B, \in \mathscr{G}$, then

$$(A \cap A') + (B \cap B') \subset (A + B) \cap (A' + B'),$$

which shows that Axiom (BF) (p. 7) is satisfied. Also easy to check is the fact that

$$\mathscr{F} \underset{R}{\sim} \mathscr{F}', \quad \mathscr{G} \underset{R}{\sim} \mathscr{G}'$$

implies

$$(\mathscr{F} + \mathscr{G}) \underset{R}{\sim} (\mathscr{F}' + \mathscr{G}').$$

Indeed, let U be two neighborhoods of 0 in E such that $V + V \subset U$. There are sets $A \in \mathscr{F}$, $A' \in \mathscr{F}'$, $B \in \mathscr{G}$, and $B' \in \mathscr{G}'$ such that

$$A - A' \subset V, \quad B - B' \subset V,$$

whence

$$(A + B) - (A' + B') \subset V + V \subset U.$$

The sum $\hat{x} + \hat{y}$ will thus be the equivalence class mod R of $\mathscr{F} + \mathscr{G}$.

(3) Topology on $\hat{E}$

Let U be an arbitrary neighborhood of zero in E. We shall set

(5.4) $\hat{U} = \{\hat{x} \in \hat{E};\ U \text{ belongs to some representative of } \hat{x}\}.$

If V is another neighborhood of 0 in E, the following inclusion is obvious:

$$\widehat{U \cap V} \subset \hat{U} \cap \hat{V}.$$

This shows that the sets $\hat{U}$ form the basis of a filter $\hat{\mathscr{F}}$ on $\hat{E}$ as U varies over the filter of neighborhoods of 0 in E. We leave it to the student to check that all properties in Theorem 3.1, (3.1)–(3.5), are satisfied when we replace, in Theorem 3.1, E by $\hat{E}$ and $\mathscr{F}$ by $\hat{\mathscr{F}}$.

(4) $\hat{E}$ Is Hausdorff

Let $\hat{x}$ be an element of E, $\hat{x} \neq 0$. This means that given any representative $\mathscr{F}$ of $\hat{x}$ and any representative $\mathscr{F}_0$ of $\hat{o}$, these two Cauchy filters are *not* equivalent modulo R. We may take as filter $\mathscr{F}_0$ the filter of all subsets of E which contain 0. Then the fact that $\mathscr{F}$ is not equivalent to $\mathscr{F}_0$ mod R means that there is some neighborhood U of 0 in E such that we cannot find $A \in \mathscr{F}$, $A_0 \in \mathscr{F}_0$ such that $A - A_0 \subset U$. In particular, since $\{0\} \in \mathscr{F}_0$, we have $A \not\subset U$ for all $A \in \mathscr{F}$, which simply means that U does not belong to $\mathscr{F}$. Let, then, V be another neighborhood of 0 such that $V + V \subset U$, and let $\mathscr{F}'$ be a Cauchy filter, equivalent to $\mathscr{F}$ modulo R (hence another representative of $\hat{x}$). I claim that V cannot belong to $\mathscr{F}'$. For otherwise, let $A \in \mathscr{F}$, $A' \in \mathscr{F}'$ be such that $A - A' \subset V$. We would have $A' \cap V \in \mathscr{F}'$, in particular $A' \cap V \neq \varnothing$, and we would have

$$A - (A' \cap V) \subset V,$$

hence

$$A \subset V + (A' \cap V) \subset V + V \subset U,$$

which is contrary to the fact $U \notin \mathscr{F}$. Thus the neighborhood of 0, V, does not belong to any representative of $\hat{x}$, which means, in view of definition (5.4), that $\hat{x} \notin \hat{V}$. This proves Statement (4).

(5) *There Is a Natural Injection of E into Ê*

The image of $x \in E$ into $\hat{E}$ is the equivalence class modulo R of the filter of neighborhoods of x; we shall denote it by $i(x)$, at least for the time being. Note that we have the following properties:

LEMMA 5.1. (a) *Two filters on E which converge to one and the same point are equivalent modulo R.*

(b) *If a filter $\mathscr{F}$ is equivalent modulo R to a filter which converges to x, then also $\mathscr{F}$ converges to x.*

The proofs are easy, and left to the student. Lemma 5.1 implies that the equivalence class modulo R of the filter of neighborhoods of x consists exactly of the filters which converge to x.

That the mapping $x \rightsquigarrow i(x)$ is linear is also very easy to check.

(6) *The Mapping i Is One-to-One, Bicontinuous, and Its Image Is Dense*

Suppose that $i(x) = i(y)$: this means that the filter of neighborhoods of x in E and the filter of neighborhoods of y in E are equivalent modulo R, therefore, in view of Lemma 5.1(b), they both converge to both points x and y. But this is impossible, unless $x = y$, in view of the uniqueness of the limit in a Hausdorff space (see p. 31). Let us now prove that i is a homeomorphism (i.e., i is continuous, and its inverse $i^{-1} : i(E) \to E$ is continuous). This means that i transforms every neighborhood U of 0 in E into a neighborhood of zero in $i(E)$, for the topology induced by $\hat{E}$, and that the preimage of such a neighborhood of zero in $i(E)$ must be a neighborhood of zero in E. Thus, first of all, we must show that if U is a neighborhood of 0 in E, $i(U)$ contains a set of the form $\hat{U}_1 \cap i(E)$, where U_1 is another neighborhood of zero in E and where

$$\hat{U}_1 = \{\hat{x} \in \hat{E}; \ U_1 \text{ belongs to some representative of } \hat{x}\}.$$

Indeed, the sets $\hat{U}_1 \cap i(E)$ form a basis of neighborhoods in the topology induced by $\hat{E}$ on $i(E)$ and the inclusion

$$(5.5) \qquad\qquad i(U) \supset \hat{U}_1 \cap i(E)$$

would imply that $i(U)$ is a neighborhood of zero in the induced topology. Conversely, we must also show that, given a neighborhood of zero, U_1, in E, there is another neighborhood of 0, U, in E, such that

$$(5.6) \qquad\qquad i(U) \subset \hat{U}_1 \cap i(E).$$

This would mean that the topology induced on $i(E)$ by $\hat{E}$ is less fine than the one carried over $i(E)$ from E by means of the one-to-one mapping i. In other words, (5.6) proves the continuity of $i : E \to i(E) \subset \hat{E}$, and (5.5) proves the continuity of $i^{-1} : i(E) \to E$.

In order to prove these facts, we shall prove the following:

(5.7) $i(\mathring{U}) \subset \mathring{U} \cap i(E) \subset i(\bar{U})$ for all neighborhoods U of 0 in E.

We recall that $\mathring{U}$ is the *interior* of U; it is, of course, a neighborhood of 0, and if $x \in \mathring{U}$, U is a neighborhood of x, which means that U belongs to a representative of $i(x)$ i.e., $i(x) \in \mathring{U}$. Suppose then that $i(x) \in \mathring{U}$; this means that U belongs to some representative of $i(x)$, in other words that U belongs to some filter converging to x. Let, then, V be an arbitrary neighborhood of 0; $V + x$ belongs to all the filters which converge to x, hence $U \cap (V + x) \neq \varnothing$, which means precisely that $x \in \bar{U}$.

(5.7) implies immediately (5.6) since $\mathring{U}$ is a neighborhood of zero. It also implies (5.5), in view of the fact (Proposition 3.1) that the closed neighborhoods of the origin in a TVS E form a basis of neighborhoods of zero in E.

This completes the proof that i is a homeomorphism, hence an isomorphism for the TVS structure. Next, we prove that the image of i, $i(E)$, is dense in $\hat{E}$.

Let $\hat{x}_0$ be an arbitrary point of $\hat{E}$; we must show that any neighborhood of $\hat{x}_0$ contains some point $i(x)$, with $x \in E$. It suffices to consider the neighborhoods of the form $\mathring{U} + \hat{x}_0$, where U is defined by (5.4). The relation $i(x) - \hat{x}_0 \in \mathring{U}$ means that U belongs to some representative of $i(x) - \hat{x}_0$, i.e., that there is a filter $\mathscr{F}$ converging to x and a filter $\mathscr{F}_0$, representing $\hat{x}_0$, some set $A \in \mathscr{F}$ and a set $A_0 \in \mathscr{F}_0$ such that

$$A - A_0 \subset U.$$

Let us then prove that such a point x exists. We select a neighborhood V of 0 in E such that $V + V \subset U$. Let $\mathscr{F}_0$ be any Cauchy filter representing x_0 and let A_0 be an element of $\mathscr{F}_0$ such that

$$A_0 - A_0 \subset V \quad \text{(remember that } \mathscr{F}_0 \text{ is a Cauchy filter!)}.$$

We choose, as point x, an arbitrary point of the set A_0. We have

$$(V + x) - A_0 \subset V + (A_0 - A_0) \subset V + V \subset U.$$

But $V + x$ belongs to any filter $\mathscr{F}$ converging to x.

(7) *Ê Is Complete*

Let $\mathscr{F}$ be a Cauchy filter on $\hat{E}$. We consider the family of subsets of E,

$$\hat{M} + \hat{U}, \qquad \hat{M} \in \mathscr{F}, \quad \hat{U}, \text{neighborhood of zero in } \hat{E}.$$

They form the basis of a filter; indeed, none of them is zero, and

$$(\hat{M} + \hat{U}) \cap (\hat{M}' + \hat{U}') \supset \hat{M} \cap \hat{M}' + \hat{U} \cap \hat{U}'.$$

(The filter $\mathscr{F}'$ generated by this basis is less fine than $\mathscr{F}$, since any element of the basis belongs to $\mathscr{F}$.) The filter $\mathscr{F}'$ is a Cauchy filter. Indeed, let $\hat{U}$ be any neighborhood of zero in $\hat{E}$. Select another neighborhood of 0 in $\hat{E}$, $\hat{V}$, such that

$$\hat{V} + \hat{V} - \hat{V} \subset \hat{U}$$

(for instance take $\hat{V}_0$ balanced, $\hat{V}_0 \subset \hat{U}$, and $\hat{V} = \frac{1}{3}\hat{V}_0$). Let $\hat{M}$ be an element of $\mathscr{F}$ such that $\hat{M} - \hat{M} \subset \hat{V}$. This implies $(\hat{M} + \hat{V}) - (\hat{M} + \hat{V}) \subset \hat{V} + \hat{V} - \hat{V} \subset \hat{U}$.

Now, let $\mathscr{F}'$ be the family of subsets of $i(E)$ of the form $\hat{A} \cap i(E)$, with $\hat{A} \in \mathscr{F}'$. These intersections contain intersections of the form

$$(\hat{M} + \hat{U}) \cap i(E), \qquad \text{with} \quad \hat{M} \in \mathscr{F}, \quad \hat{U}, \text{neighborhood of 0 in } \hat{E}.$$

As $\hat{M}$ cannot be empty, they contain intersections of the form

$$(\hat{y} + \hat{U}) \cap i(E), \qquad \text{with} \quad \hat{y} \in \hat{E}.$$

As $i(E)$ is dense in $\hat{E}$, these intersections are never empty. From there on, is is quite obvious that $\mathscr{F}'$ is a filter on $i(E)$, and in fact a Cauchy filter on $i(E)$. Using the fact that i is an isomorphism of E onto $i(E)$, we conclude that $i^{-1}(\mathscr{F}')$ is a Cauchy filter on E. Its equivalence class modulo R, $\hat{x}$, is the limit of $\mathscr{F}$. The student might try to prove this point as an exercise.

(8) *Proof of Property* (c)

It suffices to apply Theorem 5.1 with E replaced by $\hat{E}$ and A by E. We also obtain the uniqueness of the extension $\tilde{f}$, stated in Property (II).

(9) *Proof of the Uniqueness (up to Isomorphisms) of $\hat{E}$*

We prove now Property (I) of Theorem 5.2. In Diagram (5.3), we may define j as $\hat{i}_1$ with the notation of Property (c). Let, on the other hand, f be the mapping from $i_1(E)$ into $\hat{E}$ defined by

$$i_1(x) \rightsquigarrow i(x).$$

Let us apply Theorem 5.1 with $i_1(E)$ instead of A, $\hat{E}_1$ instead of E, and $\hat{E}$ instead of F. Then f has a unique extension, $\tilde{f} : \hat{E}_1 \to \hat{E}$. It is easy to check, using the density of $i(E) \subset \hat{E}$ and of $i_1(E) \subset \hat{E}_1$, and the continuity of the mappings involved, that

$$\tilde{f}(j(\hat{x})) = \hat{x} \qquad \text{for all} \quad \hat{x} \in \hat{E};$$
$$j(\tilde{f}(\hat{x}_1)) = \hat{x}_1 \qquad \text{for all} \quad \hat{x}_1 \in \hat{E}_1 .$$

This means that j and $\tilde{f}$ are the inverse of each other and that both are isomorphisms.

The proof of Theorem 5.2 is complete.

In the sequel, we shall always identify E with $i(E)$ and regard E as a (dense) vector subspace of $\hat{E}$.

Exercises

5.1. Let E, F be two TVS. Show that there is a canonical isomorphism between $\hat{E} \times \hat{F}$ and $(E \times F)^\wedge$.

5.2. Let $\mathscr{C}^1(\mathbf{R}^1)$ be the vector space of functions with complex values, defined and once continuously differentiable on the real line, and $\mathscr{C}^1_0(\mathbf{R}^1)$ the space of functions $f \in \mathscr{C}^1(\mathbf{R}^1)$ which vanish outside some finite interval $[a, b]$ $(-\infty < a < b < +\infty)$. For $\varepsilon > 0$ and $n = 1, 2, \dots$, we set

$$\mathscr{W}_{\varepsilon, n} = \{ f \in \mathscr{C}^1(\mathbf{R}^1); \sup_{|t| < n} \{ | f(t)| + | f'(t)| \} \leqslant \varepsilon \}.$$

Show that the sets $\mathscr{W}_{\varepsilon, n}$ form a basis of neighborhoods of zero for a Hausdorff topology on $\mathscr{C}^1(\mathbf{R}^1)$ compatible with the linear structure.

Prove that the TVS $\mathscr{C}^1(\mathbf{R}^1)$ is complete.

Prove that the linear subspace $\mathscr{C}^1_0(\mathbf{R}^1)$ is dense in $\mathscr{C}^1(\mathbf{R}^1)$. Does that mean that $\mathscr{C}^1(\mathbf{R}^1)$ is isomorphic to the completion of $\mathscr{C}^1_0(\mathbf{R}^1)$?

5.3. We suppose that $\mathscr{C}^1(\mathbf{R}^1)$ and $\mathscr{C}^1_0(\mathbf{R}^1)$ carry the topology defined in Exercise 5.2. Let $\mathscr{C}^0(\mathbf{R}^1)$ be the space of continuous complex functions in $\mathbf{R}^1$ equipped with the topology where a basis of neighborhoods of zero is made up by the sets

$$\{ f \in \mathscr{C}^0(\mathbf{R}^1); \sup_{|t| < n} | f(t)| < \epsilon \}, \qquad \epsilon > 0, \quad n = 1, 2, \dots .$$

Consider the mapping $f \rightsquigarrow f'$, which we denote by D. Prove that D is continuous as a map of $\mathscr{C}^1(\mathbf{R}^1)$ (resp. $\mathscr{C}^1_0(\mathbf{R}^1)$) into $\mathscr{C}^0(\mathbf{R}^1)$. Make use of the properties of the operator D in the spaces $\mathscr{C}^1(\mathbf{R}^1)$ and $\mathscr{C}^1_0(\mathbf{R}^1)$ to prove the following fact: if E and F are two TVS, $u : E \to F$ a continuous linear injection (i.e., one-to-one mapping), and $\hat{u} : \hat{E} \to \hat{F}$ the continuous extension of u to the completions, then $\hat{u}$ is not necessarily one-to-one. Prove that $\hat{u}$ is one-to-one whenever u is an isomorphism into.

5.4. Let E be a TVS and E^* its algebraic dual. Provide E^* with the topology of pointwise convergence in E. A basis of neighborhoods of zero in this topology is provided by the sets

$$\mathscr{W}(S, \varepsilon) = \{ x^* \in E^*; \sup_{x \in S} | x^*(x)| \leqslant \varepsilon \}$$

as S ranges over the family of finite subsets of E and ε over the set of numbers > 0. Prove that E^* is complete.

5.5. Let $\{E_\alpha\}$ be an arbitrary family of TVS, E their product, and p_α the coordinate projection of E onto E_α . Prove that a filter $\mathscr{F}$ on E is a Cauchy filter if and only if every $\mathscr{F}_\alpha$, image of $\mathscr{F}$ under p_α , is a Cauchy filter in E_α and that E is complete if and only if every E_α is complete (cf. Exercise 1.9).

5.6. Indicate, in the list below, which ones are the TVS that are complete and which ones are the TVS that are not complete:

(1)   the space $\mathbf{C}[[X]]$ of formal power series in one indeterminate, equipped with the topology of simple convergence of the coefficients (see p. 25 *et seq.*);

(2)   the space of finite sequences of complex numbers $s = (s_1, ..., s_\nu)$ ($\nu < \infty$ but depending on $s!$), with the topology defined by the basis of neighborhoods of zero consisting of the sets

$$\mathscr{U}(\varepsilon) = \{s = (s_j); \sup_j |s_j| \leqslant \varepsilon\}, \qquad \varepsilon > 0;$$

(3)   the space of continuous (complex) functions f on the real line, which converge to zero at infinity (i.e., $|f(t)| \to 0$ as $|t| \to \infty$), equipped with the topology of uniform convergence, i.e., the topology defined by the basis of neighborhoods of zero

$$\mathscr{U}_1(\varepsilon) = \{f; \sup_t |f(t)| \leqslant \varepsilon\}, \qquad \varepsilon > 0;$$

(4)   the space of continuous complex functions f on the closed interval $[0, 1]$, equipped with the topology defined by the basis of neighborhoods of zero

$$\mathscr{V}(\varepsilon) = \left\{f; \int_0^1 |f(t)| \, dt \leqslant \varepsilon\right\}, \qquad \varepsilon > 0.$$

5.7. Prove that the TVS $\mathscr{F}_6(\mathbf{N}, \mathbf{C})$, defined in Exercise 3.5, is complete.

6

Compact Sets

A topological space X (not necessarily the subset of a TVS) is said to be *compact* if X is Hausdorff and if every open covering $\{\Omega_i\}$ of X contains a finite subcovering. The fact that $\{\Omega_i\}$ is an open covering of X means that each Ω_i is an open subset of X and the union of the sets Ω_i is equal to X. By a finite subcovering of the covering $\{\Omega_i\}$ we mean a finite collection $\Omega_{i_1},\ldots,\Omega_{i_r}$ of sets Ω_i whose union is still equal to X. By going to the complements of the open sets Ω_i we obtain an equivalent definition of compactness: a Hausdorff space X is compact if every family of closed sets $\{F_i\}$ whose intersection is empty contains a finite subfamily whose intersection is empty.

In the sequel, we shall almost always be concerned with compact spaces which are subsets of a TVS and which carry the topology induced by the TVS in question; we shall then refer to them as *compact sets*. Let Y be a subset of a Hausdorff topological space X; a subset of Y, B, is open in the sense of the topology induced by X if and only if there is an open subset A of X such that $B = A \cap Y$. In view of this, open coverings of Y are "induced" by families of open subsets of X whose union contains Y. Thus a subset K of X is compact if every family $\{\Omega_i\}$ of open subsets of X, whose union contains K, contains a finite subfamily whose union contains K. It should be pointed out that compactness is such that many properties of compact sets are independent, to a large extent, of the surrounding space. This will become apparent soon.

We begin by stating without proof a few well-known properties of compact spaces (no linear structure is considered). If the student is not familiar with these properties, we strongly suggest that he proceed no further without having proved them by himself.

PROPOSITION 6.1.   *A closed subset of a compact space is compact.*

PROPOSITION 6.2.   *Let f be a continuous mapping of a compact space X into a Hausdorff topological space Y. Then $f(X)$ is a compact subset of Y.*

PROPOSITION 6.3. *Let f be a one-to-one continuous mapping of a compact space X onto a compact space Y. Then f is a homeomorphism (i.e., f $^{-1}$ is also continuous).*

PROPOSITION 6.4. *Let $\mathscr{T}$, $\mathscr{T}'$ be two Hausdorff topologies on a set X. Suppose that $\mathscr{T}$ is finer than $\mathscr{T}'$ and that X, equipped with $\mathscr{T}$, is compact. Then $\mathscr{T} = \mathscr{T}'$.*

Finite unions and arbitrary intersections of compact sets are compact. In a Hausdorff space X, every point is compact; every converging sequence is compact—provided that we include in it its limit point!

The student should keep in mind that compact sets can be very complicated sets. Take for instance the real line $\mathbf{R}^1$. The Borel–Lebesgue–Heine theorem says that the compact subsets of $\mathbf{R}^1$ are exactly the sets which are both closed and bounded. Note also that the Lebesgue measure of a sequence is equal to zero, and that if a set A is measurable, given any $\varepsilon > 0$, there is a compact set $K \subset A$ such that the measure of $A \cap \complement K$ is $\leqslant \varepsilon$. Take then the points x, with $0 \leqslant x \leqslant 1$, which are *nonrational*; they form a set of measure 1, since the rationals, which form a sequence, form a set of measure zero. This means that there are compact sets, contained in the interval $[0, 1]$, *which do not contain any rational number* and whose Lebesgue measure is arbitrarily close to 1. Try to draw one of them!

As the Weierstrass–Bolzano theorem shows, compact sets have interesting properties in relation with sequences of points. This extends to filters, as we are now going to see. In the immediate sequel, E is a Hausdorff topological space; when it is expressly mentioned, E is a TVS.

The following terminology is useful:

Definition 6.1. *A point x of E is called an accumulation point of a filter $\mathscr{F}$ if x belongs to the closure of every set which belongs to $\mathscr{F}$.*

Let $S = \{x_0, x_1, ...\}$ be a sequence; a point x of E is often called an accumulation point of S if every neighborhood of x contains a point of S different from x. This terminology coincides with the one introduced by Definition 6.1 if we apply the latter to the filter $\mathscr{F}_S$ associated with S (see p. 7). Let $M \in \mathscr{F}_S$ be arbitrary; M contains a subsequence of the form

$$S_n = \{x_n, x_{n+1}, ...\}.$$

If, then, x is an accumulation point of S, any neighborhood U of x contains some point x_k with k arbitrarily large, in particular $k \geqslant n$. Thus U has a nonempty intersection with M, which means that $x \in \bar{M}$. In other words, any accumulation point of the sequence S is an accumulation point of the filter $\mathscr{F}_S$. Conversely, if x is an accumulation point of the filter $\mathscr{F}_S$, x belongs to the closure of all the sets belonging to $\mathscr{F}_S$, in particular to the closures of the sets S_n, $n = 0, 1, \ldots$. This means that given any neighborhood U of x and any integer n, there is $k \geqslant n$ such that $x_k \in U$.

PROPOSITION 6.5.   *If a filter $\mathscr{F}$ converges to a point x, x is an accumulation point of $\mathscr{F}$.*

Indeed, suppose that x were not an accumulation point of $\mathscr{F}$. There would be a set $M \in \mathscr{F}$ such that $x \notin \bar{M}$. Hence the complement U of $\bar{M}$, which is open, would be a neighborhood of x, and hence should belong to $\mathscr{F}$. But then we ought to have $U \cap M \neq \emptyset$, whence a contradiction.

Of course, a filter might have more than one accumulation point. For instance, let $\mathscr{F}$ be the filter of all subsets of E containing a given subset A of E. Then every point of A is an accumulation point of $\mathscr{F}$.

PROPOSITION 6.6.   *The following two conditions are equivalent*:

(a)   *x is an accumulation point of $\mathscr{F}$;*

(b)   *there is a filter $\mathscr{F}'$ which is finer than both $\mathscr{F}$ and the filter of neighborhoods of x, $\mathscr{F}(x)$, in other words: there is a filter $\mathscr{F}'$ converging to x, which is finer than $\mathscr{F}$.*

(a) $\Rightarrow$ (b). Indeed, consider the family of subsets of E of the form $U \cap M$, where U varies over $\mathscr{F}(x)$ and M varies over $\mathscr{F}$. These sets are never empty if x is an accumulation point of $\mathscr{F}$, and they obviously have Property (BF) of p. 7, hence they generate a filter $\mathscr{F}'$ which is obviously finer than $\mathscr{F}$ and $\mathscr{F}(x)$.

(b) $\Rightarrow$ (a).   If a filter $\mathscr{F}$ is less fine than another filter $\mathscr{F}'$ and if x' is an accumulation point of $\mathscr{F}'$, then x is also an accumulation point of $\mathscr{F}$. Thus it suffices to combine (b) with Proposition 6.5.

PROPOSITION 6.7.   *If a Cauchy filter $\mathscr{F}$ on the TVS E has an accumulation point x, it converges to x.*

Let U be an arbitrary neighborhood of the origin and V another neighborhood of zero such that $V + V \subset U$. There is a set $M \in \mathscr{F}$ such that $M - M \subset V$. On the other hand, $V + x$ intersects M, hence $M - M \cap (V + x) \subset V$, or

$$M \subset V + M \cap (V + x) \subset V + V + x \subset U + x. \qquad \text{Q.E.D.}$$

PROPOSITION 6.8. *Let K be a Hausdorff topological space. The following properties are equivalent:*

(a) *K is compact;*

(b) *every filter on K has at least one accumulation point.*

(a) $\Rightarrow$ (b). Let $\mathscr{F}$ be any filter on K, and consider the family of closed sets $\bar{M}$ when M varies over $\mathscr{F}$. As no finite intersection of sets $\bar{M}$ can be empty, neither can that be true of the intersection of all of them.

(b) $\Rightarrow$ (a). Let Φ be a family of closed sets whose total intersection is empty. Suppose that Φ does not contain any *finite* subfamily whose intersection is empty. Then take the family Φ' of all the finite intersections of subsets belonging to Φ: it obviously forms a basis of a filter. This filter has an accumulation point, say x: thus x belongs to the closure of any subset belonging to the filter, in particular to any set belonging to Φ', for these are closed. In other words, x belongs to the intersection of all the sets belonging to Φ', which is the same as the intersection of all the sets belonging to Φ. But the latter was supposed to be empty!

Q.E.D.

COROLLARY 1. *A compact subset K of a Hausdorff topological space E is closed.*

Proof. Let $x \in \bar{K}$; let $\mathscr{F}(x)|K$ be the filter on K generated by the sets $U \cap K$ when U ranges over the filter of neighborhoods of x in E; that the sets $U \cap K$ form the basis of a filter means precisely that x belongs to the closure of K. In view of Proposition 6.8, $\mathscr{F}(x)|K$ must have an accumulation point $x_1 \in K$. Necessarily $x_1 = x$: otherwise we could find a neighborhood U of x whose complement in E is a neighborhood of x_1 and we could certainly not have $x_1 \in \bar{U}$, even less $x_1 \in \overline{U \cap K}$. Thus x belongs to K.

COROLLARY 2. *A compact subset of a Hausdorff TVS is complete.*

It suffices to combine Propositions 6.7 and 6.8.

COROLLARY 3. *In a compact topological space K, every sequence has an accumulation point.*

Definition 6.2. *A subset A of a topological space X is said to be relatively compact if the closure $\bar{A}$ of A is compact.*

A converging sequence (without the limit point) is a relatively compact set.

Definition 6.3. *A subset A of a Hausdorff TVS E is said to be precompact if A is relatively compact when viewed as a subset of the completion $\hat{E}$ of E.*

A Cauchy sequence in E is precompact; it is not necessarily relatively compact! For this would mean that it converges (Proposition 6.7). Another example illustrating the difference between relatively compact sets and precompact sets is the following one: let Ω be an open subset of $\mathbf{R}^n$, different from $\mathbf{R}^n$. In virtue of the Heine–Borel–Lebesgue theorem, every bounded open subset of Ω is precompact; but an open subset Ω' of Ω is relatively compact, in Ω, if and only if at the same time Ω' is bounded and its closure is contained in Ω.

A subset K of the Hausdorff TVS E is compact if and only if K is both complete and precompact. Indeed, if K is compact when viewed as subset of E, K is still compact when viewed as subset of $\hat{E}$: therefore, by Corollaries 1 and 2 of Proposition 6.8, we know that K is closed in $\hat{E}$, hence complete; of course its closure in $\hat{E}$, identical to K itself, is compact. Conversely, if K is complete, we have $\hat{K} = K$ and therefore, if $\hat{K}$ is compact, K is also compact.

Our purpose is to prove a criterion of precompactness which is to be used later. The proof of it is made very easy if we use the notion of *ultrafilter*:

Definition 6.4. *A filter* $\mathfrak{U}$ *on a set* A *is called an ultrafilter if every filter on* A *which is finer than* $\mathfrak{U}$ *is identical to* $\mathfrak{U}$.

LEMMA 6.1. *Let* $\mathscr{F}$ *be a filter on a set* A; *there is at least one ultrafilter on* A *which is finer than* $\mathscr{F}$.

Proof. Let Φ be the family of all filters on A finer than $\mathscr{F}$, ordered by the relation "to be finer than," and Φ' a subfamily of Φ totally ordered for this relation. The elements of Φ' are filters, that is to say subsets of the set of subsets $\mathfrak{P}(A)$ of A; we may therefore consider their union $\mathscr{F}'$. It is immediately seen that $\mathscr{F}'$ is a filter on A, obviously finer than $\mathscr{F}$. We may therefore apply Zorn's lemma to the family Φ, whence Lemma 6.1.

LEMMA 6.2. *Let* A *be a topological space*; *if an ultrafilter* $\mathfrak{U}$ *on* A *has an accumulation point in* A, $\mathfrak{U}$ *converges to* x.

Proof. We apply Proposition 6.6: if x is an accumulation point of $\mathfrak{U}$, there is a filter $\mathscr{F}$ which is finer than $\mathfrak{U}$ and converges to x. As $\mathfrak{U}$ is an ultrafilter, we must have $\mathscr{F} = \mathfrak{U}$.

LEMMA 6.3. *A Hausdorff topological space* K *is compact if and only if every ultrafilter on* K *converges.*

Proof. If K is compact, every filter on K has an accumulation point

(Proposition 6.8), therefore every ultrafilter converges, by Lemma 6.2. Conversely, suppose that every ultrafilter converges in K and let $\mathscr{F}$ be some filter on K. By Lemma 6.1, there is an ultrafilter $\mathfrak{U}$ on K which is finer than $\mathscr{F}$; $\mathfrak{U}$ converges to some point x. By Proposition 6.6, x is an accumulation point of $\mathscr{F}$.   Q.E.D.

We may now state and prove the announced criterion of pre-compactness:

PROPOSITION 6.9. *The following properties of a subset K of a Hausdorff TVS E are equivalent:*

(a) *K is precompact;*

(b) *given any neighborhood of the origin V in E, there is a finite family of points of K, $x_1 ,..., x_r$, such that the sets $x_i + V$ form a covering of K, i.e., such that*

$$K \subset (x_1 + V) \cup \cdots \cup (x_r + V).$$

Proof. (a) *implies* (b). Let U be an open neighborhood of zero in E, contained in V. There exists an open neighborhood of zero in $\hat{E}$, $\hat{U}$, such that $U = \hat{U} \cap E$. Consider the family of sets $x + \hat{U}$ when x varies over K. They form an open covering of the closure $\hat{K}$ of K in E. Indeed, let $\hat{y}$ be an arbitrary point of $\hat{K}$ and let $\hat{W}$ be a neighborhood of of zero in $\hat{E}$ such that $\hat{W} = -\hat{W} \subset \hat{U}$. Then there is $x \in K$ such that $x \in \hat{y} + \hat{W}$, i.e., $\hat{y} \in x + \hat{W} \subset x + \hat{U}$. This open covering of the compact set $\hat{K}$ contains a finite subcovering, $x_1 + \hat{U},..., x_r + \hat{U}$. We have:

$$K = \hat{K} \cap E \subset [(x_1 + \hat{U}) \cap E] \cap \cdots \cap [(x_r + \hat{U}) \cap E]$$
$$\subset (x_1 + V) \cup \cdots \cup (x_r + V).$$

(b) *implies* (a). If K possesses Property (b), its closure $\hat{K}$ in $\hat{E}$ possesses the same property in $\hat{E}$. Indeed, let $\hat{V}$ be an arbitrary neighborhood of zero in $\hat{E}$; let $\hat{W}$ be a *closed* neighborhood of zero in $\hat{E}$, contained in $\hat{V}$. There is a finite number of points of K, $x_1 ,..., x_r$, such that

$$K \subset (x_1 + \hat{W}) \cup \cdots \cup (x_r + \hat{W}).$$

But as the right-hand side is a closed subset of $\hat{E}$, it also contains the closure of K in $\hat{E}$, whence our assertion. In view of this, it will suffice to prove that if a *closed* subset K of a *complete* Hausdorff TVS E has Property (b), it is compact. We shall apply Lemma 6.3 and show that an arbitrary ultrafilter $\mathfrak{U}$ on K converges to a point of K.

Let V be an arbitrary neighborhood of zero in E, and $x_1, \ldots, x_r$ a finite family of points of K such that the sets $(x_j + V)$ form a covering of K. We contend that at least one of these sets $x_j + V$ belongs to the filter $\mathfrak{U}$. First of all, we show that at least one of these sets intersects every set belonging to $\mathfrak{U}$. If this were not true, for each $j = 1, \ldots, r$, we would be able to find a set $M_j \in \mathfrak{U}$ which does not intersect $x_j + V$; then the intersection $M_1 \cap \cdots \cap M_r$ would be empty, contrary to the fact that $\mathfrak{U}$ is a filter, since it would not intersect any one of the sets $x_j + V$ and since these form a covering of K. Thus one of the sets $x_j + V$, say $x_i + V$, intersects all the sets which belong to $\mathfrak{U}$. This means that, if we consider the family of subsets of K of the form $M \cap (x_i + V)$, where M runs over $\mathfrak{U}$, it is the basis of a filter on K. This filter is obviously finer than, therefore equal to, the ultrafilter $\mathfrak{U}$. In other words, the set $(x_i + V)$ belongs to $\mathfrak{U}$. If y, z are two elements of this set, we have $y - z \in V$. In other words, we have proved that, given any neighborhood of zero V in E, there is a set $M \in \mathfrak{U}$ such that $M - M \subset V$: $\mathfrak{U}$ is a Cauchy filter. But as E is complete, $\mathfrak{U}$ must converge to some point $x \in E$; as K is closed, $x \in K$. $\hspace{2em}$ Q.E.D.

Exercises

6.1. By using the compactness criterion provided by Lemma 6.3, prove the following version of *Tychonoff's theorem*:

THEOREM 6.1. *Let $\{E_i\}$ ($i \in I$) be a family of Hausdorff TVS, and $E = \prod_{i \in I} E_i$ their product (equipped with the product TVS structure). Let A_i be a subset of E_i for each index i, and $A = \prod_{i \in I} A_i$ the product of the A_i's, regarded as a subset of E.*
Then A is compact in E if and only if, for every $i \in I$, A_i is compact in E_i.

6.2. Prove that the balanced hull of a compact subset K of a Hausdorff TVS E (i.e., the smallest balanced set containing K) is compact.

6.3. Prove that a TVS E is compact if and only if it consists of a single element, 0.

6.4. Consider the TVS $\mathscr{F}_c(\mathbf{N}; \mathbf{C})$ of complex functions on the set $\mathbf{N}$ of nonnegative integers with the topology defined in Exercise 3.5. Prove that every converging sequence in $\mathscr{F}_c(\mathbf{N}; \mathbf{C})$ must be contained in some space $\mathscr{F}_c(n)$ (see Exercise 3.6). Derive from this the fact that every compact subset of $\mathscr{F}_c(\mathbf{N}, \mathbf{C})$ is contained in a finite dimensional linear subspace.

6.5. Let E be any one of the TVS, (a)–(d), of Exercise 3.4. Prove that any *infinite* dimensional linear subspace M of E contains a sequence which converges in E and which is not contained in any finite dimensional linear subspace of M.

7

Locally Convex Spaces. Seminorms

A subset K of a vector space E is convex if, whenever K contains two points x and y, K also contains the segment of straight line joining them: if $x, y \in K$ and if α, β are two numbers $\geqslant 0$ and such that $\alpha + \beta = 1$, then

$$\alpha x + \beta y \in K.$$

Let S be any subset of E. Let us call the *convex hull* of S the set of all finite linear combinations of elements of S with *nonnegative* coefficients and such, furthermore, that *the sum of the coefficients be equal to one.* Thus a set is convex if it is equal to its own convex hull. And the convex hull of a set S is the smallest convex set containing S.

Arbitrary intersections of convex sets are convex sets. Unions of convex sets are generally *not* convex. The vector sum of two convex sets is convex. The image and the preimage of a convex set under a linear map is convex.

PROPOSITION 7.1. *Let E be a TVS. The closure and the interior of convex sets are convex sets.*

The statement relative to the closure is evident. Not so the one about the interior. Let K be a convex set, $\overset{\circ}{K}$ its interior, and let x, y be any two points of $\overset{\circ}{K}$, z a point in the segment joining x to y. We know that $z \in K$ and we must show that $z \in \overset{\circ}{K}$. We have

$$z = tx + (1 - t)y \qquad \text{for some number} \quad 0 \leqslant t \leqslant 1.$$

On the other hand, there exists a neighborhood U of 0 in E such that $x + U \subset K$ and $y + U \subset K$. Then, of course, the claim is that $z + U \subset K$. This is indeed so, since any element $z + u$ of $z + U$ can be written in the form

$$tx + (1 - t)y + tu + (1 - t)u = t(x + u) + (1 - t)(y + u),$$

and since both vectors $x + u$ and $y + u$ belong to K, so does $z + u$.

Definition 7.1.   A subset T of a TVS E is called a barrel if T has the following four properties:

(1) T *is absorbing* (Definition 3.1);
(2) T *is balanced* (Definition 3.2);
(3) T *is closed;*
(4) T *is convex.*

Let U be any neighborhood of 0 in E. Let us denote by $T(U)$ the *closed convex hull* (i.e., the closure of the convex hull) of the set

$$(7.1) \qquad\qquad \bigcup_{\lambda \in \mathbf{C}, |\lambda| \leqslant 1} \lambda U.$$

Then $T(U)$ is a *barrel*. Properties (3) and (4) are evident; Property (1) holds since $U \subset T(U)$. It remains to show that $T(U)$ is balanced. It suffices to prove that the convex hull of the set (7.1) is balanced (the closure of a balanced set is obviously balanced). Any point of the convex hull can be written

$$z = tx + (1 - t)y,$$

with $x \in \lambda U$, $y \in \mu U$, for some t, λ, μ, $0 \leqslant t \leqslant 1$, $|\lambda| \leqslant 1$, $|\mu| \leqslant 1$. If $\zeta \in \mathbf{C}$, $|\zeta| \leqslant 1$, we have

$$\zeta z = t(\zeta x) + (1 - t)(\zeta y) \quad \text{and} \quad \zeta x \in \zeta \lambda U, \quad y \in \zeta \mu U.$$

Thus, every neighborhood of 0 in a TVS is contained in a neighborhood of 0 which is a barrel. But, of course, not every neighborhood of 0 contains another one which is a barrel, nor is any barrel a neighborhood of zero.

Definition 7.2.   A TVS E is said to be a locally convex space if there is a basis of neighborhoods in E consisting of convex sets.

Locally convex spaces are by far the most important class of TVS.

PROPOSITION 7.2.   *In a locally convex space E, there is a basis of neighborhoods of zero consisting of barrels.*

Let U_1 be an arbitrary neighborhood of zero in E. Since E is a TVS, U_1 contains a closed neighborhood of 0, say V (Proposition 3.1). But since, on the other hand, E is locally convex, V contains a convex neighborhood of zero, W; and finally, W contains a balanced neighborhood of 0, say U. As U is balanced, the set (7.1) associated with U is identical with U. Its convex hull is contained in W, and the closure of this convex hull, which is a barrel, is contained in V, hence in U_1.   Q.E.D.

When we deal with a family $\mathscr{B}$ of *convex balanced* absorbing subsets of a vector space E, it is very simple to ascertain that it constitutes a basis of neighborhoods of zero in a topology on E compatible with the linear structure of E (and necessarily locally convex!). Indeed, it is enough that $\mathscr{B}$ possess the following two properties:

(*)   For every pair U, $V \in \mathscr{B}$, there exists $W \in \mathscr{B}$ such that

$$W \subset U \cap V.$$

(**)   For every $U \in \mathscr{B}$ and every $\rho > 0$, there is $W \in \mathscr{B}$ such that

$$W \subset \rho U.$$

It suffices to check that the filter generated by $\mathscr{B}$ satisfies Conditions (3.1)–(3.5) of Theorem 3.1. In particular, the set of all multiples ρU of a convex balanced absorbing subset U of E form a basis of neighborhoods of 0 in a locally convex topology on E (this ceases to be true, in general, if we relax the conditions on U).

Definition 7.3.   A nonnegative function $x \rightsquigarrow p(x)$ on a vector space E is called a seminorm if it satisfies the following conditions:

(1)   *p is subadditive, i.e., for all x, $y \in E$, $p(x + y) \leqslant p(x) + p(y)$;*

(2)   *p is positively homogeneous of degree 1, i.e., for all $x \in E$ and all $\lambda \in \mathbf{C}$, $p(\lambda x) = |\lambda| \, p(x)$;*

(3)   *$p(0) = 0$ (implied by Property (2)).*

Definition 7.4.   A seminorm on a vector space E is called a norm if

$$x \in E, p(x) = 0 \qquad implies \quad x = 0.$$

Example (1).   Suppose $E = \mathbf{C}^n$ and let M be a vector subspace of E. Set $p_M(x) = $ distance from x to M, in the usual sense of the distance in $\mathbf{C}^n$. If dim $M \geqslant 1$, then p_M is a seminorm and not a norm (M is exactly the kernel of p_M). When $M = \{0\}$, p_M is the Euclidean norm.

Examples of norms in $\mathbf{C}^n$ (cf. Theorem 11.1 and Chapter 11, Example IV):

$$\zeta = (\zeta_1, ..., \zeta_n) \rightarrow |\zeta|_p = (|\zeta_1|^p + \cdots + |\zeta_n|^p)^{1/p}, \qquad 1 \leqslant p < +\infty,$$

$$\zeta \rightarrow |\zeta|_\infty = \sup_{1 \leqslant j \leqslant n} |\zeta_j|.$$

Observe that $|\zeta|_2$ is the Euclidean, or Hermitian norm; it will always be denoted by $|\zeta|$ in this text. Later on, when studying differential operators, we shall use the norm $|\;\;|_1$ on vectors whose coordinates are nonnegative integers.

Example (2). Let E be a vector space on which is defined a sesquilinear form $B(e, f)$ (*sesquilinear* means that

$$B(e_1 + e_2, f) = B(e_1, f) + B(e_2, f);$$
$$B(e, f_1 + f_2) = B(e, f_1) + B(e, f_2);$$
$$B(\lambda e, f) = \lambda\, B(e, f);$$
$$B(e, \lambda f) = \bar{\lambda}\, B(e, f);$$

and $B(e, f)$ is complex valued). Suppose that $B(e, f)$ is *Hermitian*, which means that

$$B(e, f) = \overline{B(f, e)}.$$

Observe then that, for all $e \in E$, $B(e, e)$ is a real number. Let us say that B is *nonnegative* if this number is never negative. Then it can be proved, by using the Schwarz inequality (cf. Chapter 12, Proposition 12.1 and Corollary) that

(7.2) $$e \rightsquigarrow (B(e, e))^{1/2}$$

is a seminorm on E. It is a norm on E if and only if B is *definite positive*, which means that $B(e, e) > 0$ for all $e \neq 0$.

Definition 7.5. *A vector space E over the field of complex numbers, provided with a Hermitian nonnegative form, is called a complex pre-Hilbert space.*

Example (3). Let $\mathscr{C}^0(\mathbf{R}^1)$ be the vector space (over the field of complex numbers) of complex-valued continuous functions on the real line. For any *bounded* interval $[a, b]$ $(-\infty < a < b < +\infty)$, and any function $f \in \mathscr{C}^0(\mathbf{R}^1)$, we set

$$\mathscr{P}_{[a,b]}(f) = \sup_{a \leqslant t \leqslant b} |f(t)|.$$

Then $f \rightsquigarrow \mathscr{P}_{[a,b]}(f)$ is a seminorm. It is never a norm, since f may very well vanish in the interval $[a, b]$ without being identically zero. Other seminorms are the following ones:

$$f \rightsquigarrow |f(0)|;$$

(7.3) $$f \rightsquigarrow \left(\int_a^b |f(t)|^p\, dt \right)^{1/p} \quad \text{with} \quad 1 \leqslant p < +\infty.$$

The fact that (7.3) is a seminorm is not absolutely obvious: one has to prove the triangular inequality, that is to say the subadditivity of the function (7.3). When $p < 1$, (7.3) is *not* subadditive, and therefore it cannot be a seminorm.

Example (4).   Let us denote by l^p ($1 \leqslant p < +\infty$) the vector space of of complex sequences $\{c_0, c_1, ..., c_k, ...\}$ such that

(7.4)
$$\left(\sum_{k=0}^{\infty} |c_k|^p \right)^{1/p} < +\infty.$$

Then the left-hand side of (7.4) can be regarded as the value of a *norm* on l^p. Here again, the fact that the subadditivity holds depends on p being $\geqslant 1$. One also defines l^∞ as the vector space of *bounded* complex sequences, that is to say of sequences $\{c_0, c_1, ..., c_k, ...\}$ such that

(7.5)
$$\sup_{j=0,1,...} |c_j| < +\infty;$$

then the left-hand side of (7.5) defines a *norm* on l^∞.

Definition 7.6.   *Let E be a vector space, and p a seminorm on E. The sets*

$$U_p = \{x \in E; p(x) \leqslant 1\}, \qquad \mathring{U}_p = \{x \in E; p(x) < 1\},$$

will be called, respectively, the closed and the open unit semiball of p.

PROPOSITION 7.3.   *Let E be a topological vector space, and p a seminorm on E. Then the following conditions are equivalent:*

(a)   *the open unit semiball of p is an open set;*
(b)   *p is continuous at the origin;*
(c)   *p is continuous at every point.*

(a) implies (b) since $\varepsilon \mathring{U}_p$, for $\varepsilon > 0$ arbitrary, is the preimage under p of the open interval $] - \varepsilon, \varepsilon[\subset \mathbf{R}^1$. Because of (a), $\varepsilon \mathring{U}_p$ is an open set, hence a neighborhood of zero.

(b) implies (c), since $p(x) - p(y) \leqslant p(x - y)$ (subadditivity of p).

(c) implies (a) since the preimage of an open set under a continuous mapping is open.

PROPOSITION 7.4.   *If p is a continuous seminorm on a TVS E, its closed unit semiball is a barrel.*

This is obvious in view of the definitions.

PROPOSITION 7.5. *Let E be a topological vector space, and T a barrel in E. There exists a unique seminorm p on E such that T is the closed unit semiball of p. The seminorm p is continuous if and only if T is a neighborhood of 0.*

The last part follows immediately from the fact that T is the closed unit semiball of p, and from Proposition 7.3. Let us therefore prove the first part of the statement.

We set

$$p(x) = \inf_{\lambda \geqslant 0,\, x \in \lambda T} \lambda.$$

First of all, p is indeed a nonnegative function; that is to say p is everywhere finite, because T is absorbing. That $p(\zeta x) = |\zeta| p(x)$ is pretty obvious. What has to be checked is the subadditivity of p. Let $x, y \in E$ be arbitrary. Given any $\varepsilon > 0$, there are two numbers, $\lambda, \mu \geqslant 0$, such that

$$(p(x) \leqslant) \lambda \leqslant p(x) + \varepsilon, \quad \text{and} \quad x \in \lambda T;$$

$$(p(y) \leqslant) \mu \leqslant p(y) + \varepsilon, \quad \text{and} \quad y \in \mu T.$$

Since T is convex, we have

$$\frac{\lambda}{\lambda + \mu} T + \frac{\mu}{\lambda + \mu} T \subset T,$$

whence $x + y \in (\lambda + \mu)T$, which means that

$$p(x + y) \leqslant \lambda + \mu \leqslant p(x) + p(y) + 2\varepsilon.$$

As ε is arbitrary, it shows that p is subadditive. Let us show that T is the *closed* unit semiball of p. If $p(x) \leqslant 1$, it means that, for all $\varepsilon > 0$, there exists $y_\varepsilon \in T$ such that $x = (1 + \varepsilon)y_\varepsilon$; but when $\varepsilon \to 0$, $y_\varepsilon = (1 + \varepsilon)^{-1}x$ converges to x in E (continuity of the scalar multiplication in a TVS), hence x belongs to the closure of T. But T is closed. Conversely, any vector $x \in T$ is obviously such that $p(x) \leqslant 1$. This proves that T is indeed the closed unit semiball of p.

It remains to prove the uniqueness of the seminorm p. Let p' be another seminorm on E whose closed unit semiball is identical with T. This means that $p(x) \leqslant 1$ if and only if $p'(x) \leqslant 1$. Taking in succession

$$x = y/(p(y) + \varepsilon), \qquad x = y/(p'(y) + \varepsilon),$$

with $y \in E$ and $\varepsilon > 0$ arbitrary, we see that

$$p'(y) \leqslant p(y) + \varepsilon, \qquad p(y) \leqslant p'(y) + \varepsilon.$$

As ε is arbitrary, this means that $p(y) = p'(y)$.

COROLLARY. *Let E be a locally convex space. The closed unit semiballs of the continuous seminorms on E form a basis of neighborhoods of the origin.*

Combine Proposition 7.5 with Proposition 7.2.

Definition 7.7. A family of continuous $\mathscr{P}$ seminorms on a locally convex space E will be called a basis of continuous seminorms on E if to any continuous seminorm p on E there is a seminorm q belonging to $\mathscr{P}$ and a constant $C > 0$ such that, for all $x \in E$,

$$(7.6) \qquad\qquad p(x) \leqslant C\, q(x).$$

Let us denote by U_p (resp. U_q) the closed unit semiball of p (resp. q). Then (7.6) means

$$(7.7) \qquad\qquad C^{-1} U_q \subset U_p.$$

We leave the proof of the following result to the student:

PROPOSITION 7.6. *Let $\mathscr{P}$ be a basis of continuous seminorms on the locally convex space E. Then the sets λU_p, where U_p is the closed unit semiball of p and where p varies over $\mathscr{P}$ and λ on the set of numbers >0, form a basis of neighborhoods of zero. Conversely, given any family of neighborhoods of zero, $\mathscr{B}$, consisting of barrels and such that the set λU when $U \in \mathscr{B}$ and $\lambda > 0$ form a basis of neighborhoods of 0 in E, then the seminorms whose closed unit semiballs are the barrels belonging to $\mathscr{B}$ form a basis of continuous seminorms in E.*

We shall often say that a basis of continuous seminorms on a locally convex space E *defines* the topology (or the TVS structure) of E. Thus, for instance, the seminorms $\mathscr{P}_{[a,b]}$ on $\mathscr{C}^0(R^1)$ (Example 3) define the topology of uniform convergence of continuous functions on the bounded intervals of the real line.

We shall also use the expression "*a family of seminorms on E defining the topology of E,*" in which the family under consideration, say $\{p_\alpha\}$ ($\alpha \in A$), need not be a basis of continuous seminorm. The meaning of it is the following: first, every seminorm p_α is continuous; second, the family obtained by forming the supremums of finite numbers of seminorms p_α is a basis of continuous seminorms on E. This family consists of the seminorms

$$x \rightsquigarrow p_{(B)}(x) = \sup_{\alpha \in B} p_\alpha(x),$$

where B ranges over all the finite subsets of the set of indices A of the family $\{p_\alpha\}$. Forming the supremum of a finite number of seminorms is the equivalent of forming the intersection of their closed unit semiballs and taking the "gauge" of this intersection (a seminorm p is the *gauge* of a set U if U is the closed unit semiball of p).

The following statements are obvious:

PROPOSITION 7.7.   *Let E, F be two locally convex spaces. A linear map $f : E \to F$ is continuous if and only if to every continuous seminorm q on F there is a continuous seminorm p on E such that, for all $x \in E$,*

$$q(f(x)) \leqslant p(x).$$

COROLLARY.   *A linear form f on a locally convex space E is continuous if and only if there is a continuous seminorm p on E such that, for all $x \in E$,*

$$|f(x)| \leqslant p(x).$$

Proposition 7.7 and its corollary are very often used in the following form: we are given a basis of continuous seminorms $\mathscr{P}$ (resp. $\mathscr{Q}$) on E (resp. F); then the mapping f is continuous if to every seminorm $q \in \mathscr{Q}$ there is a seminorm $p \in \mathscr{P}$ and a constant $C > 0$ such that, for all $x \in E$,

$$q(f(x)) \leqslant C\, p(x).$$

For instance, suppose that both topologies of E and F can be defined by a single (continuous) seminorm which we denote, in both spaces, by $\| \ \|$ (this notation is usual when the seminorms are norms, but this is of no importance here). Then a linear map $f : E \to F$ is continuous if and only if there is a constant $C > 0$ such that, for all $x \in E$,

$$\| f(x) \| \leqslant C \| x \|.$$

Similarly, in this case, a linear functional f on E is continuous if and only if there is a constant C such that

$$|f(x)| \leqslant C \| x \|.$$

Of course, the absolute value in $\mathbf{C}^1$ defines a continuous norm on $\mathbf{C}^1$ (for the usual topology) and constitutes, by itself, a basis of continuous seminorms in $\mathbf{C}^1$, in other words defines the topology of $\mathbf{C}^1$. The Euclidean (or Hermitian, as one prefers) norm on $\mathbf{C}^n$ (or on $\mathbf{R}^n$ if we deal with real vector spaces) defines the topology of $\mathbf{C}^n$ (or $\mathbf{R}^n$).

PROPOSITION 7.8.   *Let E be a locally convex space. Let $\mathscr{P}$ be a basis of continuous seminorms on E. A filter $\mathscr{F}$ on E converges to a point x if and* *only if to every $\varepsilon > 0$ and to every seminorm $p \in \mathscr{P}$ there is a subset M of E belonging to $\mathscr{F}$ such that, for all $y \in M$,*

$$p(x - y) < \varepsilon.$$

COROLLARY. *A sequence $\{x_1, ..., x_n, ...\}$ in E converges to x if and only if to every $\varepsilon > 0$ and to every seminorm $p \in \mathscr{P}$ there is an integer $n(p, \varepsilon)$ such that $n \geqslant n(p, \varepsilon)$ implies*

$$p(x - x_n) < \varepsilon.$$

Both statements are obvious.

PROPOSITION 7.9.   *Let E be a locally convex space, and M a linear subspace of E. Let ϕ be the canonical mapping of E onto E/M. Then the following facts are true:*

(1)   *the topology of the quotient TVS E/M is locally convex;*

(2)   *if $\mathscr{P}$ is a basis of continuous seminorms on E, let us denote by $\dot{\mathscr{P}}$ the family of seminorms on E/M consisting of the seminorms*

(7.8) $$E/M \ni \dot{x} \rightsquigarrow \dot{p}(\dot{x}) = \inf_{\phi(x) = \dot{x}} p(x).$$

Then $\dot{\mathscr{P}}$ is a basis of continuous seminorms of E/M.

The proof consists of routine checking. That $\dot{p}$, defined by (7.8), is a seminorm follows from the subadditivity of p and of the fact that ϕ is linear. In relation with (7.8), let us consider the complex two-dimensional space $\mathbf{C}^2$, playing the role of E, and its subspace

$$M = \{(\zeta_1, \zeta_2) \in \mathbf{C}^2 \mid \zeta_1 = 0\}.$$

The quotient E/M can be identified with

$$M^0 = \{(\zeta_1, \zeta_2) \mid \zeta_2 = 0\}$$

and the canonical mapping $E \rightarrow E/M$ with the *projection*

$$(\zeta_1, \zeta_2) \rightarrow (\zeta_1, 0).$$

The Euclidean norm on $\mathbf{C}^2$ (resp. M^0) defines the topology of $\mathbf{C}^2$ (resp. M^0). If for one moment we call p the Euclidean norm in $\mathbf{C}^2$ and view it as a seminorm, which it actually is, we see that

$$\dot{p}(\zeta_1, 0) = \inf_{\zeta_2 \in \mathbf{C}^1} p(\zeta_1, \zeta_2) = \inf_{\zeta_1 \in \mathbf{C}_1} (|\zeta_1|^2 + |\zeta_2|^2)^{1/2}$$
$$= |\zeta_1|.$$

This shows that we find as associated seminorm $\dot{p}$ exactly what we would expect to find.

The student should be very careful not to think that if a family of continuous seminorms defines the topology of E, without being a basis of continuous seminorms on E, then the family of continuous seminorms on E/M obtained by Formula (7.8) necessarily defines the topology of E/M. A counterexample to this is provided by $E = \mathbf{C}^2$, $M = \{(z_1, z_2) \in \mathbf{C}^2; z_1 = z_2\}$, when we take the pair of seminorms on E, $p_1 : z = (z_1, z_2) \rightsquigarrow |z_1|$ and $p_2 : z \rightsquigarrow |z_2|$. This pair obviously defines the topology of E, but

$$\dot{p}_i(\dot{z}) = \inf_{\dot{z}=z+M} p_i(z)$$

is equal to zero for $i = 1, 2$, and all $\dot{z} \in E/M$. Indeed, every equivalence class $z + M$ intersects both subspaces $z_1 = 0$ and $z_2 = 0$.

Let us go back to the general case. We call *kernel* of a seminorm p on E the set of vectors x such that $p(x) = 0$, and denote this set by Ker p. In view of the subadditivity of p and of the positive homogeneity of p, one sees immediately that Ker p is a vector subspace of E. If p is continuous, it is closed, since it is the preimage of zero when we view p as a mapping of E into the real line. In a locally convex space E, the closure of the origin is exactly the intersection

$$\bigcap_p \text{Ker } p,$$

when p runs over the family of all continuous seminorms on E. This is pretty obvious, just as the next statement is obvious:

PROPOSITION 7.10. *In a locally convex space E, the closure of $\{0\}$ is the intersection of the (closed) linear subspace Ker p, when p varies over a basis of continuous seminorms on E.*

Thus the Hausdorff space associated with an E (see the remark following corollary to Proposition 4.5) is locally convex (Proposition 7.9); it is the quotient space

$$E\Big/\Big(\bigcap_{p\in\mathscr{P}} \text{Ker } p\Big),$$

where $\mathscr{P}$ is any basis of continuous seminorms on E.

In particular, suppose that E has a basis of continuous seminorms consisting of a single seminorm p_0. Then $E/\text{Ker } p_0$ is the Hausdorff space associated with E, and its topology can be defined by the seminorm $\dot{x} \rightsquigarrow \dot{p}_0(\dot{x}) = p_0(x)$ for some $x \in E$ such that $\phi(x) = \dot{x}$. Indeed, it should be noted that the seminorm p_0 is *constant* along the submanifolds $x + \operatorname{Ker} p_0$: if $x - y \in \operatorname{Ker} p_0$, we have (by virtue of the triangular inequality)

$$| p_0(x) - p_0(y)| \leqslant p_0(x - y) = 0.$$

Now, it is evident that, if a seminorm is going to define the topology of a locally convex space, all by itself, and if this topology is Hausdorff, then the seminorm must be a norm. Thus $\dot{p}_0$ is a norm. In this context, $E/\operatorname{Ker} p_0$ is called the *normed space associated with E*.

We shall need, later on, the following result:

PROPOSITION 7.11. *Let E be a locally convex Hausdorff TVS, and K a precompact subset of E. The convex hull $\Gamma(K)$ of K is precompact.*

Proof. This consists in applying Proposition 6.9 several times. Let V be an arbitrary neighborhood of zero in E, and U a convex balanced neighborhood of zero such that $U + U \subset V$. As K is precompact, Proposition 6.9 implies that there is a finite set of points of K, $x_1 ,..., x_r$, such that $K \subset (U + x_1) \cup \cdots \cup (U + x_r)$. If we denote by S the convex hull of the finite set of points $x_1 ,..., x_r$, we see that $\Gamma(K) \subset S + U$. Observe that S is a bounded subset of a finite dimensional subspace M of E; as the topology induced by E on M is Hausdorff, this induced topology is the usual one, as we shall see in Chapter 9 (Theorem 9.1); S is bounded and closed, hence compact in view of the Borel-Lebesgue theorem. Since S is compact in E, we may apply again Proposition 6.9: there is finite set of points $y_1 ,..., y_s$ in S such that $S \subset (y_1 + U) \cup \cdots \cup (y_s + U)$, whence

$$\Gamma(K) \subset (y_1 + U + U) \cup \cdots \cup (y_s + U + U) \subset (y_1 + V) \cup \cdots \cup (y_s + V).$$

Since V is arbitrary, we see, by taking into account the implication (b) $\Rightarrow$ (a) in Proposition 6.9, that $\Gamma(K)$ is precompact.

COROLLARY. *If E is complete, the closed convex hull of a compact subset of E is compact.*

The convex hull of a compact set is not necessarily compact, even not closed.

If the surrounding space E is not complete, the closed convex hull of a compact set is not necessarily compact.

Exercises

7.1. Prove that a seminorm on a locally convex space E is continuous if and only if there is a continuous seminorm on E which is at least equal to it (at every point of E).

7.2. Prove that the product TVS of an arbitrary family of locally convex spaces is locally convex.

7.3. Prove that if a locally convex space E has at least one continuous norm it has a basis of continuous seminorms consisting of norms.

7.4. Let $\mathscr{P}$ be a family of continuous seminorms on a locally convex space E, not necessarily finite. Suppose that, for every $x \in E$,

$$p_0(x) = \sup_{p \in \mathscr{P}} p(x)$$

is finite. Prove that p_0 is a seminorm on E and that its closed unit semiball $\{x \in E; \, p_0(x) \leqslant 1\}$ is a closed subset of E, i.e., is a barrel.

7.5. Prove that if $0 < p < 1$ the function

$$\zeta = (\zeta_1, ..., \zeta_n) \leadsto |\zeta|_p = (|\zeta_1|^p + \cdots + |\zeta_n|^p)^{1/p}$$

is *not* a seminorm on $\mathbf{C}^n$.

7.6. Prove that, for any vector $\zeta \in \mathbf{C}^n$,

$$\lim_{p \to \infty} |\zeta|_p = |\zeta|_\infty \qquad \text{(see p. 59)}.$$

7.7. Let $\mathbf{C}[[X]]$ be the space of formal power series in one variable X, with complex coefficients. Construct a basis of continuous seminorms for the topology of simple convergence of the coefficients, on $\mathbf{C}[[X]]$ (see p. 25).

7.8. The convex balanced hull of a subset A of a vector space E is the smallest balanced convex set containing A.

(a) Prove that the convex balanced hull of A is the convex hull of the balanced hull of A (the latter is the smallest balanced set containing the set A).

(b) Give an example of a set A whose convex balanced hull is different from the balanced hull of its convex hull.

7.9. Let F be the space of complex valued continuous functions defined in the interval $\{t; \, 0 \leqslant t \leqslant 1\}$ of the real line; let E be the space of all mappings of F into the complex plane $\mathbf{C}$. Let us set, for each real number t, $0 \leqslant t \leqslant 1$, $\delta_t : f \leadsto f(t)$, mapping from F into $\mathbf{C}$ (thus the δ_t belong to E); let us also set $dt : f \leadsto \int_0^1 f(t) \, dt$, also a mapping of F into $\mathbf{C}$, hence an element of E. We provide the space E with the locally convex topology defined by the basis of seminorms

$$\mu \leadsto \sup_{f \in S} |\mu(f)|,$$

where S runs over the family of all finite subsets of F.

(a) Prove that when t varies over the interval $0 \leqslant t \leqslant 1$ the elements δ_t form a compact subset of E (hint: identify E to the product space

$$\mathbf{C}^F = \prod_{f \in F} \mathbf{C}_f, \qquad \mathbf{C}_f : \text{copy of the complex plane } \mathbf{C};$$

show that the topology of E is identical to the product topology on $\mathbf{C}^F$, then apply Tychonoff's theorem (Exercise 6.1, Theorem 6.1)).

(b) Prove that the mapping dt belongs to the closure of the convex hull of the set formed by the δ_t but not to the convex hull of this set (approximate the integral over $(0, 1)$ by the Riemann sums).

7.10 Give an example of a closed subset of the plane $\mathbf{R}^2$ whose convex hull is not closed.

8

Metrizable Topological Vector Spaces

A TVS E is said to be *metrizable* if it is Hausdorff and if there is a *countable* basis of neighborhoods of zero in E. The motivation for the name metrizable lies in the following fact (which we shall not prove in such a general form):

The topology of a TVS E can be defined by a metric if and only if E is Hausdorff and has a countable basis of neighborhoods of 0.

We recall that a metric d on E is a mapping $(x, y) \rightsquigarrow d(x, y)$ from $E \times E$ into the nonnegative half real line R_+ with the following properties:

(1) $d(x, y) = 0$ if and only if $x = y$ *(two points with zero distance are identical)*;

(2) $d(x, y) = d(y, x)$ for all $x, y \in E$ *(the distance is a symmetric function)*;

(3) $d(x, z) \leqslant d(x, y) + d(y, z)$ for all $x, y, z \in E$ *(triangular inequality)*.

To say then that the topology of E is *defined by the metric d* means that, for every $x \in E$, the sets

$$B_\rho(x) = \{y \in E; d(x, y) \leqslant \rho\}, \qquad \rho > 0,$$

form a basis of neighborhoods of x. The metric d is said to be *translation invariant* if the following condition is verified:

(4) $d(x, y) = d(x + z, y + z)$ for all $x, y, z \in E$.

Property (4) is equivalent with saying that, for all $x \in E$ and all $\rho > 0$,

$$B_\rho(x) = B_\rho(0) + x,$$

or that, for all pairs of points $x, y \in E$,

$$d(x, y) = d(x - y, 0).$$

70

The following can be proved: *in any metrizable TVS E, there is a translation invariant metric which defines the topology of E.*

Note that a norm on a vector space E defines a metric on E. If we denote the norm by $\| \ \|$, the metric is simply $(x, y) \leadsto \| x - y \|$. But, as we shall see, it is not true that the topology of a metrizable space can always be defined by a norm.

PROPOSITION 8.1. *Let E be a locally convex metrizable TVS, and $\{p_1, p_2, ...\}$ a nondecreasing countable basis of continuous seminorms on E. Let $\{a_1, a_2, ...\}$ be a sequence of numbers > 0, such that*

$$\sum_{j=1}^{+\infty} a_j < +\infty.$$

Then the following function on $E \times E$,

$$(x, y) \leadsto d(x, y) = \sum_{j=1}^{+\infty} a_j\, p_j(x - y)/[1 + p_j(x - y)],$$

is a translation invariant metric on E which defines the topology of E.

Proof. Let us first observe that, if E is a locally convex metrizable TVS, then there certainly exists a countable basis of continuous seminorms which is nondecreasing, meaning by this that, for all $n = 1, 2, ...$, and all $x \in E$,

$$p_n(x) \leqslant p_{n+1}(x).$$

Indeed, there is a countable basis of neighborhoods of 0 in E, U_1, U_2,..., U_n,.... . Since E is locally convex, each U_n contains a barrel, and we may therefore assume that each U_n is itself a barrel. We may then take

$$V_1 = U_1, \qquad V_2 = U_1 \cap U_2, ... \qquad V_n = U_1 \cap U_2 \cap \cdots \cap U_n, ...,$$

as a basis of neighborhoods of zero. Each V_n is a barrel, and we have $V_{n+1} \subset V_n$. If we call p_n the seminorm whose closed unit semiball is V_n, we obtain a basis of continuous seminorms on E such that $p_n \leqslant p_{n+1}$. Furthermore, as the space E is Hausdorff, we must have

$$\bigcap_{n=0}^{+\infty} \operatorname{Ker} p_n = 0.$$

This implies immediately that $d(x, y) = 0$ if and only if $x = y$. That $d(x, y) = d(y, x)$ is evident. We must therefore check the triangular inequality:

$$d(x, z) \leqslant d(x, y) + d(y, z) \qquad \text{for all} \quad x, y, z \in E.$$

This will follow if we prove, for each j, that

$$p_j(u + v)/[1 + p_j(u + v)] \leqslant p_j(u)/[1 + p_j(u)]$$
$$+ p_j(v)/[1 + p_j(v)], \qquad u, v \in E.$$

Taking then $u = x - y$, $v = y - z$ yields easily the desired result. As we know that

$$p_j(u + v) \leqslant p_j(u) + p_j(v),$$

what we have really to prove is that if a, b, c are three nonnegative numbers and if

(8.1) $$c \leqslant a + b,$$

then

(8.2) $$c/(1 + c) \leqslant a/(1 + a) + b/(1 + b).$$

If c or $a + b$ are equal to zero, there is nothing to prove so that we may assume that none of these two numbers is equal to zero. Then (8.1) is equivalent with

$$(a + b)^{-1} \leqslant 1/c,$$

which implies

$$(1 + 1/c)^{-1} \leqslant (1 + 1/(a + b))^{-1} = a/(1 + a + b) + b/(1 + a + b).$$

The left-hand side is $c/(1 + c)$; the right-hand side is obviously at most equal to

$$a/(1 + a) + b/(1 + b),$$

whence (8.2). This proves that d is indeed a metric. That it is translation invariant is obvious on the definition. What is left to prove is that the topology defined by the metric d is identical with the topology initially given on E or, which is the same, the topology defined by the seminorms p_n. We must show that every set

$$B_\rho(x) = \{y \in E; \, d(x, y) < \rho\}, \qquad \rho > 0,$$

contains some set of the form $x + \lambda V_n$, where λ is > 0 and V_n is the closed unit semiball of the seminorm p_n, and conversely that every set $x + V_n$ contains some $B_\rho(x)$. Because of the translation invariant character of d, we may of course assume $x = 0$.

Since the series of positive numbers, $\sum_{j=1}^{+\infty} a_j$, converges, we may find an integer $j(\rho) \geqslant 1$ such that

$$(8.3) \qquad \sum_{j=j(\rho)}^{+\infty} a_j < \rho/2.$$

As the sequence $\{p_1, p_2, ..., p_n, ...\}$ is nondecreasing, we have

$$(8.4) \qquad \sum_{j=1}^{j(\rho)} a_j \, p_j(x)/(1 + p_j(x)) \leqslant \left(\sum_{j=1}^{+\infty} a_j\right) p_{j(\rho)}(x).$$

Let us denote by A the sum of the series $\sum_{j=1}^{+\infty} a_j$. If the point x belongs to the set

$$(\rho/2A)V_{j(\rho)} = \{y \in E; \, p_{j(\rho)}(y) \leqslant \rho/2A\},$$

we have

$$d(x, 0) = \sum_{j=1}^{j(\rho)} a_j \, p_j(x)/(1 + p_j(x)) + \sum_{j=j(\rho)+1}^{+\infty} a_j \, p_j(x)/(1 + p_j(x)) < \rho$$

by combining (8.3) and (8.4). This shows that $(\rho/2A)V_{j(\rho)} \subset B_\rho(0)$.

In order to prove the result in the other direction, we use the fact that every number a_j is > 0. In view of this fact, we have

$$p_j(x) \leqslant a_j^{-1} \, d(x, 0) \, (1 + p_j(x)) \qquad \text{for all} \quad j = 1, 2, ... \quad \text{and all} \quad x \in E.$$

If we therefore impose upon x the condition

$$d(x, 0) < a_j/2,$$

we see that

$$p_j(x) < 1.$$

In other words,

$$B_{a_j/2}(0) \subset V_j \, .$$

The proof of Proposition 8.1 is complete.

Exercises

Let $d(x, y)$ be the metric on E defined in Proposition 8.1.

8.1. Prove that, for all $x, y \in E$,

$$d(x + y, 0) \leqslant d(x, 0) + d(y, 0).$$

Is that true of any translation invariant metric?

8.2. Prove that the nonnegative function $x \rightsquigarrow d(x, 0)$ is *not* a seminorm on E.

We go back now to the general case of a metrizable, not necessarily locally convex, TVS E. We shall prove three well-known results in the theory of metric spaces.

PROPOSITION 8.2. *A subset K of a metrizable space E is complete if and only if every Cauchy sequence in K converges to a point of K.*

In other words, in metrizable spaces, *sequentially complete implies complete*.

Proof. That every Cauchy sequence should converge in a complete set, we already know. We must prove that, if E has a countable basis of neighborhoods of zero, U_1, U_2,..., U_n,..., and if K is sequentially complete, any Cauchy filter $\mathscr{F}$ in K converges in K. To every $n = 1, 2,...,$ there is a subset M_n of K which belongs to $\mathscr{F}$ and which is such that $M_n - M_n \subset U_n$. Noting that no finite intersection of the sets M_n can be empty, since these sets belong to the same filter $\mathscr{F}$, we may choose for each n a point x_n in the set $M_1 \cap M_2 \cap \cdots \cap M_n$. It is obvious that the sequence of points x_1, x_2,..., x_n,... is a Cauchy sequence and therefore converges to some point x of K. Let us show that the filter $\mathscr{F}$ converges also to x. Let n be any integer $\geqslant 1$, and choose k such that $U_k + U_k \subset U_n$; then choose $h \geqslant k$ so that $x_h \in U_k + x$. As we have $x_h \in M_k$, we may write

$$M_k \subset U_k + x_h \subset U_k + U_k + x \subset U_n + x.$$ Q.E.D.

PROPOSITION 8.3. *A complete metrizable TVS E is a Baire space, i.e., has the property:*

  (B)   *The union of any countable family of closed sets, none of which has interior points, has no interior points.*

Remarks. 1. The union of a sequence of closed sets is not a closed set, in general.

2. The closure of the union of a sequence of closed sets may have interior points even if the space E is a Baire space: take for E the real line, with its usual topology; every point is closed. The set of rational numbers $\mathbf{Q}$ is the union of a countable family of closed sets without interior points (the rational numbers); it has no interior point, but its closure is the entire real line.

3. By going to the complements, Property (B) can be stated in the following equivalent manner:

  (B')   *The intersection of any countable family of everywhere dense open sets is an everywhere dense set.*

Indeed, the complement of a closed set without interior points is an everywhere dense open set.

4. There exist complete TVS which are not Baire spaces. The so-called *LF*-spaces will provide us with examples of such TVS.

5. There exist Baire spaces which are not metrizable, and metrizable spaces (of course, noncomplete) which are not Baire spaces. There exist noncomplete metrizable spaces which are Baire spaces.

Proof of Proposition 8.3. We shall prove that Property (B′) holds in a complete metrizable space. Let Ω_1, Ω_2,..., Ω_n,... be a sequence of dense open subsets of E. We must show that their intersection, which we denote by A, intersects every open subset of E (this means indeed that A is dense, since every neighborhood of every point contains some open set, hence some point of A). Let Ω be an arbitrary open subset of E. We are going to show that $A \cap \Omega \neq \varnothing$. Let U_1, U_2,..., U_k,... be a (countable) basis of neighborhoods of 0 in E; we may take all the sets U_k closed. Observe that $\Omega \cap \Omega_1$ is nonempty. As it is open, it contains some set of the form $x_1 + U_{k_1}$; let us call G_1 the interior of the latter set. As Ω_2 is an everywhere dense set and as G_1 is a nonempty open set, $G_1 \cap \Omega_2$ contains some set of the form $x_2 + U_{k_2}$. Choosing $k_2 > k_1$, we call G_2 the interior of the set $x_2 + U_{k_2}$. Proceeding in the indicated way, we define step by step a sequence G_1, G_2,..., G_l of open sets such that, for each l, $\bar{G}_l \subset \Omega \cap \Omega_l$; furthermore, $G_{l+1} \subset G_l$ and $G_l \subset x_l + U_{k_l}$, which implies $G_l - G_l \subset U_{k_l} - U_{k_l}$ (we have also $k_l \geqslant l$). Thus the family of sets G_1, G_2,..., G_l,... forms the basis of a Cauchy filter, which has a limit point x. Of course, we have then $x \in \bar{G}_l$ for all l, which implies

$$x \in \bigcap_{l=1}^{\infty} \bar{G}_l \subset \Omega \cap \bigcap_{l=1}^{\infty} \Omega_l = A \cap \Omega. \qquad \text{Q.E.D.}$$

The third statement is the following well-known criterion of *compactness* in metrizable spaces. In the general case, we know that a set K is compact if and only if every filter on K has an accumulation point (Proposition 6.4). By making use of the criterion of precompactness already proved (Proposition 6.9), we may prove the following:

PROPOSITION 8.4. *In a metrizable TVS E, a set K is compact if and only if every sequence in K has an accumulation point (in K).*

Proof. The necessity of the condition is true even in the absence of metrizability (Corollary 3 of Proposition 6.8). We must prove the sufficiency. In view of the general properties of precompact sets (see Definition 6.3), it is enough to show that it implies that K is complete and precompact. The completeness follows from Proposition 8.2 and from the fact that, if a Cauchy sequence has an accumulation point, it converges to it (Proposition 5.1, Proposition 6.7). As for the precompactness, we show that, if every sequence in K has an accumulation point, then Property (b) in Proposition 6.9 holds. Suppose it did not hold. Then there would exist a neighborhood V of 0 such that there is no finite covering of K by sets of the form $x + V$, $x \in K$. Choose then any point x_1 of K; since $x_1 + V$ does not cover K, there is some point $x_2 \in K$, $x_2 \notin x_1 + V$. Since $K \not\subset (x_1 + V) \cup (x_2 + V)$, there is a third point $x_3 \in K$ which does not belong to this union, etc. We construct thus, by induction, a sequence $x_1, x_2, ..., x_n, ...$ such that, given any integers n, m, we have $x_n - x_m \notin V$ (supposing, which we might, that $V = -V$). The sequence $\{x_n\}$ could certainly not have an accumulation point.

Q.E.D.

Another useful property of metrizable spaces is the equivalence of continuity with sequential continuity. In the statements below, the mappings f are not supposed to be linear.

Definition 8.1. *A mapping f of a topological space E into a topological space F is said to be sequentially continuous if, for every sequence $\{x_n\}$ which converges to a point x in E, the sequence $\{f(x_n)\}$ converges to $f(x)$ in F.*

PROPOSITION 8.5. *A mapping f (not necessarily linear) of a metrizable TVS E into a TVS F (not necessarily metrizable) is continuous if and only if it is sequentially continuous.*

Proof. If f is continuous, it is obviously sequentially continuous. Suppose, then, E to be metrizable. We show that a function $f : E \to F$ which is not continuous cannot be sequentially continuous. As f is not continuous, there is a point x^0 of E and a neighborhood V of $f(x^0)$ in F such that $f^{-1}(V)$ is *not* a neighborhood of x^0 in E. Let $U_1 \supset U_2 \supset \cdots \supset U_n \cdots$ be a countable basis of neighborhoods of zero in E. For each $n = 1, 2, ...$, we can find a point $x_n \in U_n + x^0$ which does not belong to $f^{-1}(V)$; if we could not find such a point it would mean that $U_n + x^0 \subset f^{-1}(V)$, and therefore that $f^{-1}(V)$ is a neighborhood of x^0. As we have $f(x_n) \notin V$ for all n, the sequence $\{f(x_n)\}$ does not converge to $f(x^0)$ in F. But the sequence $\{x_n\}$ does converge to x^0 in E. Therefore f is not sequentially continuous.

Q.E.D.

Exercises

Let d be a metric on a TVS E defining the topology of E. One defines the *distance of a point x to a subset A* of E as the nonnegative number

$$d(x, A) = \inf_{y \in A} d(x, y).$$

8.3. Prove that $d(x, A) = 0$ if and only if x belongs to the closure of A.

8.4. Let N be a closed vector subspace of E, and ϕ the canonical mapping of E onto E/N. Prove that the function on $(E/N) \times (E/N)$,

$$\dot{d}(\dot{x}, \dot{y}) = \inf_{\phi(x)=\dot{x}} d(x, \phi^{-1}(\dot{y})),$$

is a metric on E/N.

8.5. By using Theorem 5.1 and Exercise 5.1 prove that there is a unique metric $\hat{d}$ on the completion $\hat{E}$ of E which extends the metric d.

8.6. We use the same notation as in Exercise 8.4. Let $\hat{\phi} : \hat{E} \to (E/N)^\wedge$ be the canonical extension of the mapping ϕ (Theorem 5.1). Let $\hat{N}$ be the closure of N in $\hat{E}$.

Prove that $\hat{\phi}(\hat{x}) = 0$ if and only if $\hat{d}(\hat{x}, \hat{N}) = 0$. Derive from this that there is a canonical isomorphism of $\hat{E}/\hat{N}$ onto $(E/N)^\wedge$ and that this isomorphism transforms the quotient metric constructed out of $\hat{d}$ (Exercise 8.4) into the extension (Exercise 8.5) of the quotient metric $\dot{d}$ on E/N (Exercise 8.4).

8.7. Prove the following statements:

PROPOSITION 8.6.   *In a metrizable TVS E a point x is an accumulation point of a sequence S if and only if S contains a subsequence which converges to x.*

COROLLARY.   *A subset K of a metrizable TVS E is compact if and only if every sequence in K contains a subsequence which converges in K.*

9

Finite Dimensional Hausdorff
Topological Vector Spaces.
Linear Subspaces with Finite
Codimension. Hyperplanes

Let E be a vector space over the field of complex numbers, $\mathbf{C}$. There is equivalence between the following two properties:

(a) E is finite dimensional;

(b) there is an integer $n \geqslant 0$ such that there exists a one-to-one linear map of E onto $\mathbf{C}^n$.

Indeed, (a) means that there is some integer $n' \geqslant 0$ such that there does not exist, in E, any linearly independent set of $n' + 1$ vectors. The number n in (b) can then be taken as the smallest of those numbers n'. There exist then, in E, linearly independent sets consisting of exactly n vectors; any such set spans the whole space E, hence constitutes a *basis*. Let $(e_1, e_2, ..., e_n)$ be a basis of E. Given any vector $x \in E$ we can write

$$x = x^1 e_1 + \cdots + x^n e_n ,$$

where the "components" x^j of x are uniquely determined complex numbers. This can be precisely expressed by saying that the mapping $x \rightsquigarrow (x^1, ..., x^n)$ is an isomorphism (in the linear sense) of E onto $\mathbf{C}^n$. Then n is called the *dimension* of E; we shall denote it by dim E. If E is not finite dimensional, we say that it is *infinite dimensional*.

We are going to show that if a finite dimensional TVS E is Hausdorff, then its structure is the usual one, meaning by this that there exists an isomorphism (for the TVS structure) of E onto $\mathbf{C}^{\dim E}$. The isomorphism is in fact any one of the mappings $x \rightsquigarrow (x^1, ..., x^n)$ considered above.

THEOREM 9.1.   *Let E be a finite dimensional Hausdorff* TVS. *Then:*

(a)   *E is isomorphic, as a* TVS, *to $\mathbf{C}^d$, where $d = \dim E$. More precisely, given any basis $(e_1, \ldots, e_d)$ in E, the mapping*

(9.1) $$\mathbf{C}^d \ni (x^1, \ldots, x^d) \rightsquigarrow x^1 e_1 + \cdots + x^d e_d$$

is an isomorphism, for the TVS *structure, of $\mathbf{C}^d$ onto E.*

(b)   *Every linear functional on E is continuous.*

(c)   *Every linear map of E into any* TVS *F is continuous.*

Proof.   Observe that (9.1) is always continuous. Indeed, if all the "coordinates" x^j converge to zero, then each vector $x^j e_j$ must converge to zero and also the sum of these vectors must converge to zero (continuity of multiplication by scalars and of vector addition). What has therefore to be proven is the continuity of the inverse of (9.1).

As a first step, we prove this in dimension *one*, that is to say for $d = 1$. Let V be a balanced neighborhood of zero in E which does not contain θe_1, where θ is an arbitrary number > 0. Such a neighborhood of zero V exists because E is Hausdorff. Let x be any vector in V; we have $x = \xi e_1$ for some complex number ξ. Suppose we have $|\xi| \geqslant \theta$; then, since V is balanced and since $|\theta/\xi| \leqslant 1$, we would have

$$(\theta/\xi)x = \theta e_1 \in V, \qquad \text{contrary to our choice of } V.$$

Thus, for all $x \in V$, we have $|\xi| \leqslant \theta$ (if $x = \xi e_1$). This means precisely that the linear functional $x = \xi e_1 \rightsquigarrow \xi$ is continuous. So the bicontinuity of (9.1) is proved when $d = 1$.

Property (b) is now trivial when $d = 1$. Indeed, if f is a linear form on the one-dimensional space E, let us select a vector x_0 in E such that $f(x_0) = 1$. This is always possible when f is nonidentically zero; if $f \equiv 0$ there is nothing to prove. Then, if we write any vector x of E in the form $x = \xi x_0$, we know that $x \rightsquigarrow \xi$ is continuous. But $\xi = f(x)$.

The next step is to prove (b) in dimension $d > 1$. We assume that we have proved both (a) and (b) in all dimensions $\leqslant d - 1$. Let f be an arbitrary linear functional on E (we assume $\dim E = d$), nonidentically zero. Choose $x_0 \in E$ such that $f(x_0) = 1$. Then, given any vector $x \in E$, $x - f(x)x_0$ belongs to $\operatorname{Ker} f$. If we denote by ϕ the canonical map of E onto its quotient $E/\operatorname{Ker} f$, we see that $\phi(x) = f(x)\phi(x_0)$, in other words that $\phi(x_0)$ *spans* $E/\operatorname{Ker} f$. This simply means that $E/\operatorname{Ker} f$ *is one-dimensional*. Then the dimension of $\operatorname{Ker} f$ is $< d$ (in fact, it is exactly $d - 1$). Since we suppose that we have proved (a) in all dimensions $< d$, we conclude that $\operatorname{Ker} f$ is isomorphic to some space $\mathbf{C}^n$. In particular,

Ker f is complete, hence closed in the Hausdorff TVS E. This implies that $E/\text{Ker}\,f$ is Hausdorff (Proposition 4.5). Now let us look at the commutative diagram:

$$
\begin{array}{ccc}
E & \xrightarrow{\;\;f\;\;} & C^1 \\[2pt]
{\scriptstyle\phi}\downarrow & \nearrow {\scriptstyle\tilde{f}} & \\[2pt]
E/\text{Ker}\,f & &
\end{array}
\quad.
$$

The mapping $\tilde{f}$ is a linear functional on the one-dimensional Hausdorff TVS $E/\text{Ker}\,f$, therefore it is continuous. The canonical map ϕ is *always* continuous, therefore $f = \tilde{f} \circ \phi$ is continuous.

We have proved (b) in dimension d. It immediately implies (a) in dimension d. Indeed, we know already that the mapping (9.1) is continuous. If x converges to zero in E, each one of its components converges to zero in C^1 since they are linear functionals on E, therefore they are continuous. Thus $(x^1,\ldots,x^d) \to 0$ in C^d; this proves that the inverse of (9.1) is continuous.

Property (c) is a trivial consequence of (b). Let $e_1,\ldots,e_d$ be a basis of E, and u a linear map of E into a TVS F. If $b_j = u(e_j)$, $j = 1,\ldots,d$, the mapping u is the mapping

$$
x = \sum_{j=1}^{d} x^j e_j \rightsquigarrow \sum_{j=1}^{d} x^j b_j \,.
$$

As the forms $x \rightsquigarrow x^j$ are continuous, u is also continuous.   Q.E.D.

COROLLARY 1. *Every finite dimensional Hausdorff TVS is complete.*

Indeed, a Hausdorff TVS E of dimension $d < +\infty$ is a "copy" of C^d.

COROLLARY 2. *Every finite dimensional linear subspace of a Hausdorff TVS is closed.*

It suffices to combine Corollary 1 with Proposition 5.3.

Exercise 9.1. Show that every seminorm on a finite dimensional Hausdorff TVS is continuous.

In virtue of the Heine–Borel–Lebesgue theorem, the closures of bounded open subsets of C^d are compact; thus the origin, and consequently every point of a finite dimensional TVS, has a basis of neighborhoods consisting of compact sets. A topological space with such a property is said to be *locally compact* (this, for us, implies

Hausdorff). Locally compact spaces have remarkable properties. Locally compact groups have been the object of thorough and fruitful study and have provided a sound basis for a general theory of the Fourier transformation. One should like to know if locally compact TVS are also the receptacle of astounding properties. Indeed, they are! But these properties are nothing new to us, in view of the following theorem, due to F. Riesz:

THEOREM 9.2.   *A locally compact TVS is finite dimensional.*

Proof.   Let E be a locally compact TVS, and K a compact neighborhood of 0 in E. Since K contains a closed balanced neighborhood of zero and since a closed subset of a compact set is compact, we may assume that K is balanced. On the other hand, as K is compact and as $\frac{1}{2}K$ is a neighborhood of zero, there is a finite family of points $x_1, \ldots, x_r$ such that $K \subset (x_1 + \frac{1}{2}K) \cup \cdots \cup (x_r + \frac{1}{2}K)$. Let M be the linear subspace spanned by $x_1, \ldots, x_r$; dim M is finite, hence M is closed in E. The quotient space is Hausdorff; let ϕ be the canonical homomorphism $E \to E/M$; as we have $K \subset M + \frac{1}{2}K$, we have $\phi(K) \subset \frac{1}{2}\phi(K)$, i.e., $2\phi(K) \subset \phi(K)$. By iteration, we see that

$$\phi(2^n K) \subset \phi(K).$$

As K is balanced, we have $E = \bigcup_{n=0}^{\infty} 2^n K$. Thus $\phi(E) = E/M \subset \phi(K)$. But ϕ being continuous and E/M being Hausdorff (Proposition 4.5), $\phi(K)$ is compact. Thus E/M is a Hausdorff TVS which is compact; it must be of zero dimension, i.e., reduced to one point. Otherwise E/M would contain a subset of the form $\mathbf{R}\dot{e}$ with $\dot{e} \in E/M$, $\dot{e} \neq 0$; such a subset, necessarily closed, would be compact. But the real line is certainly not compact!

We shall now take a look at linear subspaces of a TVS E which are of *finite codimension*. We recall that the codimension of a linear subspace M of a vector space E is the dimension of the quotient space E/M. We also recall the following definition:

Definition 9.1.   A linear subspace of codimension one is called a hyperplane.

Let M be a linear subspace of a vector space E of codimension $n < +\infty$. Consider the canonical map $\phi : E \to E/M$. If $b_1, \ldots, b_n$ is a basis of the quotient space E/M, we can lift it into a linear independent set of n vectors in E, $e_1, \ldots, e_n$. Let N be the linear subspace of E spanned by $e_1, \ldots, e_n$. We claim that

(9.2) $E = M \oplus N,$

where the symbol $\oplus$ stand for *direct sum:* (9.2) means that every vector x

of E can be written $x = y + z$ with $y \in M$, $z \in N$, and that this decomposition is unique, in other words that the intersection of M and N is the set consisting of a single point, the origin. Let us prove that (9.2) indeed holds. If $x = \sum x^i e_i \in N$ belongs also to M, we have

$$0 = \phi(x) = \sum x^i b_i,$$

which is possible only if every x^i is equal to zero. This means that

$$M \cap N = \{0\}.$$

Let $x \in E$ be arbitrary. We have, for some numbers $x^1, \ldots, x^n$,

$$\phi(x) = \sum x^i b_i,$$

hence

$$\phi\left(x - \sum x^i e_i\right) = 0, \quad \text{i.e.,} \quad x - \sum x^i e_i \in M.$$

This shows that
$$E \subset M + N \quad \text{(vector addition).} \qquad \text{Q.E.D.}$$

From what we have just said, it follows that, if H is a hyperplane of a vector space E, we have

(9.3) $$E = H \oplus N,$$

where dim $N = 1$, i.e., N is a "line" (in a complex vector space, one should rather say that N is a plane, since it is a copy of the complex plane; but the general agreement is that one-dimensional linear subspaces are called *lines*). Of course, (9.2) implies that E/M is isomorphic to N, as the student can easily check (isomorphic means here isomorphic for the vector space structure: there is no topology!). Thus if E is the direct sum of its subspace M and of a subspace N of finite dimension, M is of finite codimension, exactly equal to the dimension of N. In particular, if we have (9.3) and if we know that dim $N = 1$, we know that H is a hyperplane.

PROPOSITION 9.1. *A hyperplane H in a vector space E is a maximal proper linear subspace of E.*

Trivial (that H is *proper* means that $H \neq E$).

PROPOSITION 9.2. *A hyperplane H in a TVS E either is everywhere dense or it is closed.*

Indeed, the closure $\bar{H}$ of H is a linear subspace of E (the closure of any linear subspace is a linear subspace), and, according to Proposition 9.1, we must have either $\bar{H} = E$ or $\bar{H} = H$.

PROPOSITION 9.3. *Let E be a TVS, and M a closed linear subspace of E of finite codimension. Then there is a homomorphism p of E onto M such that $p^2 = p$. We have $E = M \oplus \text{Ker } p$.*

A linear mapping p of a vector space E into itself such that $p^2 = p$, i.e., such that $p(p(x)) = p(x)$ for all $x \in E$, is called a *projection*. As immediately seen, usual projection in finite dimensional vector spaces enters in that category. The important part, in the statement above, is that p be a homomorphism, which means both *continuous* and *open*.

Proof of Proposition 9.3. Let $b_1, ..., b_n$ be a basis of E/M, which is Hausdorff since M is closed. Choose n vectors $e_1, ..., e_n$ in E such that $\phi(e_j) = b_j$ for each $j = 1, ..., n$ (ϕ is the canonical map of E onto E/M). This defines a mapping of E/M onto the vector subspace N of E spanned by the e_j's:

(9.4) $$\phi(x) = \sum \xi^i b_i \rightsquigarrow \sum \xi^i e_i \in N.$$

We know that this mapping is continuous (Theorem 9.1(c)). It is open for we know that its inverse is continuous: its inverse is the restriction of ϕ to N. As the e_j's must obviously be linearly independent, (9.4) is an isomorphism of E/M onto N. Let us call q the compose of ϕ by (9.4): it is a homomorphism of E onto N. The student may check that the mapping

$$p = I - q, \qquad I: \text{identity map of } E,$$

from E into itself, is a homomorphism of E onto M such that $p^2 = p$. We have $N = \text{Ker } p$, whence Proposition 9.3.

Exercises

9.2. Let E be a vector space, and p a linear map of E into itself such that $p^2 = p$. Let I be the identity mapping of E. Prove that $I - p$ is also a projection (i.e., $(I - p)^2 = I - p$) and that

$$E = \text{Ker } p \oplus \text{Ker}(I - p) = \text{Im } p \oplus \text{Im}(I - p)$$

(where $\oplus$ means the algebraic direct sum).

9.3. Let E be a Hausdorff TVS, and p a continuous projection of E. Prove that p is open.

Remarks. 1. Let E be a TVS, and M a linear subspace of E. Even assuming that M is closed, it is not true, in general, that there is a continuous projection p of E onto M.

2. Let E be a TVS, and A, B two closed linear subspaces of E. Suppose that E is the direct sum of A and B, in the algebraic sense: that $E = A + B$ and that $A \cap B = \{0\}$, which we have denoted by $E = A \oplus B$. It is not true in general that there is a continuous projection p of E onto A. One says that E is the *topological direct sum* of A and B if the mapping

$$(x, y) \rightsquigarrow x + y$$

from $A \times B$ into E, is one-to-one, onto and continuous both ways. Because of the continuity of vector addition, if x and y converge to zero in A and B, respectively (for the topologies induced by E), their sum $x + y$ also converges to zero. It is the converse, in the above definition, that is the nontrivial part: if $x + y$ converges to zero, both x and y must converge to zero. If E (supposed to be Hausdorff) is the topological direct sum of two linear subspaces A and B, they are automatically closed in E. Then, of course, E/A is isomorphic (for the TVS structures) with B.

Exercises

9.4. Prove the following proposition:

PROPOSITION 9.4. *Let E be a vector space, and f a linear functional on E nonidentically zero. Then $\mathrm{Ker}\, f$ is a hyperplane of E.*

Conversely, given any hyperplane H of E, there is a linear form on E, f, such that $H = \mathrm{Ker}\, f$. Any other linear form g on E such that $H = \mathrm{Ker}\, g$ is of the form λf, where λ is a complex number.

Let E be a Hausdorff TVS, H a hyperplane of E, f a linear form on E having H as kernel. Then the following properties are equivalent:

(a) H *is closed*;

(b) f *is continuous*.

9.5. Let E be a Hausdorff TVS, and A^* a subset of the algebraic dual E^* of E with the property that, for every $x \in E$, $x \neq 0$, there is $a^* \in A^*$ such that $\langle a^*, x \rangle \neq 0$. Prove that, if there is a finite number of elements of A^*, a_1^* ,..., a_r^*, such that the hyperplanes

$$\{x \in E;\ \langle a_j^*, x \rangle = 0\}, \qquad j = 1,...,r,$$

have an intersection reduced to $\{0\}$, then $\dim E \leqslant r$.

9.6. Let E be a normed space (the norm in E is denoted by $\|\ \|$). Let S be the unit sphere of E, $S = \{x \in E;\ \|x\| = 1\}$. Let E' be the dual of E, that is to say the vector space of *continuous* linear forms on E. The student is asked to admit the following result (which is a consequence of the Hahn–Banach theorem to be proved later on; see Theorem 17. 1): for every $x \in E$, $x \neq 0$, there is $x' \in E'$ such that $\langle x', x \rangle \neq 0$.

Prove that the intersection of all the closed sets $S \cap H$, where H ranges over the family of all closed hyperplanes of E, is empty. Derive from this the fact that, if E is locally compact, it must be finite dimensional (use Exercise 9.5).

10
Fréchet Spaces. Examples

A *Fréchet space* (or, in short, an *F-space*) is a TVS with the following three properties:

  (a) it is *metrizable* (in particular, it is Hausdorff);
  (b) it is *complete* (hence a Baire space, in view of Proposition 8.3);
  (c) it is *locally convex* (hence it carries a metric d of the type considered in Proposition 8.1).

Any closed subspace of an F-space is an F-space (for the induced topology). *Any product of two F-spaces is an F-space.*

The quotient of an F-space modulo a closed subspace is an F-space. (Combine Proposition 7.9 with Exercise 8.6). Hausdorff finite dimensional TVS (cf. Chapter 9), Hilbert spaces, and Banach spaces (see later on) are *F*-spaces. We shall now look at some other examples of *F*-spaces which are very important in Analysis and which do not enter in any of the latter categories (i.e., which are not Banach spaces).

Example I.
The Space of $\mathscr{C}^k$ Functions in an Open Subset Ω of $\mathbf{R}^n$

We must list the notations which we are going to use. The variable in $\mathbf{R}^n$ will be denoted by $x = (x_1,..., x_n)$, $\xi = (\xi_1,..., \xi_n)$, etc. The first-order partial differentiations with respect to the variables x_j's, will be denoted by $\partial/\partial x_j$, $j = 1,..., n$. We shall use, as differentiation indices, vectors $p = (p_1,..., p_n)$, with nonnegative integers as components, $p_1,..., p_n$, what we shall systematically call *n-tuples*, and thus we shall write

$$(\partial/\partial x)^p = (\partial/\partial x_1)^{p_1} \cdots (\partial/\partial x_n)^{p_n}.$$

We shall denote by $|p|$ the "length" of the *n*-tuple p, i.e.,

$$|p| = p_1 + \cdots + p_n.$$

(This length $|p|$ is the norm denoted by $|p|_1$ in p. 59.) The length $|p|$ is the *order* of the differentiation operator $(\partial/\partial x)^p$.

We shall be dealing with complex-valued functions $\phi(x)$ of the variables $x = (x_1, ..., x_n)$, defined in some open subset Ω of $\mathbf{R}^n$; Ω will remain fixed throughout the forthcoming description. A complex function f, defined in Ω, is said to be a $\mathscr{C}^k$ function, k being a nonnegative integer, if it is continuous and, when $k \geqslant 1$, if all the derivatives of f of order $\leqslant k$ exist (at every point of Ω) and are continuous functions in Ω. One also says that f is a *k-times continuously differentiable* function. Any $\mathscr{C}^k$ function is a $\mathscr{C}^{k-1}$ function (for $k \geqslant 1$). A function f in Ω is said to be a $\mathscr{C}^\infty$ function if it is a $\mathscr{C}^k$ function for *all* integers $k = 0, 1, 2, ...$. A $\mathscr{C}^\infty$ function is also called an *infinitely differentiable* function. The $\mathscr{C}^k$ functions in Ω $(0 \leqslant k \leqslant +\infty)$ form a vector space over the field of complex numbers, which we shall denote by $\mathscr{C}^k(\Omega)$. We shall now put a structure of topological space on $\mathscr{C}^k(\Omega)$ which will turn it into a F-space. As it is going to be a locally convex space, it suffices to define a basis of continuous seminorms (Definition 7.7). We shall choose the following seminorms:

$$|f|_{m,K} = \sup_{|p| \leqslant m} (\sup_{x \in K} |(\partial/\partial x)^p f(x)|).$$

We must say what m and K are. First of all, K is any compact subset of Ω (we recall that a compact set in $\mathbf{R}^n$ is a closed and bounded subset of $\mathbf{R}^n$; a compact subset of Ω is a compact subset of $\mathbf{R}^n$ contained in Ω). Observe that a continuous function is always bounded on a compact set. Thus, if f is a $\mathscr{C}^k$ function and if m is an integer $\leqslant k$, the quantities $|f|_{m,K}$ are finite. Thus, if k is finite, we take $m = k$; if k is infinite, we take m varying over the sequence of positive integers.

We shall provide $\mathscr{C}^k(\Omega)$ with the topology defined by the seminorms

$$f \rightsquigarrow |f|_{m,K}.$$

This topology is often referred to as the $\mathscr{C}^k$ *topology*, or as the *topology of uniform convergence on compact subsets of the functions and of their derivatives of order $\leqslant k$ (of order $\leqslant k$ is dropped when $k = +\infty$)*. The last phrase obviously describes the kind of convergence which is defined by the seminorms $|\ |_{m,K}$. The $\mathscr{C}^k$ topology turns $\mathscr{C}^k(\Omega)$ into a locally convex space. (This LC-space is evidently Hausdorff, since $|f|_{0,K} = 0$ for all compact subsets K of Ω means in particular that $|f|_{0,\{x\}} = |f(x)| = 0$ for all points x of Ω.) The next step consists of showing that the topology just defined on $\mathscr{C}^k(\Omega)$ is metrizable or, what amounts to the same, that there is a basis of continuous seminorms which is *countable*. In order to show this, let us first observe that, if $m' \geqslant m$, $K' \supset K$, we have

$$|f|_{m',K'} \geqslant |f|_{m,K}.$$

It will therefore suffice to show that there is a *sequence* of compact subsets of Ω, K_1, K_2,..., K_r,..., such that to every compact subset of Ω there is an integer j such that $K \subset K_j$. For then we will have

$$|f|_{m,K} \leqslant |f|_{m,K_j}$$

for all $\mathscr{C}^m$ functions f. As the $|\ |_{m,K}$ formed a basis of continuous semi-norms in $\mathscr{C}^k(\Omega)$ when $m = k < +\infty$ and when $k = +\infty$ for $m = 1, 2,...$, the same will be true of the $|\ |_{m,K_j}$, and the latter form a countable family.

LEMMA 10.1. *Let Ω be an open subset of $\mathbf{R}^n$. There is a sequence of compact subsets K_1, K_2,..., K_r,... of Ω with the following two properties:*

(a)   *For each $j = 1, 2,...$, K_j is contained in the interior of K_{j+1}.*
(b)   *The union of the sets K_j is equal to Ω.*

Proof.   *If $\Omega = \mathbf{R}^n$, we take*

$$K_j = B_j = \{x \in \mathbf{R}^n;\ |x| \leqslant j\},$$

where $|x|$ is the Euclidean norm on $\mathbf{R}^n$. If $\Omega \neq \mathbf{R}^n$, let us call A_j the set of points of Ω at a distance from the boundary of Ω which is $\geqslant 1/j$. The set A_j is a closed set, and A_j is contained in the interior of A_{j+1}. But of course the A_j are not bounded, in general, therefore they will not be compact. We take $K_j = A_j \cap B_j$. It is easily seen that the K_j have all the properties which we require from them.

Thus the space $\mathscr{C}^k(\Omega)$ is metrizable; we must now show that it is complete. It suffices to show that it is sequentially complete (Proposition 8.2). Let f_1, f_2,..., f_ν,... be a Cauchy sequence in $\mathscr{C}^k(\Omega)$. In order to prove that the functions f_j's converge to a function $f \in \mathscr{C}^k(\Omega)$, we use the following three facts:

(10.1)   The complex plane is complete.

(10.2)   Let A be a subset of $\mathbf{R}^n$, and $g_1, g_2,..., g_\nu,...$ a sequence of continuous functions in A, converging uniformly to a function g on A. Then g is continuous in A.

(10.3)   Let A be an open subset of $\mathbf{R}^n$, and $g_1, g_2,..., g_\nu,...$ a sequence of $\mathscr{C}^1$ functions in A. We assume that the g_ν converge uniformly in A to a function g, and that their first derivatives $\partial g_\nu/\partial x_j$ (which are continuous functions in A) converge uniformly to a function $g^{(j)}$ ($j = 1,..., n$). Then we have, for all j,

$$g^{(j)} = \partial g/\partial x_j;$$

in particular, g is $\mathscr{C}^1$.

We suppose that the student is familiar with (10.1) (observe that complete, in this statement, is equivalent with sequentially complete!). He should also be familiar with Facts (10.2) and (10.3), but we shall nevertheless recall rapidly how they are proved.

Proof of (10.2). Let x_0 be an arbitrary point of A. We should prove that to every $\varepsilon > 0$ there is $\eta > 0$ such that $|x - x_0| < \eta$ $(x \in A)$ implies $|f(x) - f(x_0)| < \varepsilon$. We have now the "three-epsilon" argument:

$$(10.4) \quad |f(x) - f(x_0)| \leqslant |f(x) - f_n(x)| + |f_n(x) - f_n(x_0)| + |f_n(x_0) - f(x_0)|.$$

We use the uniform convergence in choosing n large enough so as to have, for *all* $y \in A$, $|f(y) - f_n(y)| < \varepsilon$. Once n is chosen, and kept fixed, we use the continuity of f_n: it enables us to choose $\delta > 0$ so that $|x - x_0| < \delta$ implies $|f_n(x) - f_n(x_0)| < \varepsilon$. The point y, above, is then taken to be any point in the set

$$\{x \in A; |x - x_0| < \delta\}.$$

Taking all the properties into account in (10.4), we obtain

$$|f(x) - f(x_0)| \leqslant 3\varepsilon \qquad \text{if} \quad |x - x_0| < \delta.$$

Proof of (10.3). We shall do the reasoning in the case of one variable $x = x_1$. The extension to n variables, quite automatic, is left to the student. We may then assume that A is some open interval and we pick up any point of this interval, say a. We have then

$$(10.5) \qquad g_\nu(x) - g_\nu(a) = \int_a^x g_\nu'(t)\, dt,$$

where g_ν' is the first derivative of g_ν, and where the integral has to be taken in the usual way, which is to say that, if $x < a$, then

$$g_\nu(x) - g_\nu(a) = -\int_x^a g_\nu'(t)\, dt.$$

At any event, the g_ν' converge uniformly in A to the function $g^{(1)}$, hence $\int_a^x g_\nu'(t)\, dt$ converges to $\int_a^x g^{(1)}(t)\, dt$ (observe that $g^{(1)}$ is a continuous function in view of (10.2)). On the other hand, the g_ν converge uniformly in A. Therefore, because of (10.5), we must have

$$g(x) - g(a) = \int_a^x g^{(1)}(t)\, dt.$$

But this simply means that $g^{(1)}$ is the derivative of g.

We return then to the completeness of $\mathscr{C}^k(\Omega)$. Because of the definition of the topology of $\mathscr{C}^k(\Omega)$, for each $x \in \Omega$, the numbers $f_\nu(x)$ form a Cauchy sequence in the complex plane. Indeed, we know that, given any integer $m \leqslant k$ and any compact subset K of Ω, to every $\varepsilon > 0$ there is an $N(\varepsilon)$ such that

$$\nu, \mu \geqslant N(\varepsilon) \quad \text{implies} \quad |f_\nu - f_\mu|_{m,K} < \varepsilon.$$

It suffices to take $m = 0$ and $K = \{x\}$ to draw the conclusion which we stated. From (10.1), it follows that the complex numbers $f_\nu(x)$ have a limit, which we denote by $f(x)$. Obviously $x \rightsquigarrow f(x)$ is a function in Ω, and it is immediately seen that the functions $f_\nu(x)$ converge uniformly to f in every compact subset of Ω. By taking this subset identical with a suitable neighborhood of any point of Ω, we conclude (by (10.2)) that f is a continuous function in Ω. If $k = 0$, this finishes the argument. If $k > 0$, observe that since the f_ν form a Cauchy sequence in $\mathscr{C}^k(\Omega)$, the first derivatives

$$\partial f_\nu / \partial x_j \quad (1 \leqslant j \leqslant n)$$

form, for each j, a Cauchy sequence in $\mathscr{C}^{k-1}(\Omega)$. Suppose $k < +\infty$. Then induction on k allows us to conclude that, for each j, the $\partial f_\nu / \partial x_j$ converge to a $\mathscr{C}^{k-1}$ function, which, by (10.3), must be the derivative of f with respect to x_j. If $k = +\infty$, we have just shown that the f_ν converge in $\mathscr{C}^h(\Omega)$ to the element f of $\mathscr{C}^h(\Omega)$, whatever be the integer h, which means precisely that the f_ν converge to f in $\mathscr{C}^\infty(\Omega)$.

The last phrase is related to the fact that the topology of $\mathscr{C}^\infty(\Omega)$ is exactly the superior limit of the topologies induced by the $\mathscr{C}^k(\Omega)$: a subset U of $\mathscr{C}^\infty(\Omega)$ is a neighborhood of zero for the $\mathscr{C}^\infty$ topology if and only if there exists some *finite* integer k such that U be a neighborhood of zero for the topology induced on $\mathscr{C}^\infty(\Omega)$ by $\mathscr{C}^k(\Omega)$ (in other words, for the $\mathscr{C}^k$ topology on $\mathscr{C}^\infty(\Omega)$).

Example II.
The Space of Holomorphic Functions in an Open Subset Ω of $\mathbf{C}^n$

Let now Ω be an open subset of the complex space $\mathbf{C}^n$. The variable in $\mathbf{C}^n$ will be denoted by $z = (z_1, ..., z_n)$. For each $j = 1, ..., n$, we have $z_j = x_j + iy_j$, $i = (-1)^{1/2}$. We denote by $H(\Omega)$ the vector space of holomorphic functions in Ω. Let h be a $\mathscr{C}^1$ function in Ω ($\mathscr{C}^1$ in the sense discussed in Example I, which has a meaning if we identify $\mathbf{C}^n$ to the real vector space $\mathbf{R}^{2n}$ by way of the mapping

$$z = (z_1, ..., z_n) \rightsquigarrow (x, y) = (x_1, ..., x_n, y_1, ..., y_n).$$

Then Ω becomes an open subset of $\mathbf{R}^{2n}$; thus $\mathscr{C}^1$ means that h has first continuous derivatives with respect to the x_j's and the y_k's). We say that h is holomorphic if it satisfies the Cauchy–Riemann equations

$$\frac{\partial h}{\partial x_j} + i \frac{\partial h}{\partial y_j} = 0, \qquad j = 1,..., n,$$

in Ω (i.e., at every point of Ω). We remind the reader that this definition implies that the function h is infinitely differentiable in Ω, and not just $\mathscr{C}^1$. Let us write

$$(\partial/\partial z)^p = (\partial/\partial z_1)^{p_1} \cdots (\partial/\partial z_n)^{p_n}, \qquad p = (p_1,..., p_n),$$

where each differential operator $\partial/\partial z_j$ is defined by

$$\frac{\partial f}{\partial z_j}(x, y) = \frac{1}{2}\left(\frac{\partial f}{\partial x_j}(x, y) - i\frac{\partial f}{\partial y_j}(x, y)\right).$$

We recall how the Cauchy formulas read. Let z^0 be any point of Ω, $z^0 = (z_1^0,..., z_n^0)$. Consider the *polydisk:*

$$D(r_1,..., r_n) = \{z \in \mathbf{C}^n; \ |z_j - z_j^0| \leqslant r_j, j = 1,..., n\}.$$

Suppose that it is contained in Ω. Then, if h is holomorphic in Ω, for each $p = (p_1,..., p_n)$ we have

$$\frac{1}{p!}(\partial/\partial z)^p h(z^0) = \oint_{|z_1 - z_1^0| = r_1} \cdots \oint_{|z_n - z_n^0| = r_n} \frac{(2i\pi)^{-n} h(z)\, dz_1 \cdots dz_n}{(z_1 - z_1^0)^{p_1+1} \cdots (z_n - z_n^0)^{p_n+1}},$$

where each integral represents usual complex integration (in the complex plane). Cauchy's formula has the immediate consequence that if a sequence of holomorphic functions in Ω converges uniformly on every compact subset of Ω, then their derivatives of any order also converge uniformly on every compact subset of Ω. From the Cauchy–Riemann equations it follows that, if a function h is holomorphic in Ω,

$$\partial h/\partial z_j = \partial h/\partial x_j$$
$$\partial h/\partial z_j = -i\,\partial h/\partial y_j \qquad (1 \leqslant j \leqslant n).$$

Thus, if we view $H(\Omega)$ as a vector subspace of any one of the spaces $\mathscr{C}^k(\Omega)$, which we may since its elements are $\mathscr{C}^\infty$, we see that the induced topologies all coincide. Indeed, this is true for the least fine and for the finest of them, the $\mathscr{C}^0$ and the $\mathscr{C}^\infty$ topologies, respectively: as far as holomorphic functions are concerned, it amounts to the same to say that a sequence of functions converges uniformly on every compact subset of Ω or that the functions and all their derivatives converge uniformly on the compact subsets of Ω.

We shall provide $H(\Omega)$ with this topology. Observe that $H(\Omega)$ is a linear subspace of $\mathscr{C}^\infty(\Omega)$, and carries the induced topology. It is obvious that $H(\Omega)$ is a *closed* subspace of $\mathscr{C}^\infty(\Omega)$. Indeed, if a sequence of holomorphic functions converges in $\mathscr{C}^\infty(\Omega)$ or, for that matter, in $\mathscr{C}^1(\Omega)$, to a function f, the latter is, needless to say, a $\mathscr{C}^\infty$ function in Ω and satisfies the Cauchy–Riemann equations since its first derivatives are limits (for the uniform convergence on compact subsets) of the corresponding derivatives of functions which do satisfy those equations. As a closed subspace of a F-space, $H(\Omega)$ is itself an F-space.

Example III.
The Space of Formal Power Series in n Indeterminates

Let us denote by $\mathbf{C}[[X_1,...,X_n]]$, or shortly by $\mathbf{C}[[X]]$, the vector space of formal power series in n letters $X_1,...,X_n$, with complex coefficients, that is to say the series

$$(10.6) \qquad u = \sum u_p X^p,$$

where the summation is performed over all the vectors $p = (p_1,...,p_n)$ whose components are nonnegative integers (the set of all these vectors p will be denoted by $\mathbf{N}^n$ from now on). The coefficients u_p are complex numbers and X^p stands for the "monomial"

$$X_1^{p_1} \cdots X_n^{p_n}.$$

No condition of convergence is imposed upon the series (10.6). One can view u as a sequence depending on n indices, $p_1,...,p_n$, $\{u_p\}$, with no condition whatsoever on the complex numbers which constitute it.

We provide $\mathbf{C}[[X]]$ with the topology defined by the seminorms

$$|u|_m = \sup_{|p| \leqslant m} |u_p|, \qquad m = 0, 1,....$$

When $n = 1$, we have already considered this topology (Chapter 3, p. 28). Whatever n is, it is sometimes referred to as the topology of simple convergence of the coefficients. Provided with it, $\mathbf{C}[[X]]$ is a locally convex metrizable space. We leave to the student the proof of its completeness (one has only to use Statement (10.1) above). Thus $\mathbf{C}[[X]]$ is an F-space.

We may put on the set $\mathbf{N}^n$ of n-tuples $p = (p_1, ..., p_n)$ the discrete topology: a basis of neighborhoods of a point in $\mathbf{N}^n$ consists of the point itself. A subset of $\mathbf{N}^n$ is then compact if and only if it is finite (indeed, observe that every point is both open and closed and, for every subset of $\mathbf{N}^n$, we have an open covering just by taking the family of its points, regarded as sets). A formal power series u may then be viewed as a function on $\mathbf{N}^n$: to each $p \in \mathbf{N}^n$ it assigns its pth coefficient, u_p. On a discrete space, every function is continuous and thus $\mathbf{C}[[X]]$ may be regarded as the space of all functions, or of all continuous functions on $\mathbf{N}^n$. The topology of simple convergence of the coefficients is then nothing else but the topology of pointwise convergence in $\mathbf{N}^n$ or the one of uniform (!) convergence on the compact subsets of $\mathbf{N}^n$.

Example IV.
The Space $\mathscr{S}$ of $\mathscr{C}^\infty$ Functions in $\mathbf{R}^n$ Rapidly Decreasing at Infinity

Our last example will be an important space in the theory of distributions, in connection with Fourier transformation. It is a space of $\mathscr{C}^\infty$ functions in the whole of the Euclidean space $\mathbf{R}^n$. The functional space in question is denoted by $\mathscr{S}$: its elements are the complex-valued functions f, which are defined and infinitely differentiable in $\mathbf{R}^n$, and which have the additional property, regulating their growth (or rather, their decrease) at infinity, that all their derivatives tend to zero at infinity, faster than any power of $1/|x|$. We use here the notation

$$|x| = (x_1^2 + \cdots + x_n^2)^{1/2}.$$

This means that, given any element f of $\mathscr{S}$, any n-tuple $p = (p_1, ..., p_n) \in \mathbf{N}^n$, and any integer $k \geqslant 0$,

$$\lim_{|x| \to \infty} |x|^k |(\partial/\partial x)^p f(x)| = 0.$$

We equip $\mathscr{S}$ with the topology defined by the seminorms

$$|f|_{m,k} = \sup_{|p| \leqslant m} (\sup_{x \in \mathbf{R}^n} \{(1 + |x|)^k |(\partial/\partial x)^p f(x)|\}), \qquad m, k = 0, 1, 2, \dots.$$

Of course, $\mathscr{S}$ is metrizable. Observe that $\mathscr{S}$ is a vector subspace of

$\mathscr{C}^\infty(\mathbf{R}^n)$ (for the linear structure), but that its topology is strictly finer than the one induced by $\mathscr{C}^\infty$. A sequence of functions $f_\nu \in \mathscr{S}$ converges to zero in $\mathscr{S}$ if and only if the functions

$$(1 + | x |)^k (\partial/\partial x)^p f_\nu(x)$$

converge uniformly, *in the whole of* $\mathbf{R}^n$, to zero—for every $k = 0, 1,...$ and every $p \in \mathbf{N}^n$. In particular, for each p, the derivatives

$$(\partial/\partial x)^p f_\nu$$

must converge uniformly in $\mathbf{R}^n$ to zero. It implies immediately that the f_ν must converge to zero in $\mathscr{C}^\infty$. This enables us to show without difficulty that $\mathscr{S}$ is complete. For a Cauchy sequence $\{f_\nu\}$ in $\mathscr{S}$ is a fortiori a Cauchy sequence in $\mathscr{C}^\infty$, hence converges (in $\mathscr{C}^\infty$) to a certain $\mathscr{C}^\infty$ function f. Choose then arbitrarily k and m. There is a constant $M_{m,k}$ such that, for all ν,

(10.7) $$| f_\nu |_{m,k} \leqslant M_{m,k}.$$

(This fact will soon be generalized when we prove that a Cauchy sequence in a TVS is a bounded set; the proof of this general statement duplicates the proof that we are about to give.) Indeed, we know that there is an integer N (depending on m and k) such that, for all $\nu \geqslant N$,

$$| f_\nu - f_N |_{m,k} \leqslant 1.$$

This comes simply from the fact that we are dealing with a Cauchy sequence. We conclude that, for all ν,

$$| f_\nu |_{m,k} \leqslant 1 + \sup_{1,...,N} | f_\mu |_{m,k},$$

which proves exactly what we want.

Now, observe that (10.7) can be expressed as follows:
For every $x \in \mathbf{R}^n$,

$$\sup_{|p| \leqslant m} |(\partial/\partial x)^p f_\nu(x)| \leqslant M_{m,k}(1 + | x |)^{-k}.$$

But we know that the derivatives $(\partial/\partial x)^p f_\nu(x)$ converge uniformly in $\mathbf{R}^n$ (therefore also pointwise in $\mathbf{R}^n$) to the corresponding derivative of f, $(\partial/\partial x)^p f$. It follows that we must have, for all $x \in \mathbf{R}^n$,

$$\sup_{|p| \leqslant m} |(\partial/\partial x)^p f(x)| \leqslant M_{m,k}(1 + | x |)^{-k}.$$

This proves that f belongs indeed to $\mathscr{S}$ (until now we only knew that f was a $\mathscr{C}^\infty$ function). The last step consists in proving that f_ν converges to f in $\mathscr{S}$, and not just in $\mathscr{C}^\infty$. We might as well exchange our initial sequence $\{f_\nu\}$ with the sequence $\{f_\nu - f\}$, which is obviously also a Cauchy sequence in $\mathscr{S}$, and suppose therefore that the limit function f is zero. We are reduced to proving the following fact:

(10.8)   If a Cauchy sequence $\{f_\nu\}$ in $\mathscr{S}$ converges to zero in $\mathscr{C}^\infty(\mathbf{R}^n)$, then it also converges to zero in $\mathscr{S}$.

I think it is a good exercise for the student to try to prove (10.8).

Thus $\mathscr{S}$ is complete; it is therefore a Fréchet space. The elements of $\mathscr{S}$ are sometimes called $\mathscr{C}^\infty$ *functions rapidly decreasing at infinity*. (This implicitly means that also their derivatives are rapidly decreasing at infinity!) When we have to avoid confusion, we shall write $\mathscr{S}(\mathbf{R}^n)$ instead of $\mathscr{S}$.

Exercises

10.1. Let $\{E_n\}$ $(n = 1, 2,...)$ be a sequence of Fréchet spaces. Prove that the product TVS $E = \prod_{n=1}^\infty E_n$ is a Fréchet space.

10.2. Let K be a compact subset of $\mathbf{R}^n$, and $\mathscr{C}_c^\infty(K)$ the space of complex functions, infinitely differentiable in $\mathbf{R}^n$, having their support contained in K; we consider on $\mathscr{C}_c^\infty(K)$ the topology in which a basis of neighborhoods of zero is formed by the sets

$$\mathscr{V}(m, \varepsilon) = \left\{\phi \in \mathscr{C}_c^\infty(K); \sup_{x \in K} \sum_{|p| \leqslant m} |(\partial/\partial x)^p \phi(x)| \leqslant \varepsilon \right\}$$

as $m = 1, 2,...$ and $\varepsilon > 0$ vary in all possible ways.

Prove that $\mathscr{C}_c^\infty(K)$ is a Fréchet space.

10.3. Consider the dual E' of the space $E = \mathscr{F}_c(\mathbf{N}; \mathbf{C})$ defined in Exercise 3.5. Let us denote by B_n the subset of $\mathscr{F}_c(n)$ (Exercise 3.6) defined by the condition

$$|f(0)|^2 + \cdots + |f(n)|^2 \leqslant 1.$$

Set then, for all $n = 0, 1,...$ and all $\varepsilon > 0$,

$$\mathscr{V}(n, \varepsilon) = \{ f' \in E'; \sup_{f \in B_n} |\langle f', f \rangle| \leqslant \varepsilon\}.$$

Prove that the sets $\mathscr{V}(n, \varepsilon)$ form a basis of neighborhoods of zero for a structure of Fréchet space on E'.

11

Normable Spaces. Banach Spaces.
Examples

We shall say that a TVS E is *normable* if its topology can be defined by a norm, i.e., if there is a norm $\| \ \|$ on E such that the balls

$$B_r = \{x \in E; \| x \| \leqslant r\}, \qquad r > 0,$$

form a basis of neighborhoods of the origin. Finite-dimensional Hausdorff spaces are normable. Infinite dimensional metrizable TVS are not—in general. In this chapter and in the next one, we shall study two very important classes of normable spaces. The topology of a normable space E can be defined by many different norms. For instance, the topology of $\mathbf{C}^n$ can be defined by any one of the norms $| \ |_p$ ($1 \leqslant p \leqslant +\infty$; see Chapter 7, Example (1)).

Definition 11.1. *Let p, q be two seminorms on a vector space E. We say that p is stronger than q when there exists a constant $C > 0$ such that, for all $x \in E$,*

$$q(x) \leqslant C\,p(x).$$

We say that p and q are equivalent if each one is stronger than the other.

If p is stronger than q, the topology defined by p on E is finer than the one induced by q. If, then, q is a norm, so is p.

PROPOSITION 11.1. *If two norms define the topology of a normable space E, they are equivalent.*

Indeed, the unit ball of one of them, say p, contains a multiple of the unit ball of the other, q, which means that q is stronger than p.

COROLLARY. *Any two norms on a finite dimensional vector space are equivalent.*

Indeed, a norm on a finite dimensional space E turns E into a Hausdorff space, therefore into a TVS homeomorphic with $\mathbf{C}^{\dim E}$, which means that all the norms on E define the same topology on E.

An application of the above corollary. Let us denote by $\mathscr{P}_1^m$ the vector space of polynomials with complex coefficients, in one indeterminate X, of degree $\leqslant m$. Let $P(X)$ be such a polynomial:

$$P(X) = a_m X^m + a_{m-1} X^{m-1} + \cdots + a_0, \qquad a_0, ..., a_{m-1}, a_m \in \mathbf{C}^1.$$

Consider the following two seminorms on $\mathscr{P}_1^m$:

$$P \rightsquigarrow \| P \| = \Big(\sum_{j=0}^m | a_j |^2 \Big)^{1/2},$$

$$P \rightsquigarrow \sup_{t\, \mathrm{real},\, |t| < \varepsilon} | P(t)|, \qquad \varepsilon > 0.$$

They are both norms; it is evident as far as the first one is concerned. As for the second one, it suffices to observe that a polynomial cannot vanish in a nonempty interval of the real line without vanishing identically. We conclude that there is a constant $C_\varepsilon > 0$, depending only on m and ε, such that, for all polynomials $P \in \mathscr{P}_1^m$,

$$C_\varepsilon \sup_{t \in R^1,\, |t| < \varepsilon} | P(t)| \geqslant \| P \|.$$

It is obvious that we could have replaced the interval $| t | < \varepsilon$ of the real line by any subset of the complex plane containing at least $(m + 1)$ points.

A normed space is something different from a normable space. A *normed space* is a pair consisting of a vector space E and a norm on E. Of course, one usually puts on E the topology defined by the norm. This topology can then be defined by many other norms but, when dealing with a normed space, one should, at least in principle, continue to consider the initially given norm. If (E, p) and (F, q) are two normed spaces, an isomorphism of E into F *for the structure of normed spaces* is a *linear isometry* of E into F, that is to say a linear mapping $u : E \to F$ such that, for all $x \in E$,

$$q(u(x)) = p(x).$$

Definition 11.2.   A normed space E which is complete is called a Banach space (or a *B-space*).

The meaning of Definition 11.2 is obvious: if (E, p) is a normed space, one provides E with the topology defined by the norm p. If the TVS E

thus obtained is complete (which, in the present situation, means sequentially complete; cf. Proposition 8.2), we say that the pair (E, p) is a Banach space. Of course, one usually drops the mention of p. Since a normed space is metrizable (in particular, it is Hausdorff!), B-spaces are a particular type of Fréchet space (see Chapter 10); they are Baire's spaces.

A few words now about quotient space and completion of a normed space (E, p). If M is a closed linear subspace of E, we may turn the quotient space E/M into a normed space by equipping it with the quotient norm

$$\dot{p}(\dot{x}) = \inf_{\phi(x)=\dot{x}} p(x)$$

(ϕ, canonical map $E \to E/M$). That $\dot{p}$ is a norm is evident. It is also evident that $\dot{p}$ defines the quotient topology on E/M (cf. Proposition 7.9). The normed space $(E/M, \dot{p})$ is *the quotient modulo M of the normed space (E, p).*

As for the completion, we remark that the norm p is uniformly continuous and therefore, by Theorem 5.1, there is a unique extension $\hat{p}$ of p to the completion $\hat{E}$ of E. The student may easily check that the topology of $\hat{E}$, as it has been defined in Chapter 5, is the topology defined by the seminorm $\hat{p}$ (that $\hat{p}$ is a seminorm follows immediately by continuation of equalities and inequalities). As the topology of $\hat{E}$ is Hausdorff, $\hat{p}$ is a norm. The normed space $(\hat{E}, \hat{p})$, which, needless to say, is a Banach space, is called the *completion of the normed space* (E, p). Note that, until now, we have considered only completions of topological vector spaces. But a normed space is something more than a special type of TVS. The canonical injection of E into $\hat{E}$ is an isometry.

Example I.   Finite dimensional normed spaces.

Since any finite dimensional Hausdorff TVS is complete, finite dimensional normed spaces are Banach spaces. As a matter of fact, they are the only locally compact Banach spaces.

Example II.   The space of continuous functions on a compact set.

Let K be a compact topological space. There is no algebraic structure on K (in particular, K does not have to be a subset of a TVS). Here compact means Hausdorff plus the property that any open covering of K contains a finite subcovering (cf. Chapter 6).

We shall denote by $\mathscr{C}(K)$ the vector space of complex-valued continuous functions defined on K. We turn $\mathscr{C}(K)$ into a normed space by considering in it the norm *maximum of the absolute value*:

$$f \rightsquigarrow \|f\| = \sup_{x \in K} |f(x)|.$$

In order to verify that $\| f \|$ is a norm, which demands, in particular, that it be finite, it suffices to observe that a continuous function is always bounded on a compact set (indeed, $f(K)$ must be a compact, therefore bounded, subset of the complex plane; cf. Proposition 6.2). To say that a sequence of functions f_ν converges to a function f in the normed space $\mathscr{C}(K)$ is to say that the f_ν converge uniformly on K to f. Thus the topology of $\mathscr{C}(K)$ is the topology of *uniform convergence* on K. If then we consider a Cauchy sequence $\{f_\nu\}$ in $\mathscr{C}(K)$, we know that, for each point x of K, the complex numbers $f_\nu(x)$ form a Cauchy sequence in the complex plane $\mathbf{C}$, hence have a limit $f(x)$. Thus the f_ν converge pointwise to a function f in K. But it is easy to see that they also converge uniformly to f. Indeed, let $\varepsilon > 0$ be given arbitrarily. Let $N(\varepsilon)$ be such that $\nu, \mu \geqslant N(\varepsilon)$ implies, for all $x \in K$,

$$(11.1) \qquad\qquad |f_\nu(x) - f_\mu(x)| \leqslant \varepsilon/2$$

(we are using the fact that the f_ν form a Cauchy sequence in $\mathscr{C}(K)$). Then, for *each* $x \in K$, select $\mu_x \geqslant N(\varepsilon)$ such that

$$(11.2) \qquad\qquad |f(x) - f_{\mu_x}(x)| \leqslant \varepsilon/2.$$

Then, by combining (11.1) and (11.2) we see that, for all $\nu \geqslant N(\varepsilon)$ and all $x \in K$,

$$|f(x) - f_\nu(x)| \leqslant \varepsilon. \qquad\qquad \text{Q.E.D.}$$

But if the f_ν converge uniformly to f, then f also must be continuous, as we see by the argument already expounded in Chapter 10, Example I, Proof of (10.2), p. 88.

Thus f is an element of $\mathscr{C}(K)$ and it is the limit in this space of the Cauchy sequence f_ν, which proves that $\mathscr{C}(K)$ is complete, i.e., is a Banach space.

Example III.   The space $\mathscr{C}^k(\bar{\Omega})$, Ω: bounded open subset of $\mathbf{R}^n$.

Let Ω be an open subset of $\mathbf{R}^n$ whose closure $\bar{\Omega}$ is compact (in other words, a *bounded* open subset of $\mathbf{R}^n$). Let k be a *finite* nonnegative integer. Consider the subset of $\mathscr{C}^k(\Omega)$ consisting of the following functions (see Chapter 10, Example I): for each n-tuple $p \in \mathbf{N}^n$, such that $| p | \leqslant k$, the pth derivative of f,

$$(\partial/\partial x)^p f(x),$$

which is a continuous function in Ω, can be extended as a continuous function in the closure $\bar{\Omega}$ of Ω. We denote by $\mathscr{C}^k(\bar{\Omega})$ this set of $\mathscr{C}^k$

functions in Ω; it is a vector subspace of $\mathscr{C}^k(\Omega)$, which we turn into a normed space by considering the norm

$$f \rightsquigarrow \|f\|_k = \sup_{|p| \leqslant k} (\sup_{x \in \Omega} |(\partial/\partial x)^p f(x)|).$$

We leave to the student the proof of its completeness; in addition to (10.2), one now uses also (10.3) (p. 87). Of course, $\mathscr{C}^0(\bar{\Omega})$ is what we have denoted by $\mathscr{C}(\bar{\Omega})$ in Example II.

Before studying the next two examples, we must state and prove *Minkowski's inequalities*. In order to attain a fair amount of generality we consider a set X and a positive measure dx on it. *We shall always suppose that X is the union of a sequence of integrable subsets*, i.e., that dx is σ-finite. We shall deal with the *upperintegral* of a nonnegative function f which we denote by $\int^* f \, dx$. We recall that this is the infimum of the integrals of the countably infinite linear combinations, with nonnegative coefficients, of integrable step functions, which are $\geqslant f$. The number $\int^* f \, dx$ is equal to $+\infty$ for many an f! Two particular cases will be important in the sequel: (1) X is the set $\mathbf{N}$ of integers $\geqslant 0$, dx is the measure with mass $+1$ at every point; then functions on X are nothing else but sequences and the upper integral of a sequence with terms $\geqslant 0$ is its sum; (2) X is an open subset of the Euclidean space $\mathbf{R}^n$ and dx is the (induced) Lebesgue measure in n variables.

THEOREM 11.1.   *Let p be a real number $\geqslant 1$, and f, g, two complex functions in X. We have:*

$$\left(\int^* |f + g|^p \, dx\right)^{1/p} \leqslant \left(\int^* |f|^p \, dx\right)^{1/p} + \left(\int^* |g|^p \, dx\right)^{1/p}.$$

Proof. It is clear that we may assume that f and g are $\geqslant 0$, since $|f + g| \leqslant |f| + |g|$. We begin by studying the case where f and g are integrable step-functions, i.e., finite linear combinations (with nonnegative coefficients) of characteristic functions of integrable sets. By subdividing further, if necessary, those integrable sets, we may even assume that f and g are finite linear combinations of the *same* characteristic functions ϕ_j and that the latter have pairwise products equal to zero, i.e., are characteristic functions of disjoint sets. We see then that

$$(f)^p = \sum a_j^p \phi_j, \qquad (g)^p = \sum b_j^p \phi_j, \qquad (f + g)^p = \sum (a_j + b_j)^p \phi_j,$$

where all the summations are performed over $j = 1, ..., k$. If we set

$$A_j = a_j \left(\int \phi_j \, dx\right)^{1/p}, \qquad B_j = b_j \left(\int \phi_j \, dx\right)^{1/p},$$

we see that we are reduced to prove an inequality:

$$(11.3) \qquad \left(\sum_{j=1}^{k} (A_j + B_j)^p \right)^{1/p} \leqslant \left(\sum_{j=1}^{k} A_j^p \right)^{1/p} + \left(\sum_{j=1}^{k} B_j^p \right)^{1/p}.$$

It suffices to prove (11.3) for $k = 2$. For now suppose it has been done. We reason by induction on $k > 2$. We have

$$\left(\sum_{j=1}^{k-1} (A_j + B_j)^p + (A_k + B_k)^p \right)^{1/p}$$

$$\leqslant \left(\left[\left(\sum_{j=1}^{k-1} A_j^p \right)^{1/p} + \left(\sum_{j=1}^{k-1} B_j^p \right)^{1/p} \right]^p + (A_k + B_k)^p \right)^{1/p}.$$

From there, we derive (11.3) from the result for $k = 2$. Now, for $k = 2$, (11.3) means that the positively homogeneous function

$$(x, y) \rightsquigarrow (|x|^p + |y|^p)^{1/p}$$

is a seminorm or, which is the same, the set $\{(x, y); |x|^p + |y|^p \leqslant 1\}$ is convex. It suffices to consider the portion of this set which is contained in the region $x \geqslant 0$, $y \geqslant 0$ and therefore the piece of curve

$$y = (1 - x^p)^{1/p}, \qquad 0 \leqslant x \leqslant 1.$$

For $0 < x < 1$, the second derivative y'' is < 0, which proves what we wanted.

From this point on, the proof of Theorem 11.1 is easy to complete:

(1) We have proved it when f and g are finite linear combinations, with coefficients $\geqslant 0$, of characteristic functions of integrable sets. But then it is true if we consider *countably infinite* such linear combinations. This is immediately verified by taking the limit on increasing sequences of finite ones.

(2) Suppose now that f and g are arbitrary nonnegative functions on X; let us denote by $\sum_h$ the set of countably infinite linear combinations, with coefficients $\geqslant 0$, of integrable characteristic functions, which are $\geqslant$ than a given function $h \geqslant 0$. From the result stated in (1) for such linear combinations, we obtain

$$\inf_{u \in \Sigma_f + \Sigma_g} \left(\int |u|^p \, dx \right)^{1/p} \leqslant \inf_{v \in \Sigma_f} \left(\int |v|^p \, dx \right)^{1/p} + \inf_{w \in \Sigma_g} \left(\int |w|^p \, dx \right)^{1/p}.$$

It is immediately seen that the right-hand side is equal to

$$\left(\int^* |f|^p \, dx \right)^{1/p} + \left(\int^* |g|^p \, dx \right)^{1/p}.$$

As for the left-hand side, it is at least equal to $(\int^* (f+g)^p \, dx)^{1/p}$. This follows easily from the fact that $\Sigma_f + \Sigma_g \subset \Sigma_{f+g}$.   Q.E.D.

Example IV.   The spaces of sequences l^p $(1 \leqslant p \leqslant +\infty)$.
We denote by l^p the vector space of sequences $\sigma = (\sigma_j)$ $(j = 0, 1, ...)$ of complex numbers with the property that the quantity

$$|\sigma|_{l^p} = \left(\sum_{j=0}^{\infty} |\sigma_j|^p \right)^{1/p} \quad \text{if} \quad 1 \leqslant p < +\infty,$$

$$|\sigma|_{l^\infty} = \sup_{0 \leqslant j < +\infty} |\sigma_j| \quad \text{if} \quad p = +\infty,$$

is finite. Then, in view of Theorem 11.1, $|\sigma|_{l^p}$ is a seminorm on l^p. In fact, it is a norm and turns l^p into a normed space. This space is complete. It is a good exercise for the beginner to try to prove this fact directly. It will follow from the general Fisher–Riesz theorem proved when we discuss the next example:

Example V.   The spaces L^p $(1 \leqslant p < +\infty)$.
We deal with a set X, a positive measure dx on X; dx is σ-finite. We assume that the student is familiar with the elementary facts of integration theory. We denote by $\mathscr{F}^p$ the space of complex functions in X, f, such that

$$\int^* |f|^p \, dx < +\infty.$$

In virtue of Theorem 11.1, $f \rightsquigarrow (\int^* |f|^p \, dx)^{1/p}$ is a seminorm on $\mathscr{F}^p$. We denote then by $\mathscr{L}^p$ the closure in $\mathscr{F}^p$, in the sense of that seminorm, of the linear subspace of the integrable step-functions (finite linear combinations, with *complex* coefficients, of characteristic functions of integrable sets). When $p = 1$, $\mathscr{L}^p$ is simply the space of integrable functions. It can be proved that a function f belongs to $\mathscr{L}^p$ if and only if $f \in \mathscr{F}^p$ and if f is measurable. For integrable functions, such as $|f|^p$ when $f \in \mathscr{L}^p$, one omits the upper asterisk in the integral sign. So we set

$$\|f\|_{L^p} = \left(\int |f(x)|^p \, dx \right)^{1/p}.$$

Outside of exceptional cases, $f \rightsquigarrow \|f\|_{L^p}$ is not a norm on $\mathscr{L}^p$. We have

$\|f\|_{L^p} = 0$ if and only if $f = 0$ almost everywhere. Thus the kernel $\mathcal{N}^p$ of the seminorm $\|\ \|_{L^p}$ consists of the elements of $\mathcal{N}^p$, which vanish almost everywhere, and the associated normed space, which is denoted by L^p (often also by L_p), $L^p = \mathscr{L}^p/\mathcal{N}^p$, is a space of equivalence classes of functions modulo the relation "$f = g$ almost everywhere (a.e.)" Let $\dot{f} \in L^p$; we set

$$\|\dot{f}\|_{L^p} = \|f\|_{L^p},$$

where f is *any* representative of the class $\dot{f}$ (note indeed that any other representative g of $\dot{f}$ is such that $\|g\|_{L^p} = \|f\|_{L^p}$; this is a trivial consequence of the triangular inequality, here Theorem 11.1, as we have already pointed out on p. 67). In accordance with a well-established and convenient tradition, we shall often deal with the classes $\dot{f}$ of L^p as if they were really functions, and not simply "functions defined almost everywhere." Thus we shall drop, most of the time, the dots and write f instead of $\dot{f}$.

Next, we state and prove the classical Fischer–Riesz theorem:

THEOREM 11.2.   *Every Cauchy sequence in $\mathscr{L}^p$ converges.*

Proof.   Let $\{f_\nu\}$ be a Cauchy sequence in $\mathscr{L}^p$. We select an increasing sequence of integers ν_k such that $\nu \geqslant \nu_k$ implies

$$\|f_\nu - f_{\nu_k}\|_{L^p} \leqslant 2^{-k-1} \qquad (k = 0, 1,...).$$

We set $g_k = f_{\nu_k} - f_{\nu_{k-1}}$ for $k \geqslant 1$; $g_0 = f_{\nu_0}$. We have then

$$f_{\nu_K} = \sum_{k=0}^{K} g_k.$$

It is clear that it suffices to show that the series $\sum_{k=0}^{\infty} g_k$ converges in $\mathscr{L}^p$. Its sum will be the limit of the f_{ν_k}, hence of the f_ν (Proposition 6.7).

For $h, n = 1, 2,...$, let us denote by $N_{n,h}$ the set of points $x \in X$ such that

$$\sum_{k=h}^{\infty} |g_k(x)| \geqslant 1/n.$$

In order to estimate $\mathrm{meas}(N_{n,h})$ ($N_{n,h}$ is measurable!), we apply the following straightforward generalization of Minkowski's inequalities:

$$(11.4) \qquad \left(\int^* \left| \sum_{l=0}^{\infty} G_l(x) \right|^p dx \right)^{1/p} \leqslant \sum_{l=0}^{\infty} \left(\int^* |G_l(x)|^p \, dx \right)^{1/p}.$$

We take $G_l = g_{h+l}\chi_{N_{n,h}}$, ($\chi_A$: characteristic function of A).

We obtain

$$\frac{1}{n}(\text{meas}(N_{n,h}))^{1/p} \leqslant \sum_{k=h}^{\infty} \| g_k \|_{L^p} \leqslant 2^{-h+1}.$$

Now observe that, for fixed n, the sets $N_{n,h}$ ($h = 1, 2,...$) form a non-increasing sequence; their intersection, N_n, is obviously of measure zero; the *union* of the N_n ($n = 1, 2,...$) is therefore also of measure zero. It is immediately seen that the series $\sum_{k=0}^{\infty} g_k(x)$ converges absolutely for every $x \in X - N$; its sum will be denoted by $g(x)$. If $x \in N$, we set $g(x) = 0$. Since g is the limit, almost everywhere, of a sequence of measurable functions, it follows from Egoroff's theorem that g is measurable. On the other hand, for all $x \in X$,

$$| g(x)| \leqslant \sum_{k=0}^{\infty} | g_k(x)|.$$

By applying once more (11.4), we obtain

$$\left(\int^{*} | g(x)|^p \, dx\right)^{1/p} \leqslant \sum_{k=0}^{\infty} \| g_k \|_{L^p} \leqslant \| f_{v_0} \|_{L^p} + 1.$$

This proves that $g \in \mathscr{L}^p$. The last step consists in proving that g is indeed the sum (in $\mathscr{L}^p$) of the series $\sum g_k$. It follows from the fact that we have, for all $h \geqslant 1$ and all $x \in X - N$,

$$\left| g(x) - \sum_{k=0}^{h} g_k(x) \right| \leqslant \sum_{k=h+1}^{\infty} | g_k(x)|.$$

Raising both sides to the pth power and integrating over $X - N$ yields immediately, by application of (11.4),

$$\left\| g - \sum_{k=0}^{h} g_k \right\|_{L^p} \leqslant 2^{-h},$$

which shows what we wanted, by taking $h \to +\infty$.   Q.E.D.

COROLLARY. *L^p is a Banach space.*

We consider now the case where X is an open subset of $\mathbf{R}^n$ and dx is the induced Lebesgue measure. We observe that a *continuous* function f in X cannot vanish almost everywhere without vanishing identically: for if $f(x) \neq 0$ at some point, $f(x) \neq 0$ in some open neighborhood of that point, and such a set cannot have measure zero. This implies immediately that a class $f \in L^p$ contains at most one representative which is a continuous function; when f contains one, we say that f is a continuous function. At this stage, it is convenient to introduce the following definition, which is going to be much used in the sequel:

Definition 11.3. Let X be a topological space, E a vector space, and f a mapping of X into E. The closure of the set $\{x \in X; f(x) \neq 0\}$ is called the support of f; we denote it by supp f.

The support of f can be defined as the complement of the (open) set of points $x \in E$ with the following property: f vanishes identically in some neighborhood of x.

THEOREM 11.3. *Let X be an open subset of $\mathbf{R}^n$. If $1 \leqslant p < +\infty$, the continuous functions with compact support in X form a dense linear subspace of $\mathscr{L}^p(X)$.*[†]

We state and prove Theorem 11.3 for the Lebesgue measure. It should be pointed out, however, that it is more generally true for any Radon measure (cf. Chapter 21) on a locally compact space.

Proof. By definition of $\mathscr{L}^p(X)$, integrable step-functions are dense in it. It suffices therefore to show that every such step-function is a limit of continuous functions with compact support; but since an integrable step-function is a finite linear combination of characteristic functions of integrable sets, it suffices to show that every one of the latter can be approximated. Let $A \subset X$ be integrable, and χ_A its characteristic function. Given any $\varepsilon > 0$, there is an open subset $\Omega_\varepsilon \subset X$, a compact subset K_ε of X such that $K_\varepsilon \subset A \subset \Omega_\varepsilon$ and such that $\operatorname{meas}(K_\varepsilon) > \operatorname{meas}(A) - \varepsilon$ and $\operatorname{meas}(\Omega_\varepsilon) < \operatorname{meas}(A) + \varepsilon$. Let, then, H_ε be a compact neighborhood of K_ε contained in Ω_ε, U_ε its complement, $\delta = d(K_\varepsilon, U_\varepsilon)$ (d: Euclidean distance). Let, then, f be a continuous function on the real line such that $0 \leqslant f(t) \leqslant 1$ for all t and $f(t) = 1$ if $t \leqslant 0$, $f(t) = 0$ if $t > \frac{1}{2}$. We set $g(x) = f(1 - \delta^{-1} d(x, U_\varepsilon))$; g is continuous (cf. Lemma 16.1) with support contained in H_ε, hence compact. We have:

$$\chi_{K_\varepsilon} \leqslant g \leqslant \chi_{H_\varepsilon} \leqslant \chi_{\Omega_\varepsilon},$$

[†] When there is some risk of confusion about the set which is being considered, if this set is X, one writes $\mathscr{L}^p(X)$ rather than $\mathscr{L}^p$. If moreover we wish to make clear that we are talking about a given measure dx, we write $\mathscr{L}^p(X, dx)$, or $\mathscr{L}^p_{dx}$, or $\mathscr{L}^p_{dx}(X)$; similar remarks apply to L^p and also to the case $p = +\infty$ to be considered in the next example. When the measure dx is the Lebesgue measure on $\mathbf{R}^n$, one often reserves the notation $\mathscr{L}^p$ and L^p to the spaces $\mathscr{L}^p(\mathbf{R}^n, dx)$ and $L^p(\mathbf{R}^n, dx)$.

where we denote by χ the characteristic functions. We see immediately that

$$\int |g - \chi_A| \, dx \leqslant \varepsilon. \qquad\qquad \text{Q.E.D.}$$

COROLLARY.   *If X is open, the continuous functions with compact support form a dense subspace of $L^p(X)$ $(1 \leqslant p < +\infty)$.*

Thus, $L^p(X)$ in this case (X open in $\mathbf{R}^n$, dx the Lebesgue measure) can be regarded as a "concrete" realization of the completion of the space of continuous functions with compact support in X, equipped with the norm $\| \ \|_{L^p}$.

Example VI.   The Space L^∞.

Let X be a set, and dx a positive measure on it. We recall that dx is σ-finite, i.e., that X is the union of a sequence of dx-integrable sets. We denote by $\mathscr{L}^\infty$ the vector space of all complex-valued, measurable functions f in X such that there is a finite constant $M \geqslant 0$ with the following property:

(11.5)   *There is a subset N of X, with measure zero, such that $|f(x)| \leqslant M$ for all $x \in X - N$.*

We denote by $\|f\|_{L^\infty}$ the infimum of all numbers $M \geqslant 0$ with Property (11.5); $f \rightsquigarrow \|f\|_{L^\infty}$ is clearly a seminorm. In general, it is not a norm; its kernel is exactly the set of functions f which are equal to zero almost everywhere. Indeed, if $f = 0$ in the complement of a set N of measure zero, we may take $M = 0$ in (11.5). Conversely, suppose that to every integer $k \geqslant 1$ there is a set N_k of measure zero such that $|f(x)| \leqslant 1/k$ if $x \notin N_k$; the union of the sets N_k is a set N of measure zero and $f = 0$ in the complement of N. The normed space associated to the seminormed space $\mathscr{L}^\infty$ will be denoted by L^∞, its norm by $f \rightsquigarrow \|f\|_{L^\infty}$; one often writes L_∞ instead of L^∞. The elements of L^∞ are not functions in X but equivalence classes of functions modulo the relation "to be equal almost everywhere." However, we shall often deal with them as if they were functions. An element f of $\mathscr{L}^\infty$ (or of L^∞) is often said to be *essentially bounded* in X (with respect to the measure dx) and $\|f\|_{L^\infty}$ is called its *essential supremum*.

We have the equivalent of Theorem 11.2:

THEOREM 11.4.   *Every Cauchy sequence converges in $\mathscr{L}^\infty$.*

Proof.   The proof is a simplified version of that of Theorem 11.2.

Let $\{f_k\}$ be a Cauchy sequence in $\mathscr{L}^\infty$. First of all, we note that the sequence of numbers $\|f_k\|_{L^\infty}$ is bounded and that there is, therefore, a number $0 < A < +\infty$ and, for each k a set of measure zero N_k such that $|f_k(x)| \leqslant A$ for $x \notin N_k$; the union N of the sets N_k has measure zero and we have $|f_k(x)| \leqslant A$ for all $x \in X - N$. Next, we use the fact that to every $\nu = 1, 2, \dots$ there is k_ν such that $k, l \geqslant k_\nu$ implies $\|f_k - f_l\|_{L^\infty} \leqslant 1/\nu$: there is a set $N_{\nu,k,l}$ of measure zero such that $|f_k(x) - f_l(x)| \leqslant 2/\nu$ for all $x \notin N_{\nu,k,l}$. We denote by N' the union of all the sets $N_{\nu,k,l}$ as ν varies and so do $k, l \geqslant k_\nu$; N' has measure zero, and we have, if $k, l \geqslant k_\nu$ and $x \in X - N'$, $|f_k(x) - f_l(x)| \leqslant 2/\nu$. We derive from this that the sequence $\{f_k\}$ converges uniformly in $X - (N \cup N')$; let f be its limit there; we may extend f by zero to $N \cup N'$: we have $|f(x)| \leqslant A$ in the complement of this set; but $N \cup N'$ is of measure zero. On the other hand, f is the limit almost everywhere of the f_k's, therefore (Egoroff's theorem) is measurable, hence belongs to $\mathscr{L}^\infty$. It is evident that f is the limit of the f_k's in $\mathscr{L}^\infty$.   Q.E.D.

COROLLARY. L^∞ is a Banach space.

Let us consider the case where X is an open subset of $\mathbf{R}^n$ and dx the Lebesgue measure. The space of bounded continuous functions in X, $\mathscr{B}^0(X)$, is a linear subspace of $\mathscr{L}^\infty(X)$. We have, for $f \in \mathscr{B}^0(X)$,

$$(11.6) \qquad\qquad \|f\|_{L^\infty} = \sup_{x \in X} |f(x)|.$$

This is trivial to check. That the left-hand side, in (11.6), is at most equal to the right-hand side is evident. On the other hand, given any $\varepsilon > 0$, there is a nonempty open subset Ω of X such that $|f(y)| \geqslant \sup_{x \in X} |f(x)| - \varepsilon$ for all $y \in \Omega$. This implies immediately that the supremum of $|f(x)|$ on the complement of any set of measure zero (this complement necessarily intersects Ω) is $\geqslant \sup_{x \in X} |f(x)| - \varepsilon$. As ε is arbitrary, we derive (11.6).

Also observe that two functions belonging to $\mathscr{B}^0(X)$ cannot be equal almost everywhere without being equal everywhere. Thus (11.6) shows that the canonical homomorphism of $\mathscr{L}^\infty(X)$ onto $L^\infty(X)$ induces an *isometry* of $\mathscr{B}^0(X)$ into $L^\infty(X)$. Now, it is quite obvious that $\mathscr{B}^0(X)$ is a Banach space (cf. Example II); therefore, this isometry maps it onto a *closed* linear subspace of $L^\infty(X)$. This subspace is not the whole of $L^\infty(X)$: indeed, there are discontinuous $\mathscr{L}^\infty$ functions in X which are not equal almost everywhere to a bounded continuous function! As we see by comparing this with Theorem 11.3, the situation with respect to approximation by continuous functions is very different in the case

$p = +\infty$ from what it is in the case of p finite. This difference has far-reaching consequences.

Let E be a normed space, and $\|\ \|$ the norm on E. Let E' be the dual of E, that is to say the vector space of all *continuous* linear maps of E into the complex plane. In view of the corollary of Proposition 7.7, if $f \in E'$ there is a finite constant $C \geqslant 0$ such that, for all $x \in E$,

$$(11.7) \qquad\qquad |f(x)| \leqslant C\|x\|.$$

The infimum of the numbers C such that (11.7) holds (for *all* x) is denoted by $\|f\|$. Given any $x \in E$, we have

$$(11.8) \qquad\qquad |f(x)| \leqslant \|f\| \cdot \|x\|.$$

The student may check that we could have defined $\|f\|$ by either of the two equalities below:

$$(11.9) \qquad\qquad \|f\| = \sup_{x \in E, \|x\| \leqslant 1} |f(x)|;$$

$$(11.10) \qquad\qquad \|f\| = \sup_{x \in E, \|x\| = 1} |f(x)|.$$

Thus $\|f\|$ is the lowest upper bound of the function $x \rightsquigarrow |f(x)|$ on the unit sphere $\{x \in E; \|x\| = 1\}$ of E. From this follows immediately:

PROPOSITION 11.2.   *Let E be a normed space: $f \rightsquigarrow \|f\|$ is a norm on the dual E' of E.*

Whenever we shall be dealing with a normed space E and we refer to its dual E' as a normed space, this will mean that we consider on E' the norm defined by (11.9) or (11.10).

One should be careful not to think that there is always a point x of the unit sphere of E in which $|f(x)| = \|f\|$.

The notion of the norm of a continuous linear functional on a normed space can be immediately generalized to continuous linear maps of a normed space E into another normed space F. Let us denote by $\|\ \|$ both norms in E and F, and let $u : E \to F$ be a continuous linear map. From Proposition 7.7 it follows that there is a constant $C \geqslant 0$ such that, for all $x \in E$,

$$\|u(x)\| \leqslant C\|x\|.$$

Then again we define the norm of u, $\|u\|$, as the infimum of the constants C above. We have

$$(11.11) \qquad\qquad \|u\| = \sup_{x \in E, \|x\| \leqslant 1} \|u(x)\| = \sup_{x \in E, \|x\| = 1} \|u(x)\|.$$

The absolute value of complex numbers has been replaced here by the norm in F. Proposition 11.2 can be immediately extended, since here again $\| u \|$ is defined as the lowest upper bound of a nonnegative function, $x \rightsquigarrow \| u(x) \|$, on a set (e.g., the unit sphere of E), and that the nonnegative function in question is obviously subadditive with respect to u.

Let $L(E; F)$ be the vector space of all continuous linear maps of E into F. Assuming that E and F are both normed spaces, whenever we refer to $L(E; F)$ as a normed space, it will be implicit that it carries the norm defined by (11.11).

THEOREM 11.5. *Let E and F be two normed spaces. Suppose that F is complete. Then the normed space $L(E; F)$ is also complete.*

That the fact that E is complete or not should be irrelevant, in connection with Theorem 11.5, is obvious: indeed, any continuous linear map of E into F (assuming that F is complete) can be extended, in a unique way, into a continuous linear map of $\hat{E}$, completion of E, into F (Theorem 5.2, (c) and (II)). Thus, the extension of mappings from E to $\hat{E}$ defines an isomorphism, for the vector space structures, of $L(E; F)$ into $L(\hat{E}; F)$. We leave to the student the verification of the fact that this isomorphism is an *isometry* (we recall that an isometry is a mapping which preserves the norms).

Proof of Theorem 11.5. We must prove that $L(E; F)$ is sequentially complete. Let $\{u_\nu\}$ be a Cauchy sequence in $L(E; F)$. For every $\varepsilon > 0$, there is $N(\varepsilon)$, integer $\geqslant 0$, such that, for all ν, $\mu \geqslant N(\varepsilon)$, $\| u_\nu - u_\mu \| \leqslant \varepsilon$. Whatever be the continuous linear map $u : E \rightarrow F$, we have (cf. (11.8)),

$$(11.12) \qquad \textit{for all} \quad x \in E, \qquad \| u(x) \| \leqslant \| u \| \| x \|.$$

In particular, we shall have, for all $x \in E$ and all ν, $\mu \geqslant N(\varepsilon)$,

$$(11.13) \qquad \| u_\nu(x) - u_\mu(x) \| \leqslant \varepsilon \| x \|.$$

This means that, for fixed $x \in E$, the sequence $\{u_\nu(x)\}$ is a Cauchy sequence in F. But F is complete, hence this sequence converges to some element of F, which we shall denote by $u(x)$. This defines immediately a mapping $x \rightsquigarrow u(x)$ of E into F. Let us show that this mapping is *linear*.

Let x, y be two elements of E, arbitrary $\varepsilon > 0$, and select ν sufficiently large so as to have

$$\| u_\nu(x) - u(x) \| \leqslant \varepsilon/3, \qquad \| u_\nu(y) - u(y) \| \leqslant \varepsilon/3,$$

$$\| u_\nu(x + y) - u(x + y) \| \leqslant \varepsilon/3.$$

This is possible, since the u_ν converge pointwise to u. Combining the three preceding inequalities, we obtain that

$$\| u(x + y) - u(x) - u(y)\| \leqslant \varepsilon.$$

As ε is arbitrary, we conclude that $u(x + y) = u(x) + u(y)$. We use a similar argument in order to prove that $u(\lambda x) = \lambda\, u(x)$ for all $x \in E$, $\lambda \in \mathbf{C}$.

We prove now that u is *continuous*. Choose an integer $N(1)$ sufficiently large so as to have, for all $\nu \geqslant N(1)$,

$$\| u_\nu - u_{N(1)} \| \leqslant 1,$$

which implies, for all $x \in E$,

$$\| u_\nu(x) - u_{N(1)}(x)\| \leqslant \| x \|.$$

Choose now $\varepsilon > 0$ and x arbitrarily. There exists $\nu \geqslant N(1)$ such that

$$\| u_\nu(x) - u(x)\| \leqslant \varepsilon.$$

This implies

$$\| u(x)\| \leqslant \| u_\nu(x)\| + \varepsilon \leqslant \| u_{N(1)}(x)\| + \| x \| + \varepsilon \leqslant (\| u_{N(1)}\| + 1)\, \| x \| + \varepsilon$$

As ε is arbitrarily small, we conclude that

$$\| u(x)\| \leqslant (\| u_{N(1)}\| + 1)\, \| x \|.$$

This means that u is continuous.

It remains to prove that the u_ν converge to u in the sense of the norm of continuous linear maps. Here again, let $\varepsilon > 0$ be arbitrary; and let $N(\varepsilon)$ be as chosen in relation with (11.13). Choose arbitrarily $x \in E$, $\| x \| = 1$, and then take $\nu = \nu(x) \geqslant N(\varepsilon)$ such that $\| u_\nu(x) - u(x) \| \leqslant \varepsilon$. We have

$$\| u_{N(\varepsilon)}(x) - u(x)\| \leqslant \| u_\nu(x) - u(x)\| + \| u_\nu(x) - u_{N(\varepsilon)}(x)\| \leqslant 2\varepsilon.$$

Since x is an arbitrary point of the unit sphere of E, it means that

$$\| u_{N(\varepsilon)} - u \| \leqslant 2\varepsilon. \qquad \text{Q.E.D.}$$

COROLLARY. *Let E be a normed space. The normed space E', dual of E, is a Banach space.*

Exercises

11.1. Prove that, for $q \geqslant p$, we have $l^p \subset l^q$ and that the injection $l^p \to l^q$ is continuous and has a norm equal to one.

11.2. Let $\sigma \in l^1$. Prove that

$$| \sigma |_{l^\infty} = \lim_{r \to \infty} | \sigma |_{l^r} .$$

11.3. Let E, F be two Banach spaces (with norms denoted by $\| \ \|$), u a continuous linear map of E into a Hausdorff TVS G, and j a *continuous one-to-one* linear map of F into G. Let $E_0 = \{x \in E; u(x) \in j(F)\}$.

Prove that the norm on E_0,

$$x \rightsquigarrow \| x \| + \| j^{-1}(u(x)) \|,$$

turns E_0 into a Banach space.

11.4. Let X be a set, and $\mathscr{F}(X; \mathbf{C})$ the space of complex-valued functions in X equipped with the topology of pointwise convergence. Prove that $\mathscr{F}(X; \mathbf{C})$ is not normable unless X is finite.

11.5. Let $\mathscr{C}_\infty(\mathbf{R}^n)$ be the space of continuous functions in $\mathbf{R}^n$ which converge to zero at infinity, equipped with the topology of uniform convergence on $\mathbf{R}^n$, i.e., the topology defined by the norm

$$\phi \rightsquigarrow \sup_{x \in \mathbf{R}^n} | \phi(x) |.$$

Prove that $\mathscr{C}_\infty(\mathbf{R}^n)$ is a Banach space.

11.6. Let $E = \mathbf{C}^2$, and $\| \ \|$ the norm

$$\zeta = (\zeta_1, \zeta_2) \rightsquigarrow | \zeta |_\infty = \sup(| \zeta_1 |, | \zeta_2 |).$$

Let E_1 be the linear subspace $\{\zeta \in \mathbf{C}^2; \zeta_2 = 0\}$. For every $\zeta^0 \in E$, characterize the set of $\zeta \in E_1$ such that

$$(11.14) \qquad \qquad \| \zeta^0 - \zeta \| = \inf_{\zeta' \in E_1} \| \zeta^0 - \zeta' \|.$$

In particular, prove that, if $\zeta^0 \notin E_1$, there is an infinity of points ζ with the above property (11.14).

11.7. We keep the notation of Exercise 11.6 with one exception: the norm $\| \ \|$ now denotes a norm

$$\zeta \rightsquigarrow | \zeta |_p = (| \zeta_1 |^p + | \zeta_2 |^p)^{1/p}$$

with $p < +\infty$. Prove that there is one and only one point ζ such that (11.14) holds.

11.8. Let $E = l^\infty$, $\| \ \|$ the norm $\sigma \rightsquigarrow | \sigma |_{l^\infty}$, and l_∞ the subspace of l^∞ consisting of the sequences $\sigma = (\sigma_n)(n = 0, 1,...)$ converging to zero, i.e., such that $\sigma_n \to 0$ as $n \to +\infty$. Let B be the closed unit ball of l_∞,

$$\{\sigma \in l_\infty ; \sup_n | \sigma_n | \leqslant 1\}.$$

Let B_+ be the subset of B consisting of the sequences having all their terms $\geqslant 0$. Finally, let e be the sequence having all its terms equal to one. Prove:

(1) that B_+ is closed in l^∞;
(2) that the distance between B and e is exactly equal to one;

(3)   that the distance between e and every element of B_+ is exactly equal to one;

(4)   that the distance between e and every element of B which does not belong to B_+ is > 1.

11.9.   A TVS E is said to be *separable* if there is a dense countable set in E. Prove that, if a metrizable TVS E is separable, every noncountable subset A of E contains a converging sequence.

11.10.   Prove that $L^\infty(\mathbf{R}^1)$ is not separable (see Exercise 11.9).

12

Hilbert Spaces

Historically, the first infinite dimensional topological vector spaces whose theory has been studied and applied have been the so-called Hilbert spaces. They play a most important role in pure mathematics (e.g., in the theories of boundary value problems, of probability, of group representations), as well as in applied mathematics (e.g., in quantum mechanics and in statistics). Although many functional spaces are *not* Hilbert spaces, they can often be represented meaningfully as union of subspaces which carry a Hilbert structure (most of the time, finer than the structure induced by the surrounding space). The knowledge of the properties of these "Hilbert subspaces" may reveal important properties of the surrounding space, usually in relation with existence and uniqueness of solutions of functional equations.

The reason for the impressive success of the theory of Hilbert space is simple enough: they closely resemble finite dimensional Euclidean spaces. This, in two respects: they are complete, as all the finite dimensional TVS are; they carry an inner product, which is a positive definite sesquilinear form (see below) and which, roughly speaking, determines their properties. The usefulness of an inner product can best be emphasized by recalling how useful are orthonormal bases in the finite dimensional case (especially in relation with the diagonalization of self-adjoint matrices).

We begin by recalling what is a *sesquilinear form* on a vector space E. It is a mapping $(x, y) \to B(x, y)$ from $E \times E$ into the complex plane, $\mathbf{C}$, with the following properties:

(1) $B(x_1 + x_2, y) = B(x_1, y) + B(x_2, y);$
      $B(x, y_1 + y_2) = B(x, y_1) + B(x, y_2);$

(2) $B(\lambda x, y) = \lambda B(x, y);$

(3) $B(x, \lambda y) = \bar{\lambda} B(x, y).$

It is Property (3) which is responsible for the name *sesquilinear*: sesqui means "one time and a half" in Latin; (2) means that, for fixed y, the

112

map $x \rightsquigarrow B(x, y)$ is *linear*, whereas (3) says that, for fixed x, the map $y \rightsquigarrow B(x, y)$ is *semilinear* (if we had $B(x, \lambda y) = \lambda B(x, y)$, we would refer to B as a *bilinear* form). The importance of Condition (3) is a direct consequence of the importance of *Hermitian* forms. A Hermitian form B is a form with Properties (1) and (2), and with the additional property:

(4) $B(x, y) = \overline{B(y, x)}$.

It is then obvious that B must also have Property (3) above and thus be sesquilinear.

A sesquilinear form B on E is said to be *nondegenerate* if it has the following property:

(5) *If $x \in E$ is such that, for all $y \in E$, $B(x, y) = 0$, then $x = 0$. If $y \in E$ is such that $B(x, y) = 0$ for all $x \in E$, then $y = 0$.*

Examples in Finite Dimensional Spaces C^n

The usual Hermitian product $(\zeta, \zeta') = \zeta_1 \overline{\zeta_1'} + \cdots + \zeta_n \overline{\zeta_n'}$ is a nondegenerate Hermitian form on $\mathbf{C}^n$.
The form

$$B(\zeta, \zeta') = \zeta_1 \overline{\zeta_1'} - \zeta_2 \overline{\zeta_2'}$$

is a nondegenerate Hermitian form on $\mathbf{C}^2$ (it would be degenerate if viewed as a form on $\mathbf{C}^n$ with $n > 2$).

A sesquilinear form is Hermitian if and only if $B(x, x)$ is a real number for all $x \in E$. It is obvious one way, just by applying (4) with $y = x$. On the other hand, we have:

(12.1) $B(x + y, x + y) - B(x, x) - B(y, y) = B(x, y) + B(y, x)$.

If the left-hand side of (12.1) is real for all x, y, so must be the right-hand side, which shows that $\operatorname{Im} B(y, x) = -\operatorname{Im} B(x, y)$. Apply (12.1) with iy substituted for y. The left-hand side must again be real, and so must be the right-hand side which is now, in view of sesquilinearity,

$$i[B(x, y) - B(y, x)].$$

This shows that $\operatorname{Re} B(y, x) = \operatorname{Re} B(x, y)$, whence (4).

We shall essentially be interested, in this chapter, in *positive definite* forms. These are sequilinear forms which satisfy the following condition:

(6) *For all $x \in E$, $x \neq 0$, $B(x, x) > 0$.*

In particular, positive definite sequilinear forms are Hermitian. They are obviously nondegenerate.

We might also introduce *nonnegative* sequilinear forms, as we did in Chapter 7 (p. 60). These are forms which satisfy:

(7)   *For all $x \in E$, $B(x, x) \geqslant 0$.*

A nonnegative sesquilinear form is nondegenerate if and only if it is definite positive. One way, it is already known. The other way, our statement follows from the next result, the celebrated *Schwarz inequality* (or *Cauchy–Schwarz inequality*):

PROPOSITION 12.1.   *Let $B(\ ,\ )$ be a nonnegative sesquilinear form on E. Then, for all x and y in E, we have*

$$|B(x, y)|^2 \leqslant B(x, x)\, B(y, y).$$

Proof.   Since B is nonnegative, we have

(12.2) $0 \leqslant B(x + \lambda y, x + \lambda y) = B(x, x) + 2\,\mathrm{Re}[\bar{\lambda}\, B(x, y)] + |\lambda|^2\, B(y, y).$

It suffices then to take $\lambda = B(x, y)\, t$. The right-hand side of (12.2) becomes a polynomial in the variable t. Schwarz inequality expresses the fact that this polynomial does not have two *distinct real* roots.

COROLLARY.   *If B is nonnegative,*

(12.3) $$x \rightsquigarrow B(x, x)^{1/2}$$

is a seminorm on E. If B is positive definite, it is a norm.

Indeed, we have $B(\lambda x, \lambda x)^{1/2} = |\lambda|\, B(x, x)^{1/2}$, and

$$B(x + y, x + y)^{1/2} \leqslant (B(x, x) + 2|\,B(x, y)| + B(y, y))^{1/2}$$
$$\leqslant B(x, x)^{1/2} + B(y, y)^{1/2},$$

by Schwarz inequality.

We recall Definition 7.5: the pair consisting of a vector space E and of a nonnegative sesquilinear form B on E is called a (complex) *pre-Hilbert* space.

Definition 12.1.   *The pair consisting of a vector space E and a positive definite sesquilinear form B on E is called a complex Hausdorff pre-Hilbert space.*

Let (E, B) be a pre-Hilbert space which is not Hausdorff. Let N be the subset of E consisting of the vectors x such that $B(x, y) = 0$ for all

$y \in E$. Because of Schwarz inequality, this subset N is exactly the kernel of the seminorm (12.3). The quotient space E/N can then be regarded as a normed space. Observing that, if $x, y \in E$ and $z \in N$,

$$B(x + z, y) = B(x, y),$$

we derive that there is a canonical sesquilinear form $\dot{B}$ on E/N: if ϕ is the canonical map of E onto E/N, we have

$$\dot{B}(\dot{x}, \dot{y}) = B(x, y) \qquad \text{if} \quad \dot{x} = \phi(x), \quad \dot{y} = \phi(y).$$

Then $\dot{B}$ is positive definite, and the norm of E/N is nothing else but

$$\dot{x} \rightsquigarrow \dot{B}(\dot{x}, \dot{x})^{1/2}.$$

We say that $(E/N, \dot{B})$ *is the Hausdorff pre-Hilbert space associated with the pre-Hilbert space* (E, B).

Let (E, B) be a pre-Hilbert space; we may then regard E as a TVS: we consider on E the topology defined by the seminorm (12.3). When we speak about the topology, or the TVS structure, of a pre-Hilbert space (E, B), it will always be in this sense, unless we specify otherwise.

Definition 12.2. A Hausdorff pre-Hilbert space which is complete is called a Hilbert space.

Given a normed space $(E, \| \ \|)$, one could ask the following question: is it a (Hausdorff) pre-Hilbert space? In other words, is there a positive definite sesquilinear form $B(\ , \ )$ on E such that, for all $x \in E$, $\| x \| = B(x, x)^{1/2}$? If this is true, we say that the norm $\| \ \|$ of E is a *Hilbert norm*. The answer to our question is provided by the following result:

PROPOSITION 12.2. *The norm $\| \ \|$ on the space E is a Hilbert norm if and only if the following relation holds, for all $x, y \in E$,*

(HN) $$\| x \|^2 + \| y \|^2 = \tfrac{1}{2}(\| x + y \|^2 + \| x - y \|^2).$$

Proof. If the norm $\| \ \|$ is given by (12.3), we derive immediately (HN) from (12.1). Conversely, suppose that (HN) holds and set

$$R(x, y) = \tfrac{1}{4}(\| x + y \|^2 - \| x - y \|^2),$$

$$J(x, y) = \tfrac{1}{4}(\| x + iy \|^2 - \| x - iy \|^2).$$

Then

$$B(x, y) = R(x, y) + iJ(x, y)$$

is a sesquilinear form on E, as easily verified, and the norm $\| \ \|$ is equal to (12.3). $\qquad$ Q.E.D.

Let now (E, B) be a Hausdorff pre-Hilbert space, and $\| \ \|$ its norm (12.3). Let $\hat{E}$ be the *normed space* which is the completion of the normed space $(E, \| \ \|)$. In virtue of the continuation of the identities, (HN) holds in $\hat{E}$, hence the norm of $\hat{E}$ is a Hilbert norm; let $\hat{B}$ be the positive definite sesquilinear form on $\hat{E}$ associated to its norm. It is immediate that $\hat{B}$ extends B; $(\hat{E}, \hat{B})$ is a Hilbert space, which is called the completion of the Hausdorff pre-Hilbert space (E, B).

We could have proceeded otherwise: as easily seen, $(x, y) \rightsquigarrow B(x, y)$ is a separately uniformly continuous function on the product TVS $E \times E$, hence has a unique extension to the completion of $E \times E$, which is canonically isomorphic to $\hat{E} \times \hat{E}$; this extension is the form $\hat{B}$ and turns $\hat{E}$ into a Hilbert space.

We shall now introduce the *anti-dual* of a TVS E. It is the vector space (over the field of complex numbers) of the continuous mappings f of E into the complex plane, $\mathbf{C}$, which have the following properties:

(1) $f(x + y) = f(x) + f(y)$;

(2) $f(\lambda x) = \bar{\lambda} f(x)$.

We shall denote by $\bar{E}'$ the anti-dual of E; its elements will be called continuous *antilinear* forms (or functionals, or *semilinear* forms or functionals) in E. We underline the fact that $\bar{E}'$ is a vector space: if $f \in \bar{E}'$, the product of f by a scalar is meant in the usual sense: $(\lambda f)(x) = \lambda f(x)$. Of course, there is a canonical mapping of E' onto $\bar{E}'$, which is one-to-one, onto and antilinear: to a continuous linear functional f on E it assigns the continuous antilinear functional $x \rightsquigarrow \overline{f(x)}$ on E.

Let (E, B) be a pre-Hilbert space, not necessarily Hausdorff, not necessarily complete. Consider the following mapping:

(12.4) $x \rightsquigarrow (y \rightsquigarrow B(x, y))$.

It is a mapping of E into the anti-dual of E. Indeed, for fixed $x \in E$, the antilinear functional

(12.5) $y \rightsquigarrow B(x, y)$

is continuous, as follows immediately from Schwarz inequality. Let us denote by $\tilde{x}$ the mapping (12.5). Then (12.4) can be written $x \rightsquigarrow \tilde{x}$. This latter mapping is one-to-one if and only if B is nondegenerate, that is to say positive definite. At any event, we call it the canonical mapping of (E, B) into the anti-dual $\bar{E}'$ of E. It is one-to-one if and only if (E, B) is Hausdorff. The fundamental theorem of the theory of Hilbert spaces states that it is onto if and only if (E, B) is a Hilbert space (i.e., is Hausdorff and complete). When E is a Hausdorff pre-Hilbert space, we may regard it as a normed space (the norm is given by (12.3)), and we can also regard its dual as a normed space, which moreover is a Banach space (Corollary of Theorem 11.5). The fundamental theorem of Hilbert spaces states that $x \rightarrow \tilde{x}$ is an isometry of the Hilbert space E onto its anti-dual, $\bar{E}'$. This is the theorem that we are now going to prove, and which is often summarized (quite incorrectly) by saying that a Hilbert space *is* its own dual.

The proof is based on the following important theorem:

THEOREM 12.1.   *Let (E, B) be a Hausdorff pre-Hilbert space, and K a nonempty convex complete subset of E. To every $x \in E$, there is a unique point x_0 of K such that*

$$\| x - x_0 \| = \inf_{y \in K} \| x - y \|.$$

We have used the notation

$$\| x \| = B(x, x)^{1/2};$$

we shall do this systematically from now on.

Proof.   Let us set

$$d = \inf_{y \in K} \| x - y \|.$$

We denote by A_n $(n = 1, 2, \ldots)$ the subset of K consisting of the points y such that

$$\| x - y \| \leqslant d + 1/n.$$

By definition of inf and of d, none of these sets A_n is empty; as $A_{n+1} \subset A_n$, they form a basis of a filter on K. By using the geometry defined on E by the form $B(\quad , \quad)$ and the convexity of K, we are going to show that the filter generated by the A_n is a Cauchy filter. The completeness of K implies then that this filter has a limit, necessarily unique, which will be the point x_0 that we are seeking.

Let $\varepsilon > 0$ be given arbitrarily. We must show that there is an integer $N(\varepsilon) \geqslant 0$ such that, if $n, m \geqslant N(\varepsilon)$, given any points y_n of A_n and y_m of A_m, we must have

$$\| y_n - y_m \| \leqslant \varepsilon.$$

The argument, at this stage, is purely two-dimensional: everything takes place in a plane P containing x, y_n, y_m. And because of the properties of the sesquilinear form B, the geometry in P induced by the surrounding space E is the usual one. Consider the circumference Γ of radius d and center x. As y_n and y_m get arbitrarily close to Γ, while remaining in the exterior of it, and because of the fact that they belong to one and the same convex closed set, they *must* get arbitrarily close to each other. This argument is formalized as follows. We have

(12.6) $\| \tfrac{1}{2}[(x - y_n) + (x - y_m)]\|^2 + \| \tfrac{1}{2}(y_n - y_m)\|^2$

$$= \tfrac{1}{2}(\| x - y_n \|^2 + \| x - y_m \|^2).$$

(This follows immediately from (12.1), applied with $x - y_n$ instead of x and $x - y_m$ first, $- (x - y_m)$ next, instead of y.) Because of the convexity of K, we have $\tfrac{1}{2}(y_n + y_m) \in K$, thus

$$\| x - \tfrac{1}{2}(y_n + y_m)\| \geqslant d.$$

Taking into account how y_n and y_m were chosen, we derive, from (12.6),

$$\| y_n - y_m \|^2 \leqslant 2 \left[\left(d + \frac{1}{n} \right)^2 + \left(d + \frac{1}{m} \right)^2 \right] - 4d^2$$

$$= 8d \left(\frac{1}{n} + \frac{1}{m} \right) + 2 \left(\frac{1}{n^2} + \frac{1}{m^2} \right),$$

which easily implies what we wanted.

Definition 12.3. The point x_0 in Theorem 12.1 is called the orthogonal projection of x into the complete convex set K.

We may now prove easily the fundamental theorem of Hilbert spaces. In order to make clear the situation, we state a result about pre-Hilbert spaces which contains some of the statements on p. 116:

PROPOSITION 12.3. *Let (E, B) be a pre-Hilbert space. The mapping* (12.4),

$$x \rightsquigarrow \tilde{x} : y \rightsquigarrow B(x, y),$$

is a linear map of E into its anti-dual, $\bar{E}'$. Let us regard $\bar{E}'$ as a normed space, with the usual norm of antilinear continuous functionals (see (11.9) *and* (11.10)). *Then* (12.4) *is a continuous linear map of E into $\bar{E}'$. It is an isometry into if and only if E is Hausdorff, i.e., $B(\ ,\ )$ is nondegenerate (i.e., positive definite).*

That $\tilde{x}$ is a continuous antilinear form on E follows immediately from

Schwarz inequality, as we have already said. Also from Schwarz inequality it follows that

$$\| \tilde{x} \| = \sup_{\| y \| = 1} | B(x, y)| \leqslant \| x \|,$$

whence the continuity of $x \rightsquigarrow \tilde{x}$. Suppose now that E is Hausdorff, and let x be $\neq 0$. Take $y = x/\| x \|$. We have $\| y \| = 1$, and

$$B(x, y) = \| x \|,$$

which, by (11.10), implies that $\| \tilde{x} \| = \| x \|$. As $\bar{E}'$ is a normed space, in particular is Hausdorff, if (12.4) is an injection, E itself must be Hausdorff, hence B must be nondegenerate.

Here now is the fundamental theorem:

THEOREM 12.2. *Let (E, B) be a Hausdorff pre-Hilbert space. The canonical isometry of E into its anti-dual, $\bar{E}'$, i.e., the mapping (12.4), is onto if and only if (E, B) is a Hilbert space.*

Proof. One way, the statement is obvious. The normed space $\bar{E}'$ which, by complex conjugation of linear functionals, is nothing but a copy of the normed space E', is complete (corollary of Theorem 11.5). If E is to be an isometric image of $\bar{E}'$, E also has to be complete.

Conversely, let us assume that E is complete. By Proposition 12.3, we know that (12.4) is an isometry *into*; we must show that it is onto, in other words that, given any continuous antilinear functional f on E, there is an element x_f of E such that, for all $y \in E$,

$$f(y) = B(x_f, y).$$

Let H be the kernel of f supposed to be $\neq 0$. Since f is continuous, H is a closed hyperplane of E (Exercise 9.4, Proposition 9.4); in particular, H is a closed convex set. As E is complete so is H. Let $x \in E$, $x \notin H$. By Theorem 12.1, x has a unique orthogonal projection, x_0, in H. As $x \notin H$, we have $x \neq x_0$.

We claim that $x - x_0$ is *orthogonal* to H, that is to say that, given any $y \in H$, we have

$$B(x - x_0, y) = 0.$$

Indeed, we have, for all numbers $t > 0$,

$$0 \leqslant \| x - x_0 - ty \|^2 - \| x - x_0 \|^2 = -t2 \operatorname{Re} B(x - x_0, y) + t^2 \| y \|^2,$$

whence

$$2 \operatorname{Re} B(x - x_0 , y) \leqslant t\| y \|^2.$$

Replacing there y by $-y$, then by $\pm iy$ $(i = (-1)^{1/2})$, we obtain

$$| B(x - x_0 , y)| \leqslant t\| y \|^2.$$

Taking $t \to 0$ shows that $x_0 - x$ is orthogonal to y.

Let us go back to our functional f. We cannot have $f(x - x_0) = 0$, since $H = \operatorname{Ker} f$, and we cannot have $x - x_0 \in H$. Let us set $z = \overline{[f(x - x_0)]}^{-1}(x - x_0)$.

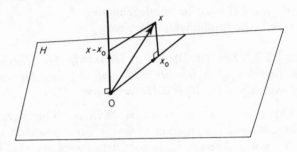

Fig. 1

Let now y be any vector in E. We may write $y = y_1 + \overline{f(y)}z$, and it is clear, since $f(z) = 1$, that

$$f(y_1) = f(y) - f(y)f(z) = 0, \quad \text{i.e.,} \quad y_1 \in H.$$

Then

$$B(z, y) = B(z, y_1) + f(y) B(z, z) = f(y) B(z, z).$$

It suffices then to take

$$x_f = (B(z, z))^{-1}z. \qquad\qquad \text{Q.E.D.}$$

Let us go back to Theorem 12.1. A closed linear subspace M of E is, in particular, a closed convex subset of E. Therefore, given any point x of E, we may consider its orthogonal projection (Definition 12.3) into M, which we denote by $P_M(x)$. We call P_M the *orthogonal projection into M*. It maps E onto M; its restriction to M is the identity mapping of M. It is of norm exactly equal to *one*. Furthermore, P_M is a projection, that is to say,

$$(12.7) \qquad P_M^2 = P_M.$$

The mapping P_M is self-adjoint, that is to say,

$$(12.8) \qquad B(P_M x, y) = B(x, P_M y).$$

(A linear map u of a pre-Hilbert space (E, B) into itself is *self-adjoint* if $B(u(x), y) = B(x, u(y))$ for all x, $y \in E$). The kernel of P_M is the orthogonal of M, M^0 (two vectors x, y of E are *orthogonal* when $B(x, y) = 0$; two *sets* A and B are *orthogonal* if, for all $x \in A$, $y \in B$, we have $B(x, y) = 0$). We have the direct sum decomposition of E,

$$(12.9) \qquad E = M \oplus M^0.$$

Here the symbol $\oplus$ means that the two subspaces which are factors in the direct sum are orthogonal; this is usually expressed by saying that it is the Hilbert direct sum, or *Hilbert sum.*

Equation (12.9) has a very important implication: in a Hilbert space E, a closed subspace M always has a *supplementary*, which means that there always is another linear subspace of E, N, also closed, such that $E = M + N$ and $M \cap N = \{0\}$: we may take $N = M^0$, orthogonal of M. This feature of Hilbert space is exceptional among TVS and even among B-spaces.

A set of vectors S is said to be *orthonormal* if $\| x \| = 1$ for all $x \in S$, and if $B(x, y) = 0$ (i.e., x and y are orthogonal) for all x, $y \in S$ such that $x \neq y$. An orthonormal set of vectors S in a pre-Hilbert space (E, B) is called an *orthonormal basis* of E if the vector space spanned by S is dense in E.[†]

Let us keep considering a Hausdorff pre-Hilbert space (E, B). We recall the following results (without proof; the student may try to prove them: they are not difficult consequences of the previous theorems):

THEOREM 12.3. *Let S be an orthonormal set in E, and V_S the closure of the linear subspace spanned by S. Then the following facts are true*:

(1) *For all $x \in E$,*

$$(12.10) \qquad \sum_{e \in S} | B(x, e)|^2 \leqslant \| x \|^2 \quad \text{(Bessel's inequality),}$$

[†] If dim $E = +\infty$, an orthonormal basis of E is *not* a basis in the algebraic sense: one cannot express every vector of E as a *finite* linear combination of vectors belonging to B.

from which it follows that the subset S_x of S consisting of the elements s of S such that $B(x, s) \neq 0$ is countable.

(2)   *For $x \in E$, the following properties are equivalent:*

   (a) $x \in V_S$;

   (b)   *in (12.10) the sign $\leqslant$ may be replaced by equality, $=$;*

   (c)   *the series $\sum_{e \in S} B(x, e)e$ converges* (with respect to the norm $\| \ \|$ of E), *and we have*

$$x = \sum_{e \in S} B(x, e)e.$$

(3)   *If V_S is complete, then, for all $x \in E$, the series $\sum_{e \in S} B(x, e)e$ converges, and we have*

$$P_{V_S}(x) = \sum_{e \in S} B(x, e)e,$$

$$\| P_{V_S}(x) \|^2 = \sum_{e \in S} | B(x, e)|^2 \quad \text{(Parseval's identity)}.$$

Let S be an arbitrary set. Let us denote by $l^2(S)$ the set of complex-valued functions λ defined in S such that

(12.11) $$\sum_{s \in S} | \lambda(s)|^2 < +\infty.$$

If a function λ on S belongs to $l^2(S)$ it vanishes identically on the complement of some countable subset of S (i.e., its support is countable; we have put on S the discrete topology so that every subset of S is closed: the support of a function is then the set in which it is different from zero). We take as norm of $\lambda \in l^2(S)$ the square root of the left-hand side of (12.11). It is easily seen that this norm, which we denote by $\| \ \|$, satisfies (HN) (Proposition 12.2); it is therefore defined by a positive definite sesquilinear form, which is in fact

$$(\lambda \mid \mu) = \sum_{s \in S} \lambda(s) \, \overline{\mu(s)}.$$

One can prove easily that $l^2(S)$ is complete, i.e., is a Hilbert space.
From Theorems 12.1 and 12.3 the next result follows easily:

THEOREM 12.4.   *In a Hilbert space (E, B), there is always an orthonormal basis. Furthermore, given any orthonormal subset L of E, there is an orthonormal basis of E containing L.*

Let S be an orthonormal basis of the Hilbert space E. To every $x \in E$ there corresponds a complex-valued function, defined in S, namely the function

$$f_x : s \leadsto B(x, s).$$

The mapping

$$x \leadsto f_x$$

is a linear isometry of E onto $l^2(S)$.

From Theorem 12.4 follows immediately that an orthonormal set S in E is an orthonormal basis of E if and only if, for any $x \in E$, $B(x, s) = 0$ for all $s \in S$ implies $x = 0$.

Examples of Hilbert Spaces

I. The space l^2 of complex sequences $\sigma = (\sigma_n)$ such that

$$\sum_{n=0}^{\infty} |\sigma_n|^2 < +\infty.$$

The *inner product* (that is to say the sesquilinear form that turns l^2 into a Hilbert space) is

$$(\sigma \mid \tau) = \sum_{n=0}^{\infty} \sigma_n \overline{\tau_n}.$$

II. The space L^2 (cf. Chapter 11, Example IV): this is the space of (classes of) square-integrable functions f (with respect to some positive measure dx on a set X). The inner product is given by

$$(f \mid g) = \int f(x) \overline{g(x)} \, dx;$$

the norm is therefore

$$\|f\|_{L^2} = \left(\int |f(x)|^2 \, dx \right)^{1/2}.$$

It is the recognized importance of the space L^2 (at first, with respect to the Lebesgue measure on open subsets of $\mathbf{R}^n$ and later on, with respect to general measures) that has been the starting point of the Hilbert space theory.

III. Any finite dimensional space $\mathbf{C}^n$ with the usual Hermitian product.

Remark 12.1. An isomorphism of a normed space (E, p) onto another normed space (F, q) is a linear mapping u of E onto F such that, for all $x \in E$,

$$q(u(x)) = p(x),$$

in other words it is a linear isometry of E onto F (such a mapping is obviously one-to-one).

Similarly, one may define an isomorphism of a pre-Hilbert space (E, B) onto another pre-Hilbert space (E_1, B_1) as a linear map u of E onto E_1, one-to-one and such that, for all $x, y \in E$,

$$B_1(u(x), u(y)) = B(x, y).$$

If now both (E, B) and (E_1, B_1) are Hausdorff, i.e., if the sesquilinear forms are positive definite, we may regard them as normed spaces. It follows immediately from Identity (HN) that an isomorphism of E onto E_1 in the sense of the normed space structure or in the sense of the Hausdorff pre-Hilbert structure are one and the same thing.

Remark 12.2. A linear subspace M of a pre-Hilbert space E is naturally equipped with a structure of pre-Hilbert space: it suffices to take the restriction $B|M$ to M of the inner product $(x, y) \rightsquigarrow B(x, y)$ which makes out of E a pre-Hilbert space. If (E, B) is Hausdorff, so is $(M, B|M)$. If (E, B) is a Hilbert space and M *closed*, then $(M, B|M)$ is also a Hilbert space.

Remark 12.3. If (E_1, B_1) and (E_2, B_2) are two pre-Hilbert spaces, one turns the product vector space $E_1 \times E_2$ into a pre-Hilbert space by considering on it the sesquilinear form

$$B((x_1, x_2), (y_1, y_2)) = B_1(x_1, y_1) + B_2(x_2, y_2).$$

The latter is called the *product* pre-Hilbert space of the two given ones.

Exercises

12.1. Let u be a continuous linear map of a Hilbert space E into itself which is a self-adjoint projection (see (12.7) and (12.8)). Prove that u is the orthogonal projection of E onto a closed linear subspace M of E.

12.2. Let E be a Hilbert space (over the field of complex numbers), not reduced to $\{0\}$. Prove that the topology of E can be defined by a norm which is *not* a Hilbert norm (cf. Proposition 12.2).

12.3. Let E, F be two Hilbert spaces. Prove that there is a unique Hilbert norm on the product $E \times F$ with the following properties:

(a)   the topology defined by this norm is the product topology;

(b)   the canonical projections on the "coordinates axes" E and F are exactly of norm one.

12.4. Let E, F be two Hilbert spaces, G an arbitrary Hausdorff TVS, $u : E \to G$ a continuous map, and $j : F \to G$ a continuous linear injection. Prove that there exists a Hilbert norm $\| \quad \|$ on the linear subspace of E,

$$H = \{x \in E; u(x) \in j(F)\},$$

with the following properties:
(a)   the natural injection of H into E is of norm $\leqslant 1$;
(b)   the mapping $j^{-1} \circ u : H \to F$ is of norm $\leqslant 1$;
(c)   with the norm $\| \quad \|$, H is a Hilbert space.

12.5. Let $(E_1 \times E_2, B)$ be the pre-Hilbert space, product of two pre-Hilbert spaces (E_1, B_1), (E_2, B_2) (see Remark 12.3). Prove that $(E_1 \times E_2, B)$ is Hausdorff if and only if both (E_j, B_j) are Hausdorff, and that it is a Hilbert space if and only if both (E_j, B_j) are Hilbert spaces ($j = 1, 2$).

12.6. Let (E, B) be a Hilbert space, and M a *closed* linear subspace of E. Show that there is a canonical structure of Hilbert space on the quotient vector space E/M. Prove that this structure has the following properties:
(a)   the quotient topology on E/M is the topology associated with the canonical Hilbert space structure;
(b)   let $\phi : E \to E/M$ be the canonical map; then the restriction of ϕ to M^0, orthogonal of M in E (equipped with the Hilbert space structure induced by E, i.e., with the restriction of B to M^0), is an isomorphism (for the Hilbert space structures) onto E/M.

12.7. A TVS E is said to be *separable* if there is a countable subset of E which is everywhere dense. A pre-Hilbert (E, B) is said to be separable if the "underlying" TVS E is separable. Prove that a Hilbert space (E, B) is separable if and only if it has a countable orthonormal basis (cf. p. 121), i.e., if and only if there is an isomorphism of (E, B) onto the Hilbert space l^2 (isomorphism for the Hilbert space structures).

12.8. Quote a theorem on Fourier series which implies that the exponentials

$$t \rightsquigarrow e^{2i\pi kt}, \qquad k = 0, \pm 1, \pm 2,...,$$

form an orthonormal basis in $L^2([0, 1])$.

12.9. For every n-tuple $p = (p_1 ,..., p_n) \in \mathbf{N}^n$, let us set

$$h_p(x) = \exp[|x|^2](\partial/\partial x_1)^{p_1} \cdots (\partial/\partial x_n)^{p_n} \exp[-\pi|x|^2].$$

Prove that, as p ranges over $\mathbf{N}^n$, the functions h_p are pairwise orthogonal in $L^2(\mathbf{R}^n)$. (One can prove that, for a suitable choice of the constants $\alpha_p > 0$, the functions $\alpha_p h_p$ form an *orthonormal basis* in $L^2(\mathbf{R}^n)$. This implies that $L^2(\mathbf{R}^n)$ is separable; cf. Exercise 12.7.)

12.10. Let (E, B) be a Hilbert space, and S an orthonormal set in E. Prove that, if $\dim E = +\infty$, there is $x \in E$ such that the series

$$\sum_{s \in S} B(x, s)\, s$$

does *not* converge absolutely in E, i.e., we have

$$\sum_{s \in S} B(x, s)\, (B(s, s))^{1/2} = +\infty.$$

(Hint: by using Theorem 12.4, prove that, if the preceding assertion were not true, we would have

$$l^2 = l^1.)$$

13

Spaces *LF*. Examples

Let E be a vector space over the field of complex numbers. Let us suppose that E is the union of an *increasing sequence* of subspaces E_n, $n = 1, 2,...$, and that on each E_n there is a structure of Fréchet space such that the natural injection of E_n into E_{n+1} (we have $E_n \subset E_{n+1}$) is an isomorphism, which means that the topology induced by E_{n+1} on E_n is identical to the topology initially given on E_n. Then we may define on E a structure of Hausdorff locally convex space, in the following way: a subset V of E, assumed to be *convex*, is a neighborhood of zero if and only if, for every $n = 1, 2,...$, $V \cap E_n$ is a neighborhood of zero in the Fréchet space E_n. When we provide E with this topology, we say that E is an *LF-space* or, equivalently, a *countable strict inductive limit* of Fréchet spaces, and that the sequence of Fréchet spaces, $\{E_n\}$ ($n = 1, 2,...$), is a *sequence of definition* of E. A space *LF* may have several, and in fact infinitely many, sequences of definition, as will soon be clear in the examples.

Let $\{E_n\}$ be a sequence of definition of an *LF*-space E. Each E_n is isomorphically embedded in the subsequent ones, $E_{n+1}, E_{n+2},....$ But a priori we do not know if E_n is isomorphically embedded in E, in other words if the topology induced by E on E_n is identical to the topology initially given on E_n. If U is a neighborhood of 0 in E, it contains a convex neighborhood of zero V in E, and $V \cap E_n$ must be a neighborhood of zero in E_n; this means that the topology induced by E on E_n is *less fine* than the original topology of E_n. That it is identical to the original topology will be a direct consequence of the following lemma:

LEMMA 13.1. *Let E be a locally convex space, E_0 a closed subspace of E, U a convex neighborhood of 0 in E_0, and x_0 a point of E which does not belong to U. Then there exists a convex neighborhood V of 0 in E, not containing x_0 and such that $V \cap E_0 = U$.*

Proof. In the statement, E_0 carries the induced topology. Therefore there is a neighborhood of zero, W, in E, such that $U = W \cap E_0$.

The trouble is that W may not be convex and that it may contain x_0; we shall modify W in such a way that this does not happen. First of all, W contains a convex neighborhood of zero W_0 in E. Let W_1 be the convex hull of $U \cup W_0$; we claim that

(13.1) $W_1 \cap E_0 = U$ (of course, we have $U \subset W_1 \cap E_0$).

Indeed, let $x \in W_1 \cap E_0$; since $x \in W_1$, we may write $x = ty + (1 - t)z$ with $0 \leqslant t \leqslant 1$, $y \in U$, and $z \in W_0$. If $t = 1$, $x = y$ belongs to U; if $t < 1$, we have $z = (1 - t)^{-1}(x - ty) \in E_0$, hence $z \in W_0 \cap E_0 \subset W \cap E_0 = U$; but as U is convex, we must then also have $x \in U$. This proves (13.1).

We must now "cut down" W_1 so that it does not contain x_0. However, if $x_0 \in E_0$, there is nothing to be done since $x_0 \notin U$: it suffices to apply (13.1). Let us therefore suppose that $x_0 \notin E_0$. Consider the quotient space E/E_0 and let ϕ be the canonical mapping of E onto E/E_0; the quotient space is a Hausdorff locally convex TVS, and we have $\phi(x_0) \neq 0$. Choose a convex neighborhood of zero in E/E_0 which does not contain $\phi(x_0)$; its preimage Ω is a convex neighborhood of E_0, and therefore also of 0, in E, which does not contain x_0; the neighborhood of zero $V = W_1 \cap \Omega$ fulfills all the requirements of the Lemma (see figure).

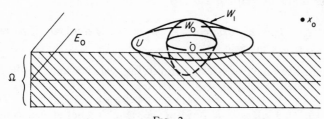

FIG. 2

Let us show now how Lemma 13.1 implies the result that E, strict inductive limit of the F-spaces E_n, induces on each E_n the initially given topology. Let U_n be an arbitrary (convex) neighborhood of 0 in E_n. There exists a convex neighborhood of zero U_{n+1} such that $U_{n+1} \cap E_n = U_n$: this follows from Lemma 13.1. By induction on n, we see that, for every $k = 1, 2, \ldots$, there exists a convex neighborhood of zero U_{n+k} in E_{n+k} such that

$$U_{n+k} \cap E_{n+k-1} = U_{n+k-1}.$$

If we set

$$U = \bigcup_{k=0}^{\infty} U_{n+k},$$

we see that $U \cap E_n = U_n$; furthermore, U is a neighborhood of zero in E since $U \cap E_m$ is a neighborhood of zero in E_m for all m.

PROPOSITION 13.1. *Let E be an LF-space, $\{E_k\}$ $(k = 0, 1,...)$ a sequence of definition of E, F an arbitrary locally convex TVS, and u a linear map of E into F. The mapping u is continuous if and only if, for each k, the restriction $u|E_k$ of u to E_k is a continuous linear map of E_k into F.*

Proof. Suppose that $u : E \to F$ is continuous. Let V be a neighborhood of zero in F; $u^{-1}(V)$ contains a convex neighborhood of zero U in E. For each k, $U_k = U \cap E_k$ is a neighborhood of zero in E_k, and we have

$$U_k = u^{-1}(V) \cap E_k = (u|E_k)^{-1}(V).$$

Suppose now that, for each k, $u|E_k : E_k \to F$ is continuous. Let V be a neighborhood of 0 in F. We use now (for the first time) the fact that F is locally convex, for we might then assume that V, hence also $u^{-1}(V)$, is convex. But, for each k,

$$u^{-1}(V) \cap E_k = (u|E_k)^{-1}(V).$$

is a neighborhood of zero in E_k. Thus $u^{-1}(V)$, being convex, must be a neighborhood of zero in E.   Q.E.D.

COROLLARY. *A linear form on E is continuous if and only if its restriction to every E_k is continuous.*

These results have a great simplifying value when applied to the theory of distributions, as will be shown.

Remark 13.1.   Unless $E = \text{ind lim}_n E_n$ is a Fréchet space, E is never a Baire space. Indeed, each E_n is a complete, therefore closed, linear subspace of E (we are using here the fact that the topology induced by E on E_n is the one initially given on E_n : otherwise we could not assert in all generality that E_n is closed in E). Thus E is a countable union of closed subsets, the E_n's: one of these, say E_{n_0}, ought to have an interior point x_0, if E were to be a Baire space.
As $x \to x - x_0$ is a homeomorphism of E onto itself, the origin should also be an interior point of E_{n_0}; in other words, E_{n_0} should be a neighborhood of the origin. As a neighborhood of zero is absorbing, E_{n_0} should be absorbing. As E_{n_0} is a linear subspace, this would imply immediately that $E_{n_0} = E$. But this would mean, in particular, that E is an F-space.

Remark 13.2.   Let E be an *LF*-space, $\{E_n\}$ a sequence of definition of E, and M a closed linear subspace of E. It is not true in general that the topology induced on M by E is the same as the inductive limit topology of the F-spaces $E_n \cap M$. One should be careful not to overlook this fact (the author has made the mistake a few times in his life and so also have a few other utilizers of the *LF*-spaces!).

THEOREM 13.1. *Any space LF is complete.*

Proof. Let E be a space *LF*, $\{E_n\}$ $(n = 1, 2,...)$ a sequence of definition of E (see p. 126), $\mathscr{F}$ a Cauchy filter on E. The collection of sets $M + V$, as M runs over $\mathscr{F}$ and V runs over the filter of neighborhoods of 0 in E, is a basis of filter on E, since

$$M \cap M' + V \cap V' \subset (M + V) \cap (M' + V');$$

let $\mathscr{G}$ be the filter that it spans. It is a Cauchy filter. Indeed, let U be an arbitrary neighborhood of 0 in E, V another neighborhood of zero such that $V + V - V \subset U$, M a set belonging to $\mathscr{F}$ such that $M - M \subset V$; then

$$(M + V) - (M + V) \subset M - M + V - V \subset U.$$

Observe that $\mathscr{F}$ is finer than $\mathscr{G}$. We shall then prove the following assertion:

(13.2)   *There is an integer $p \geqslant 1$ such that none of the sets $A \cap E_p$, as A runs over $\mathscr{G}$, is empty.*

This will imply Theorem 13.1. Indeed, if none of the sets $A \cap E_p$ is empty, they form a filter $\mathscr{G}_p$ on E_p; since the topology induced by E on E_p is identical with the original topology of E_p, $\mathscr{G}_p$ is a Cauchy filter, and since E_p is a Fréchet space, $\mathscr{G}_p$ converges to an element x in E_p. It is clear that x is an accumulation point of $\mathscr{G}$, which therefore converges to x; a fortiori, $\mathscr{F}$ converges to x.

Proof of (13.2). We shall suppose that (13.2) is false and show that this leads to a contradiction. Thus, suppose that, for every $n = 1, 2,...$, there is $A_n = M_n + V_n \in \mathscr{G}$ which does not intersect E_n. By shrinking A_n if necessary, we may assume that each neighborhood of zero V_n is convex and balanced, and that $V_n \subset V_{n-1}$ for all $n > 1$. Let, then, W_n be the convex hull of

$$V_n \cup \bigcup_{k<n} (V_k \cap E_k).$$

I contend that $M_n + W_n$ does not intersect E_n. If it did, there would be $x \in M_n$, $y \in V_n$ and $z \in E_{n-1}$ such that $x + ty + z \in E_n$ for some number t, $0 \leqslant t \leqslant 1$; but $ty \in V_n$ as this set is balanced, and $z \in E_n$;

therefore, $x + ty \in (M_n + V_n)$ would belong to E_n contrary to our choice of A_n. Now call W the convex hull of $\bigcup_{k=1}^{+\infty} (V_k \cap E_k)$. As W is convex and as $W \cap E_k$ contains $V_k \cap E_k$ for all k, W is a neighborhood of 0 in E. On the other hand, since the sequence $\{V_n\}$ is decreasing, we have, for all n,

$$V_n \cup \bigcup_{k<n} (V_k \cap E_k) \supset \bigcup_{k=1}^{\infty} (V_k \cap E_k),$$

hence $W \subset W_n$ for all n. Now, since $\mathscr{F}$ is a Cauchy filter, we may find a set $B \in \mathscr{F}$ such that $B - B \subset W \subset W_n$ (for all n). But we must have, for all n, $B \cap M_n \neq \varnothing$, hence $B - (B \cap M_n) \subset W_n$, hence

$$B \subset W_n + (B \cap M_n) \subset W_n + M_n.$$

This demands that $B \cap E_n = \varnothing$ for all n, i.e., $B = \varnothing$, which is impossible.   Q.E.D.

Example I. The space of polynomials

Let us denote by $\mathbf{C}[X]$ the vector space of polynomials in n letters $X = (X_1, ..., X_n)$ (or in n variables, if the reader prefers) with complex coefficients. This vector space has a canonical algebraic basis, the monomials

$$X^p = X_1^{p_1} \cdots X_n^{p_n}, \qquad p = (p_1, ..., p_n) \in \mathbf{N}^n.$$

(We recall that $\mathbf{N}^n$ is the set of vectors of $\mathbf{R}^n$ whose coordinates are nonnegative integers.) Any polynomial is a finite linear combination of the monomials X^p. Let $\mathscr{P}_n^m$ be the vector space spanned by the X^p such that $|p| = p_1 + \cdots + p_n \leqslant m$. The elements of $\mathscr{P}_n^m$ are the polynomials of degree $\leqslant m$. The degree of a polynomial $P(X)$ is the smallest integer m such that $P(X) \in \mathscr{P}_n^m$; we shall denote by $\deg P$ this integer. An elementary computation shows that there are exactly $\binom{m+n}{m}$ monomials X^p such that $|p| \leqslant m$, in other words

$$\dim \mathscr{P}_n^m = (m + n)!/(m!n!).$$

Let $P(x)$ be a polynomial,

$$P(X) = \sum_{|p| \leqslant \deg P} c_p X^p$$

It is obvious that P can be viewed as a function on the set $\mathbf{N}^n$: precisely the function which, to each p, assigns the value c_p if $|p| \leqslant \deg P$ and equal to zero otherwise. Note that this mapping of polynomials into functions on $\mathbf{N}^n$ does not preserve multiplication: the function corresponding to the product of the polynomials is *not* the product of the functions corresponding to each one of these polynomials (we shall see later that multiplication of polynomials is transformed into *convolution* of functions). At any event, if we put on $\mathbf{N}^n$ the discrete topology and we note that every set is closed in this topology, we see that the functions corresponding to polynomials are exactly the functions with *compact* support: a subset of $\mathbf{N}^n$ is compact if and only if it is finite. Arbitrary functions on $\mathbf{N}^n$ correspond (via the coefficients) to formal power series; it is obvious that we can regard $\mathbf{C}[X]$ as a vector subspace, or even as a subalgebra (or a subring) of the space $\mathbf{C}[[X]]$ of formal power series in n letters (with complex coefficients).

Being a finite dimensional vector space, $\mathscr{P}_n^m$ carries a unique Hausdorff topology, for which it becomes an F-space. We may then view $\mathbf{C}[X]$ as the union of the F-spaces $\mathscr{P}_n^m$ as $m = 0, 1, 2,...$, and provide it with the inductive limit topology; thus $\mathbf{C}[X]$ becomes a space *LF*. It should be noted that the topology thus defined on $\mathbf{C}[X]$ is *strictly* finer than the topology induced on $\mathbf{C}[X]$ by $\mathbf{C}[[X]]$, when the latter carries the topology of simple convergence of the coefficients (Chapter 10, Example III).

Example II.   Spaces of test functions

Let Ω be a nonempty open subset of $\mathbf{R}^n$. Let us denote by $F(\Omega)$ any one of the following spaces:

$$\mathscr{C}^k(\Omega), \quad 0 \leqslant k < \infty, \ \mathscr{C}^\infty(\Omega), \quad L^p(\Omega) \quad (1 \leqslant p \leqslant +\infty).$$

The first ones, $\mathscr{C}^k(\Omega)$, $0 \leqslant k \leqslant +\infty$, are F-spaces; the last ones, $L^p(\Omega)$, are B-spaces. The space $L^2(\Omega)$ is a Hilbert space.

Let K be a *compact* subset of Ω, which means that it is bounded and closed in $\mathbf{R}^n$ and that its closure is contained in Ω. Consider the subset of $F(\Omega)$, denoted by $F_c(K)$, consisting of the functions f whose support lies in K; it should be recalled that the *support* of f, supp f, is the closure in Ω of the subset

$$\{x \in \Omega; f(x) \neq 0\}.$$

It may of course happen that $F_c(K)$ contains only the function zero (i.e., the function identically equal to zero). This happens for instance when $F(\Omega) = \mathscr{C}^0(\Omega)$ (or, for that matter, when $F(\Omega)$ is any of the spaces considered above), and K contains a single point (or has Lebesgue measure zero). At any event, $F_c(K)$ is always a linear subspace of $F(\Omega)$ and it is easily seen that it is always *closed:* for if a sequence of functions $\{f_k\}$ converges to a function f in any of the spaces above, chosen as

$F(\Omega)$, and if the f_k all vanish in the open set $\Omega - K$, then obviously their limit must also vanish in $\Omega - K$. Thus, regarded as a subspace of $F(\Omega)$, $F_c(K)$ is an F-space, i.e., it is complete. When $F = L^p$, then $F_c(K)$ is even a B-space, and it is a Hilbert space when $F = L^2$. This turns out to be true also when $F = \mathscr{C}^k$ for $0 \leqslant k < +\infty$. Indeed, it is fairly obvious in this case that the topology of $F(K)$ can be described by the single norm

$$f \rightsquigarrow \sup_{|p| \leqslant k} (\sup_{x \in R^n} | (\partial/\partial x)^p f(x)|).$$

This is equivalent with saying that, when functions are required to have their support in some fixed compact set $K \subset \Omega$, it amounts to the same to ask that they converge uniformly on the whole space or on every compact subset of Ω.

When $F = \mathscr{C}^\infty$, it is not any more true that $F_c(K)$ is a B-space (outside of the case where it is zero); it is an F-space which, as will be seen in the following chapter, is not normable.

Notation. We shall adopt the following notation:

when $F(\Omega) = \mathscr{C}^k(\Omega)$, $0 \leqslant k \leqslant +\infty$, we write $\mathscr{C}_c^k(K)$ for $F_c(K)$;

when $F(\Omega) = L^p(\Omega)$, $1 \leqslant p \leqslant +\infty$, we write $L^p(K)$ for $F_c(K)$.

We leave to the student the verification of the fact that $L^p(K)$ is the same thing as the space normally denoted in that way, that is to say the B-space of classes of functions almost everywhere defined in K, Lebesgue measurable, and L^p. We may now consider the union of the subspaces $F_c(K)$ as K varies in all possible ways over the family of compact subsets of Ω. We denote by $F_c(\Omega)$ this union; it is a vector subspace of $F(\Omega)$, precisely the subspace consisting of all the functions belonging to $F(\Omega)$ *which have a compact support.* This is what the subscript c is meant to indicate. We shall *not* put on $F_c(\Omega)$ the topology induced by $F(\Omega)$, but a finer one, which will turn it into a space LF. We proceed as follows:

We consider a sequence of compact sets $K_1 \subset K_2 \subset \cdots \subset K_j \subset \cdots \subset \Omega$ whose union is equal to Ω. It might even be advantageous, for further purposes, that the K_j be chosen so as to be the closures of (relatively compact) open subsets of Ω, and such that K_j be contained in the interior of K_{j+1}. That such sequences of compact sets do exist has already been proved (Lemma 10.1). The space $F_c(\Omega)$ can then be regarded as the union of the spaces $F_c(K_j)$ for $j = 1, 2, \ldots$; this is simply saying that an arbitrary compact subset K of Ω is contained in K_j for sufficiently large j. Because of our way of defining the F-spaces $F_c(K)$, we see that $F_c(K_{j+1})$ induces on $F_c(K_j)$ the same topology as the one originally given on $F_c(K_j)$ (i.e., the one induced by $F(\Omega)$). Thus we may provide $F_c(\Omega)$

with the inductive limit of the topologies of the F-spaces $F_c(K_j)$. It is an easy matter to check that this topology does not really depend on the choice of the sequence of compact sets $\{K_j\}$ (provided they fill Ω). With this topology, $F_c(\Omega)$ becomes an *LF*-space.

Notation. We shall write $\mathscr{C}^k_c(\Omega)$, $\mathscr{C}^\infty_c(\Omega)$, and $L^p_c(\Omega)$ for $F_c(\Omega)$ when F is meant for $\mathscr{C}^k$, $\mathscr{C}^\infty$, and L^p, respectively. The topology just defined on $F_c(\Omega)$ will be called the *canonical LF topology*.

The space $\mathscr{C}^\infty_c(\Omega)$ plays a basic role in the theory of distributions; its elements will be called *test functions* (they are the $\mathscr{C}^\infty$ functions with compact support in Ω). A distribution in Ω is nothing else but a continuous linear functional on $\mathscr{C}^\infty_c(\Omega)$ when the latter carries the canonical *LF* topology.

PROPOSITION 13.2.   *We have the following continuous injections:*

$$\mathscr{C}^\infty_c(\Omega) \to \mathscr{C}^k_c(\Omega) \to \mathscr{C}^{k-1}_c(\Omega) \to L^\infty_c(\Omega) \qquad (0 < k < +\infty)$$

and

$$L^\infty_c(\Omega) \to L^{p+1}_c(\Omega) \to L^p_c(\Omega) \to L^1_c(\Omega) \qquad (1 < p < +\infty).$$

Proof. If we want to prove that we have a continuous injection $F_c(\Omega) \to G_c(\Omega)$, where F and G are two functions spaces of the type above, it suffices to show that we have a continuous injection $F_c(K) \to G_c(K)$ for each compact set $K \subset \Omega$, for after having shown this it will suffice to apply Proposition 13.1. The statement relative to the upper sequence becomes evident. The one about the lower sequence follows from Hölder's inequalities, which will be proved only later (Chapter 20) but which the student probably knows already. Let p, q be two real numbers, $1 \leqslant p < q \leqslant +\infty$. Let us denote by χ_K the characteristic function of the compact set K, equal to one in K and to zero everywhere else. Let $f \in L^q$ with $\operatorname{supp} f \subset K$. We may suppose at first that f is of a simple type, for instance a bounded measurable step-function. We have (cf. Lemma 20.1)

$$\int |f|^p \, dx = \int |f|^p \chi_K \, dx \leqslant \|f\|^p_{L^q} \|\chi_K\|_{L^r},$$

with $r = q/(q - p)$. We obtain thus

$$(13.3) \qquad \|f\|_{L^p} \leqslant \left(\int_K dx \right)^s \|f\|_{L^q} \qquad \left(s = \frac{1}{p} - \frac{1}{q} \right).$$

By going to the limit, e.g., along sequences of bounded measurable step-functions, we obtain easily (13.3) for all $f \in L^p(K)$, thus proving what we wanted: that $f \in L^p$ and that the natural injection of L^q into L^p is continuous.

Exercises

13.1. Let $\mathscr{P}_n$ be the space of polynomials in n variables, with complex coefficients, provided with the LF topology defined on p. 129. Prove the following two facts:

(a) The LF topology on $\mathscr{P}_n$ is the finest locally convex topology on this space.

(b) Every linear map of $\mathscr{P}_n$ into any TVS is continuous.

13.2. Let p be a positive number, $1 \leqslant p < \infty$. Let $\{\varepsilon_k\}$ $(k = 0, 1,...)$ be a decreasing sequence of numbers $\varepsilon_k > 0$, converging to zero. Prove that the subsets of $L^p_c(\mathbf{R}^n)$,

$$\mathscr{V}^p(\{\varepsilon_k\}) = \left\{ f \in L^p_c(\mathbf{R}^n); \int_{|x| \geqslant k} |f(x)|^p \, dx < \varepsilon_k \,, \, k = 0, 1,... \right\}$$

form, as the sequence $\{\varepsilon_k\}$ varies in all possible ways, a basis of neighborhoods of zero for the LF topology on $L^p_c(\mathbf{R}^n)$ (see p. 131).

13.3. Let E be a normed space, whose norm is denoted by $\| \ \|$. Let, for every $k = 0, 1, 2,..., E_k$ be a linear subspace of E of dimension k, such that $E_k \subset E_{k+1}$. Let E_∞ be the union of the subspaces E_k, equipped with the LF topology defined by means of the sequence $\{E_k\}$. Let $\{\varepsilon_k\}$ $(k = 0, 1,...)$ be a decreasing sequence, converging to 0, of numbers > 0. Set

$$\mathscr{V}(\{\varepsilon_k\}) = \{x \in E_\infty \, ; \, x \notin E_k \Rightarrow \| x \| < \varepsilon_k \,, \, k = 0, 1,...\}.$$

Prove that $\mathscr{V}(\{\varepsilon_k\})$ is *not* a neighborhood of zero in E_∞.

13.4. Let E, F be two LF-spaces, $\{E_m\}, \{F_n\}$ $(m, n = 1, 2,...)$ two sequences of definition of E and F, respectively (see p. 126), and $u : E \to F$ a continuous linear map. By using the fact that a Fréchet space is a Baire space, prove that to every m there is n such that $u(E_m) \subset F_n$.

13.5. Let $E, F, \{E_m\}$, and $\{F_n\}$ be as in Exercise 13.4. If u is an isomorphism (for the TVS structures) of E into F, prove that to every n there is m such that $u^{-1}(F_n) \subset E_m$.

13.6. Let E be a vector space, $\{E_\alpha\}$, $(\alpha \in A)$ a family of locally convex spaces, and, for each index $\alpha \in A, j_\alpha : E_\alpha \to E$ a linear map. Let $\mathscr{T}$ be the finest locally convex topology on E such that all the mappings ϕ_α be continuous. Prove that a convex subset U of E is a neighborhood of zero for $\mathscr{T}$ if and only if $\phi_\alpha^{-1}(U)$ is a neighborhood of zero in E_α for each α, but that this is not necessarily true if U is not convex. Let F be a locally convex TVS. Prove that a linear map $u : E \to F$ is continuous (when E carries the topology $\mathscr{T}$) if and only if, for every $\alpha \in A, u \circ \phi_\alpha : E_\alpha \to F$ is continuous.

13.7. Let K be a compact subset of $\mathbf{C}^n$, and Ω_k $(k = 0, 1, 2,...)$ a sequence of open sets of $\mathbf{C}^n$ containing K such that

$$\mathbf{C}^n = \Omega_0 \supset \cdots \supset \Omega_k \supset \Omega_{k+1} \supset \cdots$$

and such that any open subset of $\mathbf{C}^n$ which contains K contains some Ω_k (in other words, the Ω_k form a decreasing basis of neighborhoods of K). On the other hand, let $H(K)$ denote the vector space of functions in K which can be extended as holomorphic functions in some open set containing K. We have the natural mapping $\rho_k : H(\Omega_k) \to H(K)$, the mapping "restriction to K." Apply the scheme of Exercise 13.6 with $E = H(K)$, $A = \mathbf{N}$, $E_\alpha = H(\Omega_k)$, and $\phi_\alpha = \rho_k$ if $\alpha = k$. Prove that $H(K)$ is complete and that $H(K)$ is *not* (unless $K = \varnothing$) the strict inductive limit of the sequence of Fréchet spaces $H(\Omega_k)$.

14

Bounded Sets

Let E be a TVS (not necessarily Hausdorff nor locally convex). We wish to generalize the notion of bounded set, familiar to us in finite dimensional spaces or even in normed spaces (see Chapter 11).

Definition 14.1. *A subset B of the* TVS *E is said to be bounded if to every neighborhood of zero U in E there is a number $\lambda \geqslant 0$ such that*

$$B \subset \lambda U.$$

It may be said that a subset B of E is bounded if B can be "swallowed" by any neighborhood of zero. Of course, it suffices, in order that B be bounded, that any neighborhood in some basis of neighborhoods of zero swallows B. Since there is a basis of neighborhoods of zero in E consisting of *closed* neighborhoods of zero (Proposition 3.1), we see that *the closure of a bounded set is bounded.* It is quite obvious that finite sets, bounded subsets (in the usual sense) of finite dimensional spaces, balls with finite radii in normed spaces, are bounded sets. Also obvious are the following properties:

(1) *Finite unions of bounded sets are bounded sets* (we recall that any neighborhood of zero contains a balanced one).

(2) *Any subset of a bounded set is a bounded set.*

Notice that these properties are, in a sense, dual of the properties of neighborhoods of a point (they are also shared by the family of complements of neighborhoods of a point). This leads to the following:

Definition 14.2. *A family of bounded subsets of E, $\{B_\alpha\}$ $(\alpha \in \Omega)$, is called a basis of bounded subsets of E if to every bounded subset B of E there is an index $\alpha \in \Omega$ such that $B \subset B_\alpha$.*

A basis of neighborhoods of zero is a family of neighborhoods of 0 such that any given neighborhood of zero *contains* some neighborhood belonging to the family. A basis of bounded sets is a family of bounded sets such that any given bounded subset of E *is contained in* some bounded subset belonging to the family. As we shall see when we study the strong topology on the dual of a TVS, this "duality" between neighborhoods of zero and bounded sets has important implications.

What sets do we know to be bounded? Sets consisting of a single point are bounded: this is to be expected and it is obvious on Definition 14.1 when we take into account the fact that neighborhoods of zero are absorbing (Definition 3.1).

PROPOSITION 14.1. *Compact sets are bounded.*

Let K be a compact set, and U a neighborhood of zero in E which we can take to be open and balanced. Then we have

$$K \subset \bigcup_{n=0}^{\infty} nU = E.$$

From compactness it follows that there is a finite family of integers $n_1, \ldots, n_r$ such that

$$K \subset n_1 U \cup \cdots \cup n_r U = (\sup_{1 \leqslant \lambda \leqslant r} n_\lambda)U.$$

In finite dimensional spaces, every bounded set, provided that it is closed, is a compact set. This is not true, in general, in infinite dimensional TVS. For instance, let E be an infinite dimensional normed space. If every bounded set in E were compact, this would be true, in particular, of all the balls centered at the origin. Then E would have to be locally compact, which is impossible as dim $E = +\infty$ (Theorem 9.2). There is however an important class of infinite dimensional vector spaces, the so-called Montel spaces, in which it is true that every closed bounded set is compact. We shall study the Montel spaces later on in relation to duality. The spaces $\mathscr{C}^\infty(\Omega)$, $\mathscr{S}$, and $\mathscr{C}_c^\infty(\Omega)$ (Chapter 10 and 12) are Montel spaces (Theorem 14.4, Exercises 14.9, 14.10, p. 148).

COROLLARY 1. *Suppose that E is Hausdorff. Then precompact subsets of E are bounded in E.*

Let K be a precompact subset of E. This means that the closure $\hat{K}$ of K in the completion $\hat{E}$ of E is compact. Let U be any neighborhood of zero in E. Since the injection $E \to \hat{E}$ is an isomorphism for the TVS structure, there is a neighborhood of zero $\hat{U}$ in $\hat{E}$ such that $U = E \cap \hat{U}$. By virtue of Proposition 14.1, there is a number $\lambda > 0$ such that $\hat{K} \subset \lambda\hat{U}$, whence

$$K \subset E \cap \hat{K} \subset E \cap (\lambda\hat{U}) \subset \lambda(E \cap \hat{U}) = \lambda U.$$

COROLLARY 2. *Suppose that E is Hausdorff. The union of a converging sequence in E and of its limit is a bounded set.*

For such a union is a compact set.

COROLLARY 3. *Let E be Hausdorff. Any Cauchy sequence in E is a bounded set.*

For a Cauchy sequence is a precompact subset of E.

The Cauchy filter associated with a Cauchy sequence contains a bounded set. This is not true about a Cauchy filter in general.

PROPOSITION 14.2. *The image of a bounded set under a continuous linear mapping is a bounded set.*

Let E, F be two TVS, u a continuous linear map of E into F, B a bounded subset of E, and V an arbitrary neighborhood of zero in F. Since the preimage $u^{-1}(V)$ of V under u is a neighborhood of zero in E, there is $\lambda > 0$ such that

$$B \subset \lambda u^{-1}(V) \quad \text{which implies} \quad u(B) \subset \lambda V.$$

COROLLARY. *Let f be a continuous linear functional on E, and B a bounded subset of E. Then f is bounded on B, i.e.,*

$$\sup_{x \in B} |f(x)| < +\infty.$$

PROPOSITION 14.3. *Let E be any TVS. A subset B of E is bounded if and only if every sequence contained in B is bounded (in E).*

The necessity of the condition is obvious; let us prove its sufficiency. Suppose that B is unbounded; we shall prove that it contains a sequence of points which is also unbounded. There exists a neighborhood U of zero in E, which we might as well suppose to be balanced, which does not swallow B. In other words, for each $n = 1, 2,...$, there is a point $x_n \in B$ which does not belong to nU. The sequence of points x_n cannot be bounded.

Any ball in a normed space is a bounded set; thus we see that there exist, in normed spaces, sets which are at the same time bounded and neighborhoods of zero. This property is characteristic of normable spaces, at least among Hausdorff locally convex spaces.

PROPOSITION 14.4. *Let E be a Hausdorff locally convex space. If there is a neighborhood of zero in E which is a bounded set, then E is normable.*

Let U be a bounded neighborhood of zero in E. We may assume that U is a barrel, since it does contain a neighborhood of zero which is a barrel. We claim that, under these circumstances, the multiples $(1/n)U$

form a basis of neighborhoods of zero in E. Indeed, given any neighborhood of zero V in E, it swallows U. We may assume V balanced; hence there is an integer $n > 0$ such that $U \subset nV$, which means that $(1/n)U \subset V$. As the space E is Hausdorff, the intersection of the sets $(1/n)U$ must be equal to $\{0\}$, which means that the seminorm associated with U is a norm.   Q.E.D.

PROPOSITION 14.5. *Let E be a locally convex space. A subset B of E is bounded if and only if every seminorm, belonging to some basis of continuous seminorms of E, is bounded on B.*

To say that a seminorm p is bounded on a set B simply means that $\sup_{x \in B} p(x) < +\infty$. The proof of Proposition 14.5 is left to the student.

PROPOSITION 14.6. *Let E be an LF-space, and $\{E_n\}$ ($n = 0, 1, 2,...$) a sequence of definition of E (see Definition 12.1.). A subset B of E is bounded in E if and only if B is contained in E_n for a sufficiently large n, and if B is bounded in that F-space E_n.*

Proof. Suppose first that B is contained and bounded in some E_n. Let U be an arbitrary neighborhood of zero in E. We must show that U swallows B. As $U_n = U \cap E_n$ is a neighborhood of zero in E_n, there is a number $\lambda > 0$ such that $B \subset \lambda U_n \subset \lambda U$.

We assume now that B is bounded in E. We shall first show that B must be contained in some space E_n. We shall suppose that this is not so and show that it leads to a contradiction. For each n, there is a point $x_n \in B$, $x_n \notin E_n$. We shall construct, with the help of Lemma 13.1, a neighborhood of zero U in E which cannot swallow the sequence $\{x_n\}$, a fortiori cannot swallow B. Since $x_1 \notin E_1$, given an arbitrary convex neighborhood of zero U_1 in E_1, there is a convex neighborhood of zero U_2 in E_2, with the properties

$$U_1 = U_2 \cap E_1, \qquad x_1 \notin U_2.$$

It may of course happen that E_3 contains x_1, but, since $x_1 \notin U_2$, again in virtue of Lemma 13.1, we may find V_3', convex neighborhood of zero in E_3, such that

$$U_2 = V_3' \cap E_2, \qquad x_1 \notin V_3'.$$

On the other hand, since $x_2 \notin E_2$, hence $2x_2 \notin E_2$, we may find a convex neighborhood of zero in E_3, V_3, such that

$$U_2 = V_3 \cap E_2, \qquad 2x_2 \notin V_3.$$

We set $U_3 = V_3 \cap V_3'$. We have

$$U_2 = U_3 \cap E_2, \qquad x_1, 2x_2 \notin U_3.$$

Thus, by induction on n, we build a sequence of sets $\{U_n\}$ with the following properties:

U_n is a convex neighborhood of zero in E_n;
$U_n = U_{n+1} \cap E_n$;
the points $x_1, 2x_2, ..., nx_n$ do not belong to U_{n+1}.

Let then U be the union of those sets U_n. For each n, we have $U_n = U \cap E_n$, thus U is a neighborhood of zero in E. And obviously U cannot swallow the sequence $\{x_1, x_2, ..., x_n, ...\}$.

This proves that a bounded subset B of E must be contained in E_n for sufficiently large n. Let U_n be any neighborhood of zero in a space E_n containing B. We apply once more Lemma 13.1: there is a neighborhood of zero U in E such that $U \cap E_n \subset U_n$. By hypothesis, there is $\lambda > 0$ such that $B \subset \lambda U$. This yields

$$B = B \cap E_n \subset (\lambda U) \cap E_n = \lambda(U \cap E_n) \subset \lambda U_n.$$

Thus B is bounded in E_n.   Q.E.D.

COROLLARY 1. *Let E and the E_n be as in Proposition 14.6. A sequence $\{x_k\}$ converges in E if and only if it is contained in some subspace E_n and converges there.*

Exercises

14.0. Let E be a locally convex metrizable space. Prove that, if E is not normable, there is no countable basis of bounded sets in E.

14.1. Let Ω be an open subset of $\mathbf{R}^n$, $K_1, K_2, ..., K_\nu, ...$ a sequence of compact subsets of Ω, whose union is equal to Ω, and such that, for each ν, K_ν is contained in the interior of $K_{\nu+1}$. Let k be a nonnegative integer ($k < +\infty$). Show that, when the sequence of nonnegative numbers $\{M_\nu\}$ ($\nu = 1, 2, ...$) varies in all possible ways, the sets

$$B_k(\{M_\nu\}) = \{\phi \in \mathscr{C}^k(\Omega); \sup_{|p| \leqslant k} (\sup_{x \in K_\nu} | \partial/\partial x)^p \phi(x)|) \leqslant M_\nu\}$$

form a basis of bounded sets in $\mathscr{C}^k(\Omega)$. Conclude that (unless Ω is empty!) $\mathscr{C}^k(\Omega)$ is *not* normable.

14.2. Show that a subset B of $\mathscr{C}^\infty(\Omega)$ is bounded if and only if it is bounded in every $\mathscr{C}^k(\Omega)$, $k < +\infty$. Derive from this fact a remarkable basis of bounded sets in $\mathscr{C}^\infty(\Omega)$ using the same compact sets K_ν as in Exercise 14.1.

14.3 A subset B of $\mathscr{C}_c^k(\Omega)$ $(0 \leqslant k \leqslant +\infty)$ is bounded if and only if it consists of functions having their support contained in one and the same compact subset K of Ω, and if B is bounded in $\mathscr{C}^k(\Omega)$. Prove that, if k is finite, there is a countable basis of bounded sets in $\mathscr{C}_c^k(\Omega)$. Compare with the situation in $\mathscr{C}^k(\Omega)$.

14.4. Construct bases of bounded sets in the following locally convex TVS: the space $H(\Omega)$ of holomorphic functions in an open subset Ω of $\mathbf{C}^n$ (Chapter 10, Example II); the space of formal power series $\mathbf{C}[[X]]$ (Chapter 10, Example III); the space $\mathscr{S}$ of $\mathscr{C}^\infty$ functions rapidly decreasing at infinity in $\mathbf{R}^n$ (Chapter 10, Example IV). Prove that a subset B of the space of polynomials in n letters, with complex coefficients, $\mathbf{C}[X]$, is bounded in $\mathbf{C}[X]$ if the degrees of all the polynomials belonging to B are at most equal to some fixed integer m, and if B is bounded in $\mathbf{C}[[X]]$.

14.5. Prove the following result:

PROPOSITION 14.7. *Let E be an LF-space. A linear map of E into a locally convex space is continuous if and only if it is sequentially continuous.*

(cf. Proposition 8.5.)

14.6. Let E be an LF-space such that there is a sequence of definition $\{E_k\}(k = 0,1,2,...)$ of E consisting of *finite dimensional* (Fréchet) spaces. Prove the following facts:

(a) if F is a normed space and $u : F \to E$ is a continuous linear map, Ker u has a finite codimension (i.e., $\dim(F/\mathrm{Ker}\, u) < +\infty$);

(b) if M is a linear subspace of E such that the topology induced by E on M turns M into a normable space, then $\dim M < +\infty$.

14.7. Let E be a metrizable space, and $\{B_k\}$ $(k = 0, 1, 2,...)$ a sequence of bounded subsets of E. Prove that there is a sequence $\{\varepsilon_k\}$ of numbers > 0 such that the union

$$\bigcup_{k=0}^{\infty} \varepsilon_k B_k$$

is bounded in E.

Let E, F be two TVS, and u a linear map of E into F. Let us say that u is *bounded* if, for every bounded subset B of E, $u(B)$ is a bounded subset of F.

PROPOSITION 14.8. *Let E be a metrizable space. If a linear map of E into a TVS F is bounded, it is continuous.*

Let $f : E \to F$ be bounded. Suppose that f were not continuous. Then there would be a neighborhood of zero V in F whose preimage $f^{-1}(V)$ is not a neighborhood of 0 in E. Let us suppose that V is balanced and let $U_1 \supset U_2 \supset \cdots \supset U_m \supset \cdots$ be a totally ordered countable basis of neighborhoods of zero in E. For all m, we have

$$\frac{1}{m} U_m \not\subset f^{-1}(V);$$

in other words, there is $x_m \in (1/m)U_m$ such that $f(x_m) \notin V$. As $mx_m \in U_m$, we see that the sequence $\{mx_m\}$ converges to zero in E, in particular is bounded in E. Therefore, as u is bounded, the sequence $\{m f(x_m)\}$ must be bounded in F. This means that there is a positive number ρ such that $m f(x_m) \in \rho V$ for all m. Take $m \geqslant \rho$; we have

$$f(x_m) \in \frac{\rho}{m} V \subset V \quad \text{(since V is balanced)},$$

contrary to our assumption. We have reached a contradiction.

COROLLARY.   *A bounded linear map of a Fréchet space (resp. a normed space, resp. an LF space) into a TVS is continuous.*

The Fréchet and normed cases are trivial consequences of Proposition 14.8; the *LF* case follows from the same combined with Propositions 13.1 and 14.6.

We recall that continuous linear mappings are always bounded (Proposition 14.2).

We show now that the spaces $\mathscr{C}^\infty(\Omega)$, $\mathscr{C}^\infty_c(\Omega)$ (Ω: open subset of $\mathbf{R}^n$), and $\mathscr{S} = \mathscr{S}(\mathbf{R}^n)$ have the property that all bounded and closed subsets are compact. The proof is based on Ascoli's theorem, which we recall now. First of all, we introduce the notion of *equicontinuous* sets of functions:

Definition 14.3.   *Let X be a topological space, F a TVS, and x^0 a point of X. A set S of mappings of X into F is said to be equicontinuous at the point x^0 if, to every neighborhood of zero, V, in F, there is a neighborhood $U(x^0)$ of x^0 in X such that, for all $f \in S$,*

$$x \in U(x^0) \quad \text{implies} \quad f(x) - f(x^0) \in V.$$

The condition in Definition 14.3 implies that each mapping $f \in S$ is continuous at x^0 (but this, of course, is not enough to ensure that S is equicontinuous at x^0).

The definitions of an equicontinuous set of mappings $f : X \to F$ on a set $A \subset X$, of uniformly equicontinuous sets of mappings, etc., are obvious. Now, if X is a compact space, a set S of mappings from X into F which is equicontinuous at every point is uniformly equicontinuous (a compact space carries a canonical uniform structure; if the student does not want to hear about uniform structures, he may imagine that X is a compact subset of a TVS). In this chapter, we shall not be concerned with equicontinuous sets of *linear* mappings of a TVS E into another TVS F, but let us point out (for future purposes) that in order that such a set of mappings be equicontinuous everywhere, and also uniformly equicontinuous, it is necessary and sufficient that it be equicontinuous at the origin. If S is the set under consideration, the definition of equicontinuity reads: to every V, neighborhood of zero in F, there is U, neighborhood of zero in E, such that, for all $f \in S$,

$$f(U) \subset V, \quad \text{or} \quad U \subset f^{-1}(V).$$

Let us turn our eyes to complex-valued functions (not necessarily linear). A set S of mappings $f : X \rightsquigarrow \mathbf{C}$ is equicontinuous at x^0 if, to every $\epsilon > 0$, there is a neighborhood $U(x^0)$ of x^0 in X such that, for all $f \in S$,

$$x \in U(x^0) \quad \text{implies} \quad |f(x) - f(x^0)| < \varepsilon.$$

Now, we state and prove *Ascoli's theorem:*

THEOREM 14.1. *Let $\{f_n\}$ be a sequence of complex-valued functions defined in a compact subset of $\mathbf{R}^d$. We make the following assumptions:*

(a) *the set of functions $\{f_n\}$ is equicontinuous on K;*

(b) *there is a constant $M < +\infty$ such that, for every n and every $x \in K$,*

$$|f_n(x)| \leqslant M.$$

We conclude that the sequence $\{f_n\}$ contains a subsequence $\{f_{n_k}\}$ which converges uniformly in K.

Proof. Let $\varepsilon > 0$ be given arbitrarily. By hypothesis, to every point x^0 of K there is $\eta(x^0) > 0$ such that

$$|x - x^0| \leqslant \eta(x^0) \quad \text{implies} \quad |f_n(x) - f_n(x^0)| \leqslant \varepsilon \quad \text{for all} \quad n.$$

The compactness of K implies immediately that we may take $\eta(x^0)$ independently of x^0 (this has already been mentioned in the remarks preceding the statement of Theorem 14.1: on compact sets, equicontinuity at each point implies uniform equicontinuity). Let us then write η instead of $\eta(x^0)$.

Let us choose a finite number of points of K, $x^1, \ldots, x^m$, such that the balls $\{x \in \mathbf{R}^d \mid x - x^j \mid \leqslant \eta\}$ $(j = 1, \ldots, m)$ cover K. We focus our attention on x^1 first, and consider the set of complex numbers $\{f_n(x^1)\}$. In view of Assumption (b), this set is contained in the disk of the complex plane, $\{z; \mid z \mid \leqslant M\}$; by the theorem of Weierstrass–Bolzano we can extract a subsequence that converges; let it be $\{f_{n_\lambda}(x^1)\}$, $\lambda = 1, 2, \ldots$. Next, we repeat the same argument at the point x^2, but after having substituted the sequence $\{f_{n_\lambda}\}$ for the sequence $\{f_n\}$, which we obviously may do: if the sequence $\{f_n\}$ has Properties (a) and (b), so does every subsequence of the $\{f_n\}$. Thus we find a subsequence $\{f_{n_{\lambda_k}}\}$ of the f_{n_λ} such that the sequence of numbers $\{f_{n_{\lambda_k}}(x^2)\}$ converges; and so on. Repeating m times this procedure, we end up with a subsequence S' of the originally given sequence $\{f_n\}$, and whose elements we shall represent, for the sake of simplicity, by g_n, which has the following property: for every $j = 1,..., m$, the sequence of numbers $\{g_n(x^j)\}$ converges. Let us choose an integer $N(\varepsilon)$ so large that n_1, $n_2 \geqslant N(\varepsilon)$ implies, whatever be j, $1 \leqslant j \leqslant m$,

$$| g_{n_1}(x^j) - g_{n_2}(x^j)| \leqslant \varepsilon.$$

Let then x be an arbitrary point of K; there is some j such that $| x - x^j | < \eta$. We have, for n_1, $n_2 \geqslant N(\varepsilon)$,

$$| g_{n_1}(x) - g_{n_2}(x)| \leqslant | g_{n_1}(x) - g_{n_1}(x^j)| + | g_n(x^j) - g_{n_2}(x^j)|$$

$$+ | g_{n_2}(x^j) - g_{n_2}(x)| \leqslant 3\varepsilon.$$

In view of the properties of the sequence $\{g_{N(\varepsilon)}$, $g_{N(\varepsilon)+1}$,...$\}$, and after replacing ε by $\varepsilon/3$, we see that we have proved the following fact:

(14.1)   Let $S = \{f_n\}$ be a sequence of functions in the compact set K, having Properties (a) and (b) in Theorem 14.1. Given any $\varepsilon > 0$, there is a subsequence S' of S such that, if f_{n_1} , $f_{n_2} \in S'$,

$$\sup_{x \in K} | f_{n_1}(x) - f_{n_2}(x)| \leqslant \varepsilon.$$

From there, the proof of Theorem 14.1 follows easily: for we apply (14.1) to S with $\varepsilon = 1$, obtaining thus a subsequence of S which we shall denote by S_1; next, we apply (14.1) with S_1 instead of S and $\varepsilon = \frac{1}{2}$ and we obtain a subsequence S_2 of S_1; etc. By induction, we obtain a totally ordered sequence of sequences

$$S = S_0 \supset S_1 \supset S_2 \supset \cdots \supset S_p \supset S_{p+1} \supset \cdots$$

with the property that, if f_{n_1} , $f_{n_2} \in S_p$, then

$$\sup_{x \in K} | f_{n_1}(x) - f_{n_2}(x)| \leqslant 2^{-p}.$$

We choose arbitrarily f_{n_0} in S_0 , f_{n_1} in S_1 ,..., and f_{n_p} in S_p ,...; it is clear that the sequence $\{f_{n_k}\}$ ($k = 0, 1,...$) is a Cauchy sequence for the uniform convergence in K; it has therefore a limit, which is a continuous function.   Q.E.D.

Remark 14.1.   Consider the Banach space $\mathscr{C}(K)$ of the continuous functions in K, with the topology of uniform convergence, that is to say with the norm

$$f \rightsquigarrow \sup_{x \in K} |f(x)|.$$

Ascoli's theorem states that *a subset of $\mathscr{C}(K)$ which is both bounded and equicontinuous has a compact closure.*

Ascoli's theorem has a converse, which we are not going to use: *any subset of $\mathscr{C}(K)$ which has a compact closure is bounded and equicontinuous.*

The next result provides us with a criterion of equicontinuity; it is sometimes referred to as Arzelà's theorem. Given any bounded subset of $\mathscr{C}(K)$, we cannot assert, of course, that it is compact (otherwise the closed unit ball of $\mathscr{C}(K)$ would be compact, $\mathscr{C}(K)$ would therefore be locally compact, hence it would be finite dimensional!). But if, given a set S of functions in K, we know that it is bounded (for the maximum of the absolute value) and moreover that the functions in the set S have continuous first derivatives which are also uniformly bounded by one and the same finite constant, then we may conclude that the set S is equicontinuous, therefore relatively compact in $\mathscr{C}(K)$ ("relatively compact" means "has a compact closure"). But we must make this statement more precise: for what does it mean that a function f, defined in a compact set K, has continuous first-order derivatives? To give a meaning, we shall assume that K is contained in a bounded open subset Ω of $\mathbf{R}^n$, and we shall be concerned with the space $\mathscr{C}^1(\bar{\Omega})$ introduced in Chapter 11 (Example III, p. 98): it is the space of once continuously differentiable functions in Ω whose derivatives of order 0 and 1 can be extended as continuous functions to the closure $\bar{\Omega}$ of Ω. The space $\mathscr{C}^1(\bar{\Omega})$ carries the norm

$$\|f\|_1 = \sup_{x \in \Omega} (\sup |f(x)|, \sup_{j=1,\ldots,n} |(\partial/\partial x_j) f(x)|),$$

or any equivalent norm, like for instance

$$f \rightsquigarrow \sup_{x \in \Omega} (|f(x)| + \sum_{j=1}^{n} |(\partial/\partial x_j) f(x)|).$$

With any one of these norms, $\mathscr{C}^1(\bar{\Omega})$ is a B-space. We may now state Arzelà's theorem:

THEOREM 14.2.   *Let K be a compact subset of a bounded open subset Ω of $\mathbf{R}^n$. The restriction mapping $f \rightsquigarrow f \mid K$, which assigns to a function in Ω*

its restriction to K, transforms bounded subsets of $\mathscr{C}^1(\bar{\Omega})$ into relatively compact subsets of $\mathscr{C}(K)$.

"Relatively compact" means "has a compact closure."

Proof of Theorem 14.2. In view of Ascoli's theorem (Theorem 14.1) it suffices to prove that the restriction to K of a bounded subset of $\mathscr{C}^1(\bar{\Omega})$ is an equicontinuous set of functions in K (it is obviously bounded in $\mathscr{C}(K)$: Proposition 14.2). In fact, we might consider only balls centered at the origin, in $\mathscr{C}^1(\bar{\Omega})$.

Let f be a function belonging to $\mathscr{C}^1(\bar{\Omega})$. It follows easily, say from a Taylor expansion of order 1, that to every $x^0 \in \Omega$ there is $\rho(x^0) > 0$ such that, if $|x - x^0| \leqslant \rho(x^0)$, then $x \in \Omega$ and

$$|f(x) - f(x^0)| \leqslant \sup_{y \in \Omega} (\sup_{j=1,\ldots,n} |(\partial/\partial x_j)f(y)|) \, |x - x^0| \leqslant \|f\|_1 \, |x - x^0|.$$

This implies obviously that any ball (of finite radius), centered at the origin, in $\mathscr{C}^1(\bar{\Omega})$, is an equicontinuous set in Ω, that is to say at every point of Ω. In particular, the image of such a ball by the restriction mapping is an equicontinuous set of functions in K.   Q.E.D.

Exercise 14.8. Prove directly or derive from Theorem 14.2 the following one-dimensional version of Arzelà's theorem:

THEOREM 14.3. *Let $[a, b]$ be a closed and bounded interval of the real line, and S a bounded subset of $\mathscr{C}^1([a, b])$. The set S is relatively compact in $\mathscr{C}([a, b])$.*

We may now prove the first of the announced results (all the others follow from this first one, as can be shown):

THEOREM 14.4. *Let Ω be an open subset of $\mathbf{R}^n$. If $k < \infty$, any bounded subset of $\mathscr{C}^{k+1}(\Omega)$ is relatively compact in $\mathscr{C}^k(\Omega)$. Any bounded subset of $\mathscr{C}^\infty(\Omega)$ is relatively compact in $\mathscr{C}^\infty(\Omega)$.*

Proof. Let $K_0 \subset K_1 \subset \cdots \subset K_j \subset \cdots$ be the usual sequence of compact subsets of Ω, whose union is Ω, and such that K_j is the closure of its interior, Ω_j; for all j, $K_j \subset \Omega_{j+1}$. Since all the spaces under consideration are F-spaces, it suffices to show that any bounded sequence contains a converging subsequence (Proposition 8.4). We shall begin by considering the case where k is finite.

Let j be $\geqslant 1$. Let S be any bounded sequence in $\mathscr{C}^{k+1}(\bar{\Omega}_j)$. This means that, for each n-tuple p such that $|p| \leqslant k$, the sequence of functions

$$(\partial/\partial x)^p f, \qquad f \in S,$$

is bounded in $\mathscr{C}^1(\bar{\Omega}_j)$. In view of Arzelà's theorem, we may find a subsequence S_1 of S such that the restrictions

(14.2) $$(\partial/\partial x)^p f \mid \Omega_{j-1}, \qquad f \in S_1,$$

converge in $\mathscr{C}^1(\bar{\Omega}_{j-1})$ (to see this, order the n-tuples p, $\mid p \mid \,\leqslant k$, in any fashion: select a sequence $S_{1,1} \subset S$ such that the restrictions (14.2) converge in $\mathscr{C}^0(\bar{\Omega}_{j-1})$ as f runs over $S_{1,1}$ with the first n-tuple p; then select a subsequence $S_{1,2}$ of $S_{1,1}$ such that the restrictions (14.2) converge as f runs over $S_{1,2}$ with the second n-tuple p, and so on; in a finite number of steps, we obtain the subsequence S_1). It follows immediately from a standard argument (based on (10.2) and (10.3), p. 87) that if f^0 is the limit of the restrictions (14.2) in $\mathscr{C}^0(\bar{\Omega}_{j-1})$ when $p = 0$, then, for each p, $\mid p \mid \,\leqslant k$, $(\partial/\partial x)^p f^0$ is the limit of the restrictions (14.2) in that same space. In other words, the sequence S_1 converges in $\mathscr{C}^k(\bar{\Omega}_{j-1})$. We have proved the following fact:

(14.3) *Given any $j \geqslant 1$, and any bounded sequence S in $\mathscr{C}^{k+1}(\bar{\Omega}_j)$, there is a subsequence S_1 of S such that the restrictions of the functions $f \in S_1$ to Ω_{j-1} form a converging sequence in $\mathscr{C}^k(\bar{\Omega}_{j-1})$.*

Let now S be a bounded sequence in $\mathscr{C}^{k+1}(\Omega)$. By restriction to Ω_1, it gives rise to a bounded sequence $S|\Omega_1$ in $\mathscr{C}^{k+1}(\bar{\Omega}_1)$. By (14.3), we may find a subsequence S_1 of S such that the sequence of restrictions $S_1|\Omega_0$ to Ω_0 converges in $\mathscr{C}^k(\bar{\Omega}_0)$. But S_1 is also bounded in $\mathscr{C}^{k+1}(\Omega)$; by restriction to Ω_2, it defines a bounded sequence in $\mathscr{C}^{k+1}(\bar{\Omega}_2)$. In view of (14.3), we may find a subsequence S_2 of S_1 such that $S_2|\Omega_1$ converges in $\mathscr{C}^k(\bar{\Omega}_1)$; and so forth. We obtain in this way a totally ordered sequence of sequences

$$S = S_0 \supset S_1 \supset S_2 \supset \cdots \supset S_j \supset \cdots$$

such that, for each $j \geqslant 1$, $S_j|\Omega_{j-1}$ converges in $\mathscr{C}^k(\bar{\Omega}_{j-1})$. For each $j \geqslant 1$, let f^j be the limit of the sequence $S_j|\Omega_{j-1}$ in $\mathscr{C}^k(\bar{\Omega}_{j-1})$. Let f_j be an element of S_j such that

$$\sup_{|p| \leqslant k} \left(\sup_{x \in \Omega_{j-1}} |(\partial/\partial x)^p (f_j - f^j)| \right) \leqslant 1/j.$$

It is obvious that the sequence $S' = \{f_1, f_2, ..., f_j, ...\}$ converges in $\mathscr{C}^k(\Omega)$; its limit is the function f in Ω whose restriction to Ω_{j-1} is f^j for each $j \geqslant 1$; obviously, $S' \subset S$. Thus we have proved the result for k finite.

Let now S be a bounded sequence of $\mathscr{C}^\infty(\Omega)$; a fortiori, S is bounded in each $\mathscr{C}^k(\Omega)$, $k = 1, 2, ...$, therefore S is relatively compact in each $\mathscr{C}^{k-1}(\Omega)$. Let S_0 be a subsequence of S which converges in $\mathscr{C}^0(\Omega)$ (see Exercise 8.7), and let f be its limit. Let S_1 be a subsequence of S_0 converging in $\mathscr{C}^1(\Omega)$; the limit of S_1 must also be f (which, by way of consequence, is $\mathscr{C}^1$); let S_2 be a subsequence of S_1 converging (to f) in $\mathscr{C}^2(\Omega)$, etc. We see that f is $\mathscr{C}^\infty$ in Ω, and we have now a sequence of sequences

$$S = S_0 \supset S_1 \supset S_2 \supset \cdots \supset S_j \supset \cdots,$$

this time with the property that, for each $j \geqslant 1$, S_j converges to f in $\mathscr{C}^j(\Omega)$. For each j, we select an element $f_j \in S_j$ such that

$$\sup_{|p| \leqslant j} (\sup_{x \in \Omega_j} |(\partial/\partial x)^p (f_j - f)(x)|) \leqslant 1/j.$$

The subsequence of S, $\{f_1, f_2, ..., f_j, ...\}$, obviously converges to f in $\mathscr{C}^\infty(\Omega)$.   Q.E.D.

Exercises

14.9. Prove the following corollaries of Theorem 14.4:

COROLLARY 1.   *Any bounded subset of $C_c^{k+1}(\Omega)$ (Chapter 13, Example II), $0 \leqslant k < +\infty$, is relatively compact in $\mathscr{C}_c^k(\Omega)$.*

COROLLARY 2.   *Any closed bounded subset of $\mathscr{C}_c^\infty(\Omega)$ is compact (in $\mathscr{C}_c^\infty(\Omega)$).*

14.10. Prove the analog of Theorem 14.4 (or of its Corollary 2) for the space $\mathscr{S}$ of $\mathscr{C}^\infty$ functions rapidly decaying at infinity (Chapter 10, Example IV):

THEOREM 14.5.   *Any closed bounded subset of $\mathscr{S}$ is compact.*

We consider now the space $H(\Omega)$ of holomorphic functions in an open subset Ω of the complex n-space $\mathbf{C}^n$ (see Chapter 10, Example II, p. 89). The TVS $H(\Omega)$ carries the topology of uniform convergence of functions on every compact subset of Ω. We may identify $\mathbf{C}^n$ with $\mathbf{R}^{2n}$ via the canonical mapping

$$(x_1 + iy_1, ..., x_n + iy_n) \to (x_1, ..., x_n, y_1, ..., y_n)$$

and regard $H(\Omega)$ as a subspace of any of the spaces $\mathscr{C}^k(\Omega)$ ($0 \leqslant k \leqslant \infty$). On $H(\Omega)$ all the topologies $\mathscr{C}^k$ coincide (see Chapter 10, loc. cit.); in particular, they all coincide with the $\mathscr{C}^\infty$ topology: $H(\Omega)$ is a closed subspace of $\mathscr{C}^{\infty}(\Omega)$, meaning by this that the topology proper of $H(\Omega)$ is identical with the one induced by $\mathscr{C}^{\infty}(\Omega)$. Let now B be a closed and bounded subset of $H(\Omega)$: B is closed and bounded in $\mathscr{C}^{\infty}(\Omega)$, hence it is compact (where it is compact—in $H(\Omega)$ or in $\mathscr{C}^{\infty}(\Omega)$—is irrelevant!). Thus we have proved *Montel's theorem:*

THEOREM 14.6.   *Any closed and bounded subset of $H(\Omega)$ is compact.*

15

Approximation Procedures
in Spaces of Functions

It has been a standard tactic of the analyst, since the dawn of analysis, that, when forced to deal with a "bad" function, he should try to approximate it with "nice" ones, study the latter and prove that some of the properties in which he happens to be interested, if valid for the approximating nice functions, would carry over to their limit. Of course, the concept of a "bad" function has evolved in time, with the resulting effect that the set of functions considered "good" has steadily increased (but so has also the set of functions, or, more generally, of "function-like" objects, considered "bad"). We might imagine that Taylor and Mac Laurin felt ill at ease when confronted with analytic functions, and that is why they strove to approximate them by polynomials, whereas for our purposes here, from the local point of view, analytic functions will be regarded as the nicest type of functions (right after polynomials, which retain their supremacy); later on, nondifferentiable continuous functions and then functions which are only measurable would be regarded as bad (they still are), and approximation techniques were devised to deal with them (e.g., approximation by step functions). As we shall see in Part II, functions can become so bad as to stop being functions: they become Dirac's "function" and measures, and in distribution theory we shall be dealing with derivatives of arbitrary order of measures. In any one of these situations, it will help to have at our disposal approximation techniques, so as to approximate those objects by very smooth functions. In talking about smoothness, we deal only with the local aspect: but there is also a global aspect, not to be forgotten, for instance when considering integrals

$$\int_{\mathbf{R}^n} f(x) \, g(x) \, dx,$$

where f and g are both nice locally, and "nice" means here integrable, but where g is allowed to *grow* at infinity arbitrarily fast. Then it becomes necessary that f *decay* at infinity arbitrarily fast: the latter can have only one meaning: f must vanish outside some bounded set. Similar global requirements will force us to study the approximation of "bad" functions, measures, and distributions, by smooth functions with *compact* support. Of course, we would like to combine the global and the local aspects, and approximate by means of functions with the best possible local properties and with the best possible global ones: if it were possible, we would like to approximate by analytic functions with compact support. But nature bars such a path, since the only analytic function with compact support is the zero function: such an analytic function would have to be identically zero in the complement of a compact set, which is a big open set; this implies that the function be zero everywhere! We shall therefore deal with the two viewpoints somehow separately (although trying not to keep them too far apart): we shall approximate by analytic functions with a high order of decay at infinity, or by functions with compact support which are sufficiently smooth, say $\mathscr{C}^{\infty}$.

There is another reason to the usefulness of approximation techniques, and a reason of very fundamental importance. The objects manipulated by analysts are always extracted from topological vector spaces, spaces of functions if the objects are functions, or *duals* of spaces of functions. In fact, the latter provide us with the largest fishing ponds, as (generally speaking) the spaces of functions can be embedded in the duals of other spaces of functions. The question then arises of the mutual relation existing between those spaces or those duals, and it is important to know when inclusions $E' \subset F'$ hold (E', dual of a TVS E, F', dual of another TVS F), and more precisely to what relation between E and F could we relate such an inclusion. A much used criterion is the following one: *E' can be regarded as a vector subspace of F' if:*

(1) *as a vector space, F can be regarded as a subspace of E;*

(2) *F is a dense linear subspace of E;*

(3) *the topology of F is at least as fine as the one induced by E.*

Indeed, by (1) and (3) any continuous linear form on E defines by restriction to F a continuous linear form on F; if any two continuous linear forms on E define the same form on F, they coincide on a dense subset of E by (2), hence everywhere in E: they must be identical. Thus, to every continuous linear form on E corresponds one and only one continuous linear form on F, and this is the meaning of the inclusion $E' \subset F'$. Now, Properties (1) and (3) are always clear enough by the definition of E and F and the definition of their respective topologies (e.g., it is clear that $\mathscr{C}^{\infty}(\Omega) \subset \mathscr{C}^k(\Omega)$ and that the $\mathscr{C}^{\infty}$ topology is finer than the $\mathscr{C}^k$ one). In practice, the point to be checked is that (2) is true (when it is true!), and to that purpose one uses approximation (e.g., we shall see that every $\mathscr{C}^k$ function is the limit of a sequence of $\mathscr{C}^{\infty}$ functions—in the sense of $\mathscr{C}^k(\Omega)$). We shall see later that, instead of constructing approximating sequences or filters, one may, in certain instances, use a direct approach and functional analysis techniques (namely the theorem of Hahn–Banach), but in many of the basic instances, this leads to cruder results, and for instance does not enable us always to conclude that the object to be approximated is the limit of a *sequence*: for we must remember that the density of F in E means that every element of E is the limit of a *filter* of elements belonging to F, not necessarily of a sequence (when E is metrizable, then we know that density is equivalent with "sequential density"; it is also true in other instances, as we shall see, but it is not always true).

We begin by studying analytic functions and, as a matter of fact, *entire* functions of n variables. We recall that the holomorphic (i.e., complex analytic) functions in an open subset Ω of the complex n-space $\mathbf{C}^n$ form a vector space which is usually equipped with the topology of uniform convergence of functions on every compact subset of Ω; this turns it into a Fréchet space, which we have denoted by $H(\Omega)$. When $\Omega = \mathbf{C}^n$, we write H, if no confusion is liable to arise (see Chapter 10, Example II, and Chapter 14, Theorem 14.6).

THEOREM 15.1. *The polynomials (with respect to z) are dense in the space of entire functions in $\mathbf{C}^n$. More precisely, every entire function f is the limit in H of its finite Taylor expansions*

$$\sum_{|p| \leqslant m} 1/p!\, f^{(p)}(0)\, z^p, \qquad m = 0, 1, 2, \ldots.$$

The last part of the statement is the proof of the theorem. We have used the notation

$$p! = p_1! \cdots p_n!, \qquad f^{(p)} = (\partial/\partial z_1)^{p_1} \cdots (\partial/\partial z_n)^{p_n} f, \qquad z^p = z_1^{p_1} \cdots z_n^{p_n}.$$

When we consider open sets $\Omega \neq \mathbf{C}^n$, we must be careful. It is not always true that an holomorphic function in Ω is the limit of polynomials or of entire functions (the two facts are equivalent by Theorem 15.1).

Definition 15.1. *An open subset Ω of $\mathbf{C}^n$ is called a Runge domain if the restrictions of the entire functions to Ω are dense in $H(\Omega)$.*

The next result is almost obvious:

THEOREM 15.2.   *Open polydisks*

$$\{z \in \mathbf{C}^n; \, | \, z_j \, | < R_j \leqslant +\infty, j = 1,..., n\}$$

are Runge domains.

The following lemma is going to be used in order to derive a result of approximation by polynomials.

LEMMA 15.1.   *Let f be a continuous function in $\mathbf{R}^n$ with compact support. For each integer $k = 1, 2,...,$ the function*

$$f_k(x) = (k/\sqrt{\pi})^n \int_{\mathbf{R}^n} \exp(-k^2| \, x - y \, |^2) f(y) \, dy$$

can be extended to the complex values of the variables x as an entire function. When $k \to +\infty$, the functions f_k converge to f uniformly in $\mathbf{R}^n$.

Proof of Lemma 15.1.   The part of the statement concerning the fact that f_k can be extended to $\mathbf{C}^n$ as an entire function is trivial. It suffices to consider the "integral representation"

$$(15.1) \quad (k/\sqrt{\pi})^n \int_{\mathbf{R}^n} \exp(-k^2[(z_1 - y_1)^2 + \cdots + (z_n - y_n)^2]) f(y) \, dy,$$

where $z_1,..., z_n$ are complex variables. Observe that the integral is performed over a compact subset of $\mathbf{R}^n$, the support of f. We may apply the Cauchy–Riemann operators $\partial/\partial \bar{z}_1,..., \partial/\partial \bar{z}_n$ by differentiating under the integral sign (by Leibniz' rule), and we obtain that the function defined by (15.1) satisfies the Cauchy–Riemann equations in the whole of $\mathbf{C}^n$ (by differentiating under the integral sign with respect to the real variables Re z_j, Im $z_j, j = 1,..., n$, one sees immediately that the function (15.1) is $\mathscr{C}^\infty$ with respect to those variables; the legitimacy of applying Leibniz' rule is obvious).

Let us prove now that the f_k converge uniformly to f. By changing variables in the integrals, we see that

$$f_k(x) = (k/\sqrt{\pi})^n \int_{\mathbf{R}^n} e^{-k^2|y|^2} f(x - y) \, dy.$$

Observe then that we have, for all k,

$$(15.2) \quad (k/\sqrt{\pi})^n \int_{\mathbf{R}^n} e^{-k^2|y|^2} \, dy = 1.$$

Indeed, the left-hand side is equal to

$$\left[(k/\sqrt{\pi}) \int_{-\infty}^{+\infty} e^{-k^2 t^2} \, dt\right]^n,$$

where t is a real variable. But it is well known that

$$k \int_{-\infty}^{+\infty} e^{-k^2 t^2} \, dt = \int_{-\infty}^{+\infty} e^{-t^2} \, dt = \sqrt{\pi},$$

whence (15.2). Using this, we see that we may write

$$f_k(x) - f(x) = (k/\sqrt{\pi})^n \int_{\mathbf{R}^n} e^{-k^2 |y|^2} \left[f(x - y) - f(x)\right] dy.$$

The key feature about the functions $(k/\sqrt{\pi})^n \, e^{-k^2 |y|^2}$ is that their total mass is equal to one, according to (15.2), but tends to concentrate near the origin, as $k \to +\infty$. Notice indeed that the value of this function at zero is $(k/\sqrt{\pi})$, which converges to infinity, but that its value at any point $y \neq 0$ converges to zero quite fast. The student may plot the curves (say, when $n = 1$) and get some idea of the situation; we shall see, in due time, that these "bell-shaped" functions converge to Dirac's measure (and that any other sequence with similar properties could have been used!).

As f is a continuous function with compact support, f is uniformly continuous; in other words, to every $\varepsilon > 0$ there is $\eta > 0$ such that

$$|x - x'| < \eta \quad \text{implies} \quad |f(x) - f(x')| < \varepsilon.$$

Let us choose $k(\varepsilon)$ sufficiently large so as to have, for $k > k(\varepsilon)$,

$$(k/\sqrt{\pi})^n \int_{|y| > \eta} e^{-k^2 |y|^2} \, dy < \varepsilon.$$

We have then

$$|f_k(x) - f(x)| \leqslant \sup_{|y| < \eta} |f(x) - f(x - y)| \, (k/\sqrt{\pi})^n \int_{|y| < \eta} e^{-k^2 |y|^2} \, dy$$

$$+ 2 \sup_{y \in \mathbf{R}^n} |f(y)| \, (k/\sqrt{\pi})^n \int_{|y| > \eta} e^{-k^2 |y|^2} \, dy$$

$$\leqslant (1 + 2 \sup_{y \in \mathbf{R}^n} |f(y)|)\varepsilon. \qquad \text{Q.E.D.}$$

COROLLARY 1. *Let f be a $\mathscr{C}^m$ function with compact support in $\mathbf{R}^n$ $(0 \leqslant m \leqslant +\infty)$. For every differentiation index $p = (p_1, ..., p_n)$ such that*

$|p| < m + 1$, *the functions* $(\partial/\partial x)^p f_k$ *converge uniformly to* $(\partial/\partial x)^p f$ *in* $\mathbf{R}^n$, *as* $k \to +\infty$.

Indeed, if we make the change of variables $x - y \rightsquigarrow x$ in the integral expressing f_k, we obtain

$$f_k(x) = (k/\sqrt{\pi})^n \int_{\mathbf{R}^n} e^{-k^2|y|^2} f(x - y)\, dy.$$

By Leibniz' rule, we can differentiate "p times" under the integral sign:

$$(\partial/\partial x)^p f_k(x) = (k/\sqrt{\pi})^n \int_{\mathbf{R}^n} e^{-k^2|y|^2} [(\partial/\partial x)^p f](x - y)\, dy$$

and it suffices then to apply Lemma 15.1 with $(\partial/\partial x)^p f$ instead of f.

COROLLARY 2.   *Every function* $f \in \mathscr{C}_c^m(\mathbf{R}^n)$ *is the limit, in* $\mathscr{C}^m(\mathbf{R}^n)$, *of a sequence of polynomials.*

The idea of the proof of Corollary 2 is obvious: the function f can be approximated by entire functions in the sense of the $\mathscr{C}^m$ convergence, by Corollary 1. By Theorem 15.1, each entire function can be approximated by polynomials in the $\mathscr{C}^\infty$ sense (remember that the topology of $H(\mathbf{C}^n)$ is the topology induced by $\mathscr{C}^\infty(\mathbf{R}^{2n})$). It is then only natural that f will be the limit of a sequence of polynomials in $\mathscr{C}^m(\mathbf{R}^n)$. If the student wishes, he can work out the details, with the appropriate ε's.

We examine now the approximation by $\mathscr{C}^\infty$ functions with compact support. The student might have noticed that we have not yet exhibited any function of this type, and he may after all be uncertain as to the existence of such functions. Here is an example, which will be constantly used in the forthcoming; in particular, it will help us to prove that not only do $\mathscr{C}_c^\infty$ functions exist, but that they are everywhere dense in the spaces $\mathscr{C}^k$ ($0 \leqslant k \leqslant \infty$), L^p ($1 \leqslant p < \infty$). The function is the following one:

$$(15.3) \qquad \rho(x) = \begin{cases} a \exp\left(-\dfrac{1}{1 - |x|^2}\right) & \text{for } |x| < 1; \\ 0 & \text{for } |x| \geqslant 1. \end{cases}$$

The constant a is defined by

$$a = \left(\int_{|x| < 1} \exp\left(-\frac{1}{1 - |x|^2}\right) dx\right)^{-1},$$

so that we have

$$(15.4) \qquad \int_{\mathbf{R}^n} \rho(x)\, dx = 1.$$

The function ρ is analytic (i.e., its Taylor's series has a nonzero radius of convergence) about every point in the *open* ball $\{x; \ |x| < 1\}$; it is obviously regular in the exterior $\{x; \ |x| > 1\}$ of that ball, so that the only question about it being $\mathscr{C}^{\infty}$ could arise at the boundary of the ball, that is to say for $|x| = 1$. As ρ is rotation-invariant, it suffices to check that the function of one variable,

$$\begin{cases} \exp(-1/(1 - t^2)) & \text{for} \quad |t| \leqslant 1 \\ 0 & \text{for} \quad |t| > 1, \end{cases}$$

is $\mathscr{C}^{\infty}$ about $t = 1$, and the problem is readily reduced to proving that the function

$$\begin{cases} \exp(-1/s) & \text{for} \quad s > 0 \\ 0 & \text{for} \quad s \leqslant 0, \end{cases}$$

is $\mathscr{C}^{\infty}$, which is a well-known and evident fact (note indeed that

$$\exp(-1/(1 - t^2)) = \exp\left(-\frac{1}{2}\frac{1}{1-t}\right)\exp\left(-\frac{1}{2}\frac{1}{1+t}\right)).$$

(15.5) *Notation* (used throughout the book):

For $\varepsilon > 0$, we set $\rho_\varepsilon(x) = \varepsilon^{-n}\rho(x/\varepsilon)$.

In view of (15.4), we have, for all $\varepsilon > 0$,

$$(15.6) \qquad\qquad \int_{\mathbf{R}^n} \rho_\varepsilon(x)\,dx = 1.$$

It suffices to make the change of variables $\varepsilon x \to x$ in (15.6) to transform it into (15.4).

Observe, also, that

$$\operatorname{supp}\rho_\varepsilon = \{x \in \mathbf{R}^n; \ |x| \leqslant \varepsilon\}$$

and that

$$\rho_\varepsilon(0) = \varepsilon^{-n},$$

so that these functions ρ_ε have a lot in common with the functions

$$(15.7) \qquad\qquad (k/\sqrt{\pi})^n\, e^{-k^2|x|^2}$$

considered in relation with Lemma 15.1. Of course, the ρ_ε have compact support whereas the functions (15.7) do not: but in both cases, the total mass of the associated densities is one, and the mass tends (as $\varepsilon \to 0$ or as $k \to \infty$) to concentrate about the origin. As we have already mentioned, these features will be eventually related to Dirac's measure δ.

LEMMA 15.2.   *Let f be a continuous function with compact support in $\mathbf{R}^n$. For each $\varepsilon > 0$, the functions*

$$f_\varepsilon(x) = \int_{\mathbf{R}^n} \rho_\varepsilon(x - y) f(y) \, dy$$

are $\mathscr{C}_c^\infty$ functions in $\mathbf{R}^n$. Furthermore, the support of f_ε is contained in the neighborhood of order ε of supp f, i.e., in the set

$$\{x \in \mathbf{R}^n; d(x, \text{supp} f) \leqslant \varepsilon\}.$$

When $\varepsilon \to 0$, the functions f_ε converge uniformly to f in $\mathbf{R}^n$.

If S is a set and x a point, $d(x, S)$ means the Euclidean distance from x to S, i.e.,

$$d(x, S) = \inf_{y \in S} |x - y|.$$

Proof. of Lemma 15.2.   We can differentiate $f_\varepsilon(x)$ under the integral sign; we see immediately then that f_ε is $\mathscr{C}^\infty$. The integral expressing f_ε is performed over the set of points y such that $y \in \text{supp} f$ and that $x - y \in \text{supp} \rho_\varepsilon$, i.e., $|x - y| < \varepsilon$. If x does not belong to the neighborhood of order ε of supp f there do not exist such points y, and the integral is identically zero. Finally, we must prove that the f_ε converge uniformly to f.

Here again we use the fact that f, being continuous and identically zero outside some compact set, is uniformly continuous. Hence, to every $\eta > 0$ there is $\varepsilon > 0$ such that

$$|x - y| < \varepsilon \quad \text{implies} \quad |f(x) - f(y)| \leqslant \eta.$$

In virtue of (15.6), we have

$$f(x) - f_\varepsilon(x) = \int \rho_\varepsilon(x - y) \, [f(x) - f(y)] \, dy,$$

hence

$$|f(x) - f_\varepsilon(x)| \leqslant \sup_{|x-y| < \varepsilon} |f(x) - f(y)| \int \rho_\varepsilon(x - y) \, dy \leqslant \eta.$$

Q.E.D.

COROLLARY 1. *Let f be a $\mathscr{C}^k$ function with compact support in $\mathbf{R}^n$ $(0 \leqslant k \leqslant +\infty)$. Then, for each differentiation index $p = (p_1, ..., p_n)$ such that $|p| < k + 1$, the functions*

$$(\partial/\partial x)^p f_\varepsilon$$

converge uniformly in $\mathbf{R}^n$ to $(\partial/\partial x)^p f$ (as $\varepsilon \to 0$).

Proof. We can make the change of variables $x - y \rightsquigarrow x$ in the integral expressing f_ε, obtaining thus

$$f_\varepsilon(x) = \int_{\mathbf{R}^n} \rho_\varepsilon(y) f(x - y) \, dy.$$

By Leibniz' rule we see that

$$(\partial/\partial x)^p f_\varepsilon(x) = \int_{\mathbf{R}^n} \rho_\varepsilon(y) \left[(\partial/\partial x)^p f \right] (x - y) \, dy$$

and it suffices to apply Lemma 15.2 with $(\partial/\partial x)^p f$ instead of f.

THEOREM 15.3. *Let $0 \leqslant k \leqslant +\infty$, Ω, be an open set of $\mathbf{R}^n$. Any function $f \in \mathscr{C}^k(\Omega)$ is the limit of a sequence $\{f_j\}$ $(j = 1, 2,...)$ of $\mathscr{C}^\infty$ functions with compact support in Ω such that, for each compact subset K of Ω, the set $K \cap \operatorname{supp} f_j$ converges to $K \cap \operatorname{supp} f$.*

A sequence of sets S_j converges to a set S if to every $\varepsilon > 0$ there is an integer $J(\varepsilon)$ such that, for $j \geqslant J(\varepsilon)$, S_j is contained in the neighborhood of order ε of S, and S is contained in the neighborhood of order ε of S_j.

Proof of Theorem 15.3. We use a sequence of open subsets $\Omega_0, \Omega_1, ..., \Omega_j, ...$, whose union is equal to Ω, and such that, for each $j \geqslant 1$, $\bar{\Omega}_{j-1}$ is compact and contained in Ω_j. Let d_j be the distance from $\bar{\Omega}_{j-1}$ to the complement of Ω_j; we have $d_j > 0$ for all j. We can build a *continuous* function g_j with the following properties:

$$g_j(x) = 1 \qquad \text{if} \quad d(x, \complement \, \Omega_j) > 3d_j/4,$$

$$g_j(x) = 0 \qquad \text{if} \quad d(x, \complement \, \Omega_j) < d_j/2.$$

We choose then $\varepsilon_j = d_j/4$, and consider the function

$$h_j(x) = \int_{\mathbf{R}^n} \rho_{\varepsilon_j}(x - y) \, g_j(y) \, dy.$$

Suppose that $x \in \Omega_{j-1}$; then, if $x - y \in \operatorname{supp} \rho_{\varepsilon_j}$, we must have $d(y, \complement\Omega_j) \geqslant d(x, \complement\Omega_j) - |x - y| \geqslant d_j - d_j/4 = 3d_j/4$, which implies

$g_j(y) = 1$, hence $h_j(x) = \int \rho_{\varepsilon_j}(x - y) \, dy = 1$, in view of (15.6). Thus $h_j \equiv 1$ in Ω_{j-1}. In view of Lemma 15.2, we know that the functions h_j are $\mathscr{C}^\infty$ and have compact support, and it is obvious that h_j converges to the function identically equal to 1, that is to say $1(x)$, in $\mathscr{C}^\infty(\Omega)$. Furthermore, given any function $f \in \mathscr{C}^k(\Omega)$, we have $h_j f \in \mathscr{C}^k(\Omega)$, for the product of a $\mathscr{C}^\infty$ function with a $\mathscr{C}^k$ function is a $\mathscr{C}^k$ function; and since $h_j f = f$ in Ω_{j-1}, the functions $h_j f$ converge to f in $\mathscr{C}^k(\Omega)$. If K is a compact subset of Ω, for j large enough we have $K \subset \Omega_{j-1}$. This implies that supp $f \cap K = \mathrm{supp}(h_j f) \cap K$.

We have approximated f by $\mathscr{C}^k$ functions with compact support, namely the functions $h_j f$. We must now approximate f by $\mathscr{C}^\infty$ functions with compact support. In order to do this, it suffices to apply Corollary 1 of Lemma 15.2 to each function $h_j f$ (with the additional information about the supports contained in the statement of Lemma 15.2). Suppose for instance k *finite*. Then, by Corollary 1 of Lemma 15.2, we may find a $\mathscr{C}^\infty$ function, which we shall denote by f_j, having a compact support contained in a neighborhood of order $1/j$ of $\mathrm{supp}(h_j f)$, and such that

$$\sup_{|p| \leqslant k} (\sup_{x \in \mathbf{R}^n} |(\partial/\partial x)^p (f_j(x) - h_j(x) f(x))|) \leqslant 1/j.$$

It is trivial that the functions f_j converge to f in $\mathscr{C}^k(\Omega)$. If a compact set K is contained in some open set Ω_{j-1}, we know that $K \cap \mathrm{supp} f_j$ is contained in the neighborhood of order $1/j$ of $K \cap \mathrm{supp}(h_j f) = K \cap \mathrm{supp} f$. Conversely, if $\varepsilon > 0$ is given, let $c > 0$ be a number such that $K \cap \mathrm{supp} f$ is contained in the set

$$\{x \in K; |f(x)| \geqslant c\} + \{x \in \mathbf{R}^n; |x| \leqslant \varepsilon\}.$$

Choose j large enough so as to have $K \subset \Omega_{j-1}$ and $1/j \leqslant c/2$; then we have, in Ω_{j-1},

$$|f_j(x) - f(x)| \leqslant c/2,$$

which implies immediately that the set $\{x \in K; |f(x)| \geqslant c\}$ is contained in $K \cap \mathrm{supp} f_j$; thus $K \cap \mathrm{supp} f$ is contained in the neighborhood of order ε of $K \cap \mathrm{supp} f_j$.

We leave to the student, as an exercise, the proof for $k = \infty$.

COROLLARY 1. $\mathscr{C}^\infty_c(\Omega)$ *is dense in* $\mathscr{C}^k(\Omega)$ $(0 \leqslant k \leqslant +\infty)$.

COROLLARY 2. $\mathscr{C}^\infty_c(\Omega)$ *is sequentially dense in* $\mathscr{C}^k_c(\Omega)$ $(0 \leqslant k \leqslant +\infty)$.

COROLLARY 3.   *If* $1 \leqslant p < +\infty$, $\mathscr{C}^\infty_c(\Omega)$ *is dense in* $L^p(\Omega)$.

It suffices to combine Corollary 2 (for $k = 0$) with Theorem 11.3.

COROLLARY 4.   *The polynomials form a dense linear subspace of $\mathscr{C}^k(\Omega)$* $(0 \leqslant k \leqslant +\infty)$.

It suffices to combine Corollary 1 of Theorem 15.3 with Corollary 2 of Lemma 15.1.

Exercises

15.1. Let g be a continuous > 0 function in $\mathbf{R}^n$. Prove that there is an entire analytic function h (this means that h can be extended to $\mathbf{C}^n$ as a function everywhere holomorphic or, equivalently, that the radius of convergence of the Taylor expansion of h about any point is infinite) such that, for all $x \in \mathbf{R}^n$,

$$0 < h(x) < g(x).$$

15.2. Let $\{g_\nu\}$ $(\nu = 1, 2,...)$ be a sequence of continuous > 0 functions in $\mathbf{R}^n$. Let f be a continuous function with compact support in $\mathbf{R}^n$, ω an arbitrary open neighborhood of supp f. Prove that there is a sequence $\{h_\nu\}$ $(\nu = 1, 2,...)$ of entire functions which converge uniformly on $\mathbf{R}^n$ to f and such that, for all $x \in \mathbf{R}^n$, $x \notin \omega$,

$$| h_\nu(x)| \leqslant g_\nu(x).$$

(Use Exercise 15.1 and look at the proof of Lemma 15.1.)

15.3. Let d be a number $\geqslant 0$, and Ω an open subset of $\mathbf{R}^n$. One calls dth *Gevrey class* in Ω, and denotes by $G_d(\Omega)$, the space of $\mathscr{C}^\infty$ functions f such that, to every compact subset K of Ω, there is a constant $A(f, K) > 0$ such that, for all $p \in \mathbf{N}^n$,

$$\sup_{x \in K} |(\partial/\partial x)^p f(x)| \leqslant A(f, K)^{|p|+1} (p!)^d.$$

What are the elements of $G_0(\Omega)$ and of $G_1(\Omega)$? What are the elements of

$$\bigcup_{d < 1} G_d(\Omega)?$$

Is it true or false that every entire function belongs to this union?

15.4. Let now d be > 1. Explore the properties of the function

$$= e^{-1/t^a} \quad \text{for} \quad t > 0, \quad = 0 \quad \text{for} \quad t \leqslant 0,$$

for a suitable choice of a, so as to prove that there is a function $\sigma \in G_d(\mathbf{R}^n)$ (see Exercise 15.3) which is $\geqslant 0$ and whose support is equal to $\{x \in \mathbf{R}^n; \mid x \mid \leqslant 1\}$ (cf. construction of the function ρ, on p. 156).

15.5. We recall that $\mathscr{S} = \mathscr{S}(\mathbf{R}^n)$ is the space of $\mathscr{C}^\infty$ functions rapidly decreasing at infinity (Chapter 10, Example IV; $\mathscr{S}$ carries its Fréchet space topology). Prove:

THEOREM 15.4. $\mathscr{C}_c^\infty(\mathbf{R}^n)$ *is dense in* $\mathscr{S}$.

15.6. Prove the following density theorem:

THEOREM 15.5.   *The entire analytic functions which belong to $\mathscr{S}$ are dense in $\mathscr{S}$.*

16

Partitions of Unity

In this chapter, we shall apply some of the results of the previous one. We shall show that a $\mathscr{C}^k$ function can always be represented as a sum of $\mathscr{C}^k$ functions whose support has an arbitrarily small diameter. This is in striking contrast with the situation for analytic functions: one can certainly not represent an analytic function as a sum of other analytic functions with small support, as there are no analytic functions which have compact support (unless they are identically zero). It is obvious that, if we can represent the function identically equal to one as a sum of $\mathscr{C}^\infty$ functions $\{g^i\}$ $(i \in I)$ with arbitrarily small support, we shall have in our hands the analog representation for arbitrary $\mathscr{C}^k$ functions f by just writing $f = \sum_i (g^i f)$. A family of functions like $\{g^i\}$ is called a *partition of unity* in $\mathscr{C}^\infty$. We are going to show how to construct such partitions of unity (our definition will add the requirement that all the g^i's be nonnegative, according to a well-established custom). The fact that the supports of the g^i's are arbitrarily small is best expressed by introducing *open coverings*, as we shall now do.

Let A be an arbitrary subset of $\mathbf{R}^n$. An open covering of A is a family of open sets $\{U^i\}$ in $\mathbf{R}^n$ whose union contains A. Such a definition has the disadvantage of using the surrounding space. This is obviously unnecessary: for let A be any topological space, an open covering of A is a family of open subsets V^i of A whose ur.ion is identical to A. In the situation where $A \subset \mathbf{R}^n$, we can then take $V^i = U^i \cap A$ (A is a topological space if equipped with the induced topology). The open covering $\{V^i\}$ will be called *locally finite* if every point of A has a neighborhood which intersects only a finite number of open sets V^i. If $\{W^j\}$ is another open covering of A, one says that $\{W^j\}$ is *finer* than $\{V^i\}$ if every open set W^j is contained in some open set V^i.

THEOREM 16.1. *Let Ω be an open subset of $\mathbf{R}^n$. To every open covering $\{U^i\}$ $(i \in I)$ of Ω there is a finer open covering $\{V^j\}$ $(j \in J)$ of Ω which is locally finite.*

Proof. As usual, we select a sequence of relatively compact open subsets

161

of Ω, $\{\Omega_k\}$, $k = 0, 1, 2,...$, whose union is equal to Ω and such that, for each $k \geqslant 1$, the closure of Ω_{k-1}, $\bar{\Omega}_{k-1}$ (which is compact), is contained in Ω_k. For each $k = 0, 1,...$, we select then a *finite* family of open sets U^i, $U_k^1,..., U_k^{r_k}$, covering $\bar{\Omega}_k$. Let us then set

$$V^{(\alpha,k)} = U_k^\alpha \cap \complement \bar{\Omega}_{k-1} \quad \text{for} \quad k \geqslant 1, \quad \alpha = 1,..., r_k,$$

$$V^{(\alpha,0)} = U_0^\alpha \cap \Omega_1 \quad \text{for} \quad \alpha = 1,..., r_0.$$

For fixed k, the $V^{(\alpha,k)}$ cover $\bar{\Omega}_k \cap \complement \bar{\Omega}_{k-1}$, and the $V^{(\alpha,0)}$ cover $\bar{\Omega}_0$; this implies immediately that, as both k and α vary, the $V^{(\alpha,k)}$ cover Ω. Since every point of Ω is contained in some Ω_{k-1} and since $\Omega_{k-1} \cap V^{(\alpha,l)} = \varnothing$ as soon as $l \geqslant k$, we see that the covering $\{V^{(\alpha,k)}\}$ is locally finite; that it is finer than the covering $\{U^i\}$ is obvious, for, whatever be k and α, $V^{(\alpha,k)} \subset U_k^\alpha$.

Remark 16.1. *The open covering* $\{V^{(\alpha,k)}\}$ constructed in the proof of Theorem 17.1 is countable. It is obvious that any locally finite covering of an open set $\Omega \subset \mathbf{R}^n$ must be countable. Note that, if $\{V^j\}$ is a locally finite open covering of Ω and if K is any compact subset of Ω, K intersects only a finite number of open sets V^j.

The next theorem says that, given a locally finite open covering $\{V^j\}$ of Ω, we may shrink slightly each set V^j and still have a covering of Ω:

THEOREM 16.2. *Let $\{V^j\}$ be a locally finite open covering of Ω. To each j there is an open subset W^j of Ω such that the closure of W^j is contained in V^j and such that, when j varies, the W^j form an open covering* (necessarily locally finite) *of Ω.*

Proof. We represent Ω as the union of a sequence of relatively compact open sets Ω_k such that $\bar{\Omega}_{k-1} \subset \Omega_k$ ($k \geqslant 1$). For each k, let us consider the (finite) family of sets V^j, $j \in J$, which do intersect Ω_k; let J_k be the set of indices j such that $V^j \cap \Omega_k \neq \varnothing$. Of course, we have $\bar{\Omega}_k \subset \bigcup_{j \in J_k} V^j$. Consider then the function

$$(16.1) \quad x \rightsquigarrow \sup_{j \in J_k} d(x, \complement V^j) \quad (d(x, A): \text{distance from the point } x \text{ to the set } A).$$

This is a continuous function in $\mathbf{R}^n$, in view of the fact that the supremum of a finite number of nonnegative continuous functions is continuous, and of the following result, whose proof we leave to the reader:

LEMMA 16.1. *Let A be any subset of $\mathbf{R}^n$. The function in $\mathbf{R}^n$,*

$$(16.2) \quad x \rightsquigarrow d(x, A) = \inf_{y \in A} |x - y|,$$

is continuous.

The function (16.1) is >0 at every point of the compact set $\bar{\Omega}_k$; we conclude that there is a constant $c_k > 0$ such that (16.1) is $\geqslant c_k$ everywhere in $\bar{\Omega}_k$. For each $j \in J_k$ let

$$W_k^j = \{x \in V^j; \, d(x, \complement \, V^j) > c_k/2\}.$$

It might happen that W_k^j is empty, but in any case we set $W^j = \bigcup_{k=0}^{\infty} W_k^j$; it is easily seen that the open sets W^j satisfy the requirements in Theorem 16.2.

We begin now the construction that will lead us to partitions of unity in $\mathscr{C}^{\infty}(\Omega)$. We consider a locally finite open covering $\{V^j\}$ $(j \in J)$ of Ω. For each $k = 0, 1, 2, \ldots$, there is a finite subset J_k of indices $j \in J$ such that $V^j \cap \Omega_k \neq \varnothing$ and such that V^j does not intersect Ω_k if $j \notin J_k$ (the open sets Ω_k are the ones introduced above). Let us set $V^{k,j} = \Omega_k \cap V^j$ for $j \in J_k$. The sets $V^{k,j}$ form a locally finite open covering of Ω. We shall first apply to them Theorem 16.2 and form an open covering $\{W^{k,j}\}$ of Ω such that, for every pair (k, j), $\overline{W^{k,j}}$ is a compact subset of the (relatively compact) open set $V^{k,j}$. The idea of the proof is then the following one: we begin by constructing a *continuous* function $g^{k,j}$, identically equal to one in $W^{k,j}$ and whose support is compact and contained in $V^{k,j}$; this is easy to do and uses Lemma 16.1. Next, we take advantage of Corollary 1 of Lemma 15.2 and approximate $g^{k,j}$ by a $\mathscr{C}^{\infty}$ function $\gamma^{k,j}$ whose support lies in a sufficiently small neighborhood of supp $g^{k,j}$, in other words of $W^{k,j}$, so that the support of $\gamma^{k,j}$ is also a compact subset of $V^{k,j}$. We require that $\gamma^{k,j}$ be so close to $g^{k,j}$ so as to have $\gamma^{k,j}(x) > \frac{1}{2}$ for all $x \in W^{k,j}$. As supp $\gamma^{k,j} \subset V^{k,j}$, every point of Ω has a neighborhood which intersects only a finite number of sets supp $\gamma^{k,j}$. We may therefore form the sum $\gamma(x) = \sum_{k,j} \gamma^{k,j}(x)$ (summation performed over $k = 0, 1, \ldots$ and over $j \in J_k$), and this sum defines a $\mathscr{C}^{\infty}$ function, $x \rightsquigarrow \gamma(x)$, in Ω. We shall construct the $g^{k,j}$ so that they be everywhere nonnegative; then, inspection of the proof of Lemma 15.2 shows right away that we may also take the $\gamma^{k,j}$ so as to be nonnegative everywhere. As $\gamma^{k,j} > \frac{1}{2}$ in $W^{k,j}$ and as the $W^{k,j}$ form a covering of Ω, we have $\gamma > \frac{1}{2}$ everywhere. This implies immediately that $1/\gamma$ is also a $\mathscr{C}^{\infty}$ function in Ω, and that we may set

$$\beta^j(x) = \frac{1}{\gamma(x)} \sum_{k=0}^{\infty}{}' \gamma^{k,j}(x),$$

where the symbol $\sum'$ means that the summation is performed over those integers k such that $j \in J_k$ (simply because otherwise $\gamma^{k,j}$ is not defined). The functions β^j have the following properties:

(16.3)   for each index j, β^j is a $\mathscr{C}^{\infty}$ function in Ω;

(16.4)   for each index j, the support of β^j is contained in V^j;

(16.5)   for each index j, the function β^j is everywhere $\geqslant 0$ in Ω;

(16.6)   for all $x \in \Omega$,

$$\sum_j \beta^j(x) = 1.$$

(This follows immediately from the fact that the covering $\{V^j\}$ is locally finite and from (16.4), every point x_0 of Ω has a neighborhood in which all functions β^j vanish except a finite number of them.)

Definition 16.1.   Let $\{V^j\}$ $(j \in J)$ be a locally finite open covering of an open set $\Omega \subset \mathbf{R}^n$. A set of functions $\{\beta^j\}$, indexed by the same set of indices J as the covering $\{V^j\}$ and having Properties (16.3)–(16.6), is called a partition of unity in $\mathscr{C}^\infty(\Omega)$ subordinated to the covering $\{V^j\}$.

Observe that, if all the sets V^j were relatively compact, the function β^j would all have a compact support, in view of (16.4); i.e., we would have a partition of unity in $\mathscr{C}_c^\infty(\Omega)$. It should be kept in mind however that, if the sets V^j are not relatively compact, it will not be possible, in general, to have partitions of unity in $\mathscr{C}_c^\infty(\Omega)$ subordinated to the covering $\{V^j\}$. A trivial counterexample is obtained by taking the index set J with a single element.

The reasonings which precede Definition 16.1 constitute essentially a proof of the existence of partitions of unity in $\mathscr{C}^\infty(\Omega)$ subordinated to an arbitrary locally finite open covering $\{V^j\}$ of Ω. It will suffice to indicate how to construct the continuous functions $g^{k,j}$ by which the whole construction begins. This is easily done by considering the following function $f(t)$ of the real variable t:

$$f(t) = \begin{cases} 1 - 2t & \text{for } t \leqslant 1/2; \\ 0 & \text{for } t \geqslant 1/2. \end{cases}$$

Let then $d_{k,j}$ be the distance from $W^{k,j}$ to the complement of $V^{k,j}$. It suffices to take

$$g^{k,j}(x) = f(d(x, W^{k,j})/d_{k,j}).$$

As it is the compose of two continuous functions, by Lemma 16.1, $g^{k,j}$ is continuous; it is equal to one in $W^{k,j}$ and to zero outside the neighborhood of order $d_{k,j}/2$ of $W^{k,j}$; this neighborhood is obviously a relatively compact subset of $V^{k,j}$. The remaining details of the proof can be worked out without difficulty by the student, if he wishes to do so.

We may now state:

THEOREM 16.3. *Given an arbitrary locally finite open covering of an open subset Ω of $\mathbf{R}^n$, there is a partition of unity in $\mathscr{C}^\infty(\Omega)$ subordinated to this covering.*

COROLLARY. *Let $\{V^j\}$ be a locally finite open covering of Ω, and f a function $\in \mathscr{C}^k(\Omega)$ $(0 \leqslant k \leqslant +\infty)$. We may write*

$$f = \sum_j f^j,$$

where $f^j \subset \mathscr{C}^k(\Omega)$ and supp $f^j \subset V^j$ *for every j.*

THEOREM 16.4. *Let F be a closed subset of $\mathbf{R}^n$, and U an arbitrary open neighborhood of F. There is a function $g \in \mathscr{C}^\infty(\mathbf{R}^n)$ which is equal to one in some neighborhood of F and vanishes identically in the complement of U.*

Proof. Let V be an open neighborhood of F whose closure, $\bar{V}$, is contained in U. Let us set $W = \mathbf{R}^n - \bar{V}$. The pair (U, W) is an open covering of $\mathbf{R}^n$, obviously locally finite! In view of Theorem 16.3, there is a partition of unity in $\mathscr{C}^\infty(\mathbf{R}^n)$ subordinated to this covering. This implies, in particular, that the partition of unity in question consists of two elements, g, h. Since the support of h is contained in W, h vanishes identically in V. Since $g + h \equiv 1$, we must have $g \equiv 1$ in V. Since supp $g \subset U$, g fulfills all our requirements.

Exercises

16.1. Prove that there is an integer $\nu(n)$, depending only on the dimension n of the space $\mathbf{R}^n$, such that the following is true. There is a set of functions $\{g^i(x, t)\}(i \in I)$ of $x \in \mathbf{R}^n$ and $t > 0$ such that the following properties hold:

(a)   For each fixed t, $\{g^i(x, t)\}(i \in I)$ is a partition of unity in $\mathscr{C}^\infty_c(\mathbf{R}^n)$;

(b)   for each t, the diameter of the support of every $g^i(x, t)$, regarded as a function of $x \in \mathbf{R}^n$, is $< 1/t$;

(c)   for each t and every $i \in I$, the number of indices $i' \in I$ such that supp $g^i(\cdot, t) \cap$ supp $g^{i'}(\cdot, t) \neq \varnothing$ is at most equal to $\nu(n)$;

(d)   for each t, there is a function $g \in \mathscr{C}^\infty(\mathbf{R}^n)$ such that to every $i \in I$ corresponds one point $x^i \in \mathbf{R}^n$ such that $g^i(x, t) = g(x - x^i)$ for all x.

16.2. Let $\{U_j\}$ $(j \in J)$ be an arbitrary locally finite open covering of $\mathbf{R}^n$. Prove that there is a partition of unity $\{g_j\}(j \in J)$ subordinated to the covering $\{U_j\}$ such that, for all j, g_j belongs to the dth Gevrey class in $\mathbf{R}^n$, $G_d(\mathbf{R}^n)$ (see Exercise 15.3), provided that $d > 1$ (use Exercise 15.4).

17

The Open Mapping Theorem

This is the last section of Part I; we return, in it, to the general theory. We shall state and prove the celebrated "open mapping theorem" or *Banach theorem*. Consider two TVS E,F over the field of complex numbers, and f a linear map of E into F. We have the usual triangular diagram

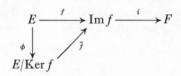

in which ϕ is the canonical mapping, i the injection of the image of f, $\operatorname{Im} f$, into F, and $\tilde{f}$ the uniquely determined linear map which makes the diagram commutative. We know that f is continuous if and only if $\tilde{f}$ is continuous. Observe that, in any case, whether f is continuous or not, $\tilde{f}$ is one-to-one, and, if viewed as a mapping of $E/\operatorname{Ker} f$ into $\operatorname{Im} f$, it is onto. By definition, f is a *homomorphism* when $\tilde{f}$ is an isomorphism of $E/\operatorname{Ker} f$ onto $\operatorname{Im} f$. One also says that f is an open mapping (assuming implicitly that we are only dealing with continuous mappings). Suppose then that F is Hausdorff. If f is continuous, $\operatorname{Ker} f$ is closed and $E/\operatorname{Ker} f$ is also Hausdorff. Suppose for a moment that $\operatorname{Im} f$ is finite dimensional; then $\operatorname{Im} f$ and $E/\operatorname{Ker} f$ have the same dimension and $\tilde{f}$ is always an isomorphism, i.e., f is a homomorphism: any continuous linear map of a TVS E into a Hausdorff TVS F, whose image is finite dimensional, is a homomorphism. We may then ask the following question: is there a class of TVS such that, if E and $\operatorname{Im} f$ both belong to that class, then f is a homomorphism as soon as it is continuous? We have just seen that this is so, whatever be E, whenever $\operatorname{Im} f$ is a finite dimensional linear subspace of a Hausdorff TVS. Of course, we are really interested in loosening the restriction on $\operatorname{Im} f$: for the condition that $\operatorname{Im} f$ be finite dimensional is a rather awkward one. We cannot hope that E will still be allowed to be any kind of TVS.

We shall have to give up some of the generality on E but, as will be shown in this chapter, it is possible to obtain a great deal of generality on Im f, while retaining a great deal of it about E, so as to have the desired property. As a matter of fact, it is possible to have the *same* conditions on E and on Im f, and these conditions are of a nature with which we are already familiar: we shall only have to ask that both E and Im f be *metrizable* and *complete* (no local convexity is involved in this problem).

Let $\mathscr{A}$, $\mathscr{B}$ be two classes of topological vector spaces. We might say that the open mapping theorem is valid for the pair $(\mathscr{A}, \mathscr{B})$ if, given any TVS $E \in \mathscr{A}$, any TVS $F \in \mathscr{B}$, and any continuous linear map of E onto F, this linear map is a homomorphism (for reasons of simplicity, we have assumed that the image of the mapping is identical with the whole space of values; we shall go on doing this, in other words we shall only consider mappings which are *onto*). The validity of the open mapping theorem has great advantages; to try to prove it is not a matter of sheer curiosity. We shall illustrate it by an example.

Let $(\mathscr{A}, \mathscr{B})$ be a pair of classes of TVS for which the open mapping theorem is valid; let E be a TVS belonging to the class $\mathscr{A}$, and F a TVS belonging to $\mathscr{B}$. Let us be given a linear map g of F into E, and suppose that we are interested in proving that g is continuous. Suppose furthermore that we have the following information about $g : g$ is *onto*; g is *one-to-one*; the inverse of g, which exists by the two preceding properties, is *continuous*. Let $f : E \to F$ be this inverse. In virtue of the open mapping theorem, we are able to conclude that f is open, i.e., that g is continuous, which is what we were seeking.

Consider the *graph* of the mapping g just introduced; the graph of g, Gr g, is the subset of $F \times E$ consisting of the pairs (x, y), $x \in F$, $y \in E$, such that $y = g(x)$. Let us go back to the basic information we have about g: that g is one-to-one, onto, and has a continuous inverse, $f : E \to F$. But let us not suppose that g is continuous. For simplicity, we shall restrict ourselves to metrizable spaces. Let (x, y) be an element of the closure of Gr g in $F \times E$; the product space $F \times E$ is also metrizable, and we can select a sequence of elements (x_n , y_n), $n = 1, 2, ...,$ of Gr g, converging to (x, y). This means that $x_n \to x$ in F, and $y_n \to y$ in E. But for each n, $y_n = g(x_n)$, i.e., $f(y_n) = x_n$, and, in view of the continuity of f, we see that $f(y_n)$ converges to $f(y)$ in F. As it also converges to x we conclude that $x = f(y)$, i.e., $y = g(x)$, i.e., $(x, y) \in$ Gr g. In other words, the graph of g is closed. If we wish then to conclude that g is continuous, we might raise the question: in what situation can we derive that a linear mapping is continuous from the fact that its graph is closed? In relation to this question, we leave to the student, as an exercise, the proof of the following statement:

PROPOSITION 17.1. *Let X, Y be two topological spaces* (not necessarily carrying an algebraic structure). *Suppose that Y is Hausdorff. Let f be a continuous mapping of X into Y. Then the graph of f in the product topological space X × Y is closed.*

Here again, we may consider the pairs of classes of TVS $(\mathscr{A}, \mathscr{B})$ such that, given any TVS $E \in \mathscr{A}$, any TVS $F \in \mathscr{B}$, and any linear map $f : E \to F$ whose graph is closed, we have the right to conclude that f is continuous. If this is so, we shall say that the *closed graph theorem* is valid for the pair $(\mathscr{A}, \mathscr{B})$. It should be expected that the validity of the closed graph theorem has advantages, as it must be easier, in many a situation, to prove that the graph of a mapping is closed than to prove directly that the mapping is continuous. This is indeed so; the reason for it lies in the following much used result:

PROPOSITION 17.2. *Let E, F, and G be three Hausdorff TVS, and j a one-to-one continuous linear map of F into G.*

Let $f : E \to F$ be a linear map such that the composed $j \circ f : E \to G$ is continuous. Then the graph of f is closed.

The mapping j has the following intuitive meaning: through j, F can be regarded as a vector subspace of G; furthermore, as j is continuous, the topology of F is finer than the topology induced on F by G.

Proof of Proposition 17.2. Let (x_0, y_0) be an element of the closure of $\operatorname{Gr} f$ in $E \times F$. There is a filter $\mathscr{F}_0$ in $\operatorname{Gr} f$, which generates a filter $\mathscr{F}$ in $E \times F$ converging to (x_0, y_0): it suffices to take as $\mathscr{F}_0$ the filter defined by the intersection of the neighborhoods of (x_0, y_0) with $\operatorname{Gr} f$. Let $\mathscr{F}_E$ and $\mathscr{F}_F$, respectively, be the images of $\mathscr{F}$ under the two coordinate projections: $(x, y) \rightsquigarrow x$ and $(x, y) \rightsquigarrow y$. To say that $\mathscr{F}$ converges to (x_0, y_0) is equivalent with saying that $\mathscr{F}_E$ converges to x_0 and $\mathscr{F}_F$ converges to y_0 (this is the definition of convergence in a product space). As the compose mapping $j \circ f$ is continuous, we derive that $j(f \mathscr{F}_E)$ is a filter in G, converging to $j(f(x_0))$; as j is continuous, we derive that $j \mathscr{F}_F$ is also a filter in G which converges; its limit is $j(y_0)$. But if we go back to the definition of the filter $\mathscr{F}_0$, precisely to the fact that it was a filter on $\operatorname{Gr} f$, we see immediately that $j(f \mathscr{F}_E)$ and $j \mathscr{F}_F$ must be one and the same filter on G. Indeed, a generic element of $\mathscr{F}_0$ is a subset of $E \times F$ consisting of elements of the form $(x, f(x))$; if we apply $j \circ f$ to the first projection of this element, we obtain $j(f(x))$, which is the same thing as if we apply j to its second projection.

Since G is Hausdorff, we conclude that $j(y_0) = j(f(x_0))$ As j is one-to-one, we may conclude that $y_0 = f(x_0)$, i.e., $(x_0, y_0) \in \operatorname{Gr} f$. Q.E.D.

The next corollary generalizes what we have said, above, about linear mappings of a metrizable space into another one:

COROLLARY.   *Let E, F be two Hausdorff TVS, and g a linear mapping of F onto E which is one-to-one and has a continuous inverse. Then the graph of g is closed.*

The student should try to prove this result by deriving it directly from Proposition 17.2.

What is the relation between the open mapping theorem and the closed graph theorem? Let $f : E \rightarrow F$ be a linear map, and let us consider its graph $\mathrm{Gr}\, f$; note that it is a linear subspace of the product vector space $E \times F$. If we put on $E \times F$ the product topology, it induces on $\mathrm{Gr}\, f$ a topology which turns $\mathrm{Gr}\, f$ into a topological vector space. Let us denote by p (resp. q) the first coordinate (resp. the second coordinate) projection restricted to $\mathrm{Gr}\, f$:

$$p(x, f(x)) = x, \qquad q(x, f(x)) = f(x).$$

We see that p is one-to-one and onto, and that, if we denote by p^{-1} its inverse, we have

$$f = q \circ p^{-1}.$$

If we have to prove that f is continuous, we may try to prove that p^{-1} is continuous. The definition of the product topology implies that both p and q are continuous. In dealing with p, we are dealing with a one-to-one mapping onto which is continuous; if we can prove that it is also open, it will follow that p^{-1} and therefore also f are continuous. We may then state:

PROPOSITION 17.3.   *Suppose that E and F are two TVS having the following property:*

(17.1)   *If G is any closed linear subspace of the product $E \times F$ and u any continuous linear map of G onto E, then u is an open mapping.*

Under this condition, if f is a linear map of E into F with a closed graph, f is continuous.

It is clear from the considerations which precede, that we could have restricted Condition (17.1) to the mappings of G into E which are onto and one-to-one. As we shall prove the open mapping theorem for metrizable and complete TVS, and that any closed subspace of the product of two metrizable and complete TVS is metrizable and complete, we shall also obtain the closed graph theorem for metrizable and complete TVS.

Finally, we state and prove the open mapping theorem:

THEOREM 17.1.   *Let E, F be two metrizable and complete* TVS. *Every continuous linear map of E onto F is a homomorphism.*

Proof.   The proof consists of two rather distinct parts. In the first one, we make use only of the fact that the mapping under consideration, $u : E \to F$, is *onto* and that F has Baire's property (p. 74). In the second part, we take advantage of the fact that both E and F can be turned into metric spaces, and that E is complete.

LEMMA 17.1.   *Let u be a linear map of a* TVS *E onto a Baire space F. Given <u>an arbitrary</u> neighborhood of zero U in E, the closure of its image $u(U)$, $\overline{u(U)}$, is a neighborhood of zero in F.*

Proof of Lemma 17.1.   The fact that F is a Baire space has the consequence that F cannot be the union of a countable family of closed sets, none of which has interior points. Let V be a balanced neighborhood of zero in E, such that $V + V \subset U$. Since V is absorbing, we have

$$E = \bigcup_{n=1}^{\infty} nV.$$

Since u is onto we have

$$F = \bigcup_{n=1}^{\infty} n\, u(V) = \bigcup_{n=1}^{\infty} n\, \overline{u(V)},$$

and at least one of the closed sets $n\,\overline{u(V)}$ must have a nonempty interior; since $x \rightsquigarrow (1/n)x$ is an isomorphism of F onto itself, $\overline{u(V)}$ must have an interior point, say x_0; let $W + x_0$ be some neighborhood of x_0 contained in $\overline{u(V)}$. Here W is a neighborhood of 0 in F; the affine mapping $x \rightsquigarrow x - x_0$ is a homomorphism of F onto itself, and maps $\overline{u(V)}$ onto $\overline{u(V)} - x_0$; as it maps $W + x_0$ onto W, we have $W \subset \overline{u(V)} - x_0$. Obviously, we have also $\overline{u(V)} - x_0 \subset \overline{u(V)} - \overline{u(V)} = \overline{u(V - V)} \subset \overline{u(U)}$ (we have $V = -V$), whence $W \subset \overline{u(U)}$, which proves that the closure of $u(U)$ is a neighborhood of zero in F.   Q.E.D.

Now we know that, because F is a Baire space and u is onto, the image of any neighborhood of zero in E is everywhere dense in a neighborhood of zero in F. The next lemma tells us that, under the assumptions of Theorem 17.1, if we enlarge a little the neighborhood of zero in E, then not only will its image be dense in a neighborhood of zero in F, but it will be itself a neighborhood of zero in F. We are going to make use of a metric in E and a metric in F, both of which we shall denote by the same letter, d, as no confusion is to be feared. If z is a point either of E or of F, we denote by $B_r(z)$ the ball with radius $r > 0$ centered at z, i.e., the set

$$\{z'; d(z, z') \leqslant r\}.$$

In the proof of Theorem 17.1, we shall need metrics on E and F which are *translation-invariant*. Every metrizable TVS carries such a metric which defines its topology (when we talk about metrics, it should always be understood that they define the underlying topology!); we have not proved this fact in general. We have only proved it for locally convex metrizable spaces; the student reluctant to take our word for the general result may confine himself to the locally convex case.

LEMMA 17.2. *Let u be a continuous linear map of a metrizable and complete TVS E into a metrizable* (not necessarily complete) *TVS F. Suppose that u has the following property:*

(17.2)   *To every number $r > 0$ there is a number $\rho > 0$ such that, for all $x \in E$,*

$$B_\rho(u(x)) \subset \overline{u(B_r(x))}.$$

Then, if r and ρ are related as in Property (17.2) *and if $a > r$, we have, for all $x \in E$,*

$$B_\rho(u(x)) \subset u(B_a(x)).$$

Proof of Lemma 17.2. We may represent a as a sum of an infinite series of positive numbers:

$$a = \sum_{n=0}^{\infty} r_n, \qquad r_0 = r, \quad r_n > 0 \quad \text{for all} \quad n.$$

Let y be an arbitrary point of $B_\rho(u(x))$. We must show that there is a point $x' \in B_a(x)$ such that $u(x') = y$. We shall define a sequence of points x_n such that $u(x_n)$ converges to y in F, and which will be a Cauchy sequence in E. Using then the completeness of E, we shall conclude that the x_n converge to an element x' of E, which necessarily satisfies $u(x') = y$. The way we define the sequence x_n will imply that $d(x, x') \leqslant a$.

We shall take advantage of Property (17.2). To each n we can find a number $\rho_n > 0$ such that $\rho_n > \rho_{n+1} \to 0$, and such that, for every $x \in E$,

$$B_{\rho_n}(u(x)) \subset \overline{u(B_{r_n}(x))}.$$

As a first step, we can select a point x_1 of $B_{r_0}(x)$ such that $d(u(x_1), y) < \rho_1$, which means that $y \in B_{\rho_1}(u(x_1))$. Therefore, we may find a point x_2 of $B_{r_1}(x_1)$ such that $d(u(x_2), y) < \rho_2$, that is to say $y \in B_{\rho_2}(u(x_2))$, and so on; for each n, we may find a point x_{n+1} of $B_{r_n}(x_n)$ such that $d(u(x_{n+1}), y) < \rho_n$. The sequence of points x_n has all the properties required: for each n, $d(x_n, x_{n+1}) < r_n$, hence the x_n form a Cauchy sequence and their limit x' satisfies $d(x, x') \leqslant \sum_{n=0}^{\infty} r_n = a$; moreover, $d(u(x_n), y) \to 0$. Q.E.D.

End of the Proof of Theorem 17.1. By Lemma 17.1 we know that to every $r > 0$ there is $\rho > 0$ such that

$$B_\rho(0) \subset \overline{u(B_r(0))}.$$

We use now the property that the metrics employed are translation-invariant: for all real numbers $\delta > 0$ and all points z,

$$B_\delta(z) = B_\delta(0) + z.$$

We conclude that Property (17.2) is valid. By Lemma 17.2, we conclude that, for $a > r$,

$$B_\rho(0) \subset u(B_a(0)).$$

This proves obviously that u transforms neighborhoods of zero into neighborhoods of zero.

COROLLARY 1. *A one-to-one continuous linear map of a metrizable and complete TVS E onto another metrizable and complete TVS F is an isomorphism* (i.e., *is bicontinuous*).

COROLLARY 2. *Let $\mathcal{T}_1$, $\mathcal{T}_2$ be two metrizable topologies on the same vector space E, both turning it into a complete TVS. Suppose that one is weaker than the other. Then they are equivalent.*

COROLLARY 3. *Let p, q be two norms on a vector space E. Suppose that both normed spaces (E, p) and (E, q) are Banach spaces, and that, for some constant $C > 0$ and all $x \in E$.*

$$p(x) \leqslant C \, q(x).$$

Then the norms p and q are equivalent, i.e., there is a constant $C' > 0$ such that, for all $x \in E$,

$$q(x) \leqslant C' \, p(x).$$

In view of Proposition 17.3 and the remark following it, we obtain the closed graph theorem for metrizable and complete TVS:

COROLLARY 4.   *Let E, F be two metrizable and complete TVS, and f a linear map of E into F. If the graph of f is closed, f is continuous.*

Exercises

17.1.   Let Ω be an open subset of $\mathbf{C}^n$, and $H(\Omega)$ the vector space of holomorphic functions in Ω. Let us identify canonically $\mathbf{C}^n$ to $\mathbf{R}^{2n}$, and regard Ω as an open subset of $\mathbf{R}^{2n}$ and $H(\Omega)$ as a vector subspace of $\mathscr{C}^k(\Omega)$ ($0 \leqslant k \leqslant +\infty$). Derive from the open mapping theorem that the topologies induced by $\mathscr{C}^k(\Omega)$ ($1 \leqslant k \leqslant +\infty$) on $H(\Omega)$ are all identical.

17.2.   Let E, F be two metrizable and complete TVS, and f a linear mapping of E into F. Prove that f is continuous if and only if, for every sequence $\{x_n\}$ converging to zero in E and such that the sequence $f(x_n)$ converges in F, the limit of the $f(x_n)$ is zero.

17.3.   Give an example of a continuous linear mapping of a Fréchet space E into another, F, which has a dense image but is not a homomorphism.

17.4.   Give an examples of two topologies on a vector space E, both metrizable and complete, one of which turns E into a TVS and is less fine than the other, but which are not identical.

17.5.   Prove the following result:

PROPOSITION 17.4.   *Any linear map f of an LF-space E into a Fréchet space F whose graph is closed is continuous.*

17.6.   Let E be a linear subspace of the space $\mathscr{C}^0([0, 1])$ of the continuous functions on the closed unit interval $[0, 1]$. Let $\| \ \|$ be a norm on E which turns E into a Banach space and defines a topology on E which is finer than the topology of pointwise convergence on $[0, 1]$. Let $\mathscr{C}^m([0, 1])$ ($0 \leqslant m \leqslant +\infty$) be the space of $\mathscr{C}^m$ functions f in the open interval $]0, 1[$ such that all the derivatives of order $< m + 1$ of f can be extended to the closure of $]0, 1[$ as continuous functions on $[0, 1]$; $\mathscr{C}^m([0, 1])$ carries the topology defined by the norms

$$f \rightsquigarrow p_k(f) = \sup_{0 < t < 1} \sum_{j=0}^{k} |f^{(j)}(t)|, \qquad k < m + 1.$$

Suppose that $\mathscr{C}^\infty([0, 1])$ is contained in E. Prove (by using Corollary 4 of Theorem 17.1, or else Exercise 17.2) that the natural injection of $\mathscr{C}^\infty([0, 1])$ into E is continuous and that, for some *finite* integer $m \geqslant 0$, it can be extended as a continuous linear injection of $\mathscr{C}^m([0, 1])$ into E. The latter injection, composed with the injection of E into $\mathscr{C}^0([0, 1])$, is equal to the natural injection of $\mathscr{C}^m([0, 1])$ into $\mathscr{C}^0([0, 1])$: in other words, $\mathscr{C}^m([0, 1]) \subset E$ and the injection is continuous (also prove these assertions).

PART II

Duality.
Spaces of Distributions

In this part, the reader will find an exposition of the main body of distribution theory and of the theory of duality between topological vector spaces. The main spaces of distributions ($\mathscr{D}'$, $\mathscr{E}'$, $\mathscr{S}'$) are defined as duals of spaces of $\mathscr{C}^\infty$ functions (of $\mathscr{C}_c^\infty$, $\mathscr{C}^\infty$, and $\mathscr{S}$, respectively). The standard operations — differentiation, multiplication by a $\mathscr{C}^\infty$ function, convolution, Fourier transformation — are systematically defined as transposes of analog operations in spaces of $\mathscr{C}^\infty$ functions. It is evident that such an approach, by duality and transposition, requires a minimum amount of knowledge about these concepts. This is provided in Chapter 18 (The Theorem of Hahn–Banach), 19 (Topologies on the Dual), and 23 (Transpose of a Continuous Linear Map). In the Chapter presenting the Hahn–Banach theorem, a few pages are devoted to showing how the theorem is used in the treatment of various problems (e.g., problems of approximation, of existence of solutions to a functional equation, and also problems of separation of convex sets). In Chapters 20, 21, and 22, examples of duals are given; Chapter 20 is entirely devoted to the duality between L^p and $L^{p'}$, the so-called Lebesgue spaces (and also between l^p and $l^{p'}$, the spaces of sequences). A proof of Hölder's inequality is given. Chapter 21 studies the dual of the space of continuous functions with compact support, which is the space of Radon measures, then the dual of the space of $\mathscr{C}^\infty$ functions with compact support, which is the space of distributions. Chapter 22 studies two cases of duality of a somewhat more abstract nature: the duality between polynomials and formal power series, and the duality between entire analytic functions in $\mathbf{C}^n$ and analytic functionals. We prove the important theorem that the Fourier–Borel transformation is an isomorphism of the space of analytic functionals onto the space of entire functions of exponential type in $\mathbf{C}^n$ (this theorem may be viewed as describing the duality between entire functions and entire functions of exponential type; this duality is closely related to the one between polynomials and power series). At the end of Chapters 20, 21, and 22, we find ourselves with a stock of spaces in duality that should provide us with a good number of examples on which to rely in the later study of duality. It should be pointed out, however, that we have at our disposal the space of all the distributions, $\mathscr{D}'$, but that we are not yet able to identify any one of its subspaces to the duals of the other spaces of functions which have

177

been introduced. In order to be able to do this, we need the notion of transpose of a linear map and the fact that the transpose is injective whenever the image of the map is dense. For then we may take advantage of the fact that the natural injection of $\mathscr{C}_c^\infty$ into L^p ($1 \leqslant p < +\infty$), $\mathscr{C}_c^k, \mathscr{C}^k$ ($0 \leqslant k \leqslant +\infty$), and $\mathscr{S}$ has a dense image. Consequently, the dual of each one of these spaces can be identified with a linear subspace of $\mathscr{D}'$, i.e., can be regarded as a space of distributions. The notion of transpose, needless to say, is important in many respects, the material treated in Chapters 23–38 bears witness to this. It is by transposition that we define the (linear partial) differential operators acting on distributions (Chapter 23), the Fourier transformation of tempered distributions (Chapter 25), and the convolution of distributions (Chapter 27). Transposition is the key to the study of the weak dual topology, as carried through in Chapter 35 (where attention is centered on the dual of a subspace and the dual of a quotient space and the related weak topologies), and to the study of reflexivity (Chapter 36: in the terms set down by Mackey and Bourbaki, with particular emphasis on reflexive Banach spaces, on one hand, and on Montel spaces, on the other). The main theorem in Chapter 37, due to S. Banach, may be regarded as the culmination of this line of thought: it shows the equivalence between the surjectivity of a continuous linear map of a Fréchet space into another Fréchet space, and the property that its transpose be one-to-one and have a weakly closed image. This theorem is complemented with a characterization of weakly closed linear subspaces in the dual of a Fréchet space, also due to Banach. In order to impress the importance of these theorems on the mind of the student, Chapter 38 (the last in Part II) shows how they can be applied to the proof of a classical theorem of E. Borel and also to the proof of one of the main results about existence of $\mathscr{C}^\infty$ solutions of linear partial differential equations (this last application is essentially due to B. Malgrange).

We have preceded these chapters by a description of the standard aspects of distribution theory: the support of a distribution is introduced in Chapter 24, where the main theorem of structure is stated and proved; the procedures of approximation of distributions by cutting and regularizing are described in Chapter 28; the Fourier transforms of distributions with compact support are characterized in Chapter 29 (this characterization forms the celebrated Paley–Wiener theorem). In Chapter 30, we show that Fourier transformation exchanges, so to speak, multiplication and convolution. We have added a section (Chapter 26) on convolution of functions, where we prove the Minkowski–Hölder–Young inequality. We have thought that it was appropriate to add also a rather lengthy section (Chapter 31) on

Sobolev's spaces: these spaces play an increasingly important role in the theory of linear (and even of nonlinear) partial differential equations, and it is mainly with the application of functional analysis to partial differential equations in mind that the material presented here has been selected.

Finally, no exposition of the theory of topological vector spaces, even admittedly succinct, could dispense with the statement and the proof of the Banach–Steinhaus theorem; we fulfill this obligation in Chapter 33. We give some of the applications of the theorem in Chapter 34. As it is a statement about equicontinuous sets of linear maps, we introduce these sets in Chapter 32 and establish their main properties. The Banach–Steinhaus theorem is extensively applied in the section on reflexivity (Chapter 36).

18

The Hahn-Banach Theorem

Let E be a topological vector space, and E' its dual, i.e., the vector space of all continuous linear functionals on E (i.e., of all continuous linear maps of E into the scalar field). Let M be a linear subspace of E; we suppose that M is equipped with the induced topology. Then M is a TVS and we may consider its dual, M'. It is clear that the restriction to M of any continuous linear functional on E defines a continuous linear functional on M. This gives a meaning to the restriction mapping

$$r_M : E' \to M', \qquad r_M : E' \ni x' \rightsquigarrow x'|M \in M'.$$

This restriction mapping (evidently linear) has no reason to be one-to-one. For, given some continuous linear functional x' on E', nonidentically zero, the subspace M may very well happen to be contained in Ker x', and thus $r_M(x')$ will be zero without x' being zero. Nor is there any a priori reason that the mapping r_M should be *onto*. Indeed, there are examples of Hausdorff TVS E on which the only continuous linear functional is the functional identically equal to zero, which means that $E' = \{0\}$. But take for M any finite dimensional subspace of E; certainly M' is not reduced to zero if $M \neq \{0\}$, for M and M' have the same dimension, and thus r_M cannot be onto. One remarkable feature of the Hausdorff TVS E whose dual is reduced to $\{0\}$ is that they are *not* locally convex; for we shall see in this chapter that, when E is locally convex, the restriction map r_M is always onto, regardless of what the subspace M is. Observe that the fact that the restriction r_M is onto means that, given any continuous linear functional on M, it can be extended as a continuous linear functional on E. We would like first to show that the fact that r_M is always onto is obvious when we are dealing with a Hilbert space E. Indeed, let y' be a continuous linear functional on M. By continuity, we may extend y' to the closure of M in E (Theorem 5.1); in other words, we might have supposed at the start that M was closed. Let then M^0 be the orthogonal of M in E; we have $E = M \oplus M^0$ ($\oplus$: Hilbert sum). Let x' be the linear functional on E which is equal to y' on M and to zero on M^0. Any element x of E can be written as $x = x_1 + x_2$ with

$x_1 \in M$, $x_2 \in M^0$, and the mapping $x \rightsquigarrow x_1$ from E onto M is continuous (it is simply the orthogonal projection onto M); thus, if x converges to zero in E, $x'(x) = y'(x_1)$ must converge to zero in the complex plane, recalling that y' is continuous. In other words, x' is a continuous linear functional in E, and obviously the restriction of x' to M is identical with y'. This proves that r_M is onto in the present context. In the general case, when E is locally convex but not a Hilbert space, most of the time we cannot represent E as a direct sum $M + N$ such that, if $x = x_1 + x_2$, $x_1 \in M$, $x_2 \in N$, the mapping $x \rightsquigarrow x_1$ is continuous. If we could, then the extension to E of continuous linear functionals (and, for that matter, of any continuous linear map) defined in M would be quite automatic. However, when E is locally convex, we need not have at our disposal a representation of E as a direct sum to be able to extend continuous linear functionals, as will be shown.

Let E be a locally convex space, M a linear subspace of E, and f a continuous linear functional defined in M. The fact that f is continuous can be expressed by saying that there is a seminorm p, defined and continuous in E, such that, for all $x \in M$,

$$|f(x)| \leqslant p(x).$$

Consider then the subset of M,

$$N = \{x \in M; f(x) = 1\}.$$

Taking any vector x_0 belonging to N, it is clear that $N - x_0$ is the kernel of f in M, which is a hyperplane of M, say M_0: $N = M_0 + x_0$ is thus the translation of a hyperplane, a linear submanifold of M which we shall call a hyperplane (to be precise, one could say *an affine hyperplane* so as to make the difference with the hyperplanes considered until now, which pass through the origin, and are simply linear subspaces of codimension one). Observe that the datum of N determines completely f in M; for we have the decomposition in direct sum

$$M = M_0 + \mathbf{C}x_0,$$

where $\mathbf{C}x_0$ is the one-dimensional linear subspace (i.e., the complex line) through x_0. In other words, every element x of M can be written in one and only one manner:

$$x = y + \lambda x_0, \qquad y \in M_0, \qquad \lambda \in \mathbf{C}.$$

As $y \in \operatorname{Ker} f$, we have

$$f(x) = \lambda f(x_0) = \lambda,$$

since $f(x_0) = 1$. Consider now the open unit semiball of p:

$$U = \{x \in E; \ p(x) < 1\}.$$

We know that U is an open convex subset of E, and it is obvious that

$$N \cap U = \varnothing.$$

Suppose we could find a *closed* (affine) hyperplane H of E with the property that

(18.1) $N \subset H, \quad H \cap U = \varnothing.$

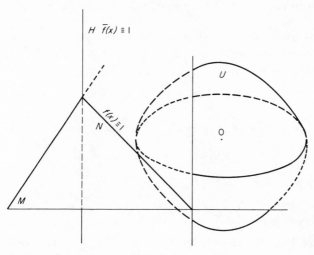

FIG. 3

Then $H - x_0$ would be the kernel of a continuous linear functional on E (Proposition 9.4) and this functional would be completely determined if we impose the condition that $\tilde{f} \equiv 1$ in H (the reasoning is identical to the one just presented, showing that the datum of N determines completely f in M). As the restriction of $\tilde{f}$ to M is equal, in N, to f, it means that the restriction of $\tilde{f}$ to M is equal to f in the whole of M. Furthermore, the fact that $H \cap U = \varnothing$ means that $\tilde{f}(x) = 1$ implies $p(x) \geqslant 1$. Let then $y \in E$ be any vector such that $\tilde{f}(y) \neq 0$. Then

$$\tilde{f}(y/\tilde{f}(y)) = 1 \quad \text{implies} \quad p(y/\tilde{f}(y)) \geqslant 1,$$

which can be rewritten into

(18.2) $|\tilde{f}(y)| \leqslant p(y).$

This remains obviously true when $\tilde{f}(y) = 0$.

Summarizing, we see that the existence of a closed hyperplane H with Properties (18.1) implies that the continuous linear form f in M can be extended as a linear form $\tilde{f}$ to the whole of E, satisfying (18.2). Condition (18.2) implies immediately that $\tilde{f}$ is continuous. In other words, we have reduced the problem of extending continuous linear functionals to the problem of *separating* by a closed hyperplane a convex open set and an affine submanifold (the image by a translation of a linear subspace) which do not intersect.

We have proved that, of the two forms of the Hahn–Banach theorem stated below, the second one, or geometric form, implies the first one, the analytic form. It will therefore suffice to prove the geometric form.

THEOREM 18.1 (Analytic form of the Hahn–Banach theorem). *Let p be a seminorm on a vector space E, M a linear subspace of E, and f a linear form in M such that*

$$|f(x)| \leqslant p(x) \quad \text{for all} \quad x \in M.$$

There exists a linear form on E, $\tilde{f}$, extending f, i.e., such that

$$\tilde{f}(x) = f(x) \quad \text{for all} \quad x \in M,$$

and such that, furthermore,

$$|\tilde{f}(x)| \leqslant p(x) \quad \text{for all} \quad x \in E.$$

THEOREM 18.2 (Geometric form of the Hahn–Banach theorem). *Let E be a topological vector space, N a linear subspace of E, and Ω a convex open subset of E such that*

$$N \cap \Omega = \varnothing.$$

There exists a closed hyperplane of E, H, such that

$$N \subset H, \quad H \cap \Omega = \varnothing.$$

It should be remarked that E does not carry any topology in Theorem 18.1, but this is somehow deluding because the datum of a semi-norm on E is equivalent to the datum of the topology defined by this semi-norm. In Theorem 18.2, E is a TVS which does not need to be Hausdorff nor locally convex.

Proof of Theorem 18.2. We assume that $\Omega \neq \varnothing$, otherwise there is nothing to prove.

The first part of the proof is quite simple and consists in a straight-forward application of Zorn's lemma: one considers the family $\mathscr{F}$ of all linear subspaces L of E such that

(18.3) $$N \subset L, \quad L \cap \Omega = \varnothing.$$

If we have a totally ordered subfamily of $\mathscr{F}$, Φ (totally ordered for the inclusion relation $L \subset L'$), it is obvious that the union of all the linear subspaces belonging to Φ is a linear subspace of E having Properties (18.3). Thus Zorn's lemma applies and we may conclude that $\mathscr{F}$ possesses maximal elements. Let H be one of them. The second part of the proof consists in showing that H is a closed hyperplane.

That H must be closed is obvious. For if H is contained in the complement of Ω, so is its closure, since Ω is open. And the closure of H is a linear subspace of E containing N. As H is maximal, it must be equal to its closure.

The fact that H is closed implies that E/H is Hausdorff (Proposition 4.5). We must show that H is a hyperplane, i.e., that dim $E/H = 1$. We shall do it in two steps.

(1)   *The scalar field is the field of real numbers.*

Let ϕ be the canonical map of E onto E/H; ϕ is a homomorphism, therefore $\phi(\Omega)$ is an open convex subset of E/H and, since $H \cap \Omega = \varnothing$, the origin of E/H is not contained in $\phi(\Omega)$. Let us set

$$A = \bigcup_{\lambda > 0} \lambda \phi(\Omega).$$

The subset A is open, convex, and it is a cone. If dim $E/H \geqslant 2$, the boundary of A must contain at least one point $x \neq 0$. It will suffice to show that, under our hypotheses, the point $-x$ cannot belong to A. But if both x and $-x$ belong to the complement of A in E/H, so also does the straight line L which these two points define; the preimage $\phi(L)$ would then be a vector subspace of E, which does not intersect Ω since $L \cap A = \varnothing$, which contains H, as $0 \in L$, but is distinct from H, as $L \neq \{0\}$. This contradicts the maximality of H.

Why cannot $-x$ belong to A? If it did, there would be a neighborhood V of $-x$ entirely contained in A. But then $-V$ is a neighborhood of x; as x is a boundary point of A, we should be able to find $y \in (-V) \cap A$. But $-y \in V \subset A$, hence, in view of the convexity of A, the whole line segment between y and $-y$ should be contained in A, in particular the origin, which is contrary to the definition of A.

(2)   *The scalar field is the field of complex numbers.*

Although the scalars are the complex numbers, we may view E as a vector space over the real numbers; it is obvious that its topology, as originally given, is compatible with its linear structure (only the continuity on $\mathbf{R} \times E$ of the scalar multiplication $(\lambda, x) \rightsquigarrow \lambda x$ has to be checked and it is obvious since the topology of $\mathbf{R}$ is the same as the topology induced on $\mathbf{R}$ by $\mathbf{C}$). Because of the result in Step 1, we know that there is a *real* hyperplane H_0 of E which contains N and does not intersect Ω. By a real hyperplane, we mean a linear subspace of E viewed as a vector space over the field of real numbers, such that $\dim_{\mathbf{R}} E/H_0 = 1$. But we must remember that complex numbers act linearly, through multiplication, on the elements of E; we have $iN = N$ ($i = (-1)^{1/2}$), hence $N \subset H_0 \cap iH_0$. But $H_0 \cap iH_0$ is a *complex* hyperplane, which does not intersect Ω. The fact that $H_0 \cap iH_0$ is a complex hyperplane is easy to check: it is obviously a complex linear subspace of E (viewed now as a complex vector space) and its real codimension is $\geqslant 1$ and $\leqslant 2$ (the intersection of two distinct hyperplanes is always a linear subspace with codimension two), hence its complex codimension is equal to one. The proof of the Hahn–Banach theorem is complete.

The Hahn–Banach theorem is frequently applied in analysis, as will be seen in the forthcoming. We shall briefly indicate three important types of problems to which it is sometimes applied: the first type are problems of approximation, the second, of existence of solutions to a functional equation, the third, of "separation" of convex sets.

(1) Problems of Approximation

Consider a locally convex space E, a closed linear subspace M of E, and a linear subspace M_0 of M. We want to show that every element of M is the limit of elements belonging to M_0. As an example of this situation, the student may think of E as a space of functions, of M as the subspace of E consisting of the solutions of some functional equation (e.g., of a partial differential equation with constant coefficients), and of M_0 as a special class of such solutions (e.g., solutions which are polynomials, or analytic functions, or $\mathscr{C}^\infty$ functions). We want to prove that M_0 is dense in M. In order to do this, we may take advantage of the following result:

COROLLARY 1.   *Let E be a locally convex space, M a closed linear subspace of E, and M_0 a linear subspace of M. Then M_0 is everywhere dense in M if and only if every continuous linear form on E vanishing identically in M_0 vanishes identically also in M.*

For suppose that $\bar{M}_0$ were $\neq M$; there would be an element x_0 of M which does not belong to $\bar{M}_0$. Consider the quotient space $E/\bar{M}_0$ and let ϕ be the canonical map of E onto $E/\bar{M}_0$; we have $\phi(x_0) \neq 0$. Let L be the one-dimensional linear subspace of $E/\bar{M}_0$ spanned by $\phi(x_0)$; every vector belonging to L is of the form $\lambda \, \phi(x_0)$. Let then $\dot{f}$ be the linear form on L, $\lambda \phi(x_0) \to \lambda$. Since $E/\bar{M}_0$ is Hausdorff, $\dot{f}$ is continuous. In virtue of the Hahn–Banach theorem, $\dot{f}$ can be extended as a continuous linear form to the whole of $E/\bar{M}_0$; we denote by $\dot{f}$ this extension. Let $f = \dot{f} \circ \phi$; f is a continuous linear form on E, which vanishes on $\bar{M}_0$, since $\bar{M}_0 = \mathrm{Ker} \, \phi$, but not on x_0, since $f(x_0) = \dot{f}(\phi(x_0)) = 1$.   Q.E.D.

In proving Corollary 1, we have proved the following result (it suffices to take $\bar{M}_0 = \{0\}$ in the proof of Corollary 1, when we deal with the quotient space $E/\bar{M}_0$ and the linear form $\dot{f}$):

COROLLARY 2.   *Let E be a Hausdorff LCS, and x_0 an element of E, $x_0 \neq 0$. There exists a continuous linear form f on E such that $f(x_0) \neq 0$.*

In particular, Corollary 2 shows that, if $E \neq \{0\}$, the dual of E cannot be reduced to $\{0\}$. On a Hausdorff locally convex space there are always nontrivial continuous linear forms.

An obvious consequence of Corollary 1 is the following one:

COROLLARY 3.   *Let M be a closed linear subspace of an LCS E. If $M \neq E$, there is a continuous linear form f, nonidentically zero but vanishing identically in M.*

(2) Problems of Existence

Let E, F be two LCS, and u a continuous linear map of E into F. Given any continuous linear functional y' on F, it is clear that the composition $y' \circ u$ is a continuous linear functional on E. This defines a mapping

$$^tu : F' \ni y' \rightsquigarrow y' \circ u \in E'$$

(E', F': duals of E and F, respectively); ^tu is called the *transpose* of u, it is obviously linear (see Chapter 19). The application of the Hahn–Banach theorem we indicate now concerns the possibility of proving the existence of a solution $y' \in F'$ to the equation

$$^tu(y') = x_0',$$

where x_0' is a given element of E'. An example of such a problem is encountered when dealing with a differential operator (in the role of u):

x_0' could then be the Dirac measure, for instance, and F', some space of distributions. In such a situation, one often applies the following consequence of Theorem 18.1:

COROLLARY 4.   *Let $E, F, u, {}^t u$, and x_0' be as said. Suppose that the linear form, defined in* Im u,

$$(18.4) \qquad\qquad u(x) \rightsquigarrow x_0'(x),$$

is continuous when we provide Im u *with the topology induced by F. Then there is a continuous linear form y' in F such that*

$$(18.5) \qquad\qquad {}^t u(y') = x_0'.$$

Proof.   If (18.4) is continuous in Im u, we can extend it as a continuous linear form to the whole of F; let y' be such an extension of (18.4). For $x \in E$, we have

$$ {}^t u(y')(x) = y'(u(x)) = x_0'(x). \qquad\qquad \text{Q.E.D.}$$

In general, the solution y' of (18.5) is not unique; if z' is any solution of the homogeneous equation

$$(18.6) \qquad\qquad {}^t u(z') = 0,$$

we see that $y' + z'$ is again a solution of (18.5). Again, from the Hahn–Banach theorem, it follows that Eq. (18.6) has nontrivial solutions if and only if Im u is nondense in F. This is a fact important enough to be stated as a corollary of Theorem 18.1:

COROLLARY 5.   *Let $E, F, u : E \to F$, and ${}^t u : F' \to E'$ be as before. The following two conditions are equivalent:*

(a)   Im u *is dense in F;*

(b) ${}^t u$ *is one-to-one.*

From Corollary 3 we derive that $\overline{\text{Im } u} \neq F$ if and only if there is a continuous linear form z' which vanishes identically on Im u but $z' \neq 0$; now, the fact that $z' = 0$ identically in Im u is obviously equivalent with ${}^t u(z') = 0$.

Observe that, when Im u is dense in F, we do not need the Hahn–Banach theorem to derive the existence of the solution y' to (18.5) from the continuity of (18.4): y' is then the unique extension of (18.4) to the whole of F; this extension is obtained by continuity. Thus, when Im u is not dense in F, we may apply the Hahn–Banach theorem to show both facts: that Eq. (18.6) has nontrivial solutions, and that the continuity of (18.4) in Im u implies the solvability of (18.5).

(3) Problems of Separation

Let E be a TVS over the field of *real* numbers; let H be a closed hyperplane of E. Then E/H is a Hausdorff one-dimensional TVS, that is to say a copy of the real line. In particular, the complement of the origin in E/H consists of two disjoint open half-lines, say D_1 and D_2. Let ϕ be the canonical homomorphism of E onto E/H; the preimages

$$\phi^{-1}(D_1) \quad \text{and} \quad \phi^{-1}(D_2)$$

are *the two open half-spaces of E determined by H*; their closures are *the two closed half-spaces determined by H*. Two subsets A and B of E are said to be *separated* (resp. *strictly separated*) by H if A is contained in one of the closed (resp. open) half-spaces determined by H, and B is contained in the other closed (resp. open) half-space determined by H.

These definitions enable us to formulate the type of problems examined in the present paragraph: can one separate, or separate strictly, two disjoint convex subsets of a TVS E? It is immediately seen that further hypotheses are necessary, on A and B, if we are to give a positive answer to this question.

PROPOSITION 18.1.   *Let E be a TVS over the real numbers, and A, B two disjoint convex subsets of E. If A is open and B is nonempty, there exists a closed hyperplane H of E separating A and B. If B is also open, the hyperplane H can be chosen so as to separate strictly A and B.*

Proof. The vector subtraction $A - B$ is an open subset of E (as it is the union of the open sets $A - y$ as y varies over B); it is convex and does not contain the origin. In view of Theorem 18.2, there is a closed hyperplane H of E which does not intersect $A - B$ (and passes through 0) or, which is equivalent, a continuous linear form f on E such that $f(A - B) > 0$, which means that $f(x) > f(y)$ for all $x \in A$ and $y \in B$. Since B is nonempty, we have $a = \inf_{x \in A} f(x) > -\infty$. The hyperplane $H_1 = \{z \in E; f(z) = a\}$ obviously separates A and B.

If now B also is open, we may find a closed hyperplane H_1 separating A and the closure $\bar{B}$ of B. It is then obvious that H_1 separates strictly A and B.   Q.E.D.

If B is not open, there will not be, in general, a closed hyperplane separating A and B strictly: even if E is finite dimensional; even if $\bar{A}$ and $\bar{B}$ do not intersect each other (see Exercise 18.2).

PROPOSITION 18.2.   *Let E be a locally convex TVS over the real numbers, and A, K two nonempty and disjoint convex subsets of E.*

   If A is closed and K is compact, there is a closed hyperplane of E which separates strictly A and K.

Proof.   Let $\mathfrak{B}$ be a basis of neighborhoods of zero in E consisting of closed convex balanced neighborhoods of zero. For each $V \in \mathfrak{B}$, let Ω_V be the complement of the closure of $A + V$. As V varies over $\mathfrak{B}$, the sets Ω_V form an open covering of K. Indeed, let x be an arbitrary point of K; the complement Ω of A is an open neighborhood of x; therefore there is $V \in \mathfrak{B}$ such that $V + V + x$ is contained in Ω. As V is balanced, this means that $V + x \subset \complement(A + V)$; the interior of $V + x$ does not intersect $A + V$, hence does not intersect its closure. We use now the compactness of K: there is a finite family of sets $V \in \mathfrak{B}$ such that the corresponding Ω_V form an open covering of K; taking W equal to the intersection of that finite family of V's, we see that $K \subset \Omega_W$. Let us then choose an open neighborhood U of zero in E, also convex and balanced, such that $U + U \subset W$. The set K is contained in the complement of the closure of $A + U + U$, therefore $K + U$ is contained in the complement of $A + U$. Since U is open, both $A + U$ and $K + U$ are open; since A, K, U are convex, $A + U$ and $K + U$ are both convex; since they are disjoint, we may apply the last part in the statement of Proposition 18.1 and conclude that there is a closed hyperplane H of E which separates strictly $A + U$ and $K + U$ and, a fortiori, A and K.

COROLLARY 1.   *In a locally convex Hausdorff TVS E over the real numbers, every closed convex set is equal to the intersection of the closed half-spaces which contain it.*

   Indeed, every point of the complement of a closed convex subset A of E is compact; by Proposition 18.2, there is a closed hyperplane H which separates strictly A and the set consisting of such a point.

COROLLARY 2.   *In a locally convex Hausdorff space E over the real numbers, the closure of a linear subspace M is the intersection of all the closed hyperplanes which contain M.*

   Let $x^0 \notin \bar{M}$; by Proposition 18.2 there is a closed hyperplane separating strictly $\bar{M}$ and x^0; let $f(x) = a$ be an equation of that hyperplane; M is contained in the set $\{x; f(x) < a\}$ and we have $f(x^0) > a$. If there were a point $y \in M$ such that $f(y) \neq 0$, then we would also have $by \in M$ with $b = a/f(y)$, which would imply $f(by) = a : M$ would not be contained in the set $\{x; f(x) < a\}$. Thus $M \subset \operatorname{Ker} f$ and $x^0 \notin \operatorname{Ker} f$.   Q.E.D.

For future purposes, it is important that we have the analogs of some of the previous results when the field of scalars is the complex field, **C**. It should be noticed, however, that the notion of separation by hyperplanes does not make any sense in a complex TVS: the complement of a closed hyperplane is always *connected* and we cannot talk about one side of the hyperplane, as we could in the real case. Of course, Corollary 2 of Proposition 18.2 still makes sense and, as a matter of fact, is still valid:

COROLLARY 3. *Let E be a locally convex Hausdorff TVS over the complex numbers, and M a linear subspace of E. The closure of M is the intersection of all the closed hyperplanes containing M.*

Proof. Let $x^0 \notin \bar{M}$; there is a closed hyperplane H of E when we regard E as a TVS over the real numbers, such that $M \subset H$ and $x^0 \notin H$. Then $H \cap iH$ is a closed hyperplane (when we regard E as a TVS over the complex numbers) which contains M, since $M = iM$, but which does not contain x^0.

The next result will be useful to us, later on.

PROPOSITION 18.3. *Let E be a vector space over the complex numbers, and $\mathcal{T}_j$ $(j = 1, 2)$ two locally convex Hausdorff topologies on E (compatible with the **C**-linear structure of E) such that the continuous linear forms on E are the same for both $\mathcal{T}_1$ and $\mathcal{T}_2$. Let A be a convex subset of E. The closures of A for $\mathcal{T}_1$ and $\mathcal{T}_2$ are identical.*

Proof. Let f be a linear map of E into the field of real numbers, **R**, i.e., an **R**-linear form on E. Let us set, for every $x \in E$,

$$g(x) = \tfrac{1}{2}(f(x) - if(ix)).$$

Let $\gamma = \alpha + i\beta$ be an arbitrary complex number. We have

$$2g(\gamma x) = f(\gamma x) - if(i\gamma x) = f(\alpha x + \beta ix) - if(\alpha ix - \beta x)$$
$$= \alpha f(x) + \beta f(ix) - i\alpha f(ix) + i\beta f(x)$$
$$= \alpha(f(x) - if(ix)) + i\beta(f(x) + i^{-1}f(ix)) = 2\gamma g(x).$$

This shows that g is a **C**-linear form. Since we have $\overline{g(x)} = \tfrac{1}{2}(f(x) + if(ix))$, remembering that f is real valued, we see that

$$f(x) = g(x) + \overline{g(x)}.$$

The definition of g and this equality show that f is continuous if and only if g is continuous.

From this fact, we derive that the continuous **R**-linear forms on E are the same for the topologies $\mathscr{T}_1$ and $\mathscr{T}_2$. This is equivalent to saying that the closed half-spaces are the same for $\mathscr{T}_1$ and $\mathscr{T}_2$. Proposition 18.3 follows then immediately from Corollary 1 of Proposition 18.2.

Exercises

18.1. Let (E, p) be a normed space, and M a linear subspace of E. Let us denote by p_M the restriction of p to M and consider the normed space (M, p_M). Let f be a continuous linear form on the normed space (M, p_M). Prove that there is a continuous linear form $\tilde{f}$ on the normed space (E, p) such that $\tilde{f}(x) = f(x)$ for all $x \in M$ and such that

$$\sup_{x \in E, p(x)=1} |\tilde{f}(x)| = \sup_{x \in M, p_M(x)=1} |f(x)|.$$

18.2. Consider the following subset of $\mathbf{R}^3$: $C = \{(x, y, z); \ x \geqslant 0, y \geqslant 0, z \geqslant 0, xy \geqslant z^2\}$. Show that the straight line $D = \{(x, y, z); x = 0, z = 1\}$ does not intersect C but that there is no plane which contains D and does not intersect C (see remark after proof of Proposition 18.1).

18.3. Let $l_{\mathbf{R}}^1$ be the Banach space of real sequences $\sigma = (\sigma_n)$, $n = 0, 1, 2,...$ such that $\sum_{n=0}^{+\infty} |\sigma_n| < +\infty$ (cf. Chapter 11, Example IV). Let us call D the one-dimensional subspace generated by the sequence $(1, 0,..., 0,...)$, i.e., the straight line $\{\sigma \in l^1; \sigma_n = 0$ for all $n > 0\}$. Let us set

$$A = \{\sigma \in l^1; \ \sigma_0 \geqslant |a_n \sigma_n - b_n|, \ n = 0, 1, 2,...\}.$$

In the definition of A, (a_n) and (b_n) $(n = 0, 1, 2,...)$ are two sequences of real numbers. Prove that these two sequences can be chosen so that $A - D$ is everywhere dense in $l_{\mathbf{R}}^1$ and that A does not intersect D. Derive from this that there is no closed hyperplane in $l_{\mathbf{R}}^1$ which separates A and D.

18.4. Let E be the vector space of polynomials in one variable with *real* coefficients (thus E is a vector space over the real). Let C be the set of polynomials $a_d X^d + a_{d-1} X^{d-1} + \cdots + a_0$ with $a_d > 0$ $(d = 0, 1, 2,...)$. Prove that C is a convex cone, that $C \cap (-C) = \{0\}$, and that $C \cup (-C) = E$. Prove also that there is no hyperplane in E, H, such that C lies on one side of H (i.e., for every hyperplane H of E, the image of C under the canonical homomorphism of E onto E/H is not contained in any "half-line"). (Note that no topology is given on E and that no condition of closedness is imposed upon H.)

18.5. A subset C of a vector space E over the field of *real* numbers is called a *cone* (cf. Exercise 18.4) if $\rho C \subset C$ for all $\rho > 0$; C is, moreover, a convex cone if C is a cone and if C is a convex set. Prove the following facts:

(a)   a subset C of E is a convex cone if and only if C is a cone and if $C + C \subset C$;

(b)   if C is a convex cone, the linear subspace of E spanned by C is equal to $C - C$;

(c)   the largest linear subspace of E contained in a convex cone C is equal to $C \cap (-C)$.

18.6. Let P be a cone in a vector space E over the field of real numbers (cf. Exercise 18.5). Let us denote by $x \geqslant y$ the relation $x - y \in P$. One says that P is *pointed* if the origin 0 of E belongs to P and that P is *salient* if the only vector subspace of E contained in P is $\{0\}$. Prove the equivalence of the following two properties:

(a) P is a pointed salient convex cone;

(b) $x \geqslant y$ is an *order relation* (i.e., is reflexive and transitive, and $x \geqslant y$ and $y \geqslant x$ implies $x = y$) *compatible with the linear structure of E* (i.e., $x \geqslant y$ implies $x + z \geqslant y + z$ and $\rho x \geqslant \rho y$ for all $z \in E$, $\rho > 0$).

Show that, if $\omega(x, y)$ is an order relation on E compatible with the linear structure of E, the set of nonnegative elements of E,

$$\Omega = \{x \in E; \ \omega(x, 0)\},$$

is a pointed salient convex cone in E. What can be said about Ω if the order defined by ω is *total* (i.e., if given any two elements x and y of E, we have either $\omega(x, y)$ or else $\omega(y, x)$)?

18.7. Let E be a vector space over the field of real numbers, E^* its algebraic dual, that is to say the vector space (over $\mathbf{R}$) of all linear mappings $E \to \mathbf{R}$. Let $x \geqslant y$ be an order relation on E, compatible with the linear structure of E; let P be the cone of nonnegative elements,

$$P = \{x \in E; \ x \geqslant 0\}.$$

A linear form $x^* \in E^*$ is called *positive* if $x^*(x) \geqslant 0$ for all $x \in P$. Prove that the set P^* of positive linear forms x^* on E is a pointed salient convex cone in E^* and that $P^* \cup (-P^*) = E^*$ if and only if dim $E = 1$.

18.8. Prove the following "complement" to the Hahn–Banach theorem:

THEOREM 18.3. *Let E be a vector space over the real numbers, M a linear subspace of E, and p a real function on E such that, for all $x, y \in E$ and all numbers $\rho > 0$,*

$$p(x + y) \leqslant p(x) + p(y), \qquad p(\rho x) = \rho \, p(x),$$

$f : M \to \mathbf{R}$ *a linear functional on M such that*

$$f(x) \leqslant p(x) \qquad \text{for all} \quad x \in M.$$

There exists a linear functional on E, $\check{f} : E \to \mathbf{R}$, such that $\check{f}(x) = f(x)$ for all $x \in M$ and such that $\check{f}(x) \leqslant p(x)$ for all $x \in E$.
(Note that p is not necessarily $\geqslant 0$.)

18.9. Let E and M be as in Theorem 18.3. Let $x \geqslant y$ be an order relation on E compatible with the linear structure of E, and $P = \{x \in E; \ x \geqslant 0\}$ the cone of nonnegative elements of E. Let f be a linear form on M which is positive (it means that $f(z) \geqslant 0$ for all $z \in M \cap P$). Suppose that, for every $x \in E$, there is $y \in M$ such that $y \geqslant x$. Prove then that the function on E,

$$x \rightsquigarrow p(x) = \inf_{y \in M, y \geqslant x} f(y),$$

is everywhere finite in E and has the following properties:

(i) $p(x + y) \leqslant p(x) + p(y)$, $p(\rho x) = \rho \, p(x)$ for all $x, y \in E$, $\rho > 0$;

(ii) $p(x) \geqslant 0$ for all $x \geqslant 0$;

(iii) $p(x) = f(x)$ for all $x \in M$.

Apply Theorem 18.3 so as to prove that there is a *positive* linear form $\check{f}$ on E, extending f.

18.10.　Prove the following theorem (due to M. Krein):

THEOREM 18.4.　*Let E be a Hausdorff TVS, $x \geqslant y$ an order relation on E, P the cone of nonnegative elements in E, M a vector subspace of E, and f a positive linear form on M. Suppose that there is at least one point $x_0 \in P \cap M$ which belongs to the interior (in E) of P. Then there is a continuous linear form $\tilde{f}$ on E which is positive and extends f.*

(Hint: show that, for every $x \in E$, there is $\rho > 0$ such that $x \leqslant \rho x_0$ and apply Exercise 18.9.)

19

Topologies on the Dual

Let E be a TVS over the field of complex numbers, and E' its dual, that is to say the vector space of all continuous linear forms (or functionals) on E (i.e., continuous linear maps of E into the complex plane, $\mathbf{C}$). If $x' \in E'$, from now on we shall denote by

$$\langle x', x \rangle$$

its value at the point x of E. Later we shall see that we also have the right to denote this value by

$$\langle x, x' \rangle.$$

Definition 19.1. *Let A be a subset of E. The subset of E',*

$$\{x' \in E'; \sup_{x \in A} |\langle x', x \rangle| \leqslant 1\},$$

is called the polar of A and denoted by A^0.

Some properties of polars:

(1) the polar A^0 of $A \subset E$ is a convex balanced subset of E';

(2) if $A \subset B$, $B^0 \subset A^0$; furthermore $(\rho A)^0 = (1/\rho)A^0$ (ρ: number >0); $(A \cup B)^0 = A^0 \cap B^0$;

(3) suppose that A is a *cone*; this means that

$$x \in A \quad \text{implies} \quad \lambda x \in A \quad \text{for all numbers} \quad \lambda > 0;$$

then, we have

$$A^0 = \{x' \in E'; \text{for all } x \in A, \langle x', x \rangle = 0\}.$$

Indeed, suppose that we have $|\langle x', x \rangle| \leqslant 1$ for all $x \in A$. Since A is a cone, we must also have $|\langle x', \lambda x \rangle| \leqslant 1$ for all $x \in A$ and all $\lambda > 0$, and this can be read as $|\langle x', x \rangle| \leqslant 1/\lambda$ for all $\lambda > 0$, which obviously means that $\langle x', x \rangle = 0$.

195

Thus, when A is a cone, in particular *when A is a vector subspace* of E, its polar A^0 is the set of all continuous linear forms on E which vanish identically in A; it is then called the *orthogonal* of A. The student should keep in mind that, in the present terminology, the orthogonal of a linear subspace of E is a subset of the dual E' of E. In fact, this subset is obviously a linear subspace of E': *if A is a cone of E, A^0 is a linear subspace of E'.*

PROPOSITION 19.1. *If B is a bounded subset of E, the polar B^0 of B is an absorbing subset of E'.*

Proof. If B is bounded, any continuous linear functional x' on E is bounded on B (corollary of Proposition 14.2), that is to say there is a constant $M(x') > 0$ such that

$$\sup_{x \in B} |\langle x', x \rangle| \leqslant M(x'),$$

and this inequality can be read as $M(x')^{-1}x' \in B^0$.   Q.E.D.

We shall define now certain topologies on the dual E' of E. We consider a family of *bounded* subsets of E, $\mathfrak{S}$, with the following two properties:

($\mathfrak{S}_1$)   *If A, $B \in \mathfrak{S}$ there is $C \in \mathfrak{S}$ such that $A \cup B \subset C$.*

($\mathfrak{S}_2$)   *If $A \in \mathfrak{S}$ and λ is a complex number, there is $B \in \mathfrak{S}$ such that $\lambda A \subset B$.*

Let us denote by $\mathfrak{S}^0$ the family of the *polars* of the sets belonging to $\mathfrak{S}$; a generic element of $\mathfrak{S}^0$ will be a subset of E' of the form A^0 where $A \in \mathfrak{S}$.

Since every subset $A \in \mathfrak{S}$ is bounded, every subset $A^0 \in \mathfrak{S}^0$ is absorbing (Proposition 19.1). If A^0, B^0 are two subsets belonging to $\mathfrak{S}^0$, the polars of A, $B \in \mathfrak{S}$, respectively, we have

$$A^0 \cap B^0 = (A \cup B)^0$$

and, as there is $C \in \mathfrak{S}$ such that $A \cup B \subset C$, we have

$$C^0 \subset (A \cup B)^0 = A^0 \cap B^0.$$

If ρ is any number > 0 and if $A \in \mathfrak{S}$, there is $B \supset \rho^{-1}A$, $B \in \mathfrak{S}$, and therefore

$$B^0 \subset \rho A^0.$$

These properties show that $\mathfrak{S}^0$ can be taken as a basis of neighborhoods of zero in a certain locally convex topology on E' (see Conditions (*) and (**) on p. 59).

Definition 19.2.   We shall call ⊝-topology on E' the locally convex topology defined by taking, as a basis of neighborhoods of zero, the family ⊝⁰ of the polars of the subsets that belong to ⊝.

PROPOSITION 19.2.   *A filter ℱ' on E' converges to an element x' of E' in the ⊝-topology on E' if ℱ' converges uniformly to x' on each subset A belonging to ⊝.*

That $\mathscr{F}'$ converges to x' uniformly on a set $A \in E$ means that, given any number $\varepsilon > 0$, there is a set M' belonging to $\mathscr{F}'$ such that

$$\sup_{x \in A} |\langle x', x \rangle - \langle y', x \rangle| \leqslant \varepsilon \qquad \text{for all } y' \in M'.$$

It is easy to see that we can take as a basis of neighborhoods of zero in the ⊝-topology the sets

$$W_\varepsilon(A) = \{x' \in E'; \sup_{x \in A} |\langle x', x \rangle| \leqslant \varepsilon\}.$$

Here ε runs over the set of numbers >0, A over ⊝. The proof of Proposition 19.2 is trivial, in view of these remarks; in virtue of it, the ⊝-topology on E' could be called the *topology of uniform convergence over the sets belonging to* ⊝. When carrying it, E' will be denoted by $E'_{\mathfrak{S}}$. We shall now introduce the main examples of ⊝-topologies.

Example I.   The weak dual topology, or weak topology on E'

This is the ⊝-topology corresponding to ⊝: family of all *finite* subsets of E; it is usually denoted by $\sigma(E', E)$, and when E' is provided with it, one writes E'_σ. Continuous linear functionals x' on E' converge *weakly* to zero if, at each point x of E, their values $\langle x', x \rangle$ converge to zero in the complex plane. In other words, the weak topology on E' is nothing else but the topology of *pointwise convergence* in E, when we look at continuous linear functionals on E as functions on E—which they are. For us, the weak topology will be most important. A basis of neighborhoods of E'_σ will be the family of sets

$$W_\varepsilon(x_1, ..., x_r) = \{x' \in E'; |\langle x', x_j \rangle| \leqslant \varepsilon, j = 1, ..., r\}.$$

Here $\{x_1, ..., x_r\}$ varies over the family of all finite subsets of E, and ε over the set of numbers >0.

Example II.   The topology of convex compact convergence

This is the ⊝-topology when we take, as family ⊝, the family of all convex compact subsets of E. We shall denote it by $\gamma(E', E)$; when provided with it, the dual of E will be denoted by E'_γ.

Example III.   The topology of compact convergence

This is the $\mathfrak{S}$-topology when $\mathfrak{S}$ is the family of all compact subsets of E; it is sometimes referred to as the topology of uniform convergence on the compact subsets of E, or compact convergence; when we put it on E', we shall write E'_c. The student should not think that it is always equivalent to the topology of convex compact convergence. We shall however see that, in one important instance, when E is a Fréchet space, these two topologies are indeed equal.

Example IV.   The strong dual topology, or strong topology on E'

With the weak topology, this one will be for us the most important $\mathfrak{S}$-topology on E'. It is defined by taking, as family $\mathfrak{S}$, the family of all bounded subsets of E. A filter in E' converges strongly to zero if it converges to zero uniformly on every bounded subset of E; this is why the strong topology on E' is sometimes called *the topology of bounded convergence*. When carrying the strong topology, E' will be called the *strong dual* of E and sometimes denoted by E'_b (b stands for bounded).

Going back to the general situation, we see that, if $\mathfrak{S}_1$, $\mathfrak{S}_2$ are two families of bounded subsets of E, satisfying $(\mathfrak{S}_1)$, $(\mathfrak{S}_2)$, and if $\mathfrak{S}_1 \supset \mathfrak{S}_2$, the $\mathfrak{S}_1$-topology is finer than the $\mathfrak{S}_2$-topology. In particular, we see that we have the following comparison relations between the four topologies on E' introduced in the above examples:

$$\sigma(E', E) \leqslant \gamma(E', E) \leqslant c(E', E) \leqslant b(E', E),$$

where c (resp. b) stands for the compact (resp. bounded) convergence topology.

PROPOSITION 19.3.   *If the union of the sets belonging to the family $\mathfrak{S}$ is dense in E, the $\mathfrak{S}$-topology on E' is Hausdorff.*

Proof.   If $x' \in E'$ is $\neq 0$, there is a point x in some set $B \in \mathfrak{S}$ such that $|\langle x', x\rangle| > 1$, therefore $x' \notin B^0$.

COROLLARY.   *The weak and strong topologies and the topologies of convex compact convergence and of compact convergence on E' are Hausdorff.*

PROPOSITION 19.4.   *Let (E, p) be a normed space, and E' the dual of E. The strong dual topology on E' may be defined by the norm*

$$p'(x') = \sup_{p(x)=1} |\langle x', x\rangle|.$$

Indeed, p' is the norm on E' whose closed unit ball is the polar of the unit ball in (E, p).

Let E and F be two TVS, and u a continuous linear map of E into F. We have already seen (Chapter 18, p. 187) that

$$^t u : F' \ni y' \rightsquigarrow y' \circ u \in E'$$

is a linear map of F' into E', called the *transpose* of u ($y' \circ u$ stands for the composition of the mappings

$$E \xrightarrow{\;u\;} F \xrightarrow{\;y'\;} \mathbf{C}.)$$

What can we say about the continuity of $^t u$ when we provide E' and F' with $\mathfrak{S}$-topologies ? In fact; suppose that we are given a family of bounded subsets of E (resp. F), $\mathfrak{S}$ (resp. $\mathfrak{H}$), having Properties $(\mathfrak{S}_1)$, $(\mathfrak{S}_2)$. Let $E'_{\mathfrak{S}}$, $F'_{\mathfrak{H}}$ be the duals of E and F, respectively, provided with the topologies defined by $\mathfrak{S}$ and $\mathfrak{H}$. That $^t u : F'_{\mathfrak{H}} \to E'_{\mathfrak{S}}$ is continuous means that, given any polar A^0 of a set $A \in \mathfrak{S}$, there is a set $B \in \mathfrak{H}$ whose polar B^0 is such that $^t u(B^0) \subset A^0$. The latter can be expressed by saying that, for all $x \in A$ and all $y' \in B^0$, we have

$$|\langle ^t u(y'), x \rangle| \leqslant 1.$$

By definition of the transpose, this is equivalent with saying that

(19.1) $$|\langle y', u(x) \rangle| \leqslant 1.$$

It is obvious that (19.1) will follow if $u(A) \subset B$. Thus we get a sufficient condition for the continuity of $^t u$:

PROPOSITION 19.5. *Let E, F be two TVS, and $\mathfrak{S}$ (resp. $\mathfrak{H}$) a family of bounded subsets of E (resp. F), having Properties $(\mathfrak{S}_1)$, $(\mathfrak{S}_2)$, (p. 196). Let u be a continuous linear map of E into F such that, to every $A \in \mathfrak{S}$, there is $B \in \mathfrak{H}$ such that $u(A) \subset B$. Then the transpose $^t u$ of u,*

$$^t u : F'_{\mathfrak{H}} \to E'_{\mathfrak{S}},$$

is continuous.

Now, observe that, if A is a finite (resp. compact and convex, resp. compact, resp. bounded) set, then $u(A)$ has the same property. This proves the following consequence of Proposition 19.5:

COROLLARY. *Let E, F be two TVS, and u a continuous linear map of E into F. Then:*

$$^t u : F' \to E'$$

is continuous when the duals E' and F' carry the weak dual topology (resp. *the topology of compact convex convergence,* resp. *the topology of compact convergence,* resp. *the strong dual topology*).

We define now the canonical map of E into the algebraic dual of its dual, which we may denote by E'^*. The image of $x \in E$ under this mapping is the linear functional in E' "*value at the point x*":

$$x' \rightsquigarrow \langle x', x \rangle.$$

Let us denote by v_x this linear form on E'. When can we say that it is continuous on $E'_\mathfrak{S}$? That v_x is continuous means that, given any $\varepsilon > 0$, there is $A \in \mathfrak{S}$ such that

$$\sup_{x' \in A^0} |v_x(x')| \leqslant \varepsilon.$$

This can be rewritten

$$\sup_{x' \in A^0} |\langle x', (1/\varepsilon)x \rangle| \leqslant 1.$$

This will certainly hold if $(1/\varepsilon)x \in A$, whence the following result:

PROPOSITION 19.6.   *Let $\mathfrak{S}$ be a family of bounded subsets of E, having Properties $(\mathfrak{S}_1)$, $(\mathfrak{S}_2)$. If $\mathfrak{S}$ is a covering of E, the canonical map of E into E'^*,*

$$x \rightsquigarrow v_x : x' \rightsquigarrow \langle x', x \rangle,$$

maps E into the dual of $E'_\mathfrak{S}$, $(E'_\mathfrak{S})'$.

From now on, we shall always suppose that $\mathfrak{S}$ is a covering of E.

In general, the canonical map of E into $(E'_\mathfrak{S})'$ will neither be onto nor one-to-one. However, the Hahn–Banach theorem has the following consequence:

PROPOSITION 19.7.   *If E is a locally convex Hausdorff TVS, the canonical map of E into the dual of $E'_\mathfrak{S}$ is one-to-one.*

Proof.   Let $x \in E$, $x \notin 0$; there is a continuous linear form x' on E such that $\langle x', x \rangle \neq 0$ (Corollary 2 of Theorem 18.1), which proves that v_x is not identically zero.   Q.E.D.

We shall see later on that, when E' carries the weak topology or the topology of compact convex convergence, the canonical mapping of E into the dual of E' is actually *onto*, which means that E can be regarded as the dual of its weak dual, E'_σ, or of E'_γ.

Exercises

In all the exercises below, E is a locally convex Hausdorff space over the field of complex numbers, **C**.

19.1. Suppose that the topology of E is the topology $\sigma(E, E^*)$, that is to say, the topology defined by the seminorms

$$p_{S^*}(x) = \sup_{x^* \in S^*} |x^*(x)|, \qquad S^*, \text{ finite subset of } E^*, \text{ algebraic dual of } E.$$

Prove the following facts:

(i)   the dual of E is identical with E^*;

(ii)   every bounded subset of E is contained in a finite dimensional vector subspace of E;

(iii)   every linear subspace of E is closed;

(iv)   every linear subspace of E has a topological supplementary.

19.2. Let E' be the dual of E, and E^* the algebraic dual of E. Prove that the completion of E', equipped with the weak topology $\sigma(E', E)$, is canonically isomorphic to E^*, equipped with the topology $\sigma(E^*, E)$ (topology defined by the seminorms

$$p_S(x^*) = \sup_{x \in S} |x^*(x)|, \qquad S, \text{ finite subset of } E).$$

19.3. Prove the equivalence of the following properties:

(a) $\dim E$ is finite;

(b) E'_σ is normable.

19.4. Prove the equivalence of the following properties:

(a) E is normable;

(b)   the strong dual E'_b of E is normable;

(c) E'_b is metrizable.

19.5. Let $\mathfrak{S}$ be a covering of E consisting of bounded subsets of E, satisfying Conditions $(\mathfrak{S}_1)$ and $(\mathfrak{S}_2)$. Prove the equivalence of following facts:

(a)   the bilinear form $(x, x') \rightsquigarrow \langle x', x \rangle$ is continuous on $E \times E'_\mathfrak{S}$;

(b) E is normable and the $\mathfrak{S}$-topology on E' is the strong dual topology.

(Hint: use Propositions 14.4 and 35.3.)

19.6. Let E be an LF-space having a sequence of definition $\{E_k\}$ $(k = 0, 1, ...)$ consisting of finite dimensional (Hausdorff) TVS. Prove the following facts:

(i)   the dual E' of E is equal to E^*, algebraic dual of E;

(ii)   on E', weak and strong topologies are equal;

(iii)   the weak (or strong) dual E' of E is a Fréchet space;

(iv)   the canonical map of E into the dual of E'_σ is onto.

19.7. Let E be an LF-space, $\{E_k\}$ $(k = 0, 1, ...)$ a sequence of definition of E. Prove the equivalence of the following properties:

(a)   the strong dual E'_b of E is a Fréchet space;

(b)   *all* the E_k are normable.

19.8. Let E be a normed space, and E' its dual. Prove that the origin in E' belongs to the closure, for the topology $\sigma(E', E)$, of the unit sphere $\{x' \in E'; \|x'\| = 1\}$ ($\| \ \|$: dual norm in E') if and only if $\dim E$ is infinite. Derive from this fact that, when $\dim E$ is infinite, the weak dual topology $\sigma(E', E)$ is strictly less fine than the strong dual topology $b(E', E)$.

20

Examples of Duals among L^p Spaces

In this chapter, we shall indicate how the duals of some of the spaces introduced in Part I can be *concretely realized*. If we wish to give a precise meaning to the latter expression, it is not difficult: let E be a topological vector space of functions, and E' its dual. Suppose that we have found a pair (F, j) consisting of a vector space F (not necessarily carrying a topology) and of a linear map j of F into E'; suppose furthermore that $j : F \to E'$ is *one-to-one* and *onto*. Then we may say that (F, j), or simply F, is a realization of the dual of E. In practice, we shall be somehow more demanding: we shall only accept realizations which are "natural," in a sense that is not easy to make precise. It will really mean that we shall select realizations which are interesting in the general context of analysis. It would be good, when E is a space of functions, to have realizations F of E' which are also spaces of functions. It will be clear, however, that this is not always possible. For instance, the dual of the space $\mathscr{C}^0(\Omega)$ of continuous functions in an open subset Ω of $\mathbf{R}^n$ cannot be naturally realized as a space of functions. Such seemingly unfavorable situations will lead us to enlarge the stock of the objects to manipulate, from functions to measures, from measures to distributions, or to analytic functionals. Let us also point out that we may look for realizations of a dual E' which are not merely algebraic, but also topological, in the sense that we may wish to find a pair (F, j), where now F is a TVS and j is an isomorphism of F onto E' in the sense of the topological vector space structures (this will require that we have put some topology on E'; the topology will usually be the strong dual one; see p. 198). When dealing with normed spaces or Hilbert spaces of functions, in which case E' will be carrying the dual Banach space or Hilbert space structure (see Proposition 11.2, corollary of Theorem 11.5 and Theorem 12.2), we might even require that F be a Banach or a Hilbert space, and that j be an isometry of F onto E'. This is indeed standard practice, and we shall see now the most important examples of such a situation.

Let us focus our attention on the case of Banach spaces. We assume that we are given two Banach spaces E, F and that we are trying to prove the existence of an isometry of E', dual of E equipped with the dual Banach space structure, onto F. This can be done if we have at our disposal a *bilinear* form on $E \times F$, which we shall denote by $\langle \ , \ \rangle$, provided with certain properties. Let us denote by $\| \ \|_E$ and $\| \ \|_F$, respectively, the norms on E and F. The first condition which should be satisfied by the bilinear form $\langle \ , \ \rangle$ is a strong form of continuity[†]:

(*) *For all* $e \in E, f \in F$, $|\langle e, f \rangle| \leqslant \| e \|_E \| f \|_F$.

If (*) holds, we are able to say the following: (1) for fixed $f \in F$,

$$e \rightsquigarrow \langle e, f \rangle$$

is a continuous linear form on E, which we shall denote by L_f; (2) we have

$$\| L_f \|_{E'} = \sup_{\| e \|_E = 1} |\langle e, f \rangle| \leqslant \| f \|_F \qquad (\| \ \|_{E'}: \text{norm in } E').$$

This means that the mapping $f \rightsquigarrow L_f$, which is obviously linear, is continuous and, furthermore, that it is a contraction, i.e., has a norm $\leqslant 1$.

The second property that the bilinear form $\langle \ , \ \rangle$ should possess is then the following one:

(**) *For each* $f \in F$ *and each* $\varepsilon > 0$, *there exists* $e \in E$ *such that*

$$\| e \|_E \leqslant 1, \qquad |\langle e, f \rangle| \geqslant \| f \|_F - \varepsilon.$$

Property (**) enables us to state that the mapping $f \rightsquigarrow L_f$ is an *isometry* on F *into* E'. Indeed, we have

$$\| f \|_F - \varepsilon \leqslant | L_f(e) | \leqslant \| L_f \|_{E'},$$

whence $\| f \|_F = \| L_f \|_{E'}$, as $\varepsilon > 0$ is arbitrary and as we already knew that $\| f \|_F \geqslant \| L_f \|_{E'}$.

The last step consists in showing that the isometry $f \rightsquigarrow L_f$ is *onto*. This is done by special techniques in each case. One considers an arbitrary continuous linear functional L on E and constructs (or ascertains the existence of) an element f of F such that $\langle e, f \rangle = L(e)$ for all $e \in E$. In some cases (e.g., the spaces l^p, $1 \leqslant p < +\infty$; see below) this is very easy and does not require any deep result; in other cases (e.g., the spaces L^p, $1 \leqslant p < +\infty$; see below), one is forced

---

† Continuity would correspond to an estimate $|\langle e, f \rangle| \leqslant \text{const} \| e \|_E \| f \|_F$.

to apply rather deep results (e.g., in the case of the L^p's, the Lebesgue–Nikodym theorem). At any event, the proofs in the present chapter all go through the three steps just described. The first step requires the proof of an estimate; this estimate will then be the one of Property (*), ensuring the continuity of the bilinear form $\langle \ , \ \rangle$. In the cases studied in this chapter, this estimate is provided by the celebrated Hölder inequalities, which we proceed now to state and prove.

We shall denote by $\mathscr{F}(X, \mathbf{C})$ the vector space of all complex-valued functions defined in a set X; we consider a seminorm $\mathfrak{p}$ on $\mathscr{F}(X, \mathbf{C})$. In order that the statements below, concerning the seminorm $\mathfrak{p}$, be true, we must allow the seminorm $\mathfrak{p}$ to take the value $+\infty$ at some elements of $\mathscr{F}(X, \mathbf{C})$. We shall furthermore assume that $\mathfrak{p}$ is *increasing*, in the following sense:

(20.1)   *If f, g are two real-valued functions in X such that*
$$f(x) \geqslant g(x) \geqslant 0 \quad \text{for all} \quad x \in X,$$
then $\mathfrak{p}(f) \geqslant \mathfrak{p}(g)$.

The Hölder inequalities, which we shall now state and prove, have a wide range of applications. However, for our limited objectives, we shall apply them with the following two choices of the seminorm $\mathfrak{p}$:

Choice 1. X is the set of nonnegative integers $j = 0, 1, 2,...$, and we identify $\mathscr{F}(X, \mathbf{C})$ with the vector space of all complex sequences $\sigma = (\sigma_j)$; the seminorm $\mathfrak{p}$ is, in this case,

$$\sigma = (\sigma_j) \rightsquigarrow |\sigma|_{l^1} = \sum_{j=0}^{\infty} |\sigma_j|;$$

of course, the set on which $\mathfrak{p}$ is finite is l^1.

Choice 2. X is an open subset of $\mathbf{R}^n$ and $\mathfrak{p}$ is given by

$$f \rightsquigarrow \int_X^* |f(x)| \, dx,$$

where $\int^*$ denotes the *upper* Lebesgue integral. We recall that the set of functions f on X such that $\mathfrak{p}(f) < +\infty$ is *not* identical with the set of Lebesgue integrable functions; the latter is smaller, it consists of the functions such that $\mathfrak{p}(f) < +\infty$ and which, furthermore, are *measurable* or, equivalently, are limits, in the sense of the seminorm $\mathfrak{p}$, of continuous functions with compact support. When f is integrable, i.e., $f \in \mathscr{L}^1$, we write $\int |f(x)| \, dx$, omitting the upper star; then $\mathrm{Re}\, f$ and $\mathrm{Im}\, f$ are also integrable and, if f is real valued, $f^+ = \sup(f, 0)$ and $f^- = \sup(-f, 0)$ are also integrable, and one defines the *Lebesgue integral* of f by

$$\int f\, dx = \int (\mathrm{Re}\, f)^+ dx - \int (\mathrm{Re}\, f)^- dx + i \int (\mathrm{Im}\, f)^+ dx - i \int (\mathrm{Im}\, f)^- dx.$$

It is well known that the two examples in Choices 1 and 2 are particular cases of a more comprehensive theory, measure theory. As a matter of fact, most of the reasonings which follow extend to the seminorm $\int_X^* |f(x)|\, dx$, where $\int_X^*$ denotes the upper integral with respect to a positive measure dx on a set X. As anybody who is familiar with this theory is also well familiar with the generalizations of the results which are going to be proved in this chapter, we shall not go into details.

We come now to Hölder's inequalities:

LEMMA 20.1. *Let X be a set, $\mathscr{F}(X, \mathbf{C})$ the space of all complex-valued functions in X, and $\mathfrak{p}$ an increasing seminorm on $\mathscr{F}(X, \mathbf{C})$ (thus $\mathfrak{p}$ has Property (20.1)). For all nonnegative functions f, g in X and all numbers $\alpha, \beta > 0$ such that $\alpha + \beta = 1$, we have*

(20.2) $\mathfrak{p}(f^\alpha g^\beta) \leqslant [\mathfrak{p}(f)]^\alpha [\mathfrak{p}(g)]^\beta.$

Proof of Lemma 20.1.   We consider in the plane, where the variable will be denoted by (s, t), the closed convex set

$$\Gamma = \{(s, t);\ s^\alpha t^\beta \geqslant 1,\ s > 0,\ t > 0\}.$$

This set Γ is equal to the intersection of the closed half-planes which contain it and whose boundaries are the straight lines tangent to the boundary of Γ, which is the curve

$$\partial\Gamma = \{(s, t);\ s^\alpha t^\beta = 1,\ s > 0,\ t > 0\}.$$

Such a half-plane can be defined by an inequality

$$as + bt \geqslant c, \qquad \text{where} \quad a, b, c \quad \text{are numbers} > 0.$$

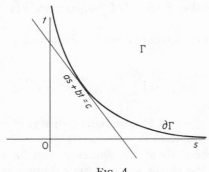

FIG. 4

Let x be an arbitrary point of X, such that

$$M(x) = [f(x)]^\alpha [g(x)]^\beta \neq 0.$$

Let us set $s_0 = f(x)/M(x)$, $t_0 = g(x)/M(x)$. We have $s_0^\alpha t_0^\beta = 1$ since $M(x)^{\alpha+\beta} = M(x)$, recalling that $\alpha + \beta = 1$. This means that, for the positive numbers a, b, c above, we have

$$c \leqslant as_0 + bt_0,$$

which reads

$$c\, M(x) \leqslant a\, f(x) + b\, g(x).$$

If we apply then the seminorm $\mathfrak{p}$ to both sides of this estimate, we obtain

$$c\, \mathfrak{p}(M) \leqslant a\, \mathfrak{p}(f) + b\, \mathfrak{p}(g).$$

Let us suppose now that $\mathfrak{p}(M) \neq 0$; the preceding inequality reads

$$c \leqslant a\, \mathfrak{p}(f)/\mathfrak{p}(M) + b\, \mathfrak{p}(g)/\mathfrak{p}(M).$$

As this must be true for all triples (a, b, c) such that the straight line $\{(s, t); as + bt = c\}$ is tangent to $\partial\Gamma$, we conclude that the point with coordinates $s = \mathfrak{p}(f)/\mathfrak{p}(M)$, $t = \mathfrak{p}(g)/\mathfrak{p}(M)$ lies in Γ. This means that $s^\alpha t^\beta \geqslant 1$, which can be rewritten

$$\mathfrak{p}(M) \leqslant [\mathfrak{p}(f)]^\alpha [\mathfrak{p}(g)]^\beta.$$

This is obviously true when $\mathfrak{p}(M) = 0$; going back to the definition of $M(x)$, we see that we have obtained (20.2).

Example I.
The Duals of the Spaces of Sequences $l^p (1 \leqslant p < +\infty)$

We recall that l^p is the space of complex sequences $\sigma = (z_j)$ $(j = 0, 1, ...)$ such that

$$\| \sigma \|_{l^p} = \left(\sum_{j=0}^\infty | z_j |^p \right)^{1/p} < +\infty.$$

With the norm $\sigma \rightsquigarrow \| \sigma \|_{l^p}$, l^p is a Banach space; l^2 is a Hilbert space (see Chapter 11, Example IV).

The duality theorem about the spaces l^p can be stated as follows: Let us denote by l_F the vector space of *finite* sequences, that is to say of sequences $\sigma = (\sigma_j)$ $(j = 0, 1,...)$ such that $\sigma_j = 0$ for large j. It is obvious that l_F is contained in each l^p; furthermore, l_F is dense in each l^p, as the student may check (remembering that p is finite!).

THEOREM 20.1.   *The bilinear form on $l_F \times l_F$,*

$$\langle \sigma, \tau \rangle = \sum_{j=0}^{\infty} \sigma_j \tau_j, \qquad \sigma = (\sigma_j), \quad \tau = (\tau_j),$$

can be extended to the product space $l^p \times l^{p'}$, where $1 \leqslant p < +\infty$ and

$$\begin{aligned} p' &= p/(p-1) \qquad \text{if} \quad p > 1, \\ p' &= +\infty \qquad\quad\; \text{if} \quad p = 1. \end{aligned}$$

This extension satisfies Hölder's inequality

(20.3) $|\langle \sigma, \tau \rangle| \leqslant \| \sigma \|_{l^p} \| \tau \|_{l^{p'}}.$

This implies that the mapping

$$\tau \rightsquigarrow (\sigma \rightsquigarrow \langle \sigma, \tau \rangle)$$

is a continuous linear map of $l^{p'}$ into the dual of l^p. In fact, it is an isometry of $l^{p'}$ onto the dual of l^p.

Concerning $l^{p'}$ when $p = 1$, we recall that l^∞ is the Banach space of bounded sequences, with the norm

$$\| \sigma \|_{l^\infty} = \sup_{0 \leqslant j \leqslant \infty} |\sigma_j|.$$

Proof.   We begin by assuming that $p > 1$; then $p' < +\infty$ and (20.3) is a trivial consequence of the general Hölder inequalities (Lemma 20.1); it shows that the mapping

(20.4) $\tau \rightsquigarrow (\sigma \rightsquigarrow \langle \sigma, \tau \rangle)$

is a continuous linear map of $l^{p'}$ into the dual of l^p, and that this map has norm $\leqslant 1$, since we derive, from (20.3),

$$\sup_{\| \sigma \|_{l^p}=1} |\langle \sigma, \tau \rangle| \leqslant \| \tau \|_{l^{p'}}.$$

Let us show that this mapping is an isometry. Let $\tau \in l^{p'}$ be arbitrary. We define a complex sequence σ by the formula

$$\begin{aligned} \sigma_j &= \bar{\tau}_j |\tau_j|^{p'-2} \qquad \text{if} \quad \tau_j \neq 0 \\ &= 0 \qquad\qquad\quad\; \text{if} \quad \tau_j = 0. \end{aligned}$$

Observe that we have

(20.5)
$$| \sigma_j |^p = | \tau_j |^{pp'-p} = | \tau_j |^{p'},$$

which implies immediately that $\sigma = (\sigma_j) \in l^p$. We have

$$\langle \sigma, \tau \rangle = \sum_{j=0}^{\infty} | \tau_j |^{p'} = \| \tau \|_{l^{p'}}^{p'}.$$

But, on the other hand,

$$\langle \sigma, \tau \rangle \leqslant \| \sigma \|_{l^p} \| \tau \|_{(l^p)'},$$

where $\| \ \|_{(l^p)'}$ is the dual norm on the dual of l^p. As we have, in view of (20.5),

$$\| \sigma \|_{l^p} = \| \tau \|_{l^{p'}}^{p'/p},$$

we conclude that

$$\| \tau \|_{l^{p'}}^{p'} \leqslant \| \tau \|_{l^{p'}}^{p'/p} \| \tau \|_{(l^p)'}.$$

But $p' - p'/p = 1$, whence our assertion.

It remains to prove that the mapping (20.4) is *onto*. In order to do this, we consider an arbitrary continuous linear functional f on l^p, and we shall show how to construct a sequence $\tau \in l^{p'}$ such that

$$f(\sigma) = \langle \sigma, \tau \rangle \quad \text{for all} \quad \sigma \in l^p.$$

Let us denote by $\sigma^{(j)}$ ($j = 0, 1,...$) the sequence (σ_k) such that $\sigma_k = 0$ if $j \neq k$, $\sigma_j = 1$. Obviously, these sequences belong to every l^p. Let us set

$$\tau_j = f(\sigma^{(j)}).$$

Let σ be a *finite* sequence; then we have $\sigma = \sum_{j=0}^{\infty} \sigma_j \sigma^{(j)}$, and

$$f(\sigma) = \sum_{j=0}^{\infty} \sigma_j \tau_j.$$

Let M be an arbitrarily large integer, and set

$$\sigma_j = \bar{\tau}_j | \tau |^{p'-2} \quad \text{if} \quad \tau_j \neq 0 \quad \text{and} \quad j \leqslant M,$$
$$= 0 \quad \text{otherwise}.$$

Also let us use the notation $\tau^{(M)}$ for the sequence whose jth entry is τ_j if $j \leqslant M$, zero otherwise. We have

$$f(\sigma) = \| \tau^{(M)} \|_{l^{p'}}^{p'}, \qquad \| \sigma \|_{l^p} = \| \tau^{(M)} \|_{l^{p'}}^{p'/p}.$$

Let us now use the fact that f is continuous; there is a constant $A > 0$ such that, for all $s \in l^p$,

$$|f(s)| \leqslant A\| s \|_{l^p},$$

whence

$$\| \tau^{(M)} \|_{l^{p'}}^{p'} = |f(\sigma)| \leqslant A\| \tau^{(M)} \|_{l^{p'}}^{p'/p}$$

and

$$\| \tau^{(M)} \|_{l^{p'}}^{p'-p'/p} \leqslant A.$$

Since $p' - p'/p = 1$, we see that $\| \tau^{(M)} \|_{l^{p'}} \leqslant A$. Since this estimate is valid for all M, it means that $\tau \in l^{p'}$. The fact that $f(\sigma) = \langle \sigma, \tau \rangle$ for all $\sigma \in l^p$ follows from the fact that this is true when σ is finite, as we have seen, and from the density of l_F in l^p for $p < +\infty$.

We must now consider the case $p = 1$, in which case $p' = +\infty$. In this case, Hőlder's inequality (20.3) is trivial, for it reads

(20.6) $$|\langle \sigma, \tau \rangle| \leqslant \{ \sup_{j=0,1,\ldots} | \tau_j |\}\| \sigma \|_{l^1},$$

and this is evident when the sequence σ is finite; in the general case, it suffices to observe that l_F is dense in l^1. This shows that, also in this case where $p = 1$, (20.4) is a continuous linear map, with norm $\leqslant 1$, of $l^{p'}$ into the dual of l^p. Let us take the sequences $\sigma^{(j)}$ introduced on p. 208; we have

$$| \tau_j | = |\langle \sigma^{(j)}, \tau \rangle| \leqslant \| \tau \|_{(l^1)'},$$

which immediately implies

$$\| \tau \|_{l^\infty} \leqslant \| \tau \|_{(l^1)'},$$

whence the equality of these two norms, in view of (20.6): the map (20.4) is an isometry of l^∞ into $(l^1)'$. It remains to show that it is an isometry *onto*.

We consider an arbitrary continuous linear functional f on l^1. We associate with it a sequence τ by setting $\tau_j = f(\sigma^{(j)})$ as before. We want to prove now that τ is bounded. This follows from the inequality

$$| \tau_j | = |f(\sigma^{(j)})| \leqslant A\| \sigma^{(j)} \|_{l^1} \leqslant A.$$

We have then, by the same argument as in the case $p > 1$, $f(\sigma) = \langle \sigma, \tau \rangle$, for all $\sigma \in l^1$.

Example II.
The Duals of the Spaces $L^p(\Omega)$ $(1 \leqslant p < +\infty)$

Let Ω be an open subset of $\mathbf{R}^n$, and dx the Lebesgue measure on $\mathbf{R}^n$. We wish to exhibit a convenient realization of the strong dual of $L^p(\Omega)$ for $p \geqslant 1$ finite. We must try to follow the direction outlined by Example I. The analogy is obvious. The norm in $L^p(\Omega)$ is

$$f \rightsquigarrow \|f\|_{L^p} = \left(\int_\Omega |f(x)|^p \, dx \right)^{1/p}.$$

With this norm, $L^p(\Omega)$ is a Banach space; $L^2(\Omega)$ is a Hilbert space (see Chapter 11, Example V). What will play the role of the vector space of finite sequences l_F is the space $\mathscr{C}_c^0(\Omega)$ of continuous functions with compact support in Ω; this vector space is dense in every $L^p(\Omega)$ for $p < +\infty$ (Theorem 11.3). We consider then the bilinear form

$$(f, g) \rightsquigarrow \langle f, g \rangle = \int f(x) g(x) \, dx$$

on $\mathscr{C}_c^0(\Omega) \times \mathscr{C}_c^0(\Omega)$. We can extend it as a bilinear form on $L^p(\Omega) \times L^{p'}(\Omega)$ if $1 < p < \infty$, $1 < p' < \infty$, $1/p + 1/p' = 1$, and this extension satisfies Hölder's inequality

$$(20.7) \qquad |\langle f, g \rangle| \leqslant \|f\|_{L^p} \|g\|_{L^{p'}}.$$

Indeed, (20.7) follows immediately from Lemma 20.1. This implies that the mapping

$$(20.8) \qquad f \rightsquigarrow (g \rightsquigarrow \langle f, g \rangle)$$

is a continuous linear map, with norm $\leqslant 1$, of $L^{p'}(\Omega)$ into the dual of $L^p(\Omega)$. The next step is to prove that this map is an isometry. We define a function g in Ω by setting

$$g(x) = \overline{f(x)} \, |f(x)|^{p'-2} \quad \text{when} \quad f(x) \neq 0;$$
$$g(x) = 0 \quad \text{otherwise.}$$

We have

$$(20.9) \qquad \|g\|_{L^p} = \|f\|_{L^{p'}}^{p'/p};$$

$$(20.10) \qquad \langle f, g \rangle = \|f\|_{L^{p'}}^{p'}$$

As we have

$$|\langle f, g \rangle| \leqslant \|f\|_{(L^p)'} \|g\|_{L^p}$$

by definition of the norm in the dual of a Banach space, by combining this with (20.9) and (20.10), we see that

$$\|f\|_{L^{p'}}^{p'-p'/p} \leqslant \|f\|_{(L^p)'},$$

which implies that the mapping (20.8) is an isometry, since $p' - p'/p = 1$.

It remains then to prove that (20.8) is *onto*. This is usually done by applying the Lebesgue–Nicodym theorem, whose statement we now recall. Anticipating the next chapter a little, we introduce *Radon measures* on the open set Ω: a Radon measure μ on Ω is a continuous linear form on the space $\mathscr{C}_c^0(\Omega)$ equipped with its natural *LF* structure (Chapter 13, Example II). The Radon measure μ is said to be *positive* if, for all functions $\varphi \in \mathscr{C}_c^0(\Omega)$ which are nonnegative everywhere, we have

$$\langle \mu, \varphi \rangle \geqslant 0.$$

An example of positive Radon measure is the *Lebesgue* measure

$$\varphi \rightsquigarrow \int_\Omega \varphi(x)\,dx.$$

We shall make use of the fact (proved in pp. 218–220) that any complex-valued Radon measure μ in Ω can be decomposed as follows:

$$\mu = \rho^+ - \rho^- + i(\sigma^+ - \sigma^-),$$

where ρ^+, ρ^-, σ^+, σ^- are *positive* Radon measures (obviously $\rho^+ - \rho^-$ is the *real part* of μ; $\sigma^+ - \sigma^-$ is the *imaginary part* of μ). We shall use the following particular case of the Radon-Nikodym theorem:

THEOREM 20.2.   *Let μ be a positive Radon measure in Ω. The following two conditions are equivalent*:

(a)   *To every nonnegative continuous function g, with compact support in Ω, and to every $\varepsilon > 0$ there is $\delta > 0$ such that the following is true: if $h \in \mathscr{C}_c^0(\Omega)$ is such that $0 \leqslant h(x) \leqslant g(x)$ for all x, then*

$$\int h(x)\,dx \leqslant \delta \quad \textit{implies} \quad \langle \mu, h \rangle \leqslant \varepsilon.$$

(b)   *There exists a function f in Ω, nonnegative almost everywhere, locally Lebesgue integrable, such that, for all $\varphi \in \mathscr{C}_c^0(\Omega)$,*

$$\langle \mu, \varphi \rangle = \int \varphi(x) f(x)\,dx.$$

That f is *locally integrable* means that f is measurable and that, whatever be the compact subset K of Ω,

$$\int_K f(x)\,dx < +\infty.$$

We recall that f is nonnegative (otherwise we should replace f by $|f|$ in the integral over K).

Let us consider a continuous linear functional on $L^p(\Omega)$, λ. There is a constant $C > 0$ such that, for all $u \in L^p(\Omega)$,

$$(20.11) \qquad |\lambda(u)| \leqslant C\|u\|_{L^p}.$$

Let now K be a compact subset of Ω, and suppose that $u \in \mathscr{C}_c^0(\Omega)$ has its support in K. We have

$$\|u\|_{L^p}^p \leqslant \mathrm{meas}(K) \cdot (\sup_x |u(x)|)^p,$$

whence

$$|\lambda(u)| \leqslant C(\mathrm{meas}(K))^{1/p} \sup_x |u(x)|.$$

We have denoted by $\mathrm{meas}(K)$ the Lebesgue measure of the set K. Thus, if u converges to zero uniformly, keeping its support in K, $\lambda(u) \to 0$; this shows that the restriction of λ to the Banach space $\mathscr{C}_c^0(K)$ is continuous (see Chapter 13, Example II). But this implies that the linear form $u \to \lambda(u)$ on $\mathscr{C}_c^0(\Omega)$ is continuous (corollary of Proposition 13.1). This linear form defines a Radon measure in Ω, which we keep denoting by λ. We shall first suppose that λ is positive. Let now g be an arbitrary nonnegative continuous function, with compact support, in Ω, and take $h \in \mathscr{C}_c^0(\Omega)$ such that $0 \leqslant h(x) \leqslant g(x)$ for all x. We have

$$\|h\|_{L^p} \leqslant (\sup_x |g(x)|^{1-1/p}) \left(\int h(x)\,dx \right)^{1/p}.$$

Setting $M = \sup_x |g(x)|^{1-1/p}$ and $\delta = (\varepsilon/CM)^p$, we see by (20.11) that

$$\int h(x)\,dx \leqslant \delta \qquad \text{implies} \quad \lambda(h) \leqslant \varepsilon.$$

We conclude that there is a locally integrable function f in Ω, $f \geqslant 0$ almost everywhere, such that

$$\lambda(\varphi) = \int \varphi(x) f(x)\,dx \qquad \text{for all} \quad \varphi \in \mathscr{C}_c^0(\Omega).$$

At this stage, we consider a sequence of relatively compact open subset $\Omega_1 \subset \Omega_2 \subset \cdots \subset \Omega_k \subset \cdots$ whose union is equal to Ω, and for each $k = 1, 2,...$ we introduce the function

$$f_k(x) = \begin{cases} f(x) & \text{if } x \in \Omega_k \text{ and if } f(x) \leqslant k; \\ 0 & \text{otherwise.} \end{cases}$$

Obviously f_k, regarded as a function in Ω, is a bounded measurable function with compact support (contained in the closure of Ω_k). As the student may know, or easily check, such a function belongs to $L^q(\Omega)$ for all $q \geqslant 1$. As $f_k \leqslant f$, we have

$$(20.12) \quad \int \varphi(x) f_k(x)\, dx \leqslant \int \varphi(x) f(x)\, dx \quad \text{for all nonnegative } \varphi \in \mathscr{C}_c^0(\Omega).$$

As $f_k \in L^{p'}(\Omega)$, we know from the first part of this proof that $u \rightsquigarrow \int u(x) f_k(x)\, dx$ is a continuous linear form on $L^p(\Omega)$; but we shall now show that the norm of this form is bounded by a constant independent of k.

If $\varphi \in \mathscr{C}_c^0(\Omega)$ is arbitrary (in particular, not necessarily $\geqslant 0$), we have

$$\left| \int \varphi(x) f_k(x)\, dx \right| \leqslant \int |\varphi(x)| f_k(x)\, dx \leqslant \int |\varphi(x)| f(x)\, dx \leqslant C \| \varphi \|_{L^p},$$

by applying (20.11) and (20.12). Using the density of $\mathscr{C}_c^0(\Omega)$ in $L^p(\Omega)$ we see that, for all $u \in L^p(\Omega)$,

$$(20.13) \quad \left| \int u(x) f_k(x)\, dx \right| \leqslant C \| u \|_{L^p}.$$

We choose then $u(x)$ such that

$$u(x) = [f_k(x)]^{p'-1} \quad \text{if } f_k(x) \neq 0,$$
$$= 0 \quad \text{otherwise.}$$

We have

$$\int u(x) f_k(x)\, dx = \| f_k \|_{L^{p'}}^{p'}, \qquad \| u \|_{L^p} = \| f_k \|_{L^{p'}}^{p'/p}.$$

Then we derive, from (20.13),

$$\| f_k \|_{L^{p'}} \leqslant C,$$

or

$$\int [f_k(x)]^{p'}\, dx \leqslant C^{p'}.$$

But $(f)^{p'}$ is the pointwise limit of the nondecreasing sequence of nonnegative integrable functions $(f_k)^{p'}$, $k = 1, 2,...$; we have, therefore,

$$\int [f(x)]^{p'} \, dx = \lim_{k \to +\infty} \int [f_k(x)]^{p'} \, dx \leqslant C^{p'}.$$

Thus $f \in L^{p'}$. The density of $\mathscr{C}_c^0(\Omega)$ in $L^p(\Omega)$ implies immediately that

$$\lambda(u) = \langle u, f \rangle = \int u(x) f(x) \, dx \qquad \text{for all} \quad u \in L^p(\Omega).$$

This was obtained in the case where the Radon measure

$$\mathscr{C}_c^0(\Omega) \ni \varphi \rightsquigarrow \lambda(\varphi)$$

is positive. If this Radon measure is arbitrary, we decompose it into its real part and imaginary part, and these into their positive and negative parts, as mentioned in the beginning. We decompose thus λ into a linear combination (with coefficients ± 1, $\pm i$) of four positive Radon measures, all of which satisfy an estimate of the kind (20.11). Applying the preceding reasonings to each of these four positive Radon measures, we conclude that there is a complex-valued function $f \in L^{p'}(\Omega)$ such that

$$\lambda(u) = \langle u, f \rangle \qquad \text{for all} \quad u \in L^p(\Omega).$$

The case $p = 1$ can be treated in a similar fashion; we leave it to the student as an exercise. We state the theorem summarizing the whole situation:

THEOREM 20.3.   *Let Ω be an open subset of $\mathbf{R}^n$. Let p be a real number, $1 \leqslant p < +\infty$. Set $p' = p/(p - 1)$ if $p > 1$, $p' = +\infty$ if $p = 1$. Then the bilinear form on $\mathscr{C}_c^0(\Omega) \times \mathscr{C}_c^0(\Omega)$,*

$$(u, v) \rightsquigarrow \langle u, v \rangle = \int u(x) \, v(x) \, dx,$$

can be extended as a bilinear form on $L^p(\Omega) \times L^{p'}(\Omega)$. This extension satisfies Hölder's inequality

$$|\langle u, v \rangle| \leqslant \| u \|_{L^p} \| v \|_{L^{p'}}.$$

The mapping

$$v \rightsquigarrow (u \rightsquigarrow \langle u, v \rangle)$$

is an isometry of $L^{p'}(\Omega)$ onto the strong dual of $L^p(\Omega)$.

Exercises

20.1. By using Hölder's inequalities, show that $l^p \subset l^q$ if $p \leqslant q$ and that the injection $l^p \to l^q$ is continuous and has a norm equal to *one*. Prove that $l^p \neq l^q$ if $p < q$ and that the topology of l^p is strictly finer, then, than the topology induced by l^q.

20.2. Relate the result for $p = 2$, stated in Theorem 20.1, to the canonical linear isometry of a Hilbert space onto its antidual (Theorem 12.2).

20.3. Let l_∞ be the space of complex sequences, converging to zero, equipped with the norm of l^∞,

$$\sigma = (\sigma_k)_{k=0,1,\dots} \rightsquigarrow |\sigma|_{l^\infty} = \sup_k |\sigma_k|.$$

Then l_∞ is a Banach space. Prove that the mapping

$$(20.14) \qquad l^1 \ni \tau = (\tau_k) \rightsquigarrow \left(\sigma = (\sigma_k) \rightsquigarrow \sum_{k=0}^{\infty} \sigma_k \tau_k \right)$$

is an isometry of l^1 onto the dual of l_∞ (carrying the dual Banach space structure). Derive from this fact that (20.14) does *not* map l^1 onto the dual of l^∞.

20.4. Can you give an example of a continuous linear form on l^∞, nonidentically zero, vanishing on l_∞ (see Exercise 20.3)?

21

Radon Measures. Distributions

In the preceding chapter, we have proved duality theorems between certain spaces of functions. As we have already said, it is not always possible to interpret "in a natural way" the dual of a functions space as a functions space. As a matter of fact, this circumstance is rather fortunate as it enables us to add new objects to our inventory. If it were not the case, the consideration of the duals would not provide us with anything but functions, which we can always consider directly. In the present chapter, we shall introduce the dual of the space of continuous functions with compact support $\mathscr{C}_c^0$ and the dual of the space of test functions (i.e., infinitely differentiable functions with compact support), $\mathscr{C}_c^\infty$. The elements of the former have already been given a name (p. 211): they are the *Radon measures*. The elements of the dual of $\mathscr{C}_c^\infty$ are the *distributions*, to the study of which much of the forthcoming is devoted. Radon measures and distributions are precisely instances of objects which cannot be naturally interpreted as functions, or at any event which it is preferable to consider in their own right. In the next chapter, we shall introduce a third example of such "new" objects: the *analytic functionals*, which are the elements of the dual of the space of holomorphic functions.

Radon Measures in an Open Subset Ω of $\mathbf{R}^n$

Let $\mathscr{C}_c^0(\Omega)$ be the space of complex-valued functions, defined and continuous in Ω, which vanish outside some compact subset of Ω. A Radon measure μ in Ω is a linear functional on $\mathscr{C}_c^0(\Omega)$ which is continuous when this space is equipped with the topology inductive limit of the spaces $\mathscr{C}_c^0(K)$. Here K is an arbitrary compact subset of Ω; $\mathscr{C}_c^0(K)$ is the space of the continuous functions in Ω which vanish outside of K; its topology is defined by the maximum norm,

$$f \rightsquigarrow \sup_{x \in \Omega} |f(x)|.$$

The student will be careful not to think that $\mathscr{C}_c^0(K)$ is the same thing as the space of complex functions, *defined in Ω with support in K and continuous in K* (space which we denote $\mathscr{C}^0(K)$). For instance, the nonzero constant functions in K, extended by zero outside of K, are not elements of $\mathscr{C}_c^0(K)$: for they are not continuous in Ω.

At any event, a Radon measure in Ω is a linear functional μ on $\mathscr{C}_c^0(\Omega)$ such that, for every compact subset K of Ω, there is a constant $C(\mu, K) > 0$ such that, for all continuous functions φ vanishing identically outside of K,

$$|\langle \mu, \varphi \rangle| \leqslant C(\mu, K) \sup_x |\varphi(x)|.$$

Examples of Radon measures: the *Dirac measure* at some point x^0,

$$\varphi \rightsquigarrow \varphi(x^0);$$

the *Lebesgue measure*, $\varphi \to \int \varphi(x)\, dx$ (which, on continuous functions, coincides with the Riemann integral); the densities

$$\varphi \rightsquigarrow \int \varphi(x)\, g(x)\, dx,$$

where g is a locally Lebesgue integrable function in Ω; this means that g is measurable and that, for any compact subset K of Ω,

$$\int_K |g(x)|\, dx < +\infty.$$

A Radon measure μ is said to be *real* if $\langle \mu, f \rangle$ is real for all real-valued functions $f \in \mathscr{C}_c^0(\Omega)$; μ is said to be *positive* if $\langle \mu, f \rangle \geqslant 0$ for all nonnegative functions $f \in \mathscr{C}_c^0(\Omega)$. Examples of positive Radon measures in the open set Ω are the Lebesgue measure dx, more generally the densities $g(x)\, dx$ when g is a locally Lebesgue integrable function in Ω almost everywhere positive; the Dirac measures δ_{x^0} at the points of Ω are also positive Radon measures.

Let μ be an arbitrary Radon measure; we can define its *complex conjugate* $\bar{\mu}$ by the following formula:

$$\langle \bar{\mu}, \varphi \rangle = \overline{\langle \mu, \bar{\varphi} \rangle},$$

then its *real part* and its *imaginary part*:

$$\operatorname{Re}\mu = \tfrac{1}{2}(\mu + \bar{\mu}), \quad \operatorname{Im}\mu = \tfrac{1}{2i}(\mu - \bar{\mu}), \quad i = (-1)^{1/2}.$$

Example: when $\mu = g(x)\,dx$, i.e., when μ is a density with respect to the Lebesgue measure,

$$\bar{\mu} = \overline{g(x)}\,dx, \qquad \operatorname{Re}\mu = (\operatorname{Re}g)(x)\,dx, \qquad \operatorname{Im}\mu = (\operatorname{Im}g)(x)\,dx.$$

Of course, we have $\mu = \operatorname{Re}\mu + i\operatorname{Im}\mu$.

We show next that every real Radon measure can be written as the difference of two positive Radon measures. We need the following result:

THEOREM 21.1.   *A positive linear functional on $\mathscr{C}_c^0(\Omega)$ is a positive Radon measure.*

We must show that a linear functional L on $\mathscr{C}_c^0(\Omega)$ which is positive is necessarily continuous (for the inductive limit topology on $\mathscr{C}_c^0(\Omega)$). Let K be an arbitrary compact subset of Ω. Let us denote by $\|\varphi\|$ the maximum norm of $\varphi \in C_c^0(K) : \|\varphi\| = \sup_x |\varphi(x)|$. Suppose that φ is real valued; then $-\|\varphi\| \leqslant \varphi(x) \leqslant \|\varphi\|$ for all $x \in \Omega$. Let $g(x)$ be a non-negative function belonging to $\mathscr{C}_c^0(\Omega)$, identically equal to one in K; we have

$$-\|\varphi\|\,g(x) \leqslant \varphi(x) \leqslant \|\varphi\|\,g(x) \qquad \text{for all } x.$$

As the linear form L is positive, we derive from there

$$-\|\varphi\|\,L(g) \leqslant L(\varphi) \leqslant \|\varphi\|\,L(g).$$

Therefore

$$|L(\varphi)| \leqslant L(g)\,\|\varphi\|.$$

This inequality is still true if φ is complex valued, as we see by applying it to $\operatorname{Re}\varphi$ and to $\operatorname{Im}\varphi$, respectively. As $L(g)$ is a nonnegative constant independent of φ, Theorem 21.1 is proved.

Let us now prove the decomposition theorem,

THEOREM 21.2.   *Every real Radon measure is equal to the difference of two positive Radon measures.*

Let μ be a real Radon measure; let φ be an arbitrary nonnegative function belonging to $\mathscr{C}_c^0(\Omega)$. Let us set

$$M(\varphi) = \sup \langle \mu, \varphi_1 \rangle,$$

where the supremum is taken over the set of all functions $\varphi_1 \in \mathscr{C}_c^0(\Omega)$ such that $0 \leqslant \varphi_1(x) \leqslant \varphi(x)$ for all x. Observe in particular that M is *positive*, i.e., that $M(\varphi) \geqslant 0$ for all $\varphi \geqslant 0$, $\varphi \in \mathscr{C}_c^0(\Omega)$. We claim that if ψ is another nonnegative continuous function with compact support in Ω,

(21.1) $$M(\varphi + \psi) = M(\varphi) + M(\psi).$$

Proof of this Statement. If $0 \leqslant \varphi_1 \leqslant \varphi$ and $0 \leqslant \psi_1 \leqslant \psi$, then $\varphi_1 + \psi_1 \leqslant \varphi + \psi$; this implies immediately

$$\sup_{0 \leqslant \varphi_1 \leqslant \varphi} \langle \mu, \varphi_1 \rangle + \sup_{0 \leqslant \psi_1 \leqslant \psi} \langle \mu, \psi_1 \rangle = \sup_{0 \leqslant \varphi_1 \leqslant \varphi, 0 \leqslant \psi_1 \leqslant \psi} \langle \mu, \varphi_1 + \psi_1 \rangle$$

$$\leqslant \sup_{0 \leqslant \chi \leqslant \varphi + \psi} \langle \mu, \chi \rangle = M(\varphi + \psi).$$

This means that

$$M(\varphi) + M(\psi) \leqslant M(\varphi + \psi).$$

We must now prove the converse inequality. It will suffice to prove that every function $\chi \in \mathscr{C}_c^0(\Omega)$ satisfying $0 \leqslant \chi \leqslant \varphi + \psi$ can be written in the form

$$\chi = \varphi_1 + \psi_1,$$

with $0 \leqslant \varphi_1 \leqslant \varphi$, $0 \leqslant \psi_1 \leqslant \psi$, $\varphi_1, \psi_1 \in \mathscr{C}_c^0(\Omega)$. In order to see this, it suffices to set

$$\varphi_1(x) = \sup(\chi(x) - \psi(x), 0) \qquad \text{for each} \quad x \in \Omega.$$

As the supremum of two continuous functions is continuous, $\varphi_1 \in \mathscr{C}_c^0(\Omega)$ and is obviously nonnegative; as $\chi \leqslant \varphi + \psi$, we have obviously $\varphi_1 \leqslant \varphi$. Set then $\psi_1 = \chi - \varphi_1$. At any point x where $\chi(x) \geqslant \psi(x)$, we have $\varphi_1(x) = \chi(x) - \psi(x)$, hence $\psi_1(x) = \psi(x)$. At a point where $\chi(x) \leqslant \psi(x)$, we have $\varphi_1(x) = 0$, hence $\psi_1(x) = \chi(x) \leqslant \psi(x)$. Thus (21.1) is proved.

The next step is to extend the functional M to functions belonging to $\mathscr{C}_c^0(\Omega)$ which are not necessarily nonnegative. This is very simply done, by observing that a real function ϕ can always be decomposed in the form $\phi = \phi_1 - \phi_2$, with ϕ_1 and ϕ_2 nonnegative continuous functions with compact support in Ω (e.g., $\phi = \sup(\phi, 0) - \sup(-\phi, 0)$). We write

$$(21.2) \qquad M(\phi) = M(\phi_1) - M(\phi_2).$$

It follows at once from (21.1) that $M(\phi)$ does not depend on the chosen decomposition of ϕ, $\phi = \phi_1 - \phi_2$, and that (21.1) is also valid when ϕ and ψ are real valued but not everywhere $\geqslant 0$. Finally, we define $M(\phi)$ for complex functions ϕ by the formula

$$(21.3) \qquad M(\phi) = M(\operatorname{Re} \phi) + iM(\operatorname{Im} \phi).$$

Equation (21.1) extends trivially to the case of ϕ, ψ complex valued.

It must now be proved that M is linear; we prove that M is **R**-linear, which implies at once, in view of (21.3), that M is **C**-linear.

The **R**-linearity of M follows from (21.1) and from the positivity of M by a standard argument. Let p, q be two positive integers; (21.1) implies immediately, for all real-valued $\varphi \subset \mathscr{C}^0_c(\Omega)$,

$$M(p\varphi) = p\, M(\varphi),$$

whence $M((1/p)\varphi) = (1/p)M(\varphi)$ by substitution of $(1/p)\varphi$ for φ; therefore we have

$$M\left(\frac{p}{q}\varphi\right) = \frac{p}{q}\, M(\varphi).$$

This shows that $M(r\varphi) = rM(\varphi)$ for all *rational* numbers r. If now λ is an arbitrary real number, and if φ is a nonnegative function, we have

$$r_1\, \varphi(x) \leqslant \lambda\, \varphi(x) \leqslant r_2\, \varphi(x)$$

for all $x \in \Omega$, and for all pairs of rational numbers r_1, r_2 such that $r_1 \leqslant \lambda \leqslant r_2$. In virtue of the positivity of M and of the linearity of M with respect to the rational numbers, we derive that

$$r_1\, M(\varphi) \leqslant M(\lambda\varphi) \leqslant r_2\, M(\varphi).$$

Taking then φ and λ fixed, and r_1 (resp. r_2) converging to λ from the left (resp. from the right), we conclude that $M(\lambda\varphi) = \lambda\, M(\varphi)$. This obviously remains true even when φ is real valued (and not necessarily nonnegative), in virtue of (21.2). As we said, it then carries over to complex-valued functions and to complex scalars λ.

Thus we have proved that M is a positive linear functional on $\mathscr{C}^0_c(\Omega)$; by Theorem 21.2, it is a (positive) Radon measure. From now on, we write μ^+ instead of M; μ^+ is called the *positive part* of μ. We define the *negative part* of μ, μ^-, as being the positive part of $-\mu$. We leave to the student the proof of the fact that

$$\mu = \mu^+ - \mu^-,$$

which completes the proof of Theorem 21.2.

Any complex measure μ can be decomposed into a linear combination of *four* positive measures, in the following manner:

$$\mu = (\mathrm{Re}\,\mu)^+ - (\mathrm{Re}\,\mu)^- + i(\mathrm{Im}\,\mu)^+ - i(\mathrm{Im}\,\mu)^-.$$

The *absolute value* of a real Radon measure μ is now easy to define: it is the positive measure

$$| \mu | = \mu^+ + \mu^-.$$

Let μ be a positive Radon measure. By definition, μ is a functional on the space of continuous functions with compact support in Ω. However, its domain of definition can be extended so as to include a larger set of functions than $\mathscr{C}_c^0(\Omega)$. The properties of such an extension constitute what is called the theory of integration of the measure μ. In this respect, Radon measures are a particular kind of *measures*; a useful theory of integration for general measures can be constructed. For further information on this subject, we refer to the treatises on integration theory.

In this chapter, we have given the definition of a Radon measure in an open subset of $\mathbf{R}^n$. The student will easily perceive that such a definition could have been given for Radon measures on any locally compact topological space. It will suffice to assume that the letter Ω stands for such a space, in the above reasonings. It should be pointed out, however, that there will be Radon measures on certain locally compact spaces which have no equivalent in others. For instance, this is true of the Lebesgue measure dx and of the densities $g(x) \, dx$. Similar measures do not exist on arbitrary locally compact spaces, although they might exist on certain types of locally compact spaces (e.g., locally compact groups: the role of the Lebesgue measure is then played by the so-called Haar measure). In the forthcoming, we shall need the following result about these densities.

THEOREM 21.3.   *Let g_1 , g_2 be two locally Lebesgue integrable functions in the open subset Ω of $\mathbf{R}^n$; the Radon measures in Ω,*

$$\varphi \rightsquigarrow \int \varphi(x) \, g_1(x) \, dx, \qquad \varphi \rightsquigarrow \int \varphi(x) \, g_2(x) \, dx,$$

are equal if and only if $g_1 = g_2$ almost everywhere in Ω.

We shall not give a proof of this theorem; the student will find a proof of it in any good book on integration theory.

At this point, an important remark is in order, namely that the set of *locally integrable* functions is the *largest* set of functions defined by local conditions involving the L^p norms. This remark will be expanded later on when we study the *local* spaces of distributions. For the time being, we shall content ourselves with a precise but limited formulation of the observed fact:

Let p be a number such that $1 \leqslant p \leqslant +\infty$.

Definition 21.1.   A function f in Ω is said to be locally L^p if f is Lebesgue measurable and if, for every compact subset K of Ω, we have

$$\int_K |f(x)|^p \, dx < +\infty.$$

THEOREM 21.4.   *Whatever be* p, $1 \leqslant p \leqslant +\infty$, *every locally* L^p *function in* Ω *is locally* L^1 *(i.e., locally integrable).*

Let f be locally L^p and K an arbitrary compact subset of Ω. Hölder's inequalities imply that

$$\int_K^* |f(x)| \, dx \leqslant \| 1_K f \|_{L^p} \| 1_K \|_{L^{p'}},$$

where 1_K is the characteristic function of K, equal to one in K and to zero everywhere else, and where p' is the number conjugate of p, $p' = +\infty$ if $p = 1$ and $p' = p/(p-1)$ if $1 < p < +\infty$. This shows immediately that

$$\int_K^* |f(x)| \, dx < +\infty;$$

therefore the restriction of f to K belongs to $L^1(K)$, as f is measurable.

Distributions in an Open Subset of $\mathbf{R}^n$

A distribution in an open subset Ω of $\mathbf{R}^n$ is a linear functional on the space of test functions $\mathscr{C}_c^\infty(\Omega)$ which is continuous when $\mathscr{C}_c^\infty(\Omega)$ carries its canonical LF topology (see Chapter 13, Example II, p. 130). We recall that the elements of $\mathscr{C}_c^\infty(\Omega)$ are the $\mathscr{C}^\infty$ (complex-valued) functions in Ω with compact support. If we consider a linear functional L on $\mathscr{C}_c^\infty(\Omega)$ we may decide whether it is a distribution by applying the corollary of Proposition 13.1, which in the present situation can be stated as follows (when combined with Proposition 8.5):

PROPOSITION 21.1.   *A linear form L on $\mathscr{C}_c^\infty(\Omega)$ is a distribution if and only if it possesses the following equivalent properties:*

(a)   *To every compact subset K of Ω there is an integer $m \geqslant 0$ and a constant $C > 0$ such that, for all $\mathscr{C}^\infty$ functions φ with support in the set K,*

$$|L(\varphi)| \leqslant C \sup_{|p| \leqslant m} (\sup_{x \in \Omega} |(\partial/\partial x)^p \varphi(x)|).$$

(b) *If a sequence of test functions* $\{\varphi_k\}$ ($k = 1, 2,...$) *converge uniformly to zero, as well as all their derivatives, and if the functions* φ_k *have their support contained in a compact subset* K *of* Ω, *independent of the index* k, *then* $L(\varphi_k) \to 0$.

A way of obtaining examples of distributions is the following one: Let E be a space of functions in Ω containing $\mathscr{C}_c^\infty(\Omega)$; suppose that E is provided with a locally convex Hausdorff topology which induces on $\mathscr{C}_c^\infty(\Omega)$ a topology less fine than the canonical LF topology on this space. Then the restriction to $\mathscr{C}_c^\infty(\Omega)$ of any continuous linear functional L on E is a continuous linear functional on $\mathscr{C}_c^\infty(\Omega)$ (equipped with its LF topology), i.e., a distribution in Ω. If we make the further requirement that any two *different* continuous linear forms L_1, L_2 on E define, in the way just described, two *different* distributions in Ω, we must impose the condition that $\mathscr{C}_c^\infty(\Omega)$ be dense in E. This is obvious: for the Hahn–Banach theorem implies that a subspace of a locally convex Hausdorff space is dense if and only if every continuous linear vanishing on the subspace is equal to zero in the whole space.

This scheme can be applied to $E = \mathscr{C}_c^0(\Omega)$. It shows immediately that every Radon measure μ in Ω defines, by restriction to $\mathscr{C}_c^\infty(\Omega)$, a distribution T_μ in Ω; if μ_1, μ_2 are two Radon measures in Ω such that $\mu_1 \neq \mu_2$, we have

$$T_{\mu_1} \neq T_{\mu_2} \qquad \text{(see Corollary 2 of Theorem 15.3).}$$

In view of these facts, we have the right to identify the distribution T_μ with the Radon measure μ. We shall therefore say that the distribution T_μ *is a Radon measure*, and we shall write μ instead of T_μ.

PROPOSITION 21.2. *A distribution* T *in* Ω *is a Radon measure if it possesses the following three equivalent properties:*

(a) *The linear form* $\varphi \rightsquigarrow \langle T, \varphi \rangle$ *is continuous on* $\mathscr{C}_c^\infty(\Omega)$ *when this space carries the topology induced by* $\mathscr{C}_c^0(\Omega)$.

(b) *To every compact subset* K *of* Ω *there is a constant* $C > 0$ *such that, for all* $\mathscr{C}^\infty$ *functions* φ *with support in the set* K,

$$|\langle T, \varphi \rangle| \leqslant C \sup_x |\varphi(x)|.$$

(c) *If a sequence of test functions* $\{\varphi_k\}$ ($k = 1, 2,...$) *converge uniformly to zero and if the functions* φ_k *have their support contained in a compact subset* K *of* Ω *independent of* k, *then* $\langle T, \varphi_k \rangle \to 0$.

The equivalence of (a), (b), and (c) follows immediately from the definition of the LF topology on $\mathscr{C}_c^0(\Omega)$ (see Chapter 13, Example II, p. 130) and from Propositions 8.5 and 13.1. We must therefore show that T is a Radon measure in Ω if and only if T has Property (a). In one direction, it is obvious: if T is a Radon measure, which is to say if T is the restriction to $\mathscr{C}_c^\infty(\Omega)$ of a Radon measure in Ω, then T is continuous for the topology induced on $\mathscr{C}_c^\infty(\Omega)$ by $\mathscr{C}_c^0(\Omega)$. Conversely, if $\varphi \rightsquigarrow \langle T, \varphi \rangle$ is continuous for this induced topology, it can be extended as a continuous linear form to the whole of $\mathscr{C}_c^0(\Omega)$: by continuity, since $\mathscr{C}_c^\infty(\Omega)$ is dense in $\mathscr{C}_c^0(\Omega)$ (Corollary 2 of Theorem 15.3). As a matter of fact, the extension of $\varphi \rightsquigarrow \langle T, \varphi \rangle$ is unique: this follows precisely from the density of $\mathscr{C}_c^\infty(\Omega)$ in $\mathscr{C}_c^0(\Omega)$, as has already been pointed out above.

A particularly important class of distributions which are Radon measures are the distributions of the form

$$\varphi \rightsquigarrow \int \varphi(x) f(x) \, dx,$$

where f is a locally integrable function in Ω. Recalling what we have just said, that two distinct Radon measures define distinct distributions, and using Theorem 21.3, we see that two locally integrable functions define the same distribution if and only if they are almost everywhere equal. This enables us to identify equivalence classes of locally integrable functions modulo the relation "to be equal almost everywhere" with the distribution defined by any one of their representative. We shall say that *a distribution T is a function* if there is a locally integrable function f such that T is the Radon measure $f(x) \, dx$; we shall then write f instead of T; it is understood that f is defined almost everywhere, or, more correctly, that f is an equivalence class of locally integrable functions modulo the relation "$f_1 = f_2$ a.e."

We shall transfer the whole terminology for functions to distributions which are functions. Thus we shall say that a distribution f is a $\mathscr{C}^k$ (resp. L^p) function ($0 \leqslant k \leqslant +\infty$) if the class denoted by f contains a representative which is a $\mathscr{C}^k$ (resp. L^p) function. We shall use such expressions as: the distribution T is a polynomial, an exponential, an analytic function, T can be extended to the complex space $\mathbf{C}^n$ as an entire analytic function, etc. We repeat: the whole terminology which is used when we deal with functions will be used when dealing with distributions which are functions.

If all distributions were Radon measures, there would not be much point in building a distributions theory. But we shall see that the space of distributions contains many more objects than Radon measures. This will become obvious as soon as we will have at our disposal the concept of a differential operator acting on distributions. We shall be able, then, to differentiate distributions as many times as we wish, and it will be obvious that differentiation of Radon measures yields distributions which, in general, will not be Radon measures.

In relation with the previous remark, it can be observed that there are many functions which are used in analysis and which are not locally integrable, which therefore do not define distributions by the formula

$$\varphi \rightsquigarrow \int \varphi(x) f(x) \, dx.$$

Such is for instance the function

$$t \rightsquigarrow t^{-k} \qquad (k = 1, 2, ...)$$

defined on the real line. Obviously, the trouble lies not with measurability since $t \rightsquigarrow t^{-k}$ is measurable (it is $\mathscr{C}^\infty$ in the complement of the origin), but with the integrals of the absolute values on bounded intervals of the real line. Indeed, we have (for $k \geqslant 1$),

$$\int_0^1 t^{-k} \, dt = +\infty.$$

There is a way, however, of defining a distribution in the real line by means of the function $t \rightsquigarrow t^{-k}$. More precisely, there is a distribution S in $\mathbf{R}^1$ such that, for any test function $\varphi \in \mathscr{C}_c^\infty(\mathbf{R}^1)$ having its support in the complement of the origin (i.e., vanishing in some neighborhood of $t = 0$),

$$\langle S, \varphi \rangle = \int \varphi(t) \frac{dt}{t}.$$

In fact, there is a standard procedure for doing this, leading to the concept of the *pseudofunctions* (in our case, the distributions Pf t^{-k}). But the student should keep in mind that ·these "extensions" S are never functions, in the sense of the expression "this or that distribution is a function." For if S were a function, i.e., if there were a locally integrable function $f(t)$ in the real line such that, for all test functions φ,

$$\langle S, \varphi \rangle = \int \varphi(t) f(t) \, dt,$$

this would imply that $f(t) = t^{-k}$ if $t \neq 0$, which in turn would mean that t^{-k} is locally integrable, which it is not.

It should also be observed that there are functions which do not define distributions, whatever procedure one tries on them. Such a function is

$$\mathbf{R}^1 \ni t \rightsquigarrow e^{1/t}.$$

There is no distribution in the real line which coincides with $t \rightsquigarrow e^{1/t}$ in the complement of the origin.

Notation 21.1. Let Ω be an open subset of $\mathbf{R}^n$. The space of distributions in Ω is denoted by $\mathscr{D}'(\Omega)$.

Exercises

In the exercises, we denote by Ω an arbitrary open subset of $\mathbf{R}^n$.

21.1. Let $x_1, x_2, ..., x_j, ...$ be an arbitrary sequence of points of Ω. Let $a_1, a_2, ..., a_j, ...$ be a sequence of complex numbers such that

$$\sum_{j=1}^{\infty} |a_j| < +\infty.$$

Prove that the functional on $\mathscr{C}_c^0(\Omega)$,

$$\varphi \rightsquigarrow \sum_{j=1}^{\infty} a_j \varphi(x_j),$$

is a Radon measure in Ω.

22.2. Let $\{x_j\}$ be a sequence of points in Ω which does not have any accumulation point in Ω. Then, prove that, for any sequence of complex numbers $\{a_j\}$, the functional on $\mathscr{C}_c^\infty(\Omega)$,

$$\varphi \rightsquigarrow \sum_{j=1}^{\infty} a_j D^j \varphi(x_j), \qquad D^j \varphi = (\partial/\partial x_1)^j \cdots (\partial/\partial x_n)^j \varphi,$$

is a distribution in Ω.

22.3. Let x^0 be an arbitrary point of Ω. Prove that the functional on $\mathscr{C}_c^\infty(\Omega)$,

$$\varphi \rightsquigarrow [(\partial/\partial x_1)\varphi](x^0),$$

is a distribution in Ω, but is not a Radon measure in Ω.

22.4. Let $|x| = (x_1^2 + \cdots + x_n^2)^{1/2}$ be the Euclidean norm on $\mathbf{R}^n$. What condition should be satisfied by the real number s in order that the measurable function $x \rightsquigarrow |x|^s$ in $\mathbf{R}^n$ define a distribution in $\mathbf{R}^n$ which is a function?

22.5. Prove that if a distribution T in the open set Ω is such that $\langle T, \phi \rangle \geq 0$ for all nonnegative $\phi \in \mathscr{C}_c^\infty(\Omega)$, then T is a positive Radon measure.

22

More Duals: Polynomials and Formal Power Series. Analytic Functionals

Polynomials and Formal Power Series

Let $\mathscr{P}_n$ be the vector space of all polynomials in n indeterminates with complex coefficients. This space is often denoted by $\mathbf{C}[X_1,...,X_n]$. We modify momentarily the notation in order to shorten it, and also in order to emphasize that we are interested not in the ring structure of the set of polynomials, but in its topological vector space structure. For $k = 0, 1, 2,...$, let $\mathscr{P}_n^k$ be the vector subspace of $\mathscr{P}_n$ consisting of the polynomials of degree $\leqslant k$; each $\mathscr{P}_n^k$ is finite dimensional, in fact its dimension is easy to compute: it is equal to

$$(k + n)!/k!n!.$$

We provide $\mathscr{P}_n$ with the locally convex topology which is the inductive limit of the topologies of the Hausdorff finite dimensional spaces $\mathscr{P}_n^k$, $k = 0, 1,...$ (see Chapter 13, Example I).

On the other hand, we consider the vector space $\mathscr{Q}_n$ of formal power series in n indeterminates, which is usually denoted by $\mathbf{C}[[X_1,...,X_n]]$. We provide $\mathscr{Q}_n$ with the topology of convergence of each coefficient. This topology is defined by the sequence of seminorms:

$$u = \sum_{p \in \mathbf{N}^n} u_p X^p \rightsquigarrow \sup_{|p| \leqslant k} |u_p|, \qquad k = 0, 1,....$$

This topology turns $\mathscr{Q}_n$ into a Fréchet space (see Chapter 10, Example III).

Now, there is a natural duality between polynomials and formal power series, which can be expressed by the bracket

$$\langle P, u \rangle = \sum_{p \in \mathbf{N}^n} P_p u_p,$$

227

where

$$P = \sum_p P_p X^p, \qquad u = \sum_p u_p X^p.$$

It should be remembered that all coefficients P_p, except possibly a finite number of them, are equal to zero; this gives a meaning to the bracket $\langle P, u \rangle$. The main result, in the present context, is the following one:

THEOREM 22.1.   (a) *The map*

(22.1) $u \rightsquigarrow (P \rightsquigarrow \langle P, u \rangle)$

is an isomorphism for the structures of topological vector spaces of the Fréchet space of formal power series $\mathscr{Q}_n$ onto the strong dual of the LF-space of polynomials, $\mathscr{P}_n$.

   (b)   *The map*

(22.2) $P \rightsquigarrow (u \rightsquigarrow \langle P, u \rangle)$

is an isomorphism of $\mathscr{P}_n$ onto the strong dual of $\mathscr{Q}_n$.

Proof.   The proof consists of a succession of very simple steps. First of all, we have, for all power series u and all polynomials P of degree $\leqslant k$,

$$|\langle P, u \rangle| \leqslant \left(\sum_{p \in N^n} |P_p| \right) \sup_{|p| \leqslant k} |u_p|.$$

This shows immediately that both maps (22.1) and (22.2) are continuous linear maps into. We must show that they are one-to-one, onto, and that their inverse is continuous.

The Maps are One-to-One

   Take for P the monomials $X^p = X_1^{p_1} \cdots X_n^{p_n}$; this means that $P_q = 0$ if $q \neq p$, $P_p = 1$. Then

$$\langle P, u \rangle = u_p \qquad \text{for any} \quad u \in \mathscr{Q}_n.$$

If the linear functional $P \rightsquigarrow \langle P, u \rangle$ were to be zero, we would have $u_p = 0$ for all p, in other words $u = 0$.

   Suppose now that, for a given polynomial P, the linear functional

$$u \rightsquigarrow \langle P, u \rangle$$

is identically zero in $\mathscr{Q}_n$; take for u the same monomial X^p as before. We obtain that $P_p = 0$ for all p, hence $P = 0$.

The Maps are Onto

Again, let us begin with (22.1). Let L be an arbitrary (continuous) linear functional on $\mathscr{P}_n$; set, for each p,

$$u_p = L(P) \qquad \text{when} \quad P(X) = X^p.$$

Then

$$u = \sum_p u_p X^p$$

is a formal power series and, obviously, we have

$$L(P) = \langle P, u \rangle \qquad \text{for all} \quad P \in \mathscr{P}_n.$$

Let us consider now (22.2). Let M be an arbitrary continuous linear form on $\mathscr{Q}_n$; by taking its value on the monomials X^p, we associate with it a formal power series

$$v = \sum_p v_p X^p.$$

From the fact that M is continuous it follows that $v_p = 0$ except possibly for a finite number of indices p. Indeed, there is a constant $C > 0$ and an integer $k \geqslant 0$ such that, for all formal power series u,

$$| M(u) | \leqslant C \sup_{|p| \leqslant k} | u_p |.$$

This means, in particular, that, for all formal power series u such that $u_p = 0$ for $| p | \leqslant k$, we have $M(u) = 0$. This applies in particular to the series $u = X^p$ for $| p | > k$. Therefore

$$v_p = M(u) = 0 \qquad \text{when} \quad u = X^p, \quad | p | > k.$$

Thus v is a polynomial, and then it becomes obvious that, for any formal power series u,

$$M(u) = \langle v, u \rangle.$$

The Inverses of Maps (22.1) and (22.2) are Continuous

Let L be a continuous linear form on $\mathscr{P}_n$, and u the associated formal power series. To say that L converges to zero in the strong dual of $\mathscr{P}_n$ is to say that, given any bounded set of polynomials, L converges uniformly to zero on this set. We may take sets consisting of a single element, in particular of the monomial X^p; thus, if L converges to zero in $\mathscr{P}'_n$, $u_p = L(X^p)$ must converge to zero for each p, which means exactly that the series u converges to zero in $\mathscr{Q}_n$. This proves that the inverse of (22.1) is continuous. Let us study now the inverse of (22.2). Let $\mathscr{V}$ be an arbitrary convex neighborhood of zero in $\mathscr{P}_n$. For each $k = 1, 2,...,$ $\mathscr{V} \cap \mathscr{P}_n^k$ contains some set

$$\mathscr{W}^k = \left\{ v = \sum_{|p| \leqslant k} v_p X^p; \sum_{|p| \leqslant k} |v_p| \leqslant \rho_k \right\}, \qquad \rho_k > 0.$$

We may assume that the ρ_k form a decreasing sequence converging to zero; the union $\mathscr{W}$ of the sets $\mathscr{W}^k$ is not, in general, a neighborhood of zero in $\mathscr{P}_n$; but its convex hull $\Gamma(\mathscr{W})$ is one. We have evidently $\Gamma(\mathscr{W}) \subset \mathscr{V}$. On the other hand, let $\mathscr{B}$ be the set of formal power series

$$\left\{ u = \sum u_p X^p; \text{ for } k = 1, 2,..., \sup_{|p| \leqslant k} |u_p| \leqslant \rho_k'^{-1} \right\},$$

where

$$\rho_k' = 2^{-k-1} \rho_k.$$

Let v be a polynomial of degree $\leqslant k$ which defines a continuous linear functional $u \rightsquigarrow \langle u, v \rangle$ on $\mathscr{Q}_n$ belonging to the polar of $\mathscr{B}$. Choose u in the following way:

$$u_p = \rho_{|p|}'^{-1} \bar{v}_p / |v_p| \qquad \text{if} \quad v_p \neq 0;$$

$$u_p = 0 \qquad \text{otherwise.}$$

We have then

(22.3) $$\langle u, v \rangle = \sum \rho_{|p|}'^{-1} |v_p| \leqslant 1,$$

since u belongs obviously to the bounded set $\mathscr{B}$. For each integer $h = 0, 1,...,$ set

$$v_h = \sum_{|p|=h} v_p X^p$$

(v_h is the homogeneous part of degree h of v). We have, in view of (22.3),

$$\sum_{|p|=h} |v_p| \leqslant \rho_h' = 2^{-h-1} \rho_h.$$

This means that $2^{h+1} v_h \in \mathscr{W}^h$; but $\sum_{h=0}^k 2^{-h-1} < 1$, hence $v = \sum_{h=0}^k v_h = \sum_{h=0}^k 2^{-h-1}(2^{h+1} v_h)$ belongs to the convex hull of the union of the $\mathscr{W}^h$, this is to say to $\Gamma(\mathscr{W}) \subset \mathscr{V}$. This shows that the image of the polar $\mathscr{B}^0$ of $\mathscr{B}$ under the canonical isomorphism of the dual of $\mathscr{Q}_n$ onto $\mathscr{P}_n$ is contained in $\mathscr{V}$, in other words that the preimage of $\mathscr{V}$ is a neighborhood of zero in the strong dual of $\mathscr{Q}_n$.   Q.E.D.

If we forget about the multiplicative structure of the sets $\mathscr{P}_n$ and $\mathscr{Q}_n$, we can regard them as sets of functions with complex values and domain of definition $\mathbf{N}^n$: this simply means that we identify a polynomial or a power series with the collection of its coefficients: instead of writing $u = \sum_{p \in \mathbf{N}^n} u_p X^p$, we write $u = (u_p)_{p \in \mathbf{N}^n}$; then $\mathscr{Q}_n$ turns out to be the space of *all* complex functions in $\mathbf{N}^n$ and $\mathscr{P}_n$ the space of those functions which vanish outside a *finite* set. Needless to say, this is the same as identifying $\mathscr{Q}_n$ with the space of arbitrary complex sequences depending on n indices, and $\mathscr{P}_n$ with the space of finite complex sequences. We may also write

$$\mathscr{Q}_n = \prod_{p \in N^n} \mathbf{C}_p, \qquad \mathbf{C}_p \cong \mathbf{C}, \text{ the complex plane.}$$

Then $\mathscr{P}_n$ can be regarded as the *direct sum* of the $\mathbf{C}_p$'s. As a matter of fact, the topology of simple convergence of the coefficients on $\mathscr{Q}_n$ is precisely the product topology of the $\mathbf{C}_p$'s, etc. Let us observe that the *LF*-space $\mathscr{P}_n$, which is canonically isomorphic with the strong dual of $\mathscr{Q}_n$, is continuously embedded in $\mathscr{Q}_n$, and is dense in $\mathscr{Q}_n$.

Analytic Functionals in an Open Subset Ω of $\mathbf{C}^n$

We denote by $H(\Omega)$ the space of holomorphic functions in Ω, equipped with the topology induced by any one of the spaces $\mathscr{C}^k(\Omega)$ $(0 \leqslant k \leqslant +\infty)$ when $\mathbf{C}^n$ is identified with $\mathbf{R}^{2n}$. For instance, we may consider that $H(\Omega)$ carries the topology of uniform convergence on compact subsets of Ω, i.e., the topology induced by $\mathscr{C}^0(\Omega)$.

Definition 22.1. The dual of $H(\Omega)$ is denoted by $H'(\Omega)$; its elements are called analytic functionals in Ω.

Observing that $H(\Omega)$ is isomorphically embedded in $\mathscr{C}^0(\Omega)$, we see, in virtue of the Hahn–Banach theorem, that any continuous linear functional L on $H(\Omega)$ can be extended as a continuous linear functional $\check{L}$ on $\mathscr{C}^0(\Omega)$, which, in turn, by restriction to $\mathscr{C}^0_c(\Omega)$, defines a Radon measure μ in Ω. As $\mathscr{C}^0_c(\Omega)$ is dense in $\mathscr{C}^0(\Omega)$, this Radon measure μ is uniquely determined by $\check{L}$; but as $\check{L}$ is not uniquely determined by L (except when $\Omega = \varnothing$), neither is μ. This is easy to understand; for let ϕ_0 be a continuous function with compact support in Ω such that there is a $\mathscr{C}^1$ function with compact support in Ω, ϕ, satisfying the equation

$$\phi_0 = \partial \phi / \partial \bar{z}_1 = \frac{1}{2} \left(\frac{\partial \phi}{\partial x_1} + i \frac{\partial \phi}{\partial y_1} \right), \qquad i = (-1)^{1/2}.$$

Then, if the Radon measure μ defines an analytic functional L in Ω, the Radon measure $\mu + \phi_0$ defines the same one, L. This simply follows from the fact that the analytic functional defined by ϕ_0 is equal to zero; this analytic functional is simply

$$H(\Omega) \ni h \rightsquigarrow \int h(x + iy) \, \phi_0(x, y) \, dx \, dy.$$

It is well defined, as ϕ_0 is continuous with compact support in Ω. By integration by parts, we see immediately that

$$\int h(x + iy) \, \phi_0(x, y) \, dx \, dy = \int h(x + iy) \, \frac{\partial \phi}{\partial z_1}(x, y) \, dx \, dy$$

$$= - \int \frac{\partial h}{\partial z_1}(x + iy) \, \phi(x, y) \, dx \, dy = 0.$$

In general, that is to say when Ω is an arbitrary open subset of $\mathbf{C}^n$, there is no natural way of interpreting as functions the analytic functionals in Ω. This is however possible when Ω is of a very simple type, for instance when Ω is a polydisk, as we are now going to show.

Notation 22.1. Let $K_1, ..., K_n$ be n numbers, $0 < K_j \leqslant +\infty$ for $j = 1, ..., n$. We denote by $\Delta(K_1, ..., K_n)$, or simply $\Delta(K)$, the open polydisk

$$\{z \in \mathbf{C}^n; \, |z_1| < K_1, ..., |z_n| < K_n\}.$$

Notation 22.2. Let $K_1, ..., K_n$ be n nonnegative finite numbers. We denote by $Exp(K_1, ..., K_n)$, or simply by $Exp(K)$, the space of entire functions of exponential type $(K_1, ..., K_n)$, i.e., the space of the entire functions f in $\mathbf{C}^n$ such that there is a constant $A(f) > 0$ such that

(22.4)   *for all* $z \in \mathbf{C}^n$, $|f(z)| \exp(-K_1|z_1| - \cdots - K_n|z_n|) \leqslant A(f).$

If $f \in Exp(K)$, the inf of the constants $A(f)$ in Property (22.4) can be taken as the *norm* of f in $Exp(K)$; that it is indeed a norm is easy to check. It induces on $Exp(K)$ a topology which is strictly finer than the one induced by $H(\mathbf{C}^n)$, the space of entire functions. Also observe that, if $K_1' \leqslant K_1, ..., K_n' \leqslant K_n$, we have $Exp(K') \subset Exp(K)$ (the two spaces are regarded, here, as subsets of $H(\mathbf{C}^n)$).

Notation 22.3. Let $K_1, ..., K_n$ be n numbers, $0 < K_j \leqslant +\infty$ for $j = 1, ..., n$. We denote by $\hat{E}xp(K)$ the union of the spaces $Exp(K')$ for all $K' = (K_1', ..., K_n')$ such that $K_1' < K_1, ..., K_n' < K_n$.

We shall not put any topology on the vector space $\tilde{E}xp(K)$. Our main result is then the following:

THEOREM 22.2.   *Let $K_1,...,K_n$ be n positive numbers, some or all of which may be infinite.*

Given any function $f \in \tilde{E}xp(K)$, the linear functional on the space $\mathscr{P}_n$ of polynomials in n indeterminates with complex coefficients (viewed as polynomials functions on $\mathbf{C}^n$, i.e., polynomials in $z_1,...,z_n$),

$$P = \sum_p P_p z^p \rightsquigarrow \langle f, P \rangle = \sum_p P_p f^{(p)}(0),$$

can be extended, in a unique way, as a continuous linear functional on $H(\Delta(K))$, i.e., as an analytic functional in the open polydisk $\Delta(K)$, μ_f. Furthermore, the mapping $f \rightsquigarrow \mu_f$ is an isomorphism (for the structures of vector spaces) of $\tilde{E}xp(K)$ onto the dual of $H(\Delta(K))$, $H'(\Delta(K))$. The inverse mapping is given by the formula

$$f(\zeta) = \langle \mu_f, e^{\langle z, \zeta \rangle} \rangle, \qquad \zeta \in \mathbf{C}^n, \qquad \langle z, \zeta \rangle = z_1 \zeta_1 + \cdots + z_n \zeta_n,$$

where μ_f operates on functions of $z \in \Delta(K)$.

Proof.   The proof is based on an estimate of the $f^{(p)}(0)$ ($f \in \tilde{E}xp(K)$) derived quite straightforwardly from Cauchy's formulas. Remembering that f is an entire function, we may write, for any set of numbers $r_1,...,r_n > 0$,

$$\frac{1}{p!} f^{(p)}(0) = (2i\pi)^{-n} \oint_{|z_1| = r_1} \cdots \oint_{|z_n| = r_n} \frac{f(z)}{z^p} \frac{dz_1}{z_1} \cdots \frac{dz_n}{z_n},$$

where $p! = p_1! \cdots p_n!$, $z^p = z_1^{p_1} \cdots z_n^{p_n}$. The Cauchy formula above implies immediately Cauchy's inequality

$$(22.5) \qquad |f^{(p)}(0)| \leqslant p! r_1^{-p_1} \cdots r_n^{-p_n} \sup_{|z_1| = r_1, ..., |z_n| = r_n} |f(z)|.$$

As $f \in \tilde{E}xp(K)$, there are numbers $A > 0$, $\varepsilon > 0$ ($\varepsilon < K_j$ for all $j = 1,...,n$), such that, for all z,

$$(22.6) \qquad |f(z)| \leqslant A \exp((K_1 - \varepsilon)|z_1| + \cdots + (K_n - \varepsilon)|z_n|).$$

Combining (22.5) and (22.6), we obtain

$$(22.7) \quad |f^{(p)}(0)| \leqslant A p! r_1^{-p_1} \cdots r_n^{-p_n} \exp((K_1 - \varepsilon)r_1 + \cdots + (K_n - \varepsilon)r_n).$$

The next step is to choose the r_j so as to minimize the right-hand side of (22.7). Consider the function of r,

$$r^{-\beta} e^{Br}, \qquad \text{where} \quad \beta \geqslant 0, \quad B > 0.$$

If $\beta = 0$, the minimum of this function is attained for $r = 0$; if $\beta \neq 0$, computation of the derivative shows that the minimum is attained for $r = \beta/B$, and there the value of the function is

$$B^{\beta}(e/\beta)^{\beta}.$$

In view of Sterling's formula, we have, with a suitable constant A_1,

$$(e/\beta)^{\beta} \leqslant A_1(\beta!)^{-1} \qquad (\beta \neq 0).$$

We see therefore that, in both cases (assuming that A_1 is sufficiently large),

$$\inf_{r>0} (r^{-\beta} e^{Br}) \leqslant A_1 B^{\beta}(\beta!)^{-1}.$$

Taking this into account in (22.7) yields (for all $p \in \mathbf{N}^n$)

$$(22.8) \qquad |f^{(p)}(0)| \leqslant A A_1 (K_1 - \varepsilon)^{p_1} \cdots (K_n - \varepsilon)^{p_n}.$$

This is the formula on which the proof of Theorem 22.2 is based. For now let g be an arbitrary entire function. Let us set

$$r_1 = K_1 - \varepsilon/2, \qquad r_n = K_n - \varepsilon/2,$$

and let us apply Estimate (22.5) with g instead of f and with this choice of the numbers r_j. We obtain

$$(22.9)$$

$$\frac{1}{p!} |g^{(p)}(0)| \leqslant (K_1 - \varepsilon/2)^{-p_1} \cdots (K_n - \varepsilon/2)^{-p_n} \sup_{|z_j| = K_j - \varepsilon/2, j=1,\ldots,n} |g(z)|.$$

Let us suppose now that g is a polynomial. We have

$$|\langle f, g \rangle| = \left| \sum_p \frac{1}{p!} f^{(p)}(0) \, g^{(p)}(0) \right| \leqslant \sum_p |f^{(p)}(0)| \frac{1}{p!} |g^{(p)}(0)|.$$

Taking into account (22.8) and (22.9), we obtain

$$|\langle f, g \rangle| \leqslant A A_1 \sum_p \theta_1^{p_1} \cdots \theta_n^{p_n} \sup_{z \in \Delta(K-\varepsilon/2)} |g(z)|,$$

where $\theta_j = (K - \varepsilon)/(K_j - \varepsilon/2) < 1, j = 1,\ldots, n$, and where we have denoted by $K - \varepsilon/2$ the set of numbers $(K_1 - \varepsilon/2,..., K_n - \varepsilon/2)$. Finally, we obtain

$$|\langle f, g \rangle| \leqslant AA_1(1 - \theta_1)^{-1} \cdots (1 - \theta_n)^{-1} \sup_{z \in \Delta(K - \varepsilon/2)} |g(z)|.$$

As the polydisk $\Delta(K - \varepsilon/2)$ is a relatively compact subset of the polydisk $\Delta(K)$, this estimate shows that the linear functional

$$g \rightsquigarrow \langle f, g \rangle$$

is continuous on the space of polynomials when this space carries the topology induced by $H(\Delta(K))$. Therefore, this linear form can be extended to the whole of $H(\Delta(K))$ in a unique manner, as the set of polynomials is dense in $H(\Delta(K))$: if $g \in H(\Delta(K))$, the finite Taylor expansions of g converge to g in $H(\Delta(K))$; cf. Theorems 15.1 and 15.2. This defines the analytic functional μ_f in the open polydisk $\Delta(K)$ corresponding to the function $f \in \tilde{E}xp(K)$.

The mapping $f \rightsquigarrow \mu_f$ is obviously *linear*. Let us show that it is *one-to-one*. It suffices to remark that $g \rightsquigarrow \langle f, g \rangle$ is a continuous linear functional on the space $\mathscr{P}_n$ of polynomials; the same that would have been denoted $g \rightsquigarrow \langle \check{f}, g \rangle$ in the duality between polynomials and formal power series (see the first part of this chapter), where $\check{f} = \sum_p f^p(0) X^p$. We know that this linear form is identically equal to zero if and only if all the coefficients of f are zero, i.e., if $f^{(p)}(0) = 0$ for all p. As f is an entire function this means that $f \equiv 0$.

It remains to show that $f \rightsquigarrow \mu_f$ is *onto*. In order to do this, we use the inverse formula given in the last part of the statement of Theorem 22.2. In fact, let μ be an arbitrary analytic functional in the open polydisk $\Delta(K)$; the restriction to $\Delta(K)$ of the entire function

$$z \rightsquigarrow e^{\langle z, \zeta \rangle}$$

belongs obviously to $H(\Delta(K))$. We have therefore the right to consider

$$\langle \mu, e^{\langle z, \zeta \rangle} \rangle;$$

this is a function of $\zeta \in \mathbf{C}^n$, which we shall denote momentarily by f. We are going to show that f is an entire function, belonging to $\tilde{E}xp(K)$, and that $\mu = \mu_f$.

The continuity of μ implies that there is a constant $C > 0$ and a relatively compact open subset U of $\Delta(K)$ such that, for all $h \in H(\Delta(K))$,

$$(22.10) \qquad |\langle \mu, h \rangle| \leqslant C \sup_{z \in U} |h(z)|.$$

We can find a number $\varepsilon > 0$ such that $U \subset \Delta(K - \varepsilon)$, where we have set $K - \varepsilon = (K_1 - \varepsilon,..., K_n - \varepsilon)$. Let p be an arbitrary n-tuple, and let us choose $h(z) = z^p$ in (22.10). We obtain

$$(22.11) \qquad |\langle \mu, z^p \rangle| \leqslant C(K_1 - \varepsilon)^{p_1} \cdots (K_n - \varepsilon)^{p_n}.$$

On the other hand, let us observe that the Taylor expansion of the function of z, $\exp(\langle z, \zeta \rangle)$, converges to this function in $H(\mathbf{C}^n)$, a fortiori in $H(\Delta(K))$. Since μ is continuous, this means that we have

$$(22.12) \qquad f(\zeta) = \langle \mu, e^{\langle z, \zeta \rangle} \rangle = \sum_{p \in \mathbf{N}^n} \frac{1}{p!} \zeta^p \langle \mu, z^p \rangle.$$

In view of (22.11), the power series at the right-hand side converges for all $\zeta \in \mathbf{C}^n$. This shows that f is an entire function. In virtue of (22.11), furthermore, we have,

$$|f(\zeta)| \leqslant C \sum_p \frac{1}{p!} (K_1 - \varepsilon)^{p_1} \cdots (K_n - \varepsilon)^{p_n} |\zeta^p|$$

$$= C \exp\{(K_1 - \varepsilon)|\zeta_1| + \cdots + (K_n - \varepsilon)|\zeta_n|\}.$$

Finally, we show that $\mu = \mu_f$; let h be an arbitrary polynomial; we have, by using (22.12),

$$\langle f, h \rangle = \sum_p \frac{1}{p!} f^{(p)}(0) h^{(p)}(0) = \sum_p \frac{1}{p!} \langle \mu, z^p \rangle h^{(p)}(0)$$

$$= \left\langle \mu, \left(\sum_p \frac{1}{p!} h^{(p)}(0) z^p \right) \right\rangle = \langle \mu, h \rangle.$$

Q.E.D.

Let now Ω be an arbitrary open subset of $\mathbf{C}^n$; if h is an entire function, the restriction of h to Ω belongs obviously to $H(\Omega)$. Given any analytic functional in Ω, we may consider its value on h, $\langle \mu, h \rangle$. Then it is evident that $h \rightsquigarrow \langle \mu, h \rangle$ is a continuous linear functional on $H(\mathbf{C}^n)$, i.e., an analytic functional in $\mathbf{C}^n$ (sometimes called simply an *analytic functional*). If μ_1, μ_2 are two analytic functionals in Ω, it may happen that $\langle \mu_1, h \rangle = \langle \mu_2, h \rangle$ for all $h \in H(\mathbf{C}^n)$, without this being true for all $h \in H(\Omega)$, i.e., without $\mu_1 = \mu_2$ being true. In view of the Hahn–Banach theorem, this will happen whenever the restriction to Ω of entire functions does not form a dense subspace of $H(\Omega)$, in other words, whenever Ω is not a Runge domain (Definition 15.1). We recall that polydisks are Runge domains. Thus, the space of analytic functionals in Ω, $H'(\Omega)$, can be canonically identified with a linear subspace of $H(\mathbf{C}^n)$ (disregarding now the question of the topologies) if and only if Ω is a Runge domain.

At any event, we may use the following terminology:

Definition 22.2. We say that an analytic functional μ in $\mathbf{C}^n$ is carried by an open set $\Omega \subset \mathbf{C}^n$ if there is a relatively compact open subset U of Ω and a constant $C > 0$ such that, for all entire functions h in $\mathbf{C}^n$,

$$|\langle \mu, h \rangle| \leqslant C \sup_{z \in U} |h(z)|.$$

In other words, μ is carried by Ω if the linear form $h \rightsquigarrow \langle \mu, h \rangle$ defined on the restriction to Ω of the entire functions can be extended as a continuous linear form to the whole of $H(\Omega)$. Furthermore, this extension of μ is unique if and only if Ω is a Runge domain.

Definition 22.3. Let μ be an analytic functional in $\mathbf{C}^n$; the function of $\zeta \in \mathbf{C}^n$,

$$\langle \mu, e^{\langle z, \zeta \rangle} \rangle,$$

will be called the Fourier–Borel transform of μ and denoted by $\hat{\mu}$.

Some authors call $\hat{\mu}$ the *Fourier–Laplace transform* of μ. With these definitions, we may restate Theorem 22.2 in the following way:

THEOREM 22.3. *The Fourier–Borel transformation is a linear isomorphism of the space of analytic functionals in $\mathbf{C}^n$ onto the space of entire functions of exponential type in $\mathbf{C}^n$.*

For every n-tuple of numbers $K_1, ..., K_n$ such that $0 < K_j \leqslant +\infty$ $(j = 1, ..., n)$, the analytic functional μ is carried by the open polydisk

$$\{z \in \mathbf{C}^n; \, |z_1| < K_1, ..., |z_n| < K_n\}$$

if and only if there are positive numbers A, ε such that, for all $z \in \mathbf{C}^n$,

$$|\hat{\mu}(\zeta)| \leqslant A \exp\{(K_1 - \varepsilon)|\zeta_1| + \cdots + (K_n - \varepsilon)|\zeta_n|\}.$$

Exercises

22.1. Let us consider $\mathscr{Q}_n$ (the Fréchet space of formal power series in n indeterminates) as the dual of the LF-space $\mathscr{P}_n$ (the space of polynomials in n indeterminates). Prove that $\mathscr{Q}_n$ is identical to the algebraic dual of $\mathscr{P}_n$ and that, on $\mathscr{Q}_n$, the weak and the strong dual topologies are identical. Furthermore, prove that the algebraic dual of $\mathscr{Q}_n$ is *not* equal to $\mathscr{P}_n$.

22.2. Let us denote by $\bar{P}$ the polynomial obtained by replacing each coefficient of P by its complex conjugate ($P \in \mathscr{P}_n$). The space $\mathscr{P}_n$ can be turned into a Hausdorff pre-Hilbert space by means of the Hermitian form

$$(P \mid Q) = \sum_{p \in \mathbf{N}^n} \frac{1}{p!} P^{(p)}(0)\, \bar{Q}^{(p)}(0).$$

Prove that the natural injection of the pre-Hilbert space $\mathscr{P}_n$ into the space $H(\mathbf{C}^n)$ of entire functions in $\mathbf{C}^n$ is continuous and can be extended as an injection of the completion $\hat{\mathscr{P}}_n$ of $\mathscr{P}_n$ onto a linear subspace, which we shall denote by Λ_n^2, of $H(\mathbf{C}^n)$. Characterize the elements of Λ_n^2 by the Taylor expansion about $z = 0$. Is it true or false that Λ_n^2 is translation invariant (a subset A of H is translation invariant if $f \in A$ implies that the function $z \rightsquigarrow f(z - z^0)$ belongs to A for all $z^0 \in \mathbf{C}^n$)?

22.3. Prove that the space $\mathrm{Exp}(K)$ (Notation 22.2), equipped with the norm

$$(22.13) \qquad \sup_{z \in \mathbf{C}^n} \{| f(z)| \exp(-(K_1|\,z_1\,| + \cdots + K_n|\,z_n\,|))\},$$

is a Banach space.

22.4. Let us consider $3n$ positive numbers K_j, K_j', K_j'' ($1 \leqslant j \leqslant n$) such that, for every j, $K_j < K_j' < K_j''$. Prove that $\mathrm{Exp}(K)$ is dense in $\mathrm{Exp}(K')$ for the topology induced by $\mathrm{Exp}(K'')$ (see Notation 22.2; the topologies are defined by the corresponding norms of the type (22.13)). Prove that the natural injection of $\mathrm{Exp}(K)$ into $\mathrm{Exp}(K')$ is continuous but that it is not an isomorphism.

22.5. Let us denote by $E\mathrm{xp}$ the vector space of all entire functions in $\mathbf{C}^n$ which are of exponential type, that is to say the union of all the spaces $\mathrm{Exp}(K)$ as $K = (K_1,...,K_n)$ ranges over the space $\mathbf{R}_+^n$ of sets of n nonnegative real numbers. We define on $E\mathrm{xp}$ the following topology: a convex subset of $E\mathrm{xp}$, U, is a neighborhood of zero if its intersection with every $\mathrm{Exp}(K)$, $K \in \mathbf{R}_+^n$, is a neighborhood of zero in this Banach space. Prove that the convex set U is a neighborhood of zero in $E\mathrm{xp}$ if and only if $U \cap \mathrm{Exp}(p)$ is a neighborhood of zero in $\mathrm{Exp}(p)$ for all n-tuple $p \in \mathbf{N}^n$.

Prove the following statements:

1. $E\mathrm{xp}$ is *not* the strict inductive limit of the spaces $\mathrm{Exp}(p)$, $p \in \mathbf{N}^n$;
2. $E\mathrm{xp}$ is complete;
3. the Fourier–Borel transformation is an isomorphism (for the TVS structures) of the strong dual H' of H onto $E\mathrm{xp}$.

22.6. Let H' be the space of analytic functionals on $\mathbf{C}^n$, and $\mathscr{F} : H' \to E\mathrm{xp}$ the Fourier–Borel transformation. For every entire analytic function h, let us set

$$\tilde{h}(z) = \overline{h(\bar{z})}, \qquad z \in \mathbf{C}^n.$$

Prove that

$$(22.14) \qquad (\mu, \nu) \rightsquigarrow \langle \mu, \tilde{\mathscr{F}}\nu \rangle,$$

where $\langle\ ,\ \rangle$ is the bracket of the duality between H and H' (using the fact that $E\mathrm{xp} \subset H$), turns H' into a Hausdorff pre-Hilbert space. Let us denote $\hat{H}'$ the Hilbert space obtained by completion of H' for the Hermitian form (22.14). Prove that the Fourier–Borel transformation $\mathscr{F} : H' \to H$ can be extended as an isometry of $\hat{H}'$ onto the Hilbert space Λ_n^2 (Exercise 22.2).

22.7.   Prove the equivalence of the following facts:

(a)   the series $\sum_{p\in\mathbf{N}^n} a_p \delta^{(p)}$ converges in H' (say strongly; $\delta^{(p)}$ is the analytic functional

$$h \rightsquigarrow (-1)^{|p|}\, h^{(p)}(0));$$

(b)   there is an entire function of exponential type, f, in $\mathbf{C}^n$, such that, for all $p \in \mathbf{N}^n$,
$p!a_p = f^{(p)}(0)$.

Prove, furthermore, that the mapping $f \rightarrow \sum_p 1/p!\, f^{(p)}(0)\, \delta^{(p)}$ is an isomorphism (for the linear structure) of $E\mathrm{xp}$ onto H'. What is the relation between this isomorphism and the Fourier–Borel transformation?

22.8.   By making use of Exercise 22.7, show that there is a sequence of elements e_p in H (resp. e_p' in H') ($p \in \mathbf{N}^n$) such that

1. $\langle e_p', e_q \rangle = 1$ if $p = q$, $= 0$ if $p \neq q$;
2.   every element e (resp. e') of H (resp. H') can be written, in one and only one way, as a convergent series

$$e = \sum_{p\in\mathbf{N}^n} a_p e_p \qquad \left(\text{resp. } e' = \sum_{p\in\mathbf{N}^n} b_p e_p'\right).$$

23

Transpose of a Continuous Linear Map

Let E, F be two TVS, and u a continuous linear map of E into F. Let y' be a continuous linear form on F, which we may regard as a continuous linear map of F into $\mathbf{C}$. We are in the situation described by the sequence

$$E \xrightarrow{\ u\ } F \xrightarrow{\ y'\ } \mathbf{C}.$$

We may form the *compose* $y' \circ u$, which is a continuous linear map of E into $\mathbf{C}$, that is to say a continuous linear form on E. Thus we end up with a mapping

$$y' \rightsquigarrow y' \circ u$$

of the dual F' of F into the dual E' of E. This mapping is called the *transpose* of u, and will always be denoted by ${}^t u$ in this book.

If x is an element of E, by using the brackets for expressing the duality between E and E', F and F', respectively, we see that

$$(y' \circ u)(x) = \langle y', u(x) \rangle.$$

As $y' \circ u$ is defined to be ${}^t u(y')$, we have the transposition formula:

$$(23.1) \qquad \langle y', u(x) \rangle = \langle {}^t u(y'), x \rangle.$$

The notion of transpose of a continuous linear map plays a central role in what follows. The reason for this is that important properties of the mapping u itself can be translated, under favorable circumstances, into properties of its transpose. As an example, let us mention the following property: we assume now, as we shall do from now on, that E and F are locally convex (so that we can apply the Hahn–Banach theorem); then the image of u is dense, i.e., $u(E)$ is dense in F, if and only if ${}^t u : F' \to E'$ is one-to-one (Corollary 5 of Theorem 18.1). Another reason for the importance of the notion of transpose lies in the fact that it enables us to extend the basic operations of analysis (differentiation, multiplication by functions, regularizing convolutions, Fourier transformation, etc.) to the new objects which have been introduced by taking into consideration the duals of the spaces of functions. For instance, as immediately seen, the multiplication by a given $\mathscr{C}^\infty$ function ψ defines a continuous linear map of $\mathscr{C}_c^\infty$ into itself; therefore, by transposition, it defines a continuous linear map of the space of distributions $\mathscr{D}'$ into itself, which may be taken as definition of the multiplication of distributions by the function ψ. In the last example, it can be seen that, when the distribution to be multiplied by ψ is a locally integrable function f, its product by ψ (defined by transposition) is equal to the ordinary product ψf. This means that we have indeed extended the operation of multiplication from functions to distributions. A similar procedure is followed—with a twist—when differentiation of distributions is defined. Another important example is Fourier transformation: it is easy to check that it is an isomorphism of the space $\mathscr{S}$ of rapidly decreasing $\mathscr{C}^\infty$ functions (see Chapter 10, Example IV) onto itself; its transpose is then an isomorphism of the dual $\mathscr{S}'$ of $\mathscr{S}$ onto itself. This transpose can then be taken as a definition of the Fourier transformation in $\mathscr{S}'$; on the other hand, $\mathscr{S}'$ can be regarded, in a canonical way, as a vector space of distributions, this is to say, as a linear subspace of $\mathscr{D}'$. We will have thus extended Fourier transformation to a class of distributions (it will be shown that this definition coincides with known ones in the cases where a classical theory of Fourier transformation already exists, for instance when the distributions are L^2 functions). These are only few examples among many which bear witness to the importance of the notion of transpose. We shall study them, and several more, soon, after a few general considerations about transposes.

We begin by a few remarks which do not involve any topology on the dual.

PROPOSITION 23.1. *If $u : E \rightarrow F$ is an isomorphism of E onto F (for the TVS structures), then the transpose of u, ${}^t u : F' \rightarrow E'$ is an isomorphism (for the vector space structures) of F' onto E'.*

Indeed, let v be the inverse of u; $v : F \rightarrow E$. The transpose of u and the one of v are inverse of each other (this means

$${}^t v \circ {}^t u = \text{identity of } F', \qquad {}^t u \circ {}^t v = \text{identity of } E').$$

But a map has an inverse (in the sense just explained) if and only if it is one-to-one and onto.

PROPOSITION 23.2. *Let E, F, and $u : E \rightarrow F$ a continuous linear map. Then we have*

(23.2) $$\text{Ker } {}^t u = (\text{Im } u)^0.$$

We recall that by A^0 we denote the *polar* of the subset A of a TVS; if A is a vector subspace, A^0 is the *orthogonal* of A, that is to say the set of all continuous linear forms on the space which vanish identically on A (then A^0 is a linear subspace of the dual; see Chapter 19, Definition 19.1). The proof of Proposition 23.2 follows immediately from Eq. (23.1): if $y' \in (\text{Im } u)^0$, the right-hand side of (23.1) is equal to zero for all $x \in E$, hence ${}^t u(y') = 0$. If $y' \in \text{Ker } {}^t u$, the right-hand side is zero for all x, hence y' is orthogonal to Im u.

By combining Proposition 23.2 with the Hahn–Banach theorem (Theorem 18.1), we easily obtain Corollary 5 of that theorem stating that ${}^t u$ is one-to-one if and only if Im u is dense in F. Indeed, we assume here that F is locally convex, and in a locally convex space a linear subspace is dense if and only if its orthogonal is reduced to $\{0\}$ (Corollary 1 of Theorem 18.1).

We shall suppose that the duals of E and F carry one of the topologies introduced in Chapter 19. Let $\mathfrak{S}$ (resp. $\mathfrak{H}$) be a family of bounded subsets of E (resp. F) having the property that the union of two subsets belonging to the family is contained in some subset belonging to the family and that, if A belongs to the family and λ is a complex number, there is a set B in the family containing λA. In practice, we shall mainly be interested in the cases where $\mathfrak{S}$ (resp. $\mathfrak{H}$) is one of the following four families of bounded subsets of E (resp. F): the family of all bounded sets, the family of all compact sets, the family of all convex compact sets, the family of all finite sets. We recall the statement of Proposition 19.5:

Let u be a continuous linear map of the TVS E into the TVS F; the transpose of u, ${}^t u$, is continuous as a linear map from $F'_{\mathfrak{H}}$ into $E'_{\mathfrak{S}}$ if to every $A \in \mathfrak{S}$ there is $B \in \mathfrak{H}$ such that $B \supset u(A)$.

As we have already pointed out after the statement of Proposition 19.5, this implies that ${}^t u$ is continuous from F'_σ into E'_σ (weak topologies), from F'_γ into E'_γ (topologies of convex compact convergence), from F'_c into E'_c (topologies of compact convergence), from F'_b into E'_b (strong dual topologies). Observe also that, if ${}^t u$ is continuous from F' into E' when these spaces carry some given topology, it remains continuous if we strengthen the topology of F' or if we weaken the one on E'.

PROPOSITION 23.3.   *If E and F are normed spaces, the norm of ${}^t u$ is equal to the norm of u.*

Proof.   Let us denote all the norms by $\| \ \|$. We must show that

$$\| {}^t u \| = \sup_{y' \in F', \, \| y' \| = 1} \| {}^t u(y') \|$$

$$= \sup_{y' \in F', \, \| y' \| = 1} \left(\sup_{x \in E, \, \| x \| = 1} |\langle y', u(x) \rangle| \right)$$

$$= \sup_{x \in E, \, \| x \| = 1} \left(\sup_{y' \in E', \, \| y' \| = 1} |\langle y', u(x) \rangle| \right)$$

is equal to $\| u \| = \sup_{x \in E, \| x \| = 1} \| u(x) \|$. This will follow at once if we show that, for every $y \in F$, $\| y \|$ is equal to the supremum of $|\langle y', y \rangle|$ over the unit ball $\{y'; \| y' \| \leqslant 1\}$ of F'. By the definition of $\| y' \|$, we have $|\langle y', y \rangle| \leqslant \| y \|$. On the other hand, by virtue of the Hahn–Banach theorem, the continuous linear form on the one-dimensional linear subspace spanned by y, $\lambda y \rightsquigarrow \lambda \| y \|$, can be extended as a continuous linear form y' on F, having the same norm, which is equal to one; thus we have $\langle y', y \rangle = \| y \|$ for some $y' \in F'$, $\| y' \| = 1$.   Q.E.D.

We consider now some examples of transposes.

Example I. Injections of Duals

We consider here a procedure for embedding the dual of a locally convex Hausdorff space E into the dual of another one, F. This procedure is standard and will be used over and over again in our discussion. The starting point is a continuous injection j of E into F; by this we mean a *one-to-one continuous* linear map of E into F. We assume furthermore that *j has a dense image*, i.e., $j(E)$ dense in F. Then, as pointed out in the remark following Proposition 23.2, the transpose of j, ${}^t j : F' \rightarrow E'$, is a one-to-one linear map. Furthermore, if we provide both E' and F' with the strong dual topology (or both with the weak dual topology), ${}^t j$ is continuous. In other words, ${}^t j$ is a continuous injection of F' into E'; for the structures of linear spaces, in particular, we may regard F' as a vector subspace of E'.

Let Ω be an open subset of $\mathbf{R}^n$. Very important examples of the situation just described are provided by the following diagram (where m is some integer $\geqslant 0$):

$$(23.3) \qquad
\begin{array}{ccccc}
\mathscr{C}_c^\infty(\Omega) & \rightarrow & \mathscr{C}_c^m(\Omega) & \rightarrow & \mathscr{C}_c^0(\Omega) \\
\downarrow & & \downarrow & & \downarrow \\
\mathscr{C}^\infty(\Omega) & \rightarrow & \mathscr{C}^m(\Omega) & \rightarrow & \mathscr{C}^0(\Omega)
\end{array}$$

The spaces carry the "natural" topologies which we have defined on them: for the first line, see Chapter 10, Example I; for the second line, see Chapter 13, Example II. Each arrow denotes the "natural" injection, which is a one-to-one continuous linear map. Each arrow has a dense image, as follows immediately from Corollaries 1 and 2 of Theorem 15.3. This means that we can reverse the arrows, replace each space by its dual, say its strong dual, and we obtain, in this way, a diagram of continuous injections. Recalling that the dual of $\mathscr{C}_c^\infty(\Omega)$ is the space of distributions in Ω, we see that the duals of all the spaces entering in

Diagram (23.3) can be regarded as linear subspaces of $\mathscr{D}'(\Omega)$. This suggests the introduction of the following general definition:

Definition 23.1. A linear subspace of $\mathscr{D}'(\Omega)$ carrying a locally convex topology finer than the one induced by the strong dual topology on $\mathscr{D}'(\Omega)$ is called a space of distributions in Ω.

With this terminology, we see that that the diagram of continuous injections with dense images (23.3) enables us to regard the duals of the intervening spaces of functions, as spaces of distributions in Ω. This, in a natural (or canonical) way. We recall that the dual of $\mathscr{C}_c^0(\Omega)$ is the space of Radon measures in Ω; we have already pointed out (using the same argument as here) that Radon measures in Ω can be viewed as a special kind of distribution in Ω. In the next chapter, we shall characterize more "concretely" the distributions which belong to the duals of the other spaces ($\mathscr{C}^m(\Omega)$, $\mathscr{C}_c^k(\Omega)$, $0 \leqslant m \leqslant +\infty$, $0 \leqslant k < \infty$).

Similar arguments can be applied to the couple

$$(23.4) \qquad \mathscr{C}_c^\infty(\Omega) \to L^p(\Omega) \qquad (1 \leqslant p < +\infty).$$

The arrow denotes the natural injection, which is continuous and has a dense image, by virtue of Corollary 3 of Theorem 15.3. Here, the fact that p is finite is essential. At any event, the strong dual of $L^p(\Omega)$ is a space of distributions in Ω. Let us then consider the following diagram:

$$(23.5)$$

$$
\begin{array}{ccc}
 & (L^p(\Omega))' & \\
 & \nearrow w \quad \searrow u & \\
 & & \mathscr{D}'(\Omega), \\
 & \nearrow v & \\
L^{p'}(\Omega) & &
\end{array}
$$

where u is the transpose of the mapping (23.4), p' is the conjugate number of p ($p' = \infty$ if $p = 1$, $p' = p/(p-1)$ if $p > 1$), w is the canonical isometry of $L^{p'}(\Omega)$ onto the dual of $L^p(\Omega)$ (Theorem 20.3), and v is the injection of elements of $L^{p'}(\Omega)$ into $\mathscr{D}'(\Omega)$ when we regard these elements as (classes of) locally integrable functions. The latter means that, if f is such an element, v assigns to f the distributions

$$\mathscr{C}_c^\infty(\Omega) \ni \varphi \rightsquigarrow \int \varphi(x) f(x) \, dx.$$

From the definition of w (Theorem 20.3), it follows immediately that the triangle (23.5) is commutative. This can be rephrased as follows: to identify elements of $L^{p'}(\Omega)$ with distributions in the manner which we have followed (which is to say as locally integrable functions) is consistent with identifying them as continuous linear functionals on $L^p(\Omega)$ in the way it is traditionally done.

Summarizing, we may say the following. By embedding each one of the spaces of Diagram (23.3), as well as the spaces $L^p(\Omega)$ $(1 \leqslant p \leqslant +\infty)$, in the space of locally integrable functions in Ω, we can regard them as spaces of distributions in Ω. Furthermore, now, by transposing the natural injection of $\mathscr{C}_c^\infty(\Omega)$ into each one of them, with the (notable) exception of $L^\infty(\Omega)$, we can also regard their duals as spaces of distributions in Ω. On the other hand, $\mathscr{C}_c^m(\Omega)$, $L^{p'}(\Omega)$, for $0 \leqslant m \leqslant +\infty$, $1 < p' \leqslant +\infty$, are contained in, or even are duals of some of the former, so that for these spaces we have at our disposal two methods of embedding them in $\mathscr{D}'(\Omega)$. What we have just said about Diagram (23.5) shows that these two embeddings are identical. Although a trivial fact, this is important to know.

Let, now, Ω be an open subset of $\mathbf{C}^n$. We may consider the space of test functions in Ω, $\mathscr{C}_c^\infty(\Omega)$. This means that we are identifying canonically $\mathbf{C}^n$ with $\mathbf{R}^{2n}$ and regarding Ω as an open subset of $\mathbf{R}^{2n}$. Observe that the intersection of $\mathscr{C}_c^\infty(\Omega)$ with $H(\Omega)$, space of holomorphic function in Ω, is reduced to the zero element. We certainly are not, therefore, in a situation where we have a natural injection of $\mathscr{C}_c^\infty(\Omega)$ in $H(\Omega)$ with dense image! It follows that we cannot identify the dual of $H(\Omega)$, $H'(\Omega)$, the space of analytic functionals in Ω, with a space of distributions in Ω, at least by the method described here (nor, as a matter of fact, by any other reasonable method).

Example II. Restrictions and Extensions

Let Ω, Ω' be two open subsets of $\mathbf{R}^n$ such that $\Omega \subset \Omega'$. Let f be a function defined in Ω, *having compact support* in Ω. By the *trivial extension of f to Ω'* we mean the function defined in Ω', equal to f in Ω and to zero in the complement of Ω with respect to Ω'.

The trivial extension to Ω' defines a continuous linear map j of $\mathscr{C}_c^\infty(\Omega)$ into $\mathscr{C}_c^\infty(\Omega')$. The transpose of this map is a continuous linear map of $\mathscr{D}'(\Omega')$ into $\mathscr{D}'(\Omega)$ (e.g., for strong dual topologies) which we shall call *restriction to Ω of the distributions in Ω'*. One sees clearly what this means: consider a distribution T in Ω'; make it operate on test functions φ with support in Ω; then $\varphi \rightsquigarrow \langle T, \varphi \rangle$ defines a distribution in Ω, which is precisely the restriction of T to Ω.

Suppose that Ω is $\neq \Omega'$. Then the following is true:

(1)   the image of $\mathscr{C}_c^\infty(\Omega)$ into $\mathscr{C}_c^\infty(\Omega')$ under the trivial extension is *not* dense;

(2)   the trivial extension is *not* an isomorphism of $\mathscr{C}_c^\infty(\Omega)$ into $\mathscr{C}_c^\infty(\Omega')$; it is of course one-to-one and continuous, but its inverse (defined on the image) is not continuous or, if one prefers, the topology of $\mathscr{C}_c^\infty(\Omega)$ is strictly finer than the one induced (via the extension mapping) by $\mathscr{C}_c^\infty(\Omega')$.

We propose the proofs of these two statements as an exercise to the student.

The two preceding statements are related by transposition to the following ones:

(1')   the restriction mapping from $\mathscr{D}'(\Omega')$ to $\mathscr{D}'(\Omega)$ is not one-to-one;

(2')   the restriction mapping from $\mathscr{D}'(\Omega')$ to $\mathscr{D}'(\Omega)$ is not onto.

Concerning (1'), observe that the restriction to Ω of a distribution which is orthogonal to the (extended) test functions with support in Ω is equal to zero. In view of (1) and of the Hahn–Banach theorem, such a distribution certainly exists (in fact, consider the Dirac measure

$$x^0 \rightsquigarrow \varphi(x^0)$$

when $x^0 \in \Omega'$, $x^0 \notin \Omega$). The proof of (2') is more complicated, and will not be given, but the student should keep in mind the two facts above.

Definition 23.2.   A distribution in Ω is said to be extendable to Ω' if it is the restriction to Ω of a distribution in Ω'. It is said to be extendable if it is extendable to $\mathbf{R}^n$.

When the boundary of Ω is sufficiently regular, it is possible to give a characterization of the distributions in Ω which are extendable.

The preceding considerations about $\mathscr{C}_c^\infty$ have obvious analogs for the spaces $\mathscr{C}_c^m, L_c^p$ ($0 \leqslant m \leqslant +\infty$, $1 \leqslant p \leqslant +\infty$) and their duals.

Let us keep Ω and Ω' as before and consider $\mathscr{C}^\infty(\Omega')$. It is obvious what we mean by the *restriction to Ω* of a function in Ω'. The restriction to Ω defines a continuous linear map of $\mathscr{C}^\infty(\Omega')$ into $\mathscr{C}^\infty(\Omega)$. Unless $\Omega = \Omega'$, this restriction mapping is neither one-to-one, nor onto. Indeed, if $\Omega \neq \Omega'$, we may find a nonidentically zero $\mathscr{C}^\infty$ function, as a matter of fact a $\mathscr{C}_c^\infty$ function φ, with support in the complement with respect to Ω of the closure of Ω'; the restriction to Ω of φ is identically zero. On the other hand, it is easy to construct a function which is $\mathscr{C}^\infty$ in Ω but which is not the restriction to Ω of a $\mathscr{C}^\infty$ function in Ω'; for instance, let x^0 be a point of Ω' which belongs to the boundary of Ω and consider the restriction to Ω of the function $| x - x^0 |^{-1}$.

Observe now that $\mathscr{C}^\infty(\Omega')$ contains, as a linear subspace, the set of $\mathscr{C}_c^\infty$ functions (defined in Ω') which have their support contained in Ω.

When we restrict to Ω such a $\mathscr{C}_c^\infty$ function, we obtain the same function, regarded as a function defined in Ω. We can say that the restriction to Ω induces the identity map of $\mathscr{C}_c^\infty(\Omega)$ onto itself. This implies that the image of $\mathscr{C}^\infty(\Omega')$ under this restriction map is dense in $\mathscr{C}^\infty(\Omega)$, as it contains $\mathscr{C}_c^\infty(\Omega)$. Therefore its transpose is a continuous injection of the dual of $\mathscr{C}^\infty(\Omega)$ into the dual of $\mathscr{C}^\infty(\Omega')$. As pointed out in Example I, the former "is" a space of distributions in Ω, the latter is a space of distributions in Ω'. For distributions in Ω which belong to the dual of $\mathscr{C}^\infty(\Omega)$, we have obtained a kind of *extension* mapping to Ω'. In the next chapter, we shall see that this mapping can be indeed regarded as an extension, very similar to the trivial extension of functions belonging to $\mathscr{C}_c^\infty(\Omega)$.

Example III. Differential Operators

As before, let Ω be an open subset of $\mathbf{R}^n$. By a differential operator in Ω, we mean here a linear map of $\mathscr{C}^\infty(\Omega)$ into itself of the form

$$(23.6) \qquad \varphi \rightsquigarrow \sum a_p(x)\, (\partial/\partial x)^p \varphi,$$

where the summation is performed over a finite set of n-tuples $p = (p_1, ..., p_n)$, where for each p the *coefficient* a_p is a complex-valued function defined as $\mathscr{C}^\infty$ in Ω and where, as usual, $(\partial/\partial x)^p$ stands for the differentiation monomial $(\partial/\partial x_1)^{p_1} \cdots (\partial/\partial x_n)^{p_n}$. In analysis, one deals with differential operators for which the condition that the coefficients a_p be $\mathscr{C}^\infty$ is considerably relaxed; for instance, one may want to consider coefficients which are just L^∞. To distinguish this wider class of differential operators from the restricted one we are considering here, one refers to the latter as *differential operators with $\mathscr{C}^\infty$ coefficients*. But as no confusion will arise for us, since we consider only the case of $\mathscr{C}^\infty$ coefficients, we shall use only the shorter name of differential operator, always meaning that the coefficients are $\mathscr{C}^\infty$. We set

$$(23.7) \qquad P(x, \partial/\partial x) = \sum a_p(x)\, (\partial/\partial x)^p,$$

so that the mapping (23.6) might be denoted $\varphi \rightsquigarrow P(x, \partial/\partial x)\varphi$. As we have said, the summation with respect to p in (23.7) is finite. There exists therefore a smallest integer $m \geqslant 0$ such that $a_p(x) \equiv 0$ for all n-tuples p such that $p_1 + \cdots + p_n = |p| > m$. The fact that m is minimum for this property means that there is an n-tuple p, $|p| = m$, such that a_p is not identically zero. This integer m is called the *order* of the differential operator $P(x, \partial/\partial x)$.

The following statement is trivial:

PROPOSITION 23.4. *Let $P(x, \partial/\partial x)$ be a differential operator in Ω. Then the mapping*

$$\varphi \leadsto P(x, \partial/\partial x)\varphi$$

is a continuous linear map of $\mathscr{C}^\infty(\Omega)$ (resp. $\mathscr{C}^\infty_c(\Omega)$) into itself.

Exercise *Prove that the LF topology on $\mathscr{C}^m_c(\Omega)$ $(0 < m \leqslant +\infty)$ is the weakest locally convex topology for which all the differential operators in Ω, of order $< m + 1$, define continuous linear maps of $\mathscr{C}^m_c(\Omega)$ into $\mathscr{C}^0_c(\Omega)$.*

By virtue of Proposition 23.4, we have the right to consider the transpose of the mapping $\varphi \leadsto P(x, \partial/\partial x) \varphi$; this transpose is a continuous linear map of the dual of $\mathscr{C}^\infty(\Omega)$ (resp. of $\mathscr{D}'(\Omega)$) into itself. Let us now consider the case of $\mathscr{D}'(\Omega)$. We observe that functions belonging to $\mathscr{C}^\infty(\Omega)$ are, in particular, locally integrable functions in Ω and therefore define distributions in Ω: $\psi \in \mathscr{C}^\infty(\Omega)$ defines the distribution

$$\varphi \leadsto \int \varphi(x) \, \psi(x) \, dx.$$

We want to find out what the effect is of applying the transpose of $P(x, \partial/\partial x)$ to ψ regarded as a distribution in Ω. In view of Eq. (23.1), we have

$$\langle {}^tP(x, \partial/\partial x) \, \psi, \varphi \rangle = \int \psi(x) \, P(x, \partial/\partial x) \, \varphi(x) \, dx.$$

$$= \sum_p \int a_p(x) \, [(\partial/\partial x)^p \, \varphi(x)] \, \psi(x) \, dx.$$

Consider an integral like

$$\int a_p(x) \, [(\partial/\partial x)^p \, \varphi(x)] \, \psi(x) \, dx.$$

We may integrate by parts. We observe that there are no boundary integrals since φ has compact support contained in the open set Ω. Therefore that integral is equal to

$$\int \varphi(x) \, (-1)^{|p|} \, (\partial/\partial x)^p \, [a_p(x) \, \psi(x)] \, dx.$$

We obtain thus

$$(23.8) \qquad \langle {}^tP(x, \partial/\partial x) \, \psi, \varphi \rangle = \left\langle \sum_p (-1)^{|p|} \, (\partial/\partial x)^p \, [a_p(x) \, \psi], \varphi \right\rangle.$$

This can be phrased as follows: the image of $\psi \in \mathscr{C}^\infty(\Omega)$, viewed as a distribution in Ω, under the transpose of the continuous linear map $\varphi \rightsquigarrow P(x, \partial/\partial x)\, \varphi$ of $\mathscr{C}_c^\infty(\Omega)$ into itself, is the distribution defined by the function

$$\sum_p (-1)^{|p|}\, (\partial/\partial x)^p\, [a_p(x)\, \psi].$$

If we now consider ψ as a function, we observe that

$$\psi \rightsquigarrow \sum_p (-1)^{|p|}\, (\partial/\partial x)^p\, [a_p(x)\, \psi]$$

is a differential operator in Ω. In fact, by using Leibniz' rule we may put it in the usual form; it suffices to observe that

$$(23.9) \qquad (\partial/\partial x)^p\, (a_p \psi) = \sum_{q \leqslant p} \binom{p}{q} [(\partial/\partial x)^{p-q}\, a_p](\partial/\partial x)^q\, \psi,$$

where the summation convention $q \leqslant p$ means that summation is performed over all n-tuples $q = (q_1, ..., q_n)$ such that $q_1 \leqslant p_1, ..., q_n \leqslant p_n$, and where the symbol $\binom{p}{q}$ stands for

$$(23.10) \qquad \binom{p_1}{q_1} \cdots \binom{p_n}{q_n}, \qquad \binom{p_j}{q_j} = p_j! / [(p_j - q_j)!\, q_j!].$$

By using (23.9) we see that we have

$$\sum_p (-1)^{|p|}\, (\partial/\partial x)^p\, [a_p(x)\, \psi] = \sum_p b_p(x)\, (\partial/\partial x)^p\, \psi,$$

where

$$(23.11) \qquad b_p(x) = \sum_{r \geqslant p} (-1)^{|r|} \binom{r}{p} (\partial/\partial x)^{r-p}\, a_r(x).$$

Definition 23.3. The differential operator $\sum_p b_p(x)\, (\partial/\partial x)^p$ with coefficients b_p given by (23.11) is called the formal transpose of the differential operator $\sum_p a_p(x)\, (\partial/\partial x)^p$.

Let us momentarily use the notation $\check{P}(x, \partial/\partial x)$ for the formal transpose of $P(x, \partial/\partial x)$. Equation (23.8) can be rewritten as

$$\langle {}^t P(x, \partial/\partial x)\, \psi, \varphi \rangle = \langle \psi, P(x, \partial/\partial x)\, \varphi \rangle = \langle \check{P}(x, \partial/\partial x)\, \psi, \varphi \rangle.$$

This justifies the following notation, which will be systematically used in the sequel:

Notation 23.1.   *The formal transpose of* $P(x, \partial/\partial x)$ *will be denoted by* ${}^{t}P(x, \partial/\partial x)$.

Observe that we have ${}^{tt}P(x, \partial/\partial x) = P(x, \partial/\partial x)$, hence

$$(23.12) \qquad \langle P(x, \partial/\partial x)\,\psi, \varphi \rangle = \langle \psi, {}^{t}P(x, \partial/\partial x)\,\varphi \rangle.$$

This suggests the following definition:

Definition 23.4.   *The transpose of the continuous linear map of* $\mathscr{C}_c^{\infty}(\Omega)$ *into itself,*

$$\varphi \rightsquigarrow {}^{t}P(x, \partial/\partial x)\varphi$$

$({}^{t}P(x, \partial/\partial x):$ *formal transpose of* $P(x, \partial/\partial x)$*), is a continuous linear map of the space of distributions* $\mathscr{D}'(\Omega)$ *into itself, which will be denoted by*

$$T \rightsquigarrow P(x, \partial/\partial x)\, T.$$

This map will be called a differential operator and denoted by $P(x, \partial/\partial x)$.

Equation (23.12) shows that, when a distribution T is defined by a $\mathscr{C}^{\infty}$ function ψ, the distribution $P(x, \partial/\partial x)T$ is defined by the $\mathscr{C}^{\infty}$ function $P(x, \partial/\partial x)\psi$ (the latter to be understood in the "classical sense"). This is what is meant by saying that differential operators, when acting on $\mathscr{C}^{\infty}$ functions in the sense of distributions, act "in the same way as in the usual sense." One can also say that a differential operator $P(x, \partial/\partial x)$ acting on distributions, as we have defined it, is an extension of the operator so denoted when acting on $\mathscr{C}^{\infty}$ functions.

Two particular cases are worth looking at:

(1)   *Differential Operators of Degree Zero*

Such a differential operator, first defined in $\mathscr{C}_c^{\infty}(\Omega)$, is nothing but the usual multiplication of test functions by a function $\alpha \in \mathscr{C}^{\infty}(\Omega)$. The differential operator is therefore the mapping $\varphi \rightsquigarrow \alpha\varphi$; its formal transpose is identical to it. In accordance with Definition 23.4, we may define αT by the formula

$$(23.13) \qquad \langle \alpha T, \varphi \rangle = \langle T, \alpha\varphi \rangle.$$

Equation (23.13) can be regarded as the definition of the multiplication of a distribution T in Ω by the $\mathscr{C}^{\infty}$ function α in Ω; $T \rightsquigarrow \alpha T$ is a continuous linear map of $\mathscr{D}'(\Omega)$ into itself. When T is a $\mathscr{C}^{\infty}$ function, αT is the $\mathscr{C}^{\infty}$ function thus denoted in the usual sense.

(2)   *Differential Operators with Constant Coefficients*

These are the finite linear combinations with complex coefficients of the differentiation monomials $(\partial/\partial x)^p$. We shall use a notation like $P(\partial/\partial x)$ for such an operator:

$$P(\partial/\partial x) = \sum_{|p| \leqslant m} a_p (\partial/\partial x)^p, \qquad a_p : \text{complex numbers}.$$

As we shall often do, we have denoted by m the order of $P(\partial/\partial x)$, this is to say the smallest integer such that $a_p = 0$ for $|p| > m$. To the operator $P(\partial/\partial x)$ we may associate the polynomial in n letters $x_1, ..., x_n$, with complex coefficients, of *degree m*,

$$P(X) = \sum_{|p| \leqslant m} a_p X^p.$$

We have used the notation $X^p = X_1^{p_1} \cdots X_n^{p_n}$ Conversely, given any polynomial in n indeterminates $X_1, ..., X_n$, with complex coefficients, we may associate with it the differential operator $P(\partial/\partial x)$ obtained by substituting $\partial/\partial x_j$ for X_j for every $j = 1, ..., n$.

What is the formal transpose of $P(\partial/\partial x)$? An immediate computation shows that it is the differential operator $P(-\partial/\partial x)$. Thus, if we want to consider the extension of the differential operator $P(\partial/\partial x)$ to distributions we have to use the formula

$$\langle P(\partial/\partial x)T, \varphi \rangle = \langle T, P(-\partial/\partial x)\varphi \rangle.$$

Take for instance the case where $P(\partial/\partial x)$ is the single partial differentiation $\partial/\partial x_j$ $(1 \leqslant j \leqslant n)$:

$$\langle \partial T/\partial x_j, \varphi \rangle = -\langle T, \partial\varphi/\partial x_j \rangle.$$

Exercises

23.1. Let $\mathscr{P}_n$ and $\mathscr{Q}_n$ be the spaces of polynomials and formal power series in n indeterminates, with complex coefficients, equipped with their LF and Fréchet topologies, respectively. Let us regard them as duals of each other by means of the bracket

$$\langle P, u \rangle = \sum_{p \in \mathbb{N}^n} \frac{1}{p!} P^{(p)}(0) u_p,$$

with $u = \sum_{p \in \mathbb{N}^n} u_p X^p \in \mathscr{Q}_n$ and $P \in \mathscr{P}_n$. Let Q be an arbitrary polynomial, belonging to $\mathscr{P}_n$. What is the transpose of the (continuous linear) mapping

(23.14) $$f \rightsquigarrow Q(\partial/\partial X)f$$

of $\mathscr{Q}_n$ (resp. $\mathscr{P}_n$) into itself?

By using Exercise 2.3, prove that, unless $Q = 0$, the mapping (23.14) of $\mathscr{P}_n$ into itself is surjective, and that this assertion is still true if we replace Q by a formal power series.

23.2. Let $\mathscr{F}$ be the Fourier–Borel transformation (Definition 22.3); $\mathscr{F} : H' \to \operatorname{Exp}$ (H' : space of analytic functionals in $\mathbf{C}^n$, Exp: space of entire functions of exponential type in $\mathbf{C}^n$). If $u : H' \to H'$ is a linear mapping, we set $\hat{u} : \mathscr{F} \circ u \circ \mathscr{F}^{-1}$. Let P be a polynomial in n letters with complex coefficients. Then we may consider the *differential polynomial* in $\mathbf{C}^n$, $P(\partial/\partial z)$,

$$H \ni h \rightsquigarrow (z \rightsquigarrow P(\partial/\partial z) \, h(z)), \qquad P(\partial/\partial z) = P(\partial/\partial z_1 , ..., \partial/\partial z_n).$$

What is $\hat{u}$ when $u : H' \to H'$ is the transpose of $P(\partial/\partial z)$?

23.3. Let E be a Hilbert space, $(\mid)_E$ the inner product in E, and F a second Hilbert space with inner product $(\mid)_F$. By using the definition of the transpose of a continuous linear map and the canonical linear isometry of a Hilbert space onto its antidual, prove that to every continuous linear map $u : E \to F$ there is a unique continuous linear map of F into E (denoted by u^* and called the *adjoint* of u) such that, for all $x \in E, y \in F$,

$$(u(x) \mid y)_F = (x \mid u^*(y))_E .$$

Prove that the mapping $u \to u^*$ is an antilinear isometry of $L(E; F)$ onto $L(F; E)$ (both spaces equipped with the operators norm; antilinear means that $(\lambda u)^* = \bar{\lambda} u^*$).

23.4. Let us denote by $\delta^{(p)}$, $p \in \mathbf{N}^n$, the analytic functional in $\mathbf{C}^n$,

$$h \rightsquigarrow (-1)^{|p|} h^{(p)}(0).$$

If P is a polynomial in n variables, with complex coefficients,

$$P(X) = \sum_p c_p X^p,$$

we set

$$P(\partial/\partial z) \, \delta^{(q)} = \sum_p c_p \delta^{(p+q)}.$$

By using the representation of analytic functionals introduced in Exercise 22.7, we extend $P(\partial/\partial z)$ to the whole of H'. On the other hand, $P(\partial/\partial z)$ operates on entire functions in the usual fashion and defines thus a continuous linear map of H into itself. Prove that the transpose of

$$P(\partial/\partial z) : H \to H$$

is equal to

$$P(-\partial/\partial z) : H' \to H'.$$

23.5. By using the fact that the ordinary multiplication of functions, $(f, g) \rightsquigarrow fg$, is a continuous bilinear map of $H \times H$ into H, prove that $(f, \mu) \rightsquigarrow f\mu$ is a separately continuous bilinear map of $H \times H'$ into H'. Suppose (cf. Exercise 22.7) that an analytic functional μ is given by

$$\mu = \sum_{p \in \mathbf{N}^n} a_p \delta^{(p)}.$$

What is the series representing $f\mu$?

23.6. Prove that, for each $m = 1, 2, ...$,

$$\phi \rightsquigarrow \frac{1}{(m-1)!} \int_{-\infty}^{+\infty} \phi^{(m)}(t) \log |t| \, dt$$

is a distribution in the real line extending the distribution in $\mathbf{R}^1 - \{0\}$ defined by the (locally L^1) function t^{-m}.

24

Support and Structure of a Distribution

Let Ω be an open subset of $\mathbf{R}^n$.

Definition 24.1. *A distribution T in Ω is said to vanish in an open subset U of Ω if $\langle T, \phi \rangle = 0$ for all functions $\phi \in \mathscr{C}_c^\infty(\Omega)$ having their support in U.*

In the terminology introduced in the preceding chapter (Chapter 23, Example II), we say that $T \in \mathscr{D}'(\Omega)$ vanishes in U, or is equal to zero in U, if the restriction of T to U is the zero distribution. As usual, when a definition such as Definition 24.1 is introduced, we ought to check that it is consistent with the terminology for functions. In the present situation, we ought to check that, if f is a locally integrable function in Ω and if f vanishes in U as a distribution, then f vanishes in U as a function, which means, in the framework of distribution theory, that f vanishes *almost everywhere* in U. But this is an immediate consequence of Theorem 21.3.

The following theorem states a very important property of distributions.

THEOREM 24.1. *Let $\{U_i\}$ $(i \in I)$ be a family of open subsets of $\mathbf{R}^n$. For each index $i \in I$, let T_i be a distribution on U_i. Suppose that, for every pair of indices $i, j \in I$, the restrictions to $U_i \cap U_j$ of T_i and T_j coincide. Then, there exists a unique distribution T on the union U of the sets U_i whose restriction to every set U_i is equal to T_i.*

The proof makes use of the existence of partitions of unity in $\mathscr{C}^\infty$ (see Chapter 16). First of all, by Theorem 16.1, we know that there is an open covering $\{V_j\}$ $(j \in J)$ of $U = \bigcup_{i \in I} U_i$ which is finer than the covering $\{U_i\}$ $(i \in I)$ and which is locally finite. Next, by Theorem 16.3, we know that there is a partition of unity in $\mathscr{C}^\infty(U)$, $\{g_j\}$ $(j \in J)$, subordinated to the covering $\{V_j\}$. Now, if $\phi \in \mathscr{C}_c^\infty(U)$, we have

$$\phi = \sum_{j \in J} g_j \phi,$$

253

where $g_j \phi$ is identically zero for all j with the possible exception of a finite number of them. In order to prove the *existence* of T, we set

$$\langle T, \phi \rangle = \sum_{j \in J} \langle T_{i(j)}, g_j \phi \rangle,$$

where $i(j)$ is an index belonging to I such that $V_j \subset U_{i(j)}$ (such an index exists since the covering $\{V_j\}$ is finer than the covering U_i). Observe that it does not make any difference what index $i(j)$ we assign to j provided that $V_j \subset U_{i(j)}$: if $V_j \subset U_i \cap U_{i'}$, we have, in view of our hypothesis, $\langle T_i, g_j \phi \rangle = \langle T_{i'}, g_j \phi \rangle$. This same hypothesis shows also that our definition of $\langle T, \phi \rangle$ is a correct one, i.e., is independent of the covering $\{V_j\}$ and of the partition of unity $\{g_j\}$. For let $\{h_k\}$ be another partition of unity in $\mathscr{C}^\infty(U)$ subordinated to some covering $\{W_k\}$ (open, locally finite, finer than $\{U_i\}$). Let us select, for each k, an index $i(k) \in I$ such that $W_k \subset U_{i(k)}$. We have

$$\sum_j \langle T_{i(j)}, g_j \phi \rangle = \sum_{j,k} \langle T_{i(j)}, g_j h_k \phi \rangle$$

$$= \sum_{j,k} \langle T_{i(k)}, g_j h_k \phi \rangle = \sum_k \langle T_{i(k)}, \sum_j g_j h_k \phi \rangle$$

$$= \sum_k \langle T_{i(k)}, h_k \phi \rangle.$$

Suppose now that ϕ converges to zero in some space $\mathscr{C}^\infty_c(K)$ (K: compact subset of U). There is a finite subset J' of J such that $g_j \phi \equiv 0$ if $j \notin J'$; J' depends only on K. If $j \in J'$, $g_j \phi \to 0$ in $\mathscr{C}^\infty_c(U_{i(j)})$, therefore $\langle T_{i(j)}, g_j \phi \rangle \to 0$. This proves the continuity on $\mathscr{C}^\infty_c(U)$ of the linear functional $\phi \leadsto \langle T, \phi \rangle$ (cf. Proposition 13.1 and corollary). Thus T, defined above, is indeed a distribution in U.

Next, suppose that the support of ϕ is contained in U_i. Then the support of $g_j \phi$ is contained in $V_j \cap U_i \subset U_{i(j)} \cap U_i$; hence

$$\langle T_{i(j)}, g_j \phi \rangle = \langle T_i, g_j \phi \rangle$$

and $\langle T, \phi \rangle = \langle T_i, \sum_j g_j \phi \rangle = \langle T_i, \phi \rangle$. Thus the restriction of T to U_i is equal to T_i.

Lastly, we prove that T is unique. If T' were a second distribution in U whose restriction to every U_i were equal to T_i, the restriction of the difference $T - T'$ to every U_i would be equal to zero. Consequently, the uniqueness of T will follow from the next statement, which can be viewed as a corollary of Theorem 24.1:

COROLLARY 1.   *If a distribution T, defined in the union of a family of open sets $\{U_i\}$ ($i \in I$) vanishes in every U_i, T vanishes in their union.*

We use the same refinement $\{V_j\}$ of $\{U_i\}$ and the same partition of unity $\{g_j\}$ subordinated to it, as in the first part of the proof. For an arbitrary $\phi \in \mathscr{C}_c^\infty(U)$, $g_j\phi = 0$ except for a finite number of indices j and therefore we have, by linearity,

$$\langle T, \phi \rangle = \sum_{j \in J} \langle T, g_j\phi \rangle.$$

But as $\operatorname{supp}(g_j\phi) \subset V_j \subset U_i$ for some i, we have $\langle T, g_j\phi \rangle = 0$. This is true for each j, whence the theorem.

COROLLARY 2.   *The union of all the open subsets of Ω in which the distribution T vanishes is an open subset of Ω in which T vanishes.*

Definition 24.2.   Let T be a distribution in Ω. The complement of the largest open subset of Ω in which T vanishes is called the support of T and will be denoted by

$$\operatorname{supp} T.$$

By complement we mean, in Definition 24.2, the complement with *respect to* Ω. This implies that supp T is a relatively closed subset of Ω; it need not be a closed subset of $\mathbf{R}^n$. Evidently, we have used the fact, stated in Corollary 2 of Theorem 24.1, that the largest open subset of Ω in which T vanishes indeed exists. This is not to say that this largest open set cannot be empty. If f is a locally integrable function in Ω, the support of f in the sense of distributions is the smallest closed subset of Ω in the complement of which f is almost everywhere equal to zero. If f is continuous, the notion of support in the sense of functions (the closure of the set of points where f is nonzero) is identical to the notion in the sense of distributions, as readily seen. The support of the Dirac measure δ_{x_0} at a point x_0 of Ω is the set $\{x_0\}$.

By virtue of the argument developed in Chapter 23, Example I, we may identify the dual of $\mathscr{C}^\infty(\Omega)$ with a linear subspace of $\mathscr{D}'(\Omega)$—in a canonical manner. We recall that this is done by transposing the canonical injection of $\mathscr{C}_c^\infty(\Omega)$ into $\mathscr{C}^\infty(\Omega)$, and by observing that this injection has a dense image. The notion of support of a distribution enables us to give a very simple characterization of the distributions which belong to the dual of $\mathscr{C}^\infty(\Omega)$. But before stating it, let us introduce the following notation, by now very standard:

Notation 24.1.   The dual of $\mathscr{C}^\infty(\Omega)$ (regarded as a space of distributions in Ω) is denoted by $\mathscr{E}'(\Omega)$.

THEOREM 24.2.   *The distributions belonging to $\mathscr{E}'(\Omega)$ are the distributions having compact support in Ω.*

The proof of Theorem 24.2 is based on the obvious remark that a distribution T belongs to $\mathscr{E}'(\Omega)$ if and only if the linear form $\phi \rightsquigarrow \langle T, \phi \rangle$ is continuous on $\mathscr{C}_c^\infty(\Omega)$ for the topology induced by $\mathscr{C}^\infty(\Omega)$. Indeed, if the form is continuous for the induced topology, it has an extension, necessarily unique, to the whole of $\mathscr{C}^\infty(\Omega)$. The statement, in the other direction, is trivial.

Proof of Theorem 24.2.   Let T be a distribution belonging to $\mathscr{E}'(\Omega)$. In view of the definition of the topology of $\mathscr{C}^\infty(\Omega)$, there is a compact subset K of Ω, an integer $m \geqslant 0$, and a constant $C > 0$ such that, for all test functions ϕ in Ω,

$$|\langle T, \phi \rangle| \leqslant C \sup_{|p| \leqslant m} (\sup_{x \in K} |(\partial/\partial x)^p \phi(x)|).$$

This implies immediately that $\langle T, \phi \rangle = 0$ whenever the support of ϕ is contained in the complement of K, which means that supp $T \subset K$.

Let, now, T be a distribution in Ω with compact support, K. Let $\alpha \in \mathscr{C}_c^\infty(\Omega)$ be equal to one in some neighborhood of K; such a function exists, by Theorem 16.4. For all test functions ϕ in Ω, we have

$$\langle T, \phi \rangle = \langle T, \alpha\phi \rangle,$$

since the support of $(1 - \alpha)\phi$ is contained in the complement of supp T. Since all the functions $\alpha\phi$ have their support contained in a fixed compact subset of Ω, namely supp α, and since on $\mathscr{C}_c^\infty(\text{supp } \alpha)$ the topologies induced by $\mathscr{C}^\infty(\Omega)$ and by $\mathscr{C}_c^\infty(\Omega)$ coincide, we observe that $\alpha\phi$ converges to zero in $\mathscr{C}_c^\infty(\Omega)$ whenever ϕ converges to zero in $\mathscr{C}^\infty(\Omega)$; in this case, therefore $\langle T, \phi \rangle \to 0$. This shows that the linear form $\phi \rightsquigarrow \langle T, \phi \rangle$ is continuous on $\mathscr{C}_c^\infty(\Omega)$ for the topology induced by $\mathscr{C}^\infty(\Omega)$, hence that $T \in \mathscr{E}'(\Omega)$.   Q.E.D.

Now that we know what the elements of $\mathscr{E}'(\Omega)$ are, dual of $\mathscr{C}^\infty(\Omega)$, we can easily interpret the transpose of the restriction mapping

$$\mathscr{C}^\infty(\Omega') \to \mathscr{C}^\infty(\Omega), \qquad \Omega \subset \Omega' \quad \text{(Chapter 23, Example II)}.$$

It is the extension mapping

$$\mathscr{E}'(\Omega) \to \mathscr{E}'(\Omega'),$$

which assigns to a distribution T with compact support in Ω the distribution in Ω' which is equal to T in Ω and to zero in the complement of supp T with respect to Ω'. This extension mapping coincides with what we have called the trivial extension of functions with compact support in Ω (see p. 245).

Exercises

24.1. Prove that, for all distributions S, T in Ω and all complex numbers $\lambda \neq 0$,

$$\text{supp}(S + T) \subset \text{supp } S \cup \text{supp } T,$$
$$\text{supp}(\lambda T) = \text{supp } T.$$

(This implies that the distributions with support in a given set $A \subset \Omega$ form a vector subspace of $\mathscr{D}'(\Omega)$.)

24.2. The vector subspace of $\mathscr{D}'(\Omega)$ consisting of all the distributions having their support in a given subset A of Ω is weakly closed in $\mathscr{D}'(\Omega)$ if and only if A is closed in Ω. Prove this statement.

24.3. Let $P(x, \partial/\partial x)$ be a differential operator in Ω. Prove that, for all distributions $T \in \mathscr{D}'(\Omega)$, we have

$$\text{supp } P(x, \partial/\partial x) \, T \subset \text{supp } T.$$

Prove that, for all $T \in \mathscr{D}'(\Omega)$ and all functions $\phi \in \mathscr{C}^\infty(\Omega)$,

$$\text{supp } (\phi T) \subset \text{supp } \phi \cap \text{supp } T.$$

Distribution theory has been built primarily in order to extend a number of basic operations of analysis, like differentiation, to functions for which these operations were not well defined in the classical framework. Needless to say, the final product of such an extended operation applied to a function for which it was not defined before will in general not be a function. It will be a rather singular distribution. But the important point is that it will be an object which we shall know how to manipulate in computations and reasonings. Since locally integrable functions are contained in the space of distributions, injectively up to equality almost everywhere, and since we have defined differential operators acting on distributions, we know how to differentiate in the sense of distributions any locally integrable function. We know also that the result of this differentiation, when applied to differentiable functions, will be the same as in the classical theory. At any event, the differential operators applied to functions yield a large class of distributions: the distributions of the form $P(x, D)f$, f locally integrable function. This raises a natural question: are there distributions which can not be represented as finite linear combinations of derivatives (in the sense of distributions!) of functions? Or to put it in different words, by introducing distributions in the manner we have, i.e., as elements of the dual of the LF-space of test functions, have we not gone beyond our scope, introducing too many objects whose interpretation may turn out to be exaggeratedly complicated? The answer to this question, as we shall now see, is no—at least in the local. It is not true, however, that globally any distribution is a finite sum of derivatives of functions, as is easily seen (see Example 24.1 below). The latter phenomenon is due to our implicit requirement (to be made explicit later) that the space of distributions be complete. The advantages deriving from the completeness of $\mathscr{D}'(\Omega)$ outnumber the ones that would follow from the "expulsion" from $\mathscr{D}'(\Omega)$ of the distributions which are not globally finite sums of derivatives of functions, especially if we take into account that the local structure of all distributions is of the kind "finite sums of derivatives of functions."

Let m be a finite nonnegative integer. We know (Corollary 2 of Theorem 15.4) that $\mathscr{C}_c^\infty(\Omega)$ is dense in $\mathscr{C}_c^m(\Omega)$. Therefore, if we transpose the natural injection of $\mathscr{C}_c^\infty(\Omega)$ into $\mathscr{C}_c^m(\Omega)$, we obtain an injection of the dual of $\mathscr{C}_c^m(\Omega)$ into $\mathscr{D}'(\Omega)$, this is to say: we obtain a "realization" of the dual of $\mathscr{C}_c^m(\Omega)$ as a space of distributions in Ω.

Definition 24.3. *The space of distributions in Ω which is the dual of $\mathscr{C}_c^m(\Omega)$ is denoted by $\mathscr{D}'^m(\Omega)$; its elements are called the* distributions of order $\leqslant m$ in Ω.

A distribution T in Ω is said to be of finite order *if there is an integer $m \geqslant 0$ such that T is of order $\leqslant m$ in Ω. The set of distributions of finite order in Ω is denoted by $\mathscr{D}'^F(\Omega)$.*

If $m' \geqslant m$, we have obviously $\mathscr{D}'^m(\Omega) \subset \mathscr{D}'^{m'}(\Omega)$. Therefore, $\mathscr{D}'^F(\Omega)$ is a vector subspace of $\mathscr{D}'(\Omega)$. We shall see in a moment that $\mathscr{D}'^F(\Omega)$ is different from $\mathscr{D}'(\Omega)$, unless $\Omega = \varnothing$.

A distribution which is of order $\leqslant m$ but which is not of order $\leqslant m - 1$ may be said to be *of order m*. The distributions of order zero in Ω are nothing else but the Radon measures in Ω.

We begin by showing that every distribution in Ω is *locally* of finite order.

THEOREM 24.3.   *Let U be a relatively compact open subset of Ω. The image of $\mathscr{D}'(\Omega)$ under the restriction mapping to U is contained in $\mathscr{D}'^F(U)$.*

Proof.   Let us denote by K the closure of U; K is a compact subset of Ω. The restriction of any distribution T in Ω to $\mathscr{C}_c^\infty(K)$ is a continuous linear form on this Fréchet space, by virtue of the definition of the LF topology on $\mathscr{C}_c^\infty(\Omega)$. But the topology of $\mathscr{C}_c^\infty(K)$ is equal to the intersection of the topologies induced by the spaces $\mathscr{C}_c^m(K)$ for $m = 0, 1, 2,\ldots.$ This means that the linear form $\phi \rightsquigarrow \langle T, \phi \rangle$ is continuous on $\mathscr{C}_c^\infty(K)$ for the topology induced by $\mathscr{C}_c^m(K)$ for some finite m. A fortiori, this form is continuous on $\mathscr{C}_c^\infty(U) \subset \mathscr{C}_c^\infty(K)$ for the topology induced by this space $\mathscr{C}_c^m(K)$. Observe then that the topology of $\mathscr{C}_c^m(U)$ is finer than the topology induced on this space by $\mathscr{C}_c^m(K)$. Therefore $\phi \rightsquigarrow \langle T, \phi \rangle$ is continuous on $\mathscr{C}_c^\infty(U)$ for the topology induced by $\mathscr{C}_c^m(U)$, which means precisely that the restriction to U of T belongs to $\mathscr{D}'^m(U)$. Q.E.D.

COROLLARY.   *Every distribution with compact support in Ω is of finite order.*

Indeed, let T belong to $\mathscr{E}'(\Omega)$; there is a relatively compact open subset U of Ω which contains supp T. The restriction of T to U is of finite order; but $T = 0$ in the complement of supp T. This implies immediately that T is of finite order in the whole of Ω.

The next result is the theorem of structure for a distribution of finite order.

THEOREM 24.4.   *Let T be a distribution of order $\leqslant m < +\infty$ in Ω, and let $S \subset \Omega$ be its support. Given any open neighborhood U of S in Ω, there is a family of Radon measures $\{\mu_p\}$ ($p \in \mathbf{N}^n$, $| p | \leqslant m$) in Ω such that*

$$T = \sum_{|p| \leqslant m} (\partial/\partial x)^p \mu_p \, ,$$

and such that supp $\mu_p \subset U$ *for every* $p \in \mathbf{N}^n$, $| p | \leqslant m$.

Proof.   Let $N = N(m, n)$ be the number of n-tuples $p = (p_1, ..., p_n)$ such that $| p | \leqslant m$. For simplicity, let us set $E_m = \mathscr{C}_c^m(\Omega)$. There is a natural injection of E_m into the product space $(E_0)^N$: it is the mapping which assigns to each $\phi \in E_m$ the set $((\partial/\partial x)^p \phi)_{(p \in \mathbf{N}^n, |p| \leqslant m)}$ of its derivatives of order $\leqslant m$. This mapping is obviously linear, obviously *not* onto. But it is an isomorphism into for the structures of TVS, as immediately seen (ϕ converges to zero in E_m if and only if every one of its derivatives of order $\leqslant m$ converges to zero in E_0). Let us denote by $\ddot{E}_m$ the image of E_m under this isomorphism. We may transfer any continuous linear functional on E_m as a continuous linear functional on $\ddot{E}_m$ and then extend the latter as a continuous linear functional on $(E_0)^N$. But the dual of a product of a finite family of TVS $(F_1, ..., F_r)$ is canonically isomorphic to the product of their duals, via the correspondence

$$F_1' \times \cdots \times F_r' \ni y' = (y_1', ..., y_r')$$

$$\rightsquigarrow \Big(y = (y_1, ..., y_r) \rightsquigarrow \langle y', y \rangle = \sum_{j=1}^{r} \langle y_j', y_j \rangle \Big) \in (F_1 \times \cdots \times F_r)'.$$

Applying this to the product $(E_0)^N$, we see that a continuous linear form on it is a set of N Radon measures $\{\lambda_p\}$ ($p \in \mathbf{N}^n$, $| \, p \, | \leqslant m$) in Ω, operating in the following manner:

$$\langle \{\lambda_p\}, \{\phi_p\} \rangle = \sum_{p \in \mathbf{N}^n, |p| \leqslant m} \langle \lambda_p, \phi_p \rangle.$$

It suffices to assume that this linear form extends the linear form $\phi \rightsquigarrow \langle T, \phi \rangle$ transferred on $\dot{E}_m$ and to take $\phi_p = (\partial/\partial x)^p \phi$ for each $p \in \mathbf{N}^n$, with $\phi \in E_m$, to see that

$$T = \sum_{|p| \leqslant m} (-1)^{|p|} (\partial/\partial x)^p \lambda_p \, .$$

We must now satisfy the condition on the supports of the measures μ_p. We take a function $g \in \mathscr{C}^\infty(\Omega)$ equal to one in a neighborhood of supp T and vanishing identically outside some closed subset contained in U. Such a function exists in view of Theorem 16.4. We consider the multiplicative product gT. First of all, $gT = T$ since for all test functions ϕ, $\langle gT, \phi \rangle = \langle T, g\phi \rangle = \langle T, \phi \rangle$, and supp$(1 - g)\phi$ is contained in the complement of supp T. Therefore we have

$$(24.1) \qquad T = gT = \sum_{|p| \leqslant m} (-1)^{|p|} g \, (\partial/\partial x)^p \lambda_p \, .$$

On the other hand, by Leibniz' formula,

$$\langle g \, (\partial/\partial x)^p \lambda_p, \phi \rangle = (-1)^{|p|} \langle \lambda_p, (\partial/\partial x)^p (g\phi) \rangle$$

$$= (-1)^{|p|} \sum_{q \leqslant p} \binom{p}{q} \langle \lambda_p, [(\partial/\partial x)^{p-q} g] (\partial/\partial x)^q \phi \rangle$$

$$= \sum_{q \leqslant p} \langle (\partial/\partial x)^q [(-1)^{|p-q|} \{(\partial/\partial x)^{p-q} g\} \lambda_p], \phi \rangle,$$

where $q \leqslant p$ means $q_1 \leqslant p_1, ..., q_n \leqslant p_n$. This means that

$$(24.2) \qquad g \, (\partial/\partial x)^p \lambda_p = \sum_{q \leqslant p} (\partial/\partial x)^q [(-1)^{|p-q|} \{(\partial/\partial x)^{p-q} g\} \lambda_p].$$

It suffices, in order to obtain the representation of T whose existence is stated in Theorem 24.4, to substitute the expressions (24.2) in (24.1), to reorder the summation and to observe that the supports of the Radon measures

$$[(\partial/\partial x)^{p-q} g] \lambda_p$$

are contained in supp g, which, in turn, is contained in U.   Q.E.D.

COROLLARY 1. *A distribution T in Ω is of order $\leqslant m$ if and only if it is equal to a finite sum of derivatives of order $\leqslant m$ of Radon measures in Ω.*

The necessity is stated in Theorem 24.4; the sufficiency is evident.

COROLLARY 2. *Let T be a distribution with compact support in Ω. There is a finite integer $m \geqslant 0$ such that, given any neighborhood U of supp T, T is equal to a finite sum of derivatives of order $\leqslant m$ of Radon measures, all of which have their support contained in U.*

Remark 24.1. In general, it is not possible to represent a distribution of order $\leqslant m$ (even if it has compact support!) as a finite sum of derivatives of Radon measures whose support is contained in its own support (even if we lift the restriction that the derivatives be of order $\leqslant m$).

As we have already stated, not every distribution in an open set $\Omega \neq \varnothing$ is of finite order.

Example 24.1. Let $\{x^k\}$ ($k = 1, 2,...$) be a sequence of points in Ω such that every compact subset of Ω contains only a finite number of them; let $\{a_k\}$ be an arbitrary sequence of complex numbers. The series

$$\sum_{k=1}^{\infty} a_k \, (\partial/\partial x_1)^k \cdots (\partial/\partial x_n)^k \, \delta_{x^k}$$

defines a distribution in Ω. However, unless the coefficients a_k are all equal to zero, with the possible exception of a finite number of them, this distribution is not of finite order in Ω.

It is clear what is meant by a *convergent series* of distributions,

$$\sum_{k=0}^{\infty} T_k \, .$$

It means that the partial sums $\sum_{k=0}^{K} T_k$ ($K = 0, 1,...$) form a sequence converging in $\mathscr{D}'(\Omega)$, say for the strong dual topology (we shall see, soon, that for sequences, strong and weak convergence are the same thing).

Exercises

24.4. Let $\{g_k\}$ ($k = 0, 1,...$) be a sequence of functions belonging to $\mathscr{C}_c^{\infty}(\Omega)$ and converging to the function 1 (i.e., the function identically equal to one in Ω) in $\mathscr{C}^{\infty}(\Omega)$. Prove that, for every distribution T in Ω, the distributions $g_k T$ converge to T in $\mathscr{D}'(\Omega)$ (for the strong dual topology).

24.5. Prove that every distribution T in Ω is equal to a convergent series $\sum_{k=0}^{+\infty} T_k$ in which each term T_k has a compact support.

24.6. Express the Dirac measure δ at the origin, on the real line $\mathbf{R}^1$, as a finite sum of derivatives of continuous functions, all of which have their support in the interval $]-\varepsilon, \varepsilon[$ ($\varepsilon > 0$ arbitrary). Prove that δ cannot be expressed as the derivative (of some order) of a single continuous function with compact support.

The fact, stated on p. 258, that every distribution is equal, at least locally, to a finite sum of derivatives of functions (i.e., locally integrable functions) is now clear, if we combine Theorem 24.4 with the following result:

THEOREM 24.5. *Let Ω be an open subset of $\mathbf{R}^n$. Every Radon measure μ in Ω is a finite sum of derivatives of order $\leqslant n$ of locally L^∞ functions in Ω.*

More precisely, given any neighborhood U of supp μ, there is a set $\{f_p\}$, $p \in \mathbf{N}^n$, $|p| \leqslant n$, of locally L^∞ functions, all of which have their support contained in U, such that

$$\mu = \sum_{|p|\leqslant n} (\partial/\partial x)^p f_p .$$

Proof. We begin by assuming that μ has compact support. Let U be an arbitrary open neighborhood of supp μ in Ω. There is a constant $C > 0$ such that, for all functions $\phi \in \mathscr{C}_c^\infty(U)$,

$$|\langle\mu, \phi\rangle| \leqslant C \sup_x |\phi(x)|.$$

For simplicity, let us use the notation

$$\triangleright = (\partial/\partial x_1) \cdots (\partial/\partial x_n).$$

We have

$$\phi(y) = \int_{-\infty}^{y_1} \cdots \int_{-\infty}^{y_n} \triangleright \phi(x) \, dx_1 \cdots dx_n ,$$

whence

$$\sup_y |\phi(y)| \leqslant \|\triangleright\phi\|_{L^1} ,$$

and

$$|\langle\mu, \phi\rangle| \leqslant C\|\triangleright\phi\|_{L^1} .$$

This means that the linear functional $\triangleright\phi \rightsquigarrow \langle\mu, \phi\rangle$ is continuous on $\triangleright \mathscr{C}_c^\infty(U)$ for the L^1 norm. By the Hahn–Banach theorem, it can be extended as a continuous linear functional on the whole of $L^1(U)$. But such a functional is of the form $g \rightsquigarrow \int f(x) g(x) \, dx$ with $f \in L^\infty(U)$ (Theorem 20.3). In particular, if we take $g = \triangleright\phi$, $\phi \in \mathscr{C}_c^\infty(U)$, we see that we have

$$\langle\mu, \phi\rangle = \langle f, \triangleright\phi\rangle = \langle(-1)^n \triangleright f, \phi\rangle,$$

that is to say

$$\mu = (-1)^n \triangleright f.$$

It remains to select a function $\alpha \in \mathscr{C}_c^\infty(U)$ equal to one in some neighborhood of supp μ, and to observe that we have

$$\mu = \alpha\mu = \sum \pm (\partial/\partial x_{i_1}) \cdots (\partial/\partial x_{i_r})\{[(\partial/\partial x_{i_{r+1}}) \cdots (\partial/\partial x_{i_n})\alpha]f\},$$

where the summation is performed over all partitions of the set of integers $(1,..., n)$ into two subsets $(i_1,..., i_r)$, $(i_{r+1},..., i_n)$. This proves the statement for μ when supp μ is compact.

In order to prove Theorem 24.5 in the general case, it suffices to make use of a locally finite open covering of Ω consisting of open sets U_j ($j \in J$) with compact closure, and of a partition of unity $\{\alpha_j\}$ subordinated to this covering; then every α_j belongs to $\mathscr{C}_c^\infty(\Omega)$. It suffices then to apply the result (in the case of compact supports) to each measure $\alpha_j\mu$. We leave the details to the reader.

COROLLARY 1.   *Every Radon measure μ in Ω is a finite sum of derivatives of order $\leqslant 2n$ of continuous functions.*

Proof.   Let f be a locally L^∞ function (or, for that matter, a locally L^1 function, which, we recall, is much less restrictive). Select an arbitrary point $x^0 = (x_1^0,..., x_n^0)$ of Ω. We have, in a neighborhood x^0,

(24.3) $$f = \triangleright F$$

where

$$F(x) = \int_{x_1^0}^{x_1} \cdots \int_{x_n^0}^{x_n} f(y)\, dy_1 \cdots dy_n.$$

One can immediately check (by using the Lebesgue–Nicodym theorem if f is only locally L^1 or directly if f is locally L^∞) that F is a continuous function of x in a sufficiently small neighborhood of x^0 (in which F is defined). Combining the representations (24.3) with Theorem 24.5, we obtain the corollary locally. By using a partition of unity in $\mathscr{C}_c^\infty(\Omega)$, we then obtain it globally.

COROLLARY 2.   *Every distribution of finite order, T, in Ω, is equal to a finite sum of derivatives of continuous functions.*

Combine Corollary 1 with Theorem 24.5.

COROLLARY 3.   *Given any relatively compact open subset Ω' of Ω, the* *restriction of a distribution T in Ω to Ω' is equal, in Ω', to a finite sum of derivatives of continuous functions in Ω'.*

Combine Corollary 2 with Theorem 24.3.

Exercises

24.7. Prove that the Dirac measure in $\mathbf{R}^n$ is equal to a derivative of order $\leqslant n$ of a bounded function (not having compact support).

24.8. Let M be a linear subspace of $\mathbf{R}^n$. Explain the meaning of the measure dx_M induced on M by the Lebesgue measure dx on $\mathbf{R}^n$.

Let us denote by $\phi \mid M$ the restriction to M of a function ϕ defined in $\mathbf{R}^n$. Prove that the Radon measure in $\mathbf{R}^n$,

$$\phi \rightsquigarrow \int_M (\phi \mid M) \, dx_M \, ,$$

is equal to a derivative of order $\leqslant n - \dim M$ of a bounded function in $\mathbf{R}^n$.

Distributions with Support at the Origin

Let $\mathscr{C}^m \, (0 \leqslant m < +\infty)$ be the space of m-times continuously differentiable functions in $\mathbf{R}^n$, with the natural $\mathscr{C}^m$ topology (of convergence of all the derivatives of order $\leqslant m$ on every compact subset of $\mathbf{R}^n$). Let us denote by N^m the closure, in $\mathscr{C}^m$, of the set of $\mathscr{C}^\infty$ functions having their support in the complement of zero. Note that this is the same as saying that N^m is the closure in $\mathscr{C}^m$ of the set of $\mathscr{C}^m$ functions with support in $\mathbf{R}^n - \{0\}$ (Corollary 1 of Theorem 15.3).

LEMMA 24.1. *N^m consists exactly of all the $\mathscr{C}^m$ functions whose derivatives of order $\leqslant m$ all vanish at the origin.*

Proof. Let $g \in \mathscr{C}_c^\infty$ be equal to 1 for $|x| \leqslant 1$ and to 0 for $|x| \geqslant 2$. Let $\phi \in \mathscr{C}^m$ have all its derivatives of order $\leqslant m$ at the origin, equal to 0. Taylor's expansion of ϕ about the origin shows that the functions $g(x/\varepsilon)\phi(x)$ converge to zero in $\mathscr{C}^m$ as $\varepsilon > 0$ converges to 0; hence $(1 - g(x/\varepsilon))\phi(x)$ converges to $\phi(x)$ in $\mathscr{C}^m$, which proves that $\phi \in N^m$. Conversely, every element of N^m is a $\mathscr{C}^m$ limit of $\mathscr{C}^m$ functions having all their derivatives of order $\leqslant m$ equal to 0 at the origin, hence has the same property.

COROLLARY. *The canonical homomorphism of $\mathscr{C}^m$ onto $\mathscr{C}^m/N^m$ induces a one-to-one linear map of the space of polynomials in n variables of degree $\leqslant m$ (with complex coefficients), $\mathscr{P}_n^m$, onto $\mathscr{C}^m/N^m$.*

Proof. Since $\mathscr{P}_n^m \cap N^m = \{0\}$, the canonical homomorphism

$\mathscr{C}^m \to \mathscr{C}^m/N^m$ restricted to $\mathscr{P}_n^m$ is one-to-one. To see that it is onto, consider an arbitrary element $f \in \mathscr{C}^m$ and its Taylor expansion of order m about $x = 0$,

$$f(x) = \sum_{|p| \leqslant m} \frac{1}{p!} [(\partial/\partial x)^p f(0)] x^p + \phi(x),$$

where the remainder ϕ is such that all its derivatives of order $\leqslant m$ vanish at the origin. In view of Lemma 24.1, $\phi \in N^m$; thus f is congruent modulo N^m to a polynomial of degree $\leqslant m$, which proves that the canonical homomorphism maps $\mathscr{P}_n^m$ onto $\mathscr{C}^m/N^m$.

Let $\mathscr{E}_0'$ be the space of distributions in $\mathbf{R}^n$ having their support at the origin (i.e., contained in the set $\{0\}$; thus the zero distribution belongs to $\mathscr{E}_0'$). By the corollary of Theorem 24.3, every distribution belonging to $\mathscr{E}_0'$ is of finite order; let $T \in \mathscr{E}_0'^m$ be the subspace of $\mathscr{E}_0'$ consisting of the distributions of order $\leqslant m$, i.e., the continuous linear forms on $\mathscr{C}^m$, which have their support as the origin.

LEMMA 24.2.   *The space $\mathscr{E}_0'^m$ is the orthogonal of N^m.*

Proof. Every distribution $T \in \mathscr{E}_0'^m$ is obviously orthogonal to all $\mathscr{C}^\infty$ functions having their support in $\mathbf{R}^n - \{0\}$ (by definition of supp T), hence to the closure of the set of these functions in $\mathscr{C}^m$. Conversely, if a distribution T of order $\leqslant m$ does not have its support at the origin, we can find a function $\phi \in \mathscr{C}^\infty$ with supp $\phi \subset \mathbf{R}^n - \{0\}$ such that $\langle T, \phi \rangle \neq 0$. This means that T does not belong to the orthogonal of N^m.

Let j be the transpose of the canonical homomorphism $\mathscr{C}^m \to \mathscr{C}^m/N^m$; j is a one-to-one linear map of the dual of $\mathscr{C}^m/N^m$ into $\mathscr{E}'^m$, the dual of $\mathscr{C}^m$ (*one-to-one* since it is the transpose of an *onto* mapping (cf. Proposition 23.2)). On the other hand, it is obvious that every continuous linear form on $\mathscr{C}^m$ which vanishes on N^m defines canonically a continuous linear form on the quotient space, $\mathscr{C}^m/N^m$, whence a one-to-one linear map k of $\mathscr{E}_0'^m$, the orthogonal of N^m (Lemma 24.2), into the dual of $\mathscr{C}^m/N^m$; it is immediately seen that the map k is the inverse of the map j. Let $N(m, n)$ be the number of n-tuples $p = (p_1, ..., p_n)$ such that $|p| = p_1 + \cdots + p_n \leqslant m$. From the corollary of Lemma 24.1, we know that

$$\dim \mathscr{C}^m/N^m = \dim \mathscr{P}_n^m = N(m, n).$$

On the other hand, $\mathscr{E}_0'^m$ contains all the derivatives of order $\leqslant m$ of the Dirac measure; these derivatives are obviously linearly independent and they number $N(m, n)$; thus

$$\dim \mathscr{E}_0'^m \geqslant N(m, n) = \dim \mathscr{C}^m/N^m.$$

But k is a one-to-one mapping of $\mathscr{E}_0'^m$ into the dual of $\mathscr{C}^m/N^m$, hence k is onto by the obvious comparison of dimensions; it is then easily seen that $k = j^{-1}$. Also we see that dim $\mathscr{E}_0'^m = N(m, n)$, in other words $\mathscr{E}_0'^m$ is spanned by the derivatives of order $\leqslant m$ of the Dirac measure. Combining all these results, we have proved:

THEOREM 24.6.   *The distributions in $\mathbf{R}^n$ which have their support at the origin are the finite linear combinations of the derivatives of the Dirac measure at 0.*

Exercises

24.9.   What is the relation between the following topologies on the space $\mathscr{E}_0'$ of distributions having their support at the origin:

(a)   the topology induced by the weak topology $\sigma(\mathscr{E}', \mathscr{C}^\infty)$ on $\mathscr{E}'$;
(b)   the topology induced by the strong dual topology on $\mathscr{E}'$;
(c)   the topology carried over from $\mathscr{P}_n$, the LF-space of polynomials in n indeterminates, with complex coefficients, via the natural isomorphism

$$\sum_p a_p X^p \rightsquigarrow \sum_p a_p \delta^{(p)} \, ?$$

24.10.   Prove that the space H' of analytic functionals in $\mathbf{C}^n$ is isomorphic to the completion of the space of distributions in $\mathbf{R}^n$, having their support at the origin, $\mathscr{E}_0'$, with respect to a certain topology (cf. Exercise 22.7). Prove also the following assertions:

1.   The natural injection of $H(\mathbf{C}^n)$ into $\mathscr{C}^\infty(\mathbf{R}^{2n})$ yields, by transposition, a homomorphism (for the strong dual structures) of $\mathscr{E}'(\mathbf{R}^{2n})$ onto $H'(\mathbf{C}^n)$; what is the kernel of this homomorphism?

2.   The mapping $H(\mathbf{C}^n) \to \mathscr{C}^\infty(\mathbf{R}^n)$, restriction to the real space $\mathbf{R}^n$ of functions defined in $\mathbf{C}^n$, yields, by transposition, a continuous injection $j : \mathscr{E}'(\mathbf{R}^n) \to H'(\mathbf{C}^n)$. Prove also that j is *not* an isomorphism into and that the image of j is dense.

25

Example of Transpose: Fourier Transformation of Tempered Distributions

As usual, x will denote the point in the Euclidean space $\mathbf{R}^n$. If we use a basis in $\mathbf{R}^n$, $(e_1, ..., e_n)$, we write $x = (x_1, ..., x_n)$, where the x_j are the coordinates of x with respect to that basis. By $\mathbf{R}_n$ we mean the dual of $\mathbf{R}^n$; elements of $\mathbf{R}_n$ will be denoted by Greek letters like ξ, η, etc.; the value of the linear form ξ at the point x will be written $\langle \xi, x \rangle$ or $\langle x, \xi \rangle$. By the dual basis of $(e_1, ..., e_n)$ we mean the basis $(e_1', ..., e_n')$ of $\mathbf{R}_n$ determined by the equations $\langle e_j', e_k \rangle = 1$ if $j = k$, $=0$ if $j \neq k$. If $\xi_1, ..., \xi_n$ are the coordinates of $\xi \in \mathbf{R}_n$ in this dual basis, we have

$$\langle x, \xi \rangle = x_1 \xi_1 + \cdots + x_n \xi_n.$$

We shall make use of the Lebesgue measure in $\mathbf{R}^n$; it will be denoted by dx. The student ought to keep in mind that the Lebesgue measure in $\mathbf{R}^n$ is determined up to a constant factor. We assume that we have somehow made a choice of a particular one, for instance by selecting a basis $(e_1, ..., e_n)$ in $\mathbf{R}^n$ and by requiring that the measure of the hypercube $\{x; 0 \leqslant x_j \leqslant 1, j = 1, ..., n\}$ be equal to one. Such a choice determines immediately a Lebesgue measure $d\xi$ in $\mathbf{R}_n$: we take the dual basis and require that the measure of the hypercube $\{\xi; 0 \leqslant \xi_j \leqslant 1, j = 1, ..., n\}$ be one. If we perform a change of basis in $\mathbf{R}^n$ and if we denote by $y_1, ..., y_n$ the coordinates in the new basis and by dy the Lebesgue measure such that the measure of the set $\{y; 0 \leqslant y_j \leqslant 1, 1 \leqslant j \leqslant n\}$ is one, we have $dy = t\, dx$ for some positive number t. Using the coordinates η_j in $\mathbf{R}_n$ with respect to the basis which is the dual of the new basis in $\mathbf{R}_n$, it is easy to check that $d\eta = t\, d\xi$.

We recall now the definition of the space $\mathscr{S}(\mathbf{R}^n)$. It is the space of $\mathscr{C}^\infty$ functions φ in $\mathbf{R}^n$ such that, for all pairs P, Q of polynomials in n indeterminates, with complex coefficients,

$$\sup_{x \in \mathbf{R}^n} | P(x) Q(\partial/\partial x) \varphi(x) | < +\infty;$$

267

the topology of $\mathscr{S}(\mathbf{R}^n)$ is defined by the seminorms

$$\varphi \rightsquigarrow \sup_x |\, P(x)\, Q(\partial/\partial x)\, \varphi(x)|.$$

We have seen that $\mathscr{S}(\mathbf{R}^n)$ is a Fréchet space in which every closed bounded set is compact (see Chapter 10, Example IV; also Theorem 14.5). The student will perceive that the definition of the topological vector space $\mathscr{S}(\mathbf{R}^n)$ is independent of the choice of a particular basis in $\mathbf{R}^n$.

Also observe that, for all $\xi \in \mathbf{R}_n$, the function of x,

$$\exp(-2i\pi \langle x, \xi \rangle)\, \varphi(x)$$

belongs to $\mathscr{S}(\mathbf{R}^n)$ as soon as $\varphi \in \mathscr{S}(\mathbf{R}^n)$. Such a function decreases so fast at infinity that the integral of its absolute value, i.e., $\int |\, \varphi(x)|\, dx$, is always finite.

Definition 25.1.   The Fourier transform of $\varphi \in \mathscr{S}(\mathbf{R}^n)$ is the function of $\xi \in \mathbf{R}_n$,

$$\int \exp(-2i\pi \langle x, \xi \rangle)\, \varphi(x)\, dx.$$

We denote it by $\mathscr{F}\varphi(\xi)$, or by $\hat{\varphi}(\xi)$ when no confusion is to be feared.

Very often, the Fourier transform of φ is defined to be equal to the integral

$$\int \exp(-i \langle x, \xi \rangle)\, \varphi(x)\, dx.$$

We choose to put the factor 2π in the exponential, following Schwartz, so as to avoid factors 2π that appear when computing the inverse transformation.

THEOREM 25.1.   *The Fourier transformation is an isomorphism of $\mathscr{S}(\mathbf{R}^n)$ onto $\mathscr{S}(\mathbf{R}_n)$ (for the structures of topological vector spaces). The inverse mapping is the mapping*

$$\mathscr{F} : \mathscr{S}(\mathbf{R}_n) \ni \psi \rightsquigarrow \int \exp(2i\pi \langle x, \xi \rangle)\, \psi(\xi)\, d\xi.$$

It is understood that we are using here the Lebesgue measure $d\xi$ associated canonically with dx (as indicated above).

Proof of Theorem 25.1.   (1) The Fourier transformation is a continuous linear map of $\mathscr{S}(\mathbf{R}^n)$ into $\mathscr{S}(\mathbf{R}_n)$.

We have, for any pair of polynomials $P, Q \in \mathbf{C}[X_1, ..., X_n]$,

$$Q(\partial/\partial\xi)\, \hat{\varphi}(\xi) = \int \exp(-2i\pi \langle x, \xi \rangle)\, Q(-2i\pi x)\, \varphi(x)\, dx;$$

$$P(\xi)\, \hat{\varphi}(\xi) = \int \exp(-2i\pi \langle x, \xi \rangle)\, P((2i\pi)^{-1}\, \partial/\partial x)\, \varphi(x)\, dx.$$

The first identity is obvious in view of Leibniz' rule of differentiation under the integral sign (keeping in mind that the product of φ with any polynomial decreases "very" fast at infinity). The second formula is obvious if we perform an integration by parts in the integral, at the right-hand side. If, now, we combine the two formulas, we obtain

$$P(\xi)\,Q(\partial/\partial\xi)\,\hat{\varphi}(\xi) = \int \exp(-2i\pi\,\langle x,\,\xi\rangle)\,P((2i\pi)^{-1}\,\partial/\partial x)\,[Q(-2i\pi x)\,\varphi(x)]\,dx,$$

whence, for all $\xi \in \mathbf{R}_n$,

$$|\,P(\xi)\,Q(\partial/\partial\xi)\,\hat{\varphi}(\xi)| \leqslant \int |\,P((2i\pi)^{-1}\,\partial/\partial x)\,[Q(-2i\pi x)\,\varphi(x)]|\,dx$$

$$\leqslant \{\sup_{x\in\mathbf{R}^n}(1 + |\,x\,|)^{n+1}|\,P((2i\pi)^{-1}\,\partial/\partial x)\,[Q(-2i\pi x)\varphi(x)]|\,\} \int (1 + |\,x\,|)^{-n-1}\,dx.$$

Our statement follows immediately from this inequality.

(2) $\mathscr{F}$ *is a continuous linear map of* $\mathscr{S}(\mathbf{R}_n)$ *into* $\mathscr{S}(\mathbf{R}^n)$.

This is obvious since $\mathscr{F}$ is defined exactly as $\mathscr{F}$ except for the sign in the exponent of the exponential.

(3) $\mathscr{F}$ *is the inverse of* $\mathscr{F}$.

If we prove (3), the theorem will be proved since we shall then be dealing with two continuous linear maps which are the inverse of each other. We must prove that

$$\mathscr{F} \circ \mathscr{F} = \text{identity of } \mathscr{S}(\mathbf{R}^n); \qquad \mathscr{F} \circ \mathscr{F} = \text{identity of } \mathscr{S}(\mathbf{R}_n).$$

In view of the symmetry in the definitions of $\mathscr{F}$ and $\mathscr{F}$, it suffices to prove the first one of these two equalities.

Let now f be some function belonging to $\mathscr{S}(\mathbf{R}_n)$. We have, with obvious notations,

$$\int f(\xi)\hat{\phi}(\xi)\exp(2i\pi\,\langle x,\,\xi\rangle)\,d\xi = \int\int f(\xi)\,\phi(y)\exp(2i\pi\,\langle x - y,\,\xi\rangle)\,dy\,d\xi$$

$$= \int \phi(y)\,\mathscr{F}f(x - y)\,dy$$

$$= \int \phi(x - y)\,\mathscr{F}f(y)\,dy.$$

We were able to interchange the order of integrations, in the double integral, because of the obvious fact that the function $(y, \xi) \leadsto \phi(y) f(\xi)$ is Lebesgue integrable in $\mathbf{R}^n \times \mathbf{R}_n$. At any event, we have obtained

$$(25.1) \qquad \int f(\xi) \hat{\phi}(\xi) \exp(2i\pi \langle x, \xi \rangle) \, d\xi = \int \phi(x - y) \mathscr{F} f(y) \, dy.$$

The proof will be complete if we can make f vary over a sequence of functions of ξ, belonging to $\mathscr{S}(\mathbf{R}_n)$, such that f converges to the function identically equal to 1 in $\mathbf{R}_n$—e.g., in the sense of the weak convergence in $L^2(\mathbf{R}_n)$—whereas $\mathscr{F}f$ converges to the Dirac measure at the origin in $\mathscr{S}'$, dual of $\mathscr{S}$ (anticipating slightly on what follows, we are interpreting $\mathscr{S}'$ as a space of distributions). To find such a sequence is easy; it is enough, for instance, to take the sequence

$$e^{-|\xi|^2/k}, \qquad k = 1, 2, \dots \text{ (see Exercise 25.1 below).} \qquad \text{Q.E.D.}$$

THEOREM 25.2. (*Plancherel–Parseval*). *Let ϕ, ψ be any two functions belonging to $\mathscr{S}(\mathbf{R}^n)$. We have*

$$(25.2) \qquad \int \phi(x) \, \overline{\psi(x)} \, dx = \int \hat{\phi}(\xi) \, \overline{\hat{\psi}(\xi)} \, d\xi,$$

$$(25.3) \qquad \int |\phi(x)|^2 \, dx = \int |\hat{\phi}(\xi)|^2 \, d\xi.$$

The bars in (25.2) mean the complex conjugates.

Proof of Theorem 25.2. In Eq. (25.1), we take $x = 0$ and make the change of variables $y \leadsto -y$ in the integral of the right-hand side. This yields

$$\int \hat{\phi}(\xi) f(\xi) \, d\xi = \int \phi(y) \, \mathscr{F} f(-y) \, dy.$$

We then choose $f \in \mathscr{S}(\mathbf{R}_n)$ in such a way that $\mathscr{F} f(-y) = \overline{\psi(y)}$. This is possible by virtue of Theorem 25.1. It is then trivial to check that $f(\xi) = \overline{\hat{\psi}(\xi)}$.

COROLLARY 1. *The Fourier transformation $\mathscr{F} : \mathscr{S}(\mathbf{R}^n) \to \mathscr{S}(\mathbf{R}_n)$ can be extended as an isometry of $L^2(\mathbf{R}^n)$ onto $L^2(\mathbf{R}_n)$.*

Observe that $\mathscr{S}$, which contains $\mathscr{C}_c^\infty$, is dense in L^2 (Corollary 3 of Theorem 15.3); Corollary 1 follows then immediately from (25.3).

Notation 25.1. *We shall denote by $\mathscr{F}$ the isometry of $L^2(\mathbf{R}^n)$ onto $L^2(\mathbf{R}_n)$ extending the Fourier transformation in $\mathscr{S}(\mathbf{R}^n)$ and by $\bar{\mathscr{F}}$ the inverse isometry.*

These isometries will be called respectively *Fourier transformation* and *inverse transformation* in L^2.

COROLLARY 2.   *If u, v are any two elements of $L^2(\mathbf{R}^n)$, we have*

(25.4) $(u \mid v)_{L^2(\mathbf{R}^n)} = (\mathscr{F}u \mid \mathscr{F}v)_{L^2(\mathbf{R}_n)}$ (*Parseval formula*);

(25.5) $\| u \|_{L^2(\mathbf{R}^n)} = \| \mathscr{F}u \|_{L^2(\mathbf{R}_n)}$ (*Plancherel formula*).

We have denoted by (|) the inner product in L^2 and by ‖ ‖ the norm. Observe that (25.4) (resp. (25.5)) follows by continuity from (25.2) (resp. (25.3)). As a matter of fact, (25.5) is a restatement of Corollary 1, essentially; (25.4) and (25.5) are equivalent by polarization.

Exercises

25.1.   Compute the Fourier transform of the function $x \to e^{-|x|^2/k}$ ($k = 1, 2, \ldots$).

25.2.   Let $\varepsilon > 0$ be arbitrary. Prove that the Fourier transform of the function $x \rightsquigarrow \varepsilon^{-n} \varphi(x/\varepsilon)$, $\varphi \in \mathscr{S}(\mathbf{R}^n)$, is the function $\xi \to \hat{\varphi}(\varepsilon\xi)$. Using this fact, show how to construct sequences in $\mathscr{S}(\mathbf{R}^n)$ which converge to the Dirac measure in $\mathscr{D}'(\mathbf{R}^n)$ and whose Fourier transforms converge to the function one in $\mathscr{C}^\infty(\mathbf{R}_n)$.

25.3.   Let $\mathscr{C}_\infty(\mathbf{R}_n)$ be the space of continuous complex-valued functions in $\mathbf{R}_n$ which converge to zero at infinity, equipped with the norm of uniform convergence in the whole of $\mathbf{R}_n$,

$$\varphi \rightsquigarrow \sup_{\xi \in \mathbf{R}_n} | \varphi(\xi)|.$$

Prove that $\mathscr{C}_\infty(\mathbf{R}_n)$ is a Banach space. Then prove the following (important) theorem of Lebesgue:

THEOREM 25.3.   *The Fourier transformation $\mathscr{F} : \mathscr{S}(\mathbf{R}^n) \to \mathscr{S}(\mathbf{R}_n)$ can be extended as a one-to-one continuous linear map of $L^1(\mathbf{R}^n)$ into $\mathscr{C}_\infty(\mathbf{R}_n)$.*

25.4.   Prove that the Fourier transformation, extended from $\mathscr{S}(\mathbf{R}^n)$ to $L^1(\mathbf{R}^n)$, is neither a mapping of $L^1(\mathbf{R}_n)$ *onto*, nor an *isomorphism into* $\mathscr{C}_\infty(\mathbf{R}_n)$.

We proceed to define the Fourier transformation in the dual $\mathscr{S}'$ of $\mathscr{S}$. But before doing this, we wish to show that $\mathscr{S}'$ is a space of distributions in $\mathbf{R}^n$, intermediary between the space $\mathscr{E}'$ of distributions with compact support and the space of all distributions, $\mathscr{D}'$; (we omit the mention $(\mathbf{R}^n)$ because all functions and distributions considered here are defined in the whole Euclidean space $\mathbf{R}^n$). We shall also give a characterization of the elements of $\mathscr{S}'$ (regarded as distributions).

Consider the sequence

$$\mathscr{C}^\infty_c \to \mathscr{S} \to \mathscr{C}^\infty.$$

The arrows indicate the natural injections; they are continuous linear mappings, in view of the definitions of the topologies in the three spaces under consideration, and they have dense images (Corollary 1 of Theorem 15.3; Theorem 15.4). By transposing the above sequence, we obtain a new sequence of continuous injections,

$$\mathscr{E}' \to \mathscr{S}' \to \mathscr{D}',$$

which enables us to regard the (strong) dual $\mathscr{S}'$ of $\mathscr{S}$ as a space of distributions in $\mathbf{R}^n$ (containing $\mathscr{E}'$), just as announced.

Definition 25.2. *The distributions belonging to $\mathscr{S}'(\mathbf{R}^n)$ are called the tempered (or temperate) distributions in $\mathbf{R}^n$.*

The name *tempered* is motivated by the following structure theorem for distributions belonging to $\mathscr{S}'$ (it shows, among other things, that all tempered distributions are of finite order):

THEOREM 25.4.   *A distribution in $\mathbf{R}^n$ is tempered if and only if it is a finite sum of derivatives of continuous functions growing at infinity slower than some polynomial.*

Proof. The proof is very similar to the arguments used in the preceding chapter to prove the various structure theorems in $\mathscr{D}'^F$. Of course, there are a few slight differences.

First of all, a distribution T is tempered if and only if the linear form $\varphi \rightsquigarrow \langle T, \varphi \rangle$ is continuous on $\mathscr{C}_c^\infty$ for the topology induced by $\mathscr{S}$. This is a restatement of the definition of the "natural" injection of $\mathscr{S}'$ into $\mathscr{D}'$. From it, the sufficiency of the condition stated in Theorem 25.4 is evident. We need only prove its necessity.

To every tempered distribution T there are two integers $m, h \geqslant 0$ and a positive constant C such that, for all $\varphi \in \mathscr{C}_c^\infty$,

$$|\langle T, \varphi \rangle| \leqslant C \sup_{|p| \leqslant m} \sup_{x \in \mathbf{R}^n} |(1 + |x|^2)^h (\partial/\partial x)^p \varphi(x)|.$$

Let us set

$$\varphi_h(x) = (1 + |x|^2)^h \varphi(x).$$

Of course, φ_h is also a test function. In fact, $\varphi \rightsquigarrow \varphi_h$ is a one-to-one linear map of $\mathscr{C}_c^\infty$ onto itself. Furthermore, as immediately proved by induction on $h = 0, 1, \ldots,$

$$|(\partial/\partial x)^p \varphi(x)| \leqslant C_{p,h}(1 + |x|^2)^{-h} \sum_{q \leqslant p} |(\partial/\partial x)^q \varphi_h(x)|,$$

where $q \leqslant p$ means $q_1 \leqslant p_1 ,..., q_n \leqslant p_n$ and where the constant $C_{p,h}$ depends only on the n-tuple p and on the integer h. We see therefore immediately that

$$(25.6) \qquad |\langle T, \varphi \rangle| \leqslant C' \sup_{|p| \leqslant m} \sup_x |(\partial/\partial x)^p \varphi_h(x)|.$$

Let us introduce once more the differentiation monomial

$$\triangleright = (\partial/\partial x_1) \cdots (\partial/\partial x_n).$$

We have (cf. p. 262)

$$\sup_x | \varphi(x)| \leqslant \| \triangleright \varphi \|_{L^1} .$$

Therefore, combining this with (25.6),

$$(25.7) \qquad |\langle T, \varphi \rangle| \leqslant C'' \sup_{|p| \leqslant m+n} \|(\partial/\partial x)^p \varphi_h \|_{L^1} .$$

Let then N be the number of n-tuples p such that $| p | \leqslant m + n$; we consider the product space $L^1 \times \overset{N.}{\cdots} \times L^1 = (L^1)^N$ and the injection

$$J : \psi \rightsquigarrow ((\partial/\partial x)^p \, \psi)_{|p| \leqslant m+n}$$

of $\mathscr{C}_c^\infty$ into $(L^1)^N$. Estimate (25.7) can be read as saying that the linear functional $J\varphi_h \rightsquigarrow \langle T, \varphi \rangle$ is continuous on $J\mathscr{C}_c^\infty$ for the topology induced by $(L^1)^N$. Therefore, by the Hahn–Banach theorem, it can be extended as a continuous linear form in the whole of $(L^1)^N$. But the dual of $(L^1)^N$ is canonically isomorphic with $(L^\infty)^N$ (cf. p. 259 and Theorem 20.3), therefore there exist N L^∞ functions $h_p(| p | \leqslant m + n)$ such that

$$\langle T, \varphi \rangle = \sum_{|p| \leqslant m+n} \langle h_p , (\partial/\partial x)^p \varphi_h \rangle$$

Recalling the expression of φ_h in terms of φ, we see that this simply means that

$$T = \sum_{|p| \leqslant m+n} (1 + | x |^2)^h (-1)^{|p|} (\partial/\partial x)^p h_p .$$

For each p, we set

$$g_p(x) = \int_0^{x_1} \cdots \int_0^{x_n} h_p(t_1 ,..., t_n) \, dt_1 \cdots dt_n .$$

Since h_p is L^∞, we see that g_p is a *continuous* function in $\mathbf{R}^n$ and that

$$| g_p(x) | \leqslant | x_1 | \cdots | x_n | \| h_p \|_{L^\infty} .$$

Furthermore, we have

$$h_p = \rhd g_p \, ,$$

and consequently

(25.8)
$$T = \sum_{|p| \leqslant m+2n} (1 + | x |^2)^h \, (\partial / \partial x)^p \, \tilde{g}_p \, ,$$

with an obvious definition of the $\tilde{g}_p$. One sees then easily, by induction on h, that

$$(1 + | x |^2)^h \, (\partial / \partial x)^p \, \tilde{g}_p(x) = \sum_{q \leqslant p} (\partial / \partial x)^q \, [P_{p,q}(x) \, \tilde{g}_p(x)],$$

where the $P_{p,q}$ are polynomials depending only on h, p, and q. Putting this back into (25.8), we obtain the desired expression of $T \in \mathscr{S}'$.   Q.E.D.

Examples of Tempered Distributions

1. All distributions with compact support (in particular, all functions with compact support) are tempered distributions.

2. All continuous functions which grow at infinity slower than some polynomial are tempered distributions.

3. All (classes of) functions belonging to some L^p ($1 \leqslant p \leqslant +\infty$) are tempered distributions (Proof: if $f \in L^p$,

$$\varphi \rightsquigarrow \int \varphi(x) f(x) \, dx$$

is continuous on $\mathscr{C}_c^\infty$ for the topology induced by $\mathscr{S}$).

Multiplication by polynomials, differential operators in $\mathbf{R}^n$ with coefficients which are polynomials define continuous linear mappings of $\mathscr{S}'$ into itself. We are supposing here that these mappings are already defined in the whole of $\mathscr{D}'$ and we are taking their restrictions to $\mathscr{S}'$. These restrictions can be obtained also as the transposes of similar mappings defined in $\mathscr{S}$. These definitions are consistent, as one sees by observing that the mappings in question, defined in $\mathscr{S}$, when restricted to $\mathscr{C}_c^\infty$, are continuous on the latter for the topology induced by $\mathscr{S}$, etc.

Concerning the multiplication by $\mathscr{C}^\infty$ functions, the following theorem can be proved:

THEOREM 25.5.   *Let α be a $\mathscr{C}^\infty$ function. The following three properties are equivalent:*

(1)   *The multiplication mapping $S \rightsquigarrow \alpha S$ from $\mathscr{S}'$ into $\mathscr{D}'$ is a continuous linear map of $\mathscr{S}'$ into itself.*

(2)   *The multiplication mapping $\varphi \rightsquigarrow \alpha\varphi$ of $\mathscr{S}$ into $\mathscr{C}^\infty$ is a continuous linear map of $\mathscr{S}$ into itself.*

(3)   *For every n-tuple p, there is a polynomial P_p in $\mathbf{R}^n$ such that, for all $x \in \mathbf{R}^n$,*

$$|(\partial/\partial x)^p \, \alpha(x)| \leqslant |P_p(x)|.$$

Definition 25.3.   The space of $\mathscr{C}^\infty$ functions α in $\mathbf{R}^n$ having the equivalent properties (1), (2), and (3) of Theorem 25.5 is denoted by $\mathcal{O}_M$.

The letters $\mathcal{O}_M$ stand for *multiplication operators* (in $\mathscr{S}$ or $\mathscr{S}'$), referring to Properties (1) and (2) in Theorem 25.5. The functions $\alpha \in \mathcal{O}_M$ are often called *$\mathscr{C}^\infty$ functions slowly increasing at infinity*, referring then to Property (3).

Exercises

25.5.   Prove Theorem 25.5.

25.6.   Give an example of a continuous function f in $\mathbf{R}^n$ with the following two properties:

(1)   there is no polynomial P in $\mathbf{R}^n$ such that

$$|f(x)| \leqslant |P(x)| \qquad \text{for all} \quad x \in \mathbf{R}^n;$$

(2)   the distribution $\varphi \to \int \varphi(x) f(x) \, dx$ is tempered.

27.7.   The Radon measure on the real line,

$$\varphi \rightsquigarrow \sum_{k=1}^\infty a_k \, \varphi(k),$$

is a tempered distribution on $\mathbf{R}^1$ if and only if there is an integer $m \geqslant 0$ and a constant $C > 0$ such that, for all $k = 1, 2,...,$

$$|a_k| \leqslant Ck^m.$$

Prove this statement.

We define now the Fourier transformation in the space $\mathscr{S}'$ of tempered distributions as the transpose of the Fourier transformation in the space $\mathscr{S}$ of $\mathscr{C}^\infty$ functions rapidly decaying at infinity.

Definition 25.4.   The transpose of the continuous linear map

$$\mathscr{F} : \mathscr{S}(\mathbf{R}^n) \ni \varphi \rightsquigarrow (\xi \rightsquigarrow \int \exp(-2i\pi \, \langle x, \, \xi \rangle) \, \varphi(x) \, dx) \in \mathscr{S}(\mathbf{R}_n)$$

is called the Fourier transformation in $\mathscr{S}'(\mathbf{R}_n)$ and is denoted by $\mathscr{F}$.

Of course, in all these statements, one may interchange $\mathbf{R}^n$ and $\mathbf{R}_n$.
Theorem 25.1 yields immediately, by transposition (see Proposition 23.1):

THEOREM 25.6.   *The Fourier transformation is an isomorphism (for the structures of topological vector spaces) of $\mathscr{S}'(\mathbf{R}^n)$ onto $\mathscr{S}'(\mathbf{R}_n)$.*

Let us show rapidly that the Fourier transformation in $\mathscr{S}'$ does indeed extend the Fourier transformation in the spaces of functions, say in $L^2(\mathbf{R}^n)$. Let $f \in L^2(\mathbf{R}^n)$, $\varphi \in \mathscr{S}(\mathbf{R}_n)$, be arbitrary. Observe that we have

$$(25.9) \qquad \overline{\mathscr{F}\varphi} = \overline{\mathscr{F}}\bar\varphi,$$

where the bars stand for complex conjugate ($\overline{\mathscr{F}}$ is the inverse Fourier transformation). By Parseval's formula, we have

$$\int f(x)\,\overline{\overline{\mathscr{F}\varphi(x)}}\,dx = \int \mathscr{F}f(\xi)\,\overline{\overline{\mathscr{F}\mathscr{F}\varphi}(\xi)}\,d\xi.$$

But in view of (25.9), $\mathscr{F}\,\overline{\mathscr{F}\varphi} = \bar\varphi$, whence

$$\langle f, \mathscr{F}\varphi \rangle = \langle \mathscr{F}f, \varphi \rangle,$$

which proves our assertion.

In Chapter 29, we shall study the Fourier transformation in the space $\mathscr{E}'$ of distributions with compact support.

Exercises

25.8.   Does the function $e^{|x|}$ have a Fourier transform in the sense of distributions ?

25.9.   Compute the Fourier transform of the function $e^{i|x|^2}$.

25.10.   Compute the Fourier transforms of $\sin x$ and $\cos x$ ($x \in \mathbf{R}^1$).

25.11.   For what values of the real number s does the function $x \rightsquigarrow |x|^s$ define a tempered distribution in $\mathbf{R}^n$ which is a function ? Compute the Fourier transform of this distribution.

25.12.   Compute the Fourier transform in $\mathscr{S}'(\mathbf{R}^1)$ of the Heaviside function

$$Y(x) = \begin{cases} 1 & \text{for } x > 0 \\ 0 & \text{for } x < 0. \end{cases}$$

25.13.   Compute the Fourier transform in $\mathscr{S}'(\mathbf{R}^1)$ of the distribution

$$(d/dx)\log|x|.$$

25.14.   We recall that multiplication by a polynomial and differential operators with constant coefficients define continuous linear maps of $\mathscr{S}$ (resp. $\mathscr{S}'$) into itself. Prove the following theorem:

THEOREM 25.7. *Let P be a polynomial in n variables, with complex coefficients. Let S be a tempered distribution in $\mathbf{R}^n$. We have*

$$(25.10) \qquad \mathscr{F}[P(x)\,S] = P(-(2i\pi)^{-1}\,\partial/\partial\xi)\,\mathscr{F}S;$$

$$(25.11) \qquad \mathscr{F}[P(\partial/\partial x)\,S] = P(2i\pi\xi)\,\mathscr{F}S.$$

25.15. Compute the Fourier transform of a polynomial function in $\mathbf{R}^n$.

25.16. Let T be an arbitrary distribution in $\mathbf{R}^n$. Prove that the following properties of a point ξ of $\mathbf{R}_n$ are equivalent:

(25.12)   there exists $\eta \in \mathbf{R}_n$ such that

$$\exp(-2\pi\,\langle\xi + i\eta,\, x\rangle)\,T \in \mathscr{S}'_x\,;$$

(25.13)   for all $\eta \in \mathbf{R}_n$, $T\exp(-2\pi\,\langle\xi + i\eta,\, x\rangle) \in \mathscr{S}'_x$.

Let us denote by Γ_T the set of points having Properties (25.12) and (25.13). Give a characterization (of the type of Theorem 25.4) of the distributions T such that Γ_T is nonempty.

25.17. Let T and Γ_T be as in Exercise 25.16. What is the set Γ_T when $T \in \mathscr{E}'$, i.e., has a compact support? What is the *interior* of Γ_T when T is a tempered distribution whose support is contained in some convex salient cone $\Gamma \subset \mathbf{R}_n$? (A cone is salient if it does not contain any straight line.)

25.18. Let T and Γ_T be as in Exercise 25.16. Prove that Γ_T is convex.

25.19. Let T and Γ_T be as in Exercise 25.16, and $\mathring{\Gamma}_T$ the interior of Γ_T. Prove that

$$\exp(-2\pi\,\langle\xi + i\eta,\, x\rangle)\,T \in (\mathscr{O}'_C)_x \qquad \text{(Definition 30.1)}$$

for all $\xi \in i\eta \in \mathring{\Gamma}_T + i\mathbf{R}_n$.

25.20. We use the same notation as in Exercise 25.19. Let the Fourier transformation, $\mathscr{F}$, operate from distributions with respect to the variable x, into distributions with respect to $\eta \in \mathbf{R}_n$. Prove that

$$\mathscr{F}(\exp(-2\pi\,\langle\xi,\, x\rangle)\,T)$$

belongs to $(\mathscr{O}_M)_\eta$ for all $\xi \in \mathring{\Gamma}_T$ and that it is a *holomorphic* function $F(\xi + i\eta)$ of $\xi + i\eta = \zeta$ in $\mathring{\Gamma}_T + i\mathbf{R}_n$ (the function $\mathscr{L}T(\zeta) = F((1/2\pi)\zeta)$ is called the *Laplace transform* of T).

25.21. Compute the Laplace transforms of the following distributions on the real line (Y denotes the Heaviside function, equal to 1 for $x > 0$ and to zero for $x < 0$):

(i) $Y(x)\,(x^k/k!),\ k = 0, 1,\ldots$;

(ii) $\delta^{(k)},\ k = 0, 1,\ldots$;

(iii) $Y(x)\,e^{\zeta x},\ \zeta \in \mathbf{C}^1$;

(iv) $Y(x)\cos x,\ Y(x)\sin x$;

(v) $Y(x)\cosh x,\ Y(x)\sinh x$.

26

Convolution of Functions

Let f, g be two continuous complex-valued functions in $\mathbf{R}^n$, having compact support. The convolution of f and g is the function, also defined in $\mathbf{R}^n$,

$$(26.1) \qquad (f * g)(x) = \int_{\mathbf{R}^n} f(x - y) g(y) \, dy = \int_{\mathbf{R}^n} f(y) g(x - y) \, dy.$$

It is clear, on this definition, that there is no need for *both* f and g to have compact support: the integrals defining $f * g$ have a meaning if either one of the "factors" f or g has compact support while the other has an arbitrary support. Under this assumption, it is immediately seen that the function $f * g$ is continuous. Furthermore, if both supp f and supp g are compact, so is supp$(f * g)$. If now we release the requirement that f and/or g have compact support, it is clear that we must impose a condition ensuring that, for every $x \in \mathbf{R}^n$, the function of y, $f(y) g(x - y)$, be integrable. This demands that the two functions f and g be *locally* integrable and that the growth at infinity of one of them be "matched" by the decay of the other. What we have said earlier is an illustration of this situation: if the growth at infinity of f, say, is arbitrary, then the decay of g must also be arbitrarily fast, which can only mean that g vanishes outside a compact set. But other conditions, less "lax" on f and less restrictive on g, may easily be imagined. Suppose for instance that f grows slowly at infinity, i.e., slower than some polynomial, and that g decreases rapidly at infinity, i.e., faster than any power of $1/|x|$. Then obviously $f * g$ is well defined by Eq. (26.1). As will now be shown, one may strengthen the condition on f and weaken the one on g in such a way that the two conditions come to coincide, and in fact are also shared by the resulting function $f * g$; and the conditions are nothing else but that f and g be integrable!

THEOREM 26.1. *Let p, q, and r be three numbers such that $1 \leqslant p$, q, $r \leqslant +\infty$ and such that*

$$(26.2) \qquad 1/r = (1/p) + (1/q) - 1.$$

278

Then, for all pairs of continuous functions with compact support in $\mathbf{R}^n$, f, g, we have

(26.3) $$\left(\int_{\mathbf{R}^n} |f * g|^r \, dx\right)^{1/r} \leqslant \left(\int_{\mathbf{R}^n} |f|^p \, dx\right)^{1/p} \left(\int_{\mathbf{R}^n} |g|^q \, dx\right)^{1/q}.$$

Proof. Set $h(x) = (f * g)(x) = \int_{\mathbf{R}^n} f(x - y)g(y) \, dy$. Set $s = p(1 - 1/q)$. By Hölder's inequalities (Theorem 20.3), we obtain

$$|h(x)| \leqslant \left\{\int |f(x - y)|^{(1-s)q} |g(y)|^q \, dy\right\}^{1/q} \| |f|^s \|_{L^{q'}},$$

where q' is the conjugate of q, i.e., $q'^{-1} = 1 - q^{-1}$. Observing that $sq' = p$, we obtain

(26.4) $$|h(x)|^q \leqslant \|f\|_{L^p}^{sq} \int |f(x - y)|^{(1-s)q} |g(y)|^q \, dy.$$

At this point, we make use of the following general fact: let $t \leadsto \mathbf{f}(t)$ be a continuous function from $\mathbf{R}^n$ into some Banach space E, which has compact support. We may then define its integral $\int_{\mathbf{R}^n} \mathbf{f}(t) \, dt$, say by considering Riemann sums; the value of this integral is an element $\mathbf{e}$ of E. We have

(26.5) $$\| \mathbf{e} \| \leqslant \int_{\mathbf{R}^n} \| \mathbf{f}(t) \| \, dt.$$

We have denoted by $\| \ \|$ the norm in E. Our statement follows immediately from the "triangular" inequality for the norm applied to the finite Riemann sums approaching the integral (also observe that $t \leadsto \| \mathbf{f}(t) \|$ is a nonnegative continuous function with compact support in $\mathbf{R}^n$). We apply this to the function

$$y \leadsto |f(x - y)|^{(1-s)q} |g(y)|^q,$$

regarded as a function of $y \in \mathbf{R}^n$, obviously continuous with compact support, into some space L^α (with respect to the variable x). By applying (26.5) together with (26.4), observing that $h(x)$ is a continuous function with compact support in $\mathbf{R}^n$, we obtain

$$\| |h|^q \|_{L^\alpha} \leqslant \|f\|_{L^p}^{sq} \| |f|^{(1-s)q} \|_{L^\alpha} \| |g|^q \|_{L^1}.$$

But this can be rewritten as

(26.6) $$\| h \|_{L^{\alpha q}}^q \leqslant \|f\|_{L^p}^{sq} \|f\|_{L^{\alpha(1-s)q}}^{(1-s)q} \| g \|_{L^q}^q.$$

We choose $\alpha = r/q$ and take the qth root of both sides of (26.6). We obtain

$$\| h \|_{L^r} \leqslant \| f \|_{L^p}^s \| f \|_{L^{(1-s)r}}^{1-s} \| g \|_{L^q} .$$

This is nothing else but (26.3) if we observe that

$$(1 - s)r = \left(1 - p + \frac{p}{q}\right) r = pr \left(\frac{1}{p} + \frac{1}{q} - 1\right) = p,$$

by (26.2).   Q.E.D.

By using the density of the space $\mathscr{C}_c^0(\mathbf{R}^n)$ of continuous functions with compact support in the spaces $L^\alpha(\mathbf{R}^n)$, $\alpha < +\infty$, one may then prove the following consequence of Theorem 26.1 (when $r < +\infty$; when $r = +\infty$, one applies Hölder's inequalities):

COROLLARY 1.   *If $f \in L^p$, $g \in L^q$, then*

$$\int_{\mathbf{R}^n} f(x - y) g(y) \, dy$$

defines an element of $L^r (p, q, r$ as in Theorem 26.1), denoted by $f * g$; we have

(26.7) $\| f * g \|_{L^r} \leqslant \| f \|_{L^p} \| g \|_{L^q} .$

From this one, the following results are easily derived:

COROLLARY 2.   *Let p be a number such that $1 \leqslant p \leqslant +\infty$. For every $f \in L^1$, the mapping*

$$g \rightsquigarrow f * g$$

is a continuous linear map of L^p into itself, with norm $\leqslant \| f \|_{L^1}$.

COROLLARY 3.   *The convolution*

$$(f, g) \rightsquigarrow f * g$$

is a bilinear mapping of $L^1 \times L^1$ into L^1; we have:

$$\| f * g \|_{L^1} \leqslant \| f \|_{L^1} \| g \|_{L^1} .$$

One often summarizes the content of Corollary 3 by saying that L^1 is a *convolution algebra*.

Let A be an arbitrary subset of $\mathbf{R}^n$, and f and g two continuous functions in $\mathbf{R}^n$, one of which has a compact support. Let f_A be the function equal to f in $A - \operatorname{supp} g$ and to zero everywhere else. Then we have obviously, for all $x \in A$,

$$(26.8) \qquad (f * g)(x) = \int f(x - y) g(y)\, dy = \int f_A(x - y) g(y)\, dy$$
$$= (f_A * g)(x).$$

We may summarize this as follows:

PROPOSITION 26.1. *Let f, g be continuous functions in $\mathbf{R}^n$, one of which has a compact support. Let A be a subset of $\mathbf{R}^n$. The values of the convolution $f * g$ in the set A do not depend on the values of f in the complement of the set*

$$A - \operatorname{supp} g = \{x \in \mathbf{R}^n; x = x' - x'' \text{ for some } x' \in A', x'' \in \operatorname{supp} g\}.$$

We have already taken advantage of the fact that, if both f and g have compact support, so does $f * g$. This may be regarded as a corollary of the following result:

PROPOSITION 26.2. *Let f, g be as in Proposition 26.1. We have*

$$(26.9) \qquad \operatorname{supp}(f * g) \subset \operatorname{supp} f + \operatorname{supp} g \qquad \text{(vector sum)}.$$

Proof. Let x belong to the complement of $\operatorname{supp} f + \operatorname{supp} g$; then, for $y \in \operatorname{supp} g$, $x - y$ belongs to the complement of $\operatorname{supp} f$, hence

$$(f * g)(x) = \int f(x - y) g(y)\, dy = 0.$$

COROLLARY. *If both $\operatorname{supp} f$ and $\operatorname{supp} g$ are compact, $\operatorname{supp}(f * g)$ is also compact.*

Indeed, if K and H are compact subsets of a Hausdorff TVS E (here $\mathbf{R}^n$), $K + H$ is also a compact subset of E, as it is the image of the product $K \times H$, compact subset of $E \times E$, under the continuous mapping $(x, y) \rightsquigarrow x + y$.

We proceed now to study convolution from the viewpoint of differentiability. Suppose that f and g are two $\mathscr{C}^1$ functions, one of which has compact support in $\mathbf{R}^n$. Then it follows immediately from Leibniz' rule for differentiation under the integral sign that

$$(f * g)(x) = \int f(x - y) g(y)\, dy$$

is a differentiable function (at every point x) and that

$$(\partial/\partial x_j)(f * g) = ((\partial/\partial x_j)f) * g = f * ((\partial/\partial x_j)g), \qquad j = 1,...,n.$$

In fact, combining this with the fact that the convolution of two continuous functions, one of which has compact support, is a continuous function, we see that $f * g$ is a $\mathscr{C}^1$ function in $\mathbf{R}^n$.

Furthermore, if we apply (26.8) with $\partial f/\partial x_j$ instead of f, we derive the fact that, for all $x \in A$,

$$|(\partial/\partial x_j)(f * g)(x)| \leqslant \sup_{y \in B} |[(\partial/\partial x_j)f](y)| \, \| g \|_{L^1} ,$$

where $B = A - \operatorname{supp} g$. Of course, the right-hand side of the above inequality may very well be infinite. But it will be finite whenever supp g and A are both compact sets. This implies easily the following result, which is a particular case of a more general fact:

PROPOSITION 26.3. *Let m be an integer, $0 \leqslant m \leqslant +\infty$. If $g \in L^1$ has compact support, the convolution*

$$f \rightsquigarrow f * g$$

is a continuous linear map of $\mathscr{C}^m(\mathbf{R}^n)$ into itself.

COROLLARY. *Let m, g be as in Proposition 26.3. The convolution $f \rightsquigarrow f * g$ is a continuous linear map of $\mathscr{C}_c^m(\mathbf{R}^n)$ into itself.*

Proof of Corollary. In view of the general properties of LF-spaces, it suffices to prove that, for every compact subset K of $\mathbf{R}^n$, the restriction of the mapping $f \rightsquigarrow f * g$ to $\mathscr{C}_c^m(K)$ is continuous, as a map of $\mathscr{C}_c^m(K)$ into $\mathscr{C}^m(\mathbf{R}^n)$ (see Proposition 13.1). But in view of Proposition 26.2, it maps $\mathscr{C}_c^m(K)$ into $\mathscr{C}_c^m(K + \operatorname{supp} g)$, and on both these spaces the topology induced by $\mathscr{C}_c^m(\mathbf{R}^n)$ is the same as the one induced by $\mathscr{C}^m(\mathbf{R}^n)$. The corollary follows then directly from Proposition 26.3.

Exercises

26.1. Let μ be a Radon measure, and f a continuous function with compact support in $\mathbf{R}^n$. Prove that

$$(\mu * f)(x) = \int f(x - y) \, d\mu(y)$$

is a continuous function in $\mathbf{R}^n$.

26.2. Let f, g be two *locally* L^1 functions in the real line which vanish identically in the negative open half-line $\{t \in \mathbf{R}^1; t < 0\}$. Prove that

$$(f * g)(x) = \int_0^s f(x - t) g(t)\, dt$$

is a locally L^1 function in $\mathbf{R}^1$ vanishing for $x < 0$.

26.3. Compute the Fourier transform of $f * g$ when $f, g \in \mathscr{S}(\mathbf{R}^n)$.

26.4. Let f, g be two continuous functions in $\mathbf{R}^n$, one of which has compact support. Prove Leibniz' formula:

$$x^p(f * g) = \sum_{q \leqslant p} \binom{p}{q} (x^q f) * (x^{p-q} g),$$

where $p, q \in \mathbf{N}^n$ and $q \leqslant p$ means $q_j \leqslant p_j$ for all $j = 1, ..., n$.

26.5. Let f be a continuous function with compact support in $\mathbf{R}^n$. Prove that, if g is a polynomial (resp. an exponential $x \leadsto \exp(\langle x, \zeta \rangle)$, $\zeta \in \mathbf{C}^n$; resp. a function which can be continued to the complex space $\mathbf{C}^n$ as an entire analytic function), the same is true of the convolution $f * g$.

27

Example of Transpose:
Convolution of Distributions

In this chapter, we shall define the convolution of two distributions S, T in $\mathbf{R}^n$ under the condition that one of the two have compact support; for example, if supp S is compact, T may be arbitrary.

Later on, this condition on the supports will have to be relaxed; but of course this will not be possible unless we relate somehow the growth of the two factors, S, T. We know already, for instance, that, if S and T are both L^1 functions, their supports can be arbitrary (needless to say, the convolution of two distributions will have to coincide with the convolution defined in Chapter 26, when these distributions are functions!). To give another important example, we shall want to define the convolution operators on the space $\mathscr{S}'$ of tempered distributions, that is to say to find out what are the distributions T such that $S * T$ (suitably defined) is a tempered distribution for all $S \in \mathscr{S}'$. We shall see that rather restrictive conditions of decay at infinity must be imposed upon T. The situation here is similar to the one which we would encounter if we tried to define the convolution $f * g$ of two functions f, g, one of which is a polynomial: g will have to decrease rapidly at infinity (i.e., more rapidly than any power of $1/|x|$).

We are going to need a very general and very simple result about functions with values in a TVS (defined in some open subset of $\mathbf{R}^n$).

THEOREM 27.1. *Let Ω be an open subset of $\mathbf{R}^n$, and $x \rightsquigarrow \varphi(x)$ a function defined in Ω with values in a Hausdorff TVS E. Let e' be an arbitrary continuous linear functional on E. Then*

(a) *If the function φ is continuous, the complex-valued function*

(27.1) $$\Omega \ni x \rightsquigarrow \langle e', \varphi(x) \rangle$$

is continuous.

(b) *If φ is continuous and if φ is differentiable at some point x^0 of Ω, then (27.1) is differentiable at x^0 and we have*

(27.2) $$(\partial/\partial x_j)\langle e', \varphi(x)\rangle \Big|_{x=x^0} = \langle e', (\partial\varphi/\partial x_j)(x^0)\rangle.$$

(c) *Let k be either a positive finite integer or $+\infty$. If φ is a $\mathscr{C}^k$ function in Ω with values in E, the function (27.1) is a complex-valued $\mathscr{C}^k$ function in Ω and we have, for all $p \in \mathbf{N}^n$ such that $|p| < k+1$, and all $x \in \Omega$,*

(27.3) $$(\partial/\partial x)^p \langle e', \varphi(x)\rangle = \langle e', (\partial/\partial x)^p \varphi(x)\rangle.$$

Theorem 27.1 is a kind of rule of differentiation under the integral sign. Its proof is trivial, as soon as we know what we mean by a differentiable function with values in a TVS. First of all, let $t \rightsquigarrow f(t)$ be a function from some open interval $]t_0 - \varepsilon, t_0 + \varepsilon[$ of the real line into the Hausdorff TVS E. We say that the *derivative of $f(t)$ at t_0 exists*, or that f is *differentiable at t_0* if

$$h^{-1}[f(t_0 + h) - f(t_0)]$$

converges to some element of E, denoted then by $f'(t_0)$, as the real number $h \neq 0$ converges to zero. If we study now a function of several variables, like φ defined in $\Omega \subset \mathbf{R}^n$, it is clear what we mean when we say that *the partial derivatives of φ exist at a point x^0 of Ω*. We say then that φ is differentiable at x^0 if

$$\varphi(x) - \varphi(x^0) - \sum_{j=1}^{n} (x_j - x_j^0)(\partial\varphi/\partial x_j)(x^0)$$

converges to zero in E when $x \to x^0$ in Ω.

Proof of Theorem 27.1. Part (a) is obvious since the mapping (27.1) is nothing else but $e' \circ \varphi : \Omega \to \mathbf{C}$; (c) follows by combining (a) and (b). As for (b), it is simply a matter of combining the continuity of e' with the differentiability of φ as we have defined it. The student may work out the details if he likes to.

We shall apply Theorem 27.1 in the following two situations:

(1) $E = \mathscr{C}_c^\infty(\mathbf{R}^n)$; the elements of E are functions of the variable in $\mathbf{R}^n$ which, for reasons of clarity, we shall denote by $y = (y_1,...,y_n)$. On the other hand, Ω will be the whole space $\mathbf{R}^n$; the variable point in Ω will be denoted by x.

(2) $E = \mathscr{C}^\infty(\mathbf{R}^n)$, the rest staying as in (1).

The function φ in Theorem 27.1 will be, in both these situations,

(27.4) $$x \rightsquigarrow (y \rightsquigarrow \varphi(x - y)),$$

where φ is a given element of E (which is to say, either of $\mathscr{C}_c^\infty(\mathbf{R}^n)$ or of $\mathscr{C}^\infty(\mathbf{R}^n)$, depending on whether we are in Situation (1) or (2)).

LEMMA 27.1.   *Suppose that $\varphi \in \mathscr{C}_c^\infty(\mathbf{R}^n)$ (resp. $\mathscr{C}^\infty(\mathbf{R}^n)$). Then (27.4) is a differentiable function of $x \in \mathbf{R}^n$ with values in $\mathscr{C}_c^\infty(\mathbf{R}_y^n)$ (resp. $\mathscr{C}^\infty(\mathbf{R}_y^n)$). Its first partial derivative with respect to x_j ($1 \leqslant j \leqslant n$) is the function*

(27.5) $$x \rightsquigarrow (y \rightsquigarrow (\partial/\partial x_j)\, \varphi(x - y)).$$

Proof.   We shall prove the result only in the case of $\mathscr{C}_c^\infty(\mathbf{R}^n)$. As in the other case, when the space under consideration is $\mathscr{C}^\infty(\mathbf{R}^n)$, the proof is very similar and, as a matter of fact, somewhat simpler.

Let x^0 be an arbitrary point of $\mathbf{R}^n$ and let $U = \{x \in \mathbf{R}^n;\ |x - x^0| < 1\}$. When x varies in U, the function $y \rightsquigarrow \varphi(x - y)$ keeps its support in the set $U - \mathrm{supp}\, \varphi$, which is an open subset of $\mathbf{R}^n$ whose closure, K, is compact. We shall prove that (27.4) is a function in U, with values in $\mathscr{C}_c^\infty(K_y)$ when the latter carries the topology induced by $\mathscr{C}_c^\infty(\mathbf{R}_y^n)$ (the subscript y signifies that the elements of these spaces are functions of y), that (27.4) is then differentiable at x^0, and that its partial derivative with respect to x_j at x^0 is equal to the value of (27.5) at x^0. As x^0 is arbitrary, this will obviously prove the lemma.

The topology of $\mathscr{C}_c^\infty(K_y)$ is defined by the seminorms

$$\psi \rightsquigarrow \sup_{y \in \mathbf{R}^n} |(\partial/\partial y)^q\, \psi(y)|, \qquad q \in \mathbf{N}^n.$$

Let us denote by q_q this seminorm, by φ the function of x, (27.4), and by φ^j the function (27.5). We must prove that

$$\mathrm{q}_q \left(\varphi(x) - \varphi(x_0) - \sum_{j=1}^{n} (x_j - x_j^0)\, \varphi^j(x^0) \right)$$

converges to zero when $x \in U$ converges to x^0. But this only means that

$$(\partial/\partial y)^q\, \varphi(x - y) - (\partial/\partial y)^q\, \varphi(x^0 - y) - \sum_{j=1}^{n} (x_j - x_j^0)(\partial/\partial y)^q(\partial/\partial x_j)\, \varphi(x^0 - y)$$

converges uniformly to zero with respect to $y \in U - \mathrm{supp}\, \varphi$ as $x \in U$ converges to x^0, which is trivially true.

COROLLARY. *Suppose that* $\varphi \in \mathscr{C}_c^\infty(\mathbf{R}^n)$ *(resp.* $\mathscr{C}^\infty(\mathbf{R}^n)$*). The function* (27.4) *is a* $\mathscr{C}^\infty$ *function of* $x \in \mathbf{R}^n$ *with values in* $\mathscr{C}_c^\infty(\mathbf{R}_y^n)$ *(resp.* $\mathscr{C}^\infty(\mathbf{R}_y^n)$*). Its partial derivative of order* $p = (p_1, ..., p_n) \in \mathbf{N}^n$ *is equal to*

$$(27.6) \qquad x \rightsquigarrow (y \rightsquigarrow (\partial/\partial x)^p \, \varphi(x - y)).$$

Proof. Set $E = \mathscr{C}_c^\infty(\mathbf{R}_y^n)$ (resp. $\mathscr{C}^\infty(\mathbf{R}_y^n)$).

Since (27.4) is differentiable at every point, it is continuous everywhere. Its partial derivative with respect to x_j $(1 \leqslant j \leqslant n)$ is given by (27.5), which is just the same as (27.4) except that φ has been replaced by $(\partial/\partial x_j)\varphi$, therefore it is a continuous function of x with values in E. Thus (27.4) is a $\mathscr{C}^1$ function of x with values in E. Suppose then that we have proved that (27.4) is a $\mathscr{C}^k$ function of x for some $k = 0, 1, 2,...$; but this also applies to the first derivatives of (27.4) with respect to the variables x_j, since they have the same form as (27.4) (except that φ has to be replaced by $(\partial/\partial x_j)\varphi$), we conclude that these first derivatives are also $\mathscr{C}^k$ functions of x with values in E, which proves that (27.4) is a $\mathscr{C}^{k+1}$ function of x. This immediately implies the corollary.

THEOREM 27.2. *Let* φ *be a* $\mathscr{C}^\infty$ *function, and* T *a distribution in* $\mathbf{R}^n$. *Suppose that at least one of the two sets,* supp φ, supp T, *is compact.* *Then*

$$x \rightsquigarrow \langle T_y, \varphi(x - y) \rangle$$

is a $\mathscr{C}^\infty$ *function in* $\mathbf{R}^n$. *For all n-tuples* $p = (p_1, ..., p_n)$, *we have*

$$(27.7) \qquad (\partial/\partial x)^p \langle T_y, \varphi(x - y) \rangle = \langle T_y, (\partial/\partial x)^p \, \varphi(x - y) \rangle.$$

The notation T_y means that the distribution T acts on a function $\psi(x - y)$ when the latter is regarded as a function of the variable y.

Proof. It suffices to combine the corollary of Lemma 27.1 with Part (c) of Theorem 27.1 (remembering that distributions with compact support are continuous linear forms on $\mathscr{C}^\infty(\mathbf{R}^n)$).

Definition 27.1.   *The function*

$$x \rightsquigarrow \langle T_y, \varphi(x - y) \rangle$$

is called the convolution of φ *and* T *and denoted by* $T * \varphi$ *or* $\varphi * T$.

When the distribution T is a locally integrable function f, we have

$$\langle T_y, \varphi(x - y) \rangle = \int f(y) \, \varphi(x - y) \, dy,$$

which shows that Definition 27.1 agrees, in this case, with the notion of convolution introduced in Chapter 26 (Eq. (26.1)).

The convolution

$$(\varphi, T) \rightsquigarrow \varphi * T$$

is a bilinear map from $\mathscr{C}_c^\infty \times \mathscr{D}'$ (resp. $\mathscr{C}^\infty \times \mathscr{E}'$) into $\mathscr{C}^\infty$. We state and prove now some of its elementary properties.

Observe that Eq. (27.7) can be read

$$(\partial/\partial x)^p(T * \varphi) = T * [(\partial/\partial x)^p\varphi].$$

But if we observe that

$$(\partial/\partial x)^p \varphi(x - y) = (-1)^{|p|}(\partial/\partial y)^p \varphi(x - y),$$

the right-hand side of (27.7) can also be read

$$\langle(\partial/\partial y)^p T_y, \varphi(x - y)\rangle = \{[(\partial/\partial x)^p T] * \varphi\}(x),$$

so that we can state:

PROPOSITION 27.1.   *Let φ be a $\mathscr{C}^\infty$ function, and T a distribution in $\mathbf{R}^n$. Suppose that at least one of the two sets, supp φ, supp T, is compact. Then, for all n-tuples p,*

$$(\partial/\partial x)^p(T * \varphi) = [(\partial/\partial x)^p T] * \varphi = T * [(\partial/\partial x)^p\varphi].$$

Another important property is the following one:

PROPOSITION 27.2.   *Let T and φ be as in Proposition 27.1. Then*

(27.8) $\mathrm{supp}(\varphi * T) \subset \mathrm{supp}\,\varphi + \mathrm{supp}\,T$ (vector sum).

Proof.   As one of the two sets supp φ or supp T is compact, their vector sum is closed, according to an elementary result of point-set topology. Suppose for instance that supp φ is compact. Let U be an open set whose closure $\bar{U}$ is compact and contained in the complement of supp φ + supp T. The compact set

$$\bar{U} - \mathrm{supp}\,\varphi$$

does not intersect supp T; therefore we may find a $\mathscr{C}^\infty$ function g in $\mathbf{R}^n$, equal to one in a neighborhood of supp T and to zero in some neighborhood of $\bar{U}$ − supp φ (Theorem 16.4). In particular, $gT = T$ and we have, therefore,

$$(T * \varphi)(x) = \langle T_y, \varphi(x - y)\rangle = \langle g(y)T_y, \varphi(x - y)\rangle$$
$$= \langle T_y, g(y)\,\varphi(x - y)\rangle.$$

But if $x \in U$, the function $y \rightsquigarrow g(y)\,\varphi(x - y)$ is identically zero. Indeed, if it were not zero at some point y, it would mean that $y \in$ supp g and that $x - y \in$ supp φ; the latter means that $y \in U -$ supp φ. But this is impossible as g vanishes in a neighborhood of $\bar{U} -$ supp φ.   Q.E.D.

We study now the continuity of the mapping $\varphi \rightsquigarrow T * \varphi$.

THEOREM 27.3.   *Let T be a distribution (resp. a distribution with compact support) in $\mathbf{R}^n$. The convolution*

$$\varphi \rightsquigarrow T * \varphi$$

is a continuous linear map of $\mathscr{C}_c^\infty(\mathbf{R}^n)$ (resp. $\mathscr{C}^\infty(\mathbf{R}^n)$) into $\mathscr{C}^\infty(\mathbf{R}^n)$.

Proof.   We shall treat only the case of an arbitrary distribution T and prove that $\varphi \rightsquigarrow T * \varphi$ is a continuous linear map $\mathscr{C}_c^\infty \rightsquigarrow \mathscr{C}^\infty$. The other case is simpler to handle (and, in fact, follows quite easily from this one).

In view of a general property of LF-spaces (Proposition 13.1), it will suffice to prove that, for every compact subset K of $\mathbf{R}^n$, $\varphi \rightsquigarrow T * \varphi$ is a continuous linear map of $\mathscr{C}_c^\infty(K)$ into $\mathscr{C}^\infty(\mathbf{R}^n)$. The topology of $\mathscr{C}^\infty(\mathbf{R}^n)$ is defined by the seminorms

$$\varphi \rightsquigarrow \sup_{x \in H} |\,(\partial/\partial x)^p\,\varphi(x)|, \qquad H, \text{ compact subset of } \mathbf{R}^n, \quad p \in \mathbf{N}^n.$$

When x varies in H and supp φ is contained in K, the function

$$y \rightsquigarrow (\partial/\partial x)^p\,\varphi(x - y)$$

varies in $\mathscr{C}_c^\infty(H - K)$. The restriction of T to $\mathscr{C}_c^\infty(H - K)$ is a continuous linear form on this space. Therefore, there is a constant $M > 0$ and an integer k such that we have, for all $\psi \in \mathscr{C}_c^\infty(H - K)$,

$$|\langle T, \psi \rangle| \leqslant M \sup_{|q| \leqslant k} \left(\sup_{y \in \mathbf{R}^n} |(\partial/\partial y)^q\,\psi(y)| \right).$$

We replace $\psi(y)$ by $(\partial/\partial x)^p \varphi(x - y)$ with $x \in H$, supp $\varphi \subset K$, and we apply Eq. (27.7) (Theorem 27.2). It yields

$$\sup_{x \in H} |(\partial/\partial x)^p \langle T_y, \varphi(x - y) \rangle|$$

$$\leqslant M \sup_{|q| \leqslant k} \left(\sup_{x \in H, y \in \mathbf{R}^n} |(\partial/\partial x)^p(\partial/\partial y)^q\,\varphi(x - y)| \right)$$

$$\leqslant M \sup_{r \leqslant k + |p|} \left(\sup_{\xi \in \mathbf{R}^n} |(\partial/\partial \xi)^r\,\varphi(\xi)| \right).$$

This proves the asserted continuity.

COROLLARY. *Let T be a distribution with compact support in $\mathbf{R}^n$. The convolution*

$$\varphi \rightsquigarrow T * \varphi$$

is a continuous linear map of $\mathscr{C}_c^\infty(\mathbf{R}^n)$ into itself.

Proof. It suffices to prove that, for every compact subset K of $\mathbf{R}^n$, $\varphi \rightsquigarrow T * \varphi$ is a continuous linear map of $\mathscr{C}_c^\infty(K)$ into $\mathscr{C}_c^\infty(\mathbf{R}^n)$. But we know that it is continuous from $\mathscr{C}_c^\infty(K)$ into $\mathscr{C}^\infty(\mathbf{R}^n)$, by Theorem 27.3; and by Proposition 27.2, it maps $\mathscr{C}_c^\infty(K)$ into $\mathscr{C}_c^\infty(K + \operatorname{supp} T)$. On the latter space, the topologies induced by $\mathscr{C}^\infty(\mathbf{R}^n)$ and by $\mathscr{C}_c^\infty(\mathbf{R}^n)$ coincide, whence the corollary.

By using the corollary of Theorem 27.3, we may now easily define the convolution of an arbitrary distribution S with a distribution T having compact support. But so as to provide a motivation for the definition, we shall first consider the case where S is a locally integrable function f and T an L^1 function with compact support, g. Then we know what the convolution $f * g$ must be; it is the locally L^1 function

$$(f * g)(x) = \int f(x - y) g(y) \, dy = \int f(y) g(x - y) \, dy.$$

Let, then, φ be a test function; considering $f * g$ as a distribution, we observe that we have

$$\langle f * g, \varphi \rangle = \iint f(x - y) g(y) \, \varphi(x) \, dx \, dy$$

$$= \int \left[\int f(x - y) \, \varphi(x) \, dx \right] g(y) \, dy = \langle g, \check{f} * \varphi \rangle,$$

where we have set $\check{f}(x) = f(-x)$. Similarly,

$$\langle f * g, \varphi \rangle = \int f(y) \left[\int g(x - y) \, \varphi(x) \, dx \right] dy = \langle f, \check{g} * \varphi \rangle.$$

Now, the operation $f \rightsquigarrow \check{f}$ is easily extendable to distributions, if we note that

$$\langle \check{f}, \varphi \rangle = \int f(-x) \, \varphi(x) \, dx = \int f(x) \, \varphi(-x) \, dx = \langle f, \check{\varphi} \rangle.$$

Definition 27.2. *Let T be a distribution in $\mathbf{R}^n$. By $\check{T}$ we denote the distribution defined by*

$$\langle \check{T}, \varphi \rangle = \langle T, \check{\varphi} \rangle, \qquad \varphi \in \mathscr{C}_c^\infty(\mathbf{R}^n).$$

The operation $T \rightsquigarrow \check{T}$ is often referred to as *the symmetry with respect to the origin* ($\check{}$: tchetch).

We now focus our attention on the following two maps. Let S, T be two distributions in $\mathbf{R}^n$, S having compact support. Consider

$$(27.9) \qquad \mathscr{C}_c^\infty \ni \varphi \rightsquigarrow \varphi * \check{S} \in \mathscr{C}_c^\infty,$$

$$(27.10) \qquad \mathscr{C}_c^\infty \ni \varphi \rightsquigarrow \varphi * \check{T} \in \mathscr{C}^\infty.$$

Both are continuous linear mappings, and we may introduce their transposes. The transpose of (27.9) is a continuous linear map of $\mathscr{D}'$ into itself, whereas the transpose of (27.10) is a continuous linear map of $\mathscr{E}'$ into $\mathscr{D}'$. We have then the following commutativity result:

THEOREM 27.4. *Let S, T be two distributions in $\mathbf{R}^n$, S having compact support.*

The image of T under the transpose of (27.9) is equal to the image of S under the transpose of (27.10).

Proof. For simplicity, we shall denote by $S * T$ (resp. $T * S$) the image of T (resp. S) under the transpose of (27.9) (resp. (27.10)). We must prove that $S * T = T * S$.

1. *Case Where S and T Are Locally Integrable Functions*

Suppose that S is an L^1 function f, having compact support, and that T is a locally integrable function g. Let φ be an arbitrary element of $\mathscr{C}_c^\infty$. We have, according to the definitions,

$$\langle S * T, \varphi \rangle = \langle T, \check{S} * \varphi \rangle = \langle T, \int f(y - x)\, \varphi(y)\, dy \rangle$$

$$= \int g(x) \left[\int f(y - x)\, \varphi(y)\, dy \right] dx.$$

As both f and φ have compact support, we may obviously interchange the order of integration. This yields

$$\langle S * T, \varphi \rangle = \langle f * g, \varphi \rangle.$$

Similarly,

$$\langle T * S, \varphi \rangle = \langle g * f, \varphi \rangle.$$

We have used the notation

$$(f * g)(x) = \int f(x - y)\, g(y)\, dy, \qquad (g * f)(x) = \int g(x - y)\, f(y)\, dy,$$

but, of course, $f * g = g * f$, whence the result in this case.

2. Case Where Both S and T Have Compact Support

In view of Proposition 27.1, we have, whatever the n-tuple p,

(27.11)
$$(\partial/\partial x)^p(S * T) = [(\partial/\partial x)^p S] * T = S * [(\partial/\partial x)^p T],$$
$$(\partial/\partial x)^p(T * S) = [(\partial/\partial x)^p T] * S = T * [(\partial/\partial x)^p S].$$

Indeed, it suffices to remark that, for all distributions U,

$$\overset{\smile}{\widehat{(\partial/\partial x)^p U}} = (-1)^{|p|}(\partial/\partial x)^p \overset{\smile}{U}.$$

Let us now assume that both S and T have compact support. Then they are both equal to a finite sum of derivatives of L^1 functions with compact support (Corollary 2 of Theorem 24.4, Theorem 24.5):

$$S = \sum_{|p| \leqslant m_1} (\partial/\partial x)^p f_p, \qquad T = \sum_{|q| \leqslant m_2} (\partial/\partial x)^q g_q.$$

In view of Eqs. (27.11), we have

$$S * T = \sum_{|p| \leqslant m_1, |q| \leqslant m_2} (\partial/\partial x)^{p+q}(f_p * g_q).$$

It suffices to observe that $f_p * g_q = g_q * f_p$ for all pairs (p, q); this implies obviously that $S * T = T * S$ in this case.

3. General Case

Let φ be an arbitrary element of $\mathscr{C}_c^\infty$. Let $f \in \mathscr{C}_c^\infty$ be equal to one in a neighborhood of supp S, and $g \in \mathscr{C}_c^\infty$ equal to one in a neighborhood of supp $\varphi -$ supp f. We have

$$\langle T * S, \varphi \rangle = \langle S, \varphi * \overset{\smile}{T} \rangle = \langle S, f(\varphi * \overset{\smile}{T}) \rangle, \qquad \text{since} \quad fS = S.$$

On the other hand,

$$f(x)(\varphi * \overset{\smile}{T})(x) = \langle T_y, f(x)\, \varphi(x + y) \rangle = \langle T_y, f(x)\, g(y)\, \varphi(x + y) \rangle,$$

since $g(y) = 1$ when $y \in$ supp $\varphi -$ supp f. In other words,

$$f(\varphi * \overset{\smile}{T}) = f(\varphi * \overset{\smile}{(gT)}).$$

Consequently

(27.12) $\langle T * S, \varphi \rangle = \langle S, f(\varphi * \overset{\smile}{(gT)}) \rangle = \langle S, \varphi * \overset{\smile}{(gT)} \rangle = \langle (gT) * S, \varphi \rangle.$

We have also

$$(27.13) \quad \langle S*T,\varphi \rangle = \langle T,\varphi * \check{S} \rangle = \langle T, g(\varphi * \check{S}) \rangle = \langle gT,\varphi * \check{S} \rangle = \langle S*(gT),\varphi \rangle.$$

We have used the fact that $g = 1$ in a neighborhood of $\mathrm{supp}(\varphi * \check{S})$ (cf. Proposition 27.2; note that $\mathrm{supp}\ \check{S} = -\mathrm{supp}\ S$).

Since gT has compact support, we know by Part 2 in the proof of Theorem 27.4 that

$$S*(gT) = (gT)*S.$$

Combining this with (27.12) and (27.13), we obtain the desired result.

The statement and the proof of Theorem 27.4 contain practically all the information we need. First of all, we may define the convolution of two distributions, one of which has a compact support.

Definition 27.3. Let S and T be two distributions. Suppose that S has compact support. The convolution of S and T, denoted by $S*T$ or $T*S$, is the image of S under the transpose of the continuous linear map $\varphi \rightsquigarrow \varphi * \check{T}$ of $\mathscr{C}_c^\infty(\mathbf{R}^n)$ into $\mathscr{C}^\infty(\mathbf{R}^n)$, or equivalently, the image of T under the transpose of the continuous linear map $\varphi \rightsquigarrow \varphi * \check{S}$ if $\mathscr{C}_c^\infty(\mathbf{R}^n)$ into itself.

Indeed, these two images are equal, in view of Theorem 27.4.

The beginning of the proof of Theorem 27.4 shows that, when S and T are locally integrable functions (S having compact support), the convolution $S*T$ in the sense of distributions is equal to the convolution in the sense of functions. Keeping this in mind, we may show that Definitions 27.1 and 27.3 are consistent, that is to say:

THEOREM 27.5. *Let S, T be two distributions, one of which has compact support. Suppose furthermore that T is a $\mathscr{C}^\infty$ function, φ. Then the distribution $S*T$ (Definition 27.3) is the $\mathscr{C}^\infty$ function*

$$x \rightsquigarrow \langle S_y , \varphi(x-y) \rangle.$$

Proof. The statement is true when S is a locally integrable function (Part 1 in the proof of Theorem 27.4). Suppose now that S is an arbitrary distribution with compact support; let us write

$$S = \sum_p (\partial/\partial x)^p g_p(x),$$

where the summation is finite and where the g_p are L^1 functions with compact support. We have, by (27.11),

$$S*T = \sum_p [(\partial/\partial x)^p g_p]*T = \sum_p g_p * [(\partial/\partial x)^p T],$$

but

$$g_p * [(\partial/\partial x)^p T]$$

is the function

$$x \rightsquigarrow \int g_p(y)\,(\partial/\partial x)^p\,\varphi(x-y)\,dy = \int g_p(y)\,(-1)^{|p|}(\partial/\partial y)^p\,\varphi(x-y)\,dy$$

$$= \langle (\partial/\partial y)^p\,g(y),\,\varphi(x-y)\rangle,$$

which means exactly that $S * T$ is the function $x \rightsquigarrow \langle S_y,\,\varphi(x-y)\rangle$.

Suppose now that the support of S is not compact, in which case the support of φ must be compact. Let U be an arbitrary bounded open subset of $\mathbf{R}^n$. Let g be a $\mathscr{C}^\infty$ function equal to one in the open set $U - \operatorname{supp}\varphi$. In the open set U, we have

$$S * T = (gS) * T \qquad \text{(in the sense of distributions in } U\text{)},$$

as immediately seen. From the first part of the proof, we know that $(gS) * T$ is a $\mathscr{C}^\infty$ function, equal to $x \rightsquigarrow \langle S_y,\,g(y)\,\varphi(x-y)\rangle$. If $x \in U$, we have $g(y) \equiv 1$ in a neighborhood of the support of the function $y \rightsquigarrow \varphi(x-y)$, hence, for $x \in U$,

$$\langle S_y,\,g(y)\,\varphi(x-y)\rangle = \langle S_y,\,\varphi(x-y)\rangle,$$

which implies that $S * T$ is (in U) the function $x \rightsquigarrow \langle S_y,\,\varphi(x-y)\rangle$. We know that the latter function is a $\mathscr{C}^\infty$ function (Theorem 27.2). Since U is arbitrary, this proves Theorem 27.5. By applying Theorem 27.3 (combined with the fact that the transpose of a continuous linear map is a continuous linear map—if the duals of the spaces involved carry the strong dual topology), we obtain immediately:

THEOREM 27.6.   *The convolution*

$$(S, T) \rightsquigarrow S * T$$

is a separately continuous bilinear map of $\mathscr{E}' \times \mathscr{D}'$ into $\mathscr{D}'$.

Separately continuous means here that, if S is kept fixed, the linear map $T \rightsquigarrow S * T$ of $\mathscr{D}'$ into $\mathscr{D}'$ is continuous, and if T is kept fixed, the linear map $S \rightsquigarrow S * T$ of $\mathscr{E}'$ into $\mathscr{D}'$ is continuous.

We have Eq. (27.11):

PROPOSITION 27.3.   *Let $S \in \mathscr{E}'$, $T \in \mathscr{D}'$. For all n-tuples p,*

$$(\partial/\partial x)^p(S * T) = [(\partial/\partial x)^p S] * T = S * [(\partial/\partial x)^p T].$$

Also:

PROPOSITION 27.4.   *Let* $S \in \mathscr{E}'$, $T \in \mathscr{D}'$. *We have*

$$\text{supp}(S * T) \subset \text{supp } S + \text{supp } T.$$

Proof.   Let $\varphi \in \mathscr{C}_c^\infty$ have its support in the complement of the closed set $\text{supp } S + \text{supp } T$; then $\text{supp } \varphi - \text{supp } S$ is a compact subset of the complement of $\text{supp } T$. In view of Proposition 27.2,

$$\text{supp}(\check{S} * \varphi) \subset \text{supp } \varphi + \text{supp } \check{S};$$

but as $\text{supp } \check{S} = -\text{supp } S$, we see that $\langle T, \check{S} * \varphi \rangle = 0$.   Q.E.D.

COROLLARY.   *If both S and T have compact support, so does $S * T$.*

The corollary shows that $\mathscr{E}'$, space of distributions with compact support in $\mathbf{R}^n$, is a ring for the operations addition and convolution (in fact, it is an algebra if we consider its vector space structure). In fact, we have:

THEOREM 27.7.   *The space $\mathscr{E}'$ of distributions with compact support in $\mathbf{R}^n$ is a commutative convolution algebra with the Dirac measure as unit element.*

The commutativity of the convolution in $\mathscr{E}'$ is just a restatement of Theorem 27.4 when both S and T have compact support. That the Dirac measure is the unity for convolution follows from the more general fact:

PROPOSITION 27.5.   *If T is any distribution in $\mathbf{R}^n$,*

$$T * \delta = T.$$

Proof.   We have, for all test functions φ,

$$\langle T * \delta, \varphi \rangle = \langle T, \varphi * \delta \rangle, \qquad \text{since, obviously, } \check{\delta} = \delta.$$

But

$$(\varphi * \delta)(x) = \langle \delta_y, \varphi(x - y) \rangle = \varphi(x).$$ Q.E.D.

COROLLARY.   *If $T \in \mathscr{D}'$, $p \in \mathbf{N}^n$,*

$$(\partial/\partial x)^p T = [(\partial/\partial x)^p \delta] * T.$$

Proof.   Combine Propositions 27.3 and 27.5.

We define now the *translation* of a distribution. Let a be any vector in $\mathbf{R}^n$; let us define the translation by a, $\tau_a f$, of a function f by the formula

$$\tau_a f(x) = f(x - a), \qquad x \in \mathbf{R}^n.$$

Suppose that f is locally integrable; if φ is a test function, we have

$$\int f(x - a)\, \varphi(x)\, dx = \int f(x)\, \varphi(x + a)\, dx,$$

which can be read

$$\langle \tau_a f, \varphi \rangle = \langle f, \tau_{-a}\varphi \rangle.$$

This motivates the following definition:

Definition 27.4. *Let T be a distribution, and a a vector belonging to $\mathbf{R}^n$. The translation of T by a is the distribution $\tau_a T$ defined by*

$$\langle \tau_a T, \varphi \rangle = \langle T, \tau_{-a}\varphi \rangle, \qquad \varphi \in \mathscr{C}_c^\infty.$$

Exercise 27.1. Prove the following Proposition and its corollary:

PROPOSITION 27.6. *Let S, T be two distributions in $\mathbf{R}^n$, one of which at least has compact support; let $a \in \mathbf{R}^n$. Then*

$$\tau_a(S * T) = (\tau_a S) * T = S * (\tau_a T).$$

COROLLARY. *Let T be any distribution in $\mathbf{R}^n$, $a \in \mathbf{R}^n$. Then*

$$\tau_a T = \delta_a * T, \qquad \delta_a : \text{Dirac measure at the point } a.$$

We should also mention that the convolution of a finite number of distributions, all of which, except at most one, have compact support, is *associative*.

Exercises

27.2. Let T be a distribution with compact support, and f a function in $\mathbf{R}^n$. Prove the following statements:

(1) if f is a polynomial, so is $T * f$;
(2) if f is an exponential $\exp\langle x, \zeta \rangle$ ($\zeta \in \mathbf{C}^n$),

$$(T * f)(x) = C(\zeta) \exp\langle x, \zeta \rangle;$$

what is the value of the constant $C(\zeta)$?

(3) if f is an analytic function, so is $T * f$;
(4) if f is the restriction to $\mathbf{R}^n$ of an entire analytic function in $\mathbf{C}^n$, so is $T * f$;
(5) if f is the restriction of an entire function of exponential type in $\mathbf{C}^n$, so is $T * f$.

27.3. Prove that, given any distribution T with compact support in $\mathbf{R}^n$,

$$(\partial/\partial x_j)T = \lim \frac{1}{h}(\tau_{h_j}T - T),$$

where h_j is the vector in $\mathbf{R}^n$ whose kth component is equal to zero if $k \neq j$ and whose jth component is equal to a number $h > 0$, and where the limit is to be understood in the sense of $\mathscr{D}'(\mathbf{R}^n)$ as $h \to 0$.

27.4. Let f, g be two $\mathscr{C}^\infty$ functions with compact support in the real line; let Y be the Heaviside function, equal to 1 for $x > 0$ and to zero for $x < 0$. Verify the formula

$$(d/dx)^p(f * (Yg)) = f * [(dx)^p(Yg)].$$

27.5. Let Γ be a *closed convex salient* cone in $\mathbf{R}^n$, i.e., $\bar{\Gamma} = \Gamma$, $\rho\Gamma \subset \Gamma$ for all $\rho > 0$, $\Gamma + \Gamma \subset \Gamma$, Γ containing no straight line. Let us denote by $\mathscr{D}'(\Gamma)$ the subspace of $\mathscr{D}'(\mathbf{R}^n)$ consisting of the distributions in $\mathbf{R}^n$ having their support in Γ. Let us denote by $\mathscr{E}'(\Gamma)$ the intersection $\mathscr{D}'(\Gamma) \cap \mathscr{E}'(\mathbf{R}^n)$, that is to say the space of distributions with compact support contained in Γ.

Prove that to every pair of distributions $U, V \in \mathscr{D}'(\Gamma)$ there is a unique distribution $W \in \mathscr{D}'(\Gamma)$ with the following property:

For every $\rho > 0$, if $U', V' \in \mathscr{E}'(\Gamma)$ are equal to U and V, respectively, in the open ball $\{x \in \mathbf{R}^n; |x| < \rho\}$, then

$$W = U' * V' \quad \text{in the open ball} \quad \{x; |x| < \rho\}.$$

Prove that, by setting $W = U * V$, we turn $\mathscr{D}'(\Gamma)$ into a commutative convolution algebra which contains $\mathscr{E}'(\Gamma)$ (with the convolution induced by $\mathscr{E}'(\mathbf{R}^n)$) as a subalgebra.

Apply this to $n = 1$ and to $\Gamma = \{t \in \mathbf{R}^1; t \geqslant 0\}$. Compute the convolution of two locally integrable functions in $\mathbf{R}^1$ which vanish identically for $t < 0$. (The space $\mathscr{D}'(\Gamma)$ for this particular choice of Γ is usually denoted by $\mathscr{D}'_+$.)

27.6. Let $\mathscr{D}'_+$ be the space of distributions in $\mathbf{R}^1$ having their support in the closed positive half-line $\{t; t \geqslant 0\}$. In this exercise, we regard $\mathscr{D}'_+$ as a convolution algebra for the convolution defined in Exercise 27.5.

Prove the following facts:

(1)   the subspace $\mathscr{D}'_0$ of $\mathscr{D}'_+$, consisting of the distributions with support at the origin, is a *subfield* of the convolution algebra $\mathscr{D}'_+$;

(2)   the subspace of $\mathscr{D}'_+$ consisting of the distributions with support in a closed half-line $\{t; t \geqslant a\}$, $a > 0$, is an ideal in $\mathscr{D}'_+$.

27.7. We keep considering the convolution algebra $\mathscr{D}'_+$ of Exercise 27.6.

Prove that the distributions $Y(t)(t^k/k!)$ $(k = 0, 1, 2,...)$ are invertible in $\mathscr{D}'_+$ and compute their inverses (Y is the Heaviside function, equal to 1 for $t > 0$ and to zero for $t < 0$).

Give an example of a continuous function in $\mathbf{R}^1$, f, such that $f(t) > 0$ for $t > 0$, $f(t) = 0$ for $t < 0$, and which does not have any inverse in the convolution algebra $\mathscr{D}'_+$.

27.8. Let A be a subset of $\{t; t \geqslant 0\}$, and $\mathscr{D}'(A)$ the space of distributions in $\mathbf{R}^1$ having their support in A. Give a necessary and sufficient condition on A so that $\mathscr{D}'(A)$ be an ideal of the convolution algebra $\mathscr{D}'_+$ (cf. Exercises 27.6 and 27.7). Give an example of an ideal of $\mathscr{D}'_+$ which is not of the form $\mathscr{D}'(A)$.

28

Approximation of Distributions by Cutting and Regularizing

In this chapter, we are going to show that every distribution T in an open set Ω of $\mathbf{R}^n$ is the limit of a sequence of functions belonging to $\mathscr{C}_c^\infty(\Omega)$ and, furthermore, that there is a standard procedure for constructing this sequence from T.

Let $\{\Omega_k\}$ $(k = 0, 1,...)$ be a sequence of open subsets of Ω whose union is equal to Ω and such that $\Omega_{k-1} \subset \Omega_k$ $(k = 1, 2,...)$. For each k, select a function $g_k \in \mathscr{C}^\infty(\Omega)$ which is equal to one in Ω_k. Now, given any distribution T in Ω, it is clear that the distributions $g_k T$ converge to T in $\mathscr{D}'(\Omega)$, say for the strong dual topology, although this is not important, as strong and weak convergences in $\mathscr{D}'(\Omega)$ are one and the same thing for sequences, as we shall soon see. Anyway, if $\mathscr{B}$ is a bounded subset of $\mathscr{C}_c^\infty(\Omega)$, there is a compact subset K of Ω such that $\mathscr{B} \subset \mathscr{C}_c^\infty(K)$ (Proposition 14.6); there is an integer $k(K)$ such that $K \subset \Omega_k$ for all $k \geqslant k(K)$, therefore such that $g_k \varphi = \varphi$ for all $k \geqslant k(K)$ and all $\varphi \in \mathscr{B}$. Then, for all $\varphi \in \mathscr{B}$,

$$\langle g_k T, \varphi \rangle = \langle T, g_k \varphi \rangle = \langle T, \varphi \rangle,$$

which proves that $g_k T \to T$ in $\mathscr{D}'(\Omega)$. The fact that the functions g_k are identically equal to one in open sets which form an expanding sequence is not at all necessary to reach the conclusion that $g_k T \to T$. In connection with this, we propose to the student the following exercise:

Exercise 28.1. *Let $\{g_k\}$ be a sequence in $\mathscr{C}^\infty(\Omega)$ which converges to the function identically one in Ω. Prove that, given any distribution T in Ω, the sequence of distributions $g_k T$ converges to T in $\mathscr{D}'(\Omega)$, and that, if T has compact support, $g_k T$ converges to T in $\mathscr{E}'(\Omega)$.*

Going back to the considerations above, we can take the $\bar{\Omega}_k$ to be compact and the $\mathscr{C}^\infty$ functions g_k with compact support. We may then state:

THEOREM 28.1. *Let T be a distribution in Ω. There is a sequence of distributions with compact support, $\{T_k\}$ $(k = 0, 1,...)$, such that, given any relatively compact open subset Ω' of Ω, there is an integer $k(\Omega') \geqslant 0$ such that, for all $k \geqslant k(\Omega')$, the restriction of T_k to Ω', $T_k \mid \Omega'$, is equal to the restriction of T to Ω', $T \mid \Omega'$.*

Let T and the T_k be as in Theorem 28.1. If K is any compact subset of Ω, there is an integer $k(K)$ such that, for $k \geqslant k(K)$, T and T_k are equal in some neighborhood of K and, in particular,

$$K \cap \operatorname{supp} T = K \cap \operatorname{supp} T_k.$$

Remark 28.1. The operation just described, of multiplying a distribution by $\mathscr{C}^\infty$ functions with compact support, g_k, equal to one in relatively compact open subsets of Ω which expand as $k \to \infty$ and ultimately fill Ω, is the extension to distributions of the "cutting operation" on functions. In the latter case, if we deal with some function f in Ω, we regard f as the limit of the functions f_k equal to f in Ω_k and to zero outside Ω_k; then obviously the f_k converge to f uniformly on every compact subset of Ω. Note that f_k is the product of f by the characteristic function φ_{Ω_k} of Ω_k. Of course, we cannot multiply a distribution by φ_{Ω_k} since this function is not smooth; thus we must multiply the distribution by a $\mathscr{C}^\infty$ function (with compact support) which coincides with φ_{Ω_k} in Ω_k, that is to say, which is equal to one in Ω_k.

We have approximated an arbitrary distribution in Ω by a sequence of distributions in Ω which have compact support. The next step is to approximate any distribution with compact support by a sequence of test functions. This is done by convoluting the given distributions with $\mathscr{C}_c^\infty$ functions which converge to the Dirac measure δ. We shall therefore use the properties of the convolution of distributions, established in the preceding chapter (Chapter 27).

We begin by considering the convolution $T * \varphi$ of a distribution T with a $\mathscr{C}^\infty$ function φ, one of the two having compact support. We may regard φ as a distribution, in which case $T * \varphi$ is the distribution

$$\mathscr{C}_c^\infty \ni \psi \rightsquigarrow \langle T * \varphi, \psi \rangle = \langle T, \check{\varphi} * \psi \rangle,$$

where $\check{\varphi}(x) = \varphi(-x)$. But $T * \varphi$ is, in fact, a $\mathscr{C}^\infty$ function, precisely the function

$$x \rightsquigarrow \langle T_y, \varphi(x - y) \rangle \qquad \text{(Theorem 27.5)}.$$

The approximation result which we are seeking will be a consequence of the following lemma:

LEMMA 28.1.   *Let ρ be the function defined by*

$$\rho(x) = \begin{cases} a \ \exp\left(-\dfrac{1}{1-|x|^2}\right) & \text{if} \ \ |x| < 1, \\ 0 & \text{if} \ \ |x| \geqslant 0, \end{cases}$$

where

$$a^{-1} = \int_{|x|<1} \exp\left(-\frac{1}{1-|x|^2}\right) dx.$$

For $\varepsilon > 0$, call ρ_ε the function $x \rightsquigarrow \varepsilon^{-n}\rho(x/\varepsilon)$. Then the sequence $\{\rho_{1/j}\}$ ($j = 1, 2, ...$) converges to the Dirac measure δ in the space $\mathscr{E}'$ of distributions with compact support in $\mathbf{R}^n$.

About the function ρ_ε, see Chapter 15, p. 155.

Proof of Lemma 28.1.   We advise the student to take a look at the proof of Lemma 15.2; the arguments here and there are closely related. Let f be a $\mathscr{C}^\infty$ function in $\mathbf{R}^n$. We have

$$\left| f(0) - \int \rho_\varepsilon(x) f(x)\, dx \right| \leqslant \int \rho_\varepsilon(x) \,|f(x) - f(0)|\, dx$$

$$\leqslant \sup_{|x|\leqslant\varepsilon} |f(x) - f(0)| \int \rho_\varepsilon(x)\, dx$$

$$\leqslant \sup_{|x|\leqslant\varepsilon} |f(x) - f(0)|.$$

We may assume that $\varepsilon < 1$. Then, if $|x| < \varepsilon$,

$$|f(x) - f(0)| \leqslant C|x| \sup_{|x|<1} \sum_{j=1}^{n} |(\partial/\partial x_j) f(x)|;$$

therefore,

$$\left| f(0) - \int \rho_\varepsilon(x) f(x)\, dx \right| \leqslant C\varepsilon \sup_{|x|<1} \sum_{j=1}^{n} |(\partial/\partial x_j) f(x)|.$$

This shows that, if f remains in a bounded set of $\mathscr{C}^\infty(\mathbf{R}^n)$, in fact, in a bounded set of $\mathscr{C}^1(\mathbf{R}^n)$, $\langle \rho_\varepsilon, f \rangle$ converges to $f(0)$.   Q.E.D.

Observe that we do not need the precise information which we have about the functions ρ_ε. In connection with this, we propose the following exercise to the student:

Exercise 28.2.   *Let $\{\mu_k\}$ be a sequence of Radon measures in $\mathbf{R}^n$, having the following properties:*

(1) *supp μ_k is contained in a ball of radius r_k centered at $x = 0$ such that $r_k \to 0$ as $k \to +\infty$;*

(2) *the numbers* $\langle \mu_k, 1 \rangle$ (1 : function identically equal to one in $\mathbf{R}^n$) *converge to one as* $k \to +\infty$;

(3) *there is a constant* $C > 0$ *such that, for all* $k = 1, 2,...,$ *and all functions* $f \in \mathscr{C}_c^0(\mathbf{R}^n)$,

$$|\langle \mu_k, f \rangle| \leqslant C \sup_{x \in \mathbf{R}^n} |f(x)|.$$

Under these hypotheses, prove that μ_k *converge to* δ *in* $\mathscr{E}'(\mathbf{R}^n)$.

What modifications could be made on the hypotheses if we were only to require that the μ_k *converge to* δ *in* $\mathscr{D}'(\mathbf{R}^n)$?

We may now easily prove the following result:

THEOREM 28.2. *Let* T *be a distribution in the open set* Ω. *There is a sequence of functions* $\varphi_k \in \mathscr{C}_c^\infty(\Omega)$ $(k = 1, 2,...)$ *with the following properties:*

(i) φ_k *converges to* T *in* $\mathscr{D}'(\Omega)$ *and, if the support of* T *is compact,* φ_k *converges to* T *in* $\mathscr{E}'(\Omega)$;

(ii) *for every compact subset* K *of* Ω, $K \cap \operatorname{supp} \varphi_k$ *converges to* $K \cap \operatorname{supp} T$ *and, if* $\operatorname{supp} T$ *is compact,* $\operatorname{supp} \varphi_k$ *converges to* $\operatorname{supp} T$.

A sequence $\{A_k\}$ of subsets of a metric space E (with metric $(x, y) \rightsquigarrow d(x, y)$) is said to converge to $A \subseteq E$ if, to every $\varepsilon > 0$, there is $k(\varepsilon)$ such that, for every $k \geqslant k(\varepsilon)$,

$$A_k \subseteq \{x \in E; d(x, A) < \varepsilon\},$$

$$A \subseteq \{x \in E; d(x, A_k) < \varepsilon\}.$$

If $B \subseteq E$, we have set $d(x, B) = \inf_{y \in B} d(x, y)$.

Proof of Theorem 28.2. We begin by selecting a sequence of distributions T_k $(k = 0, 1,...)$ with compact support, as in Theorem 28.1; for every relatively compact open subset Ω' of Ω there is $k(\Omega')$ such that, if $k \geqslant k(\Omega')$, $T_k | \Omega' = T | \Omega'$. Next we consider a sequence of functions

$$\rho_{1/j} * T_k,$$

observing that

$$\operatorname{supp}(\rho_{1/j} * T_k) \subseteq \operatorname{supp} \rho_{1/j} + \operatorname{supp} T_k \qquad \text{(Proposition 27.5)}.$$

For each k we select j according to two requirements: $j \geqslant k$ (in order to ensure that $j \to +\infty$); j sufficiently large so that the neighborhood of order $1/j$ of $\operatorname{supp} T_k$ is a compact subset of Ω. If we then call j_k the integer thus selected, we contend that the test functions $\varphi_k = \rho_{1/j_k} * T_k$ converge to T in $\mathscr{D}'(\Omega)$.

Indeed, let $\mathscr{B}$ be a bounded subset of $\mathscr{C}_c^\infty(\Omega)$; there is a compact subset K of Ω such that supp $\varphi \subset K$ for all functions $\varphi \in \mathscr{B}$. Let Ω' be a relatively compact open subset of Ω containing K; for $k \geqslant k(\Omega')$, we have $T_k | \Omega' = T | \Omega'$. On the other hand, for $1/k < d(K, \complement \Omega')$,

$$\text{supp}(\rho_{1/j_k} * \varphi) \subset \Omega' \qquad \text{for all} \quad \varphi \in \mathscr{B}.$$

We recall that $j_k \geqslant k$ and, also, that $\check{\rho} = \rho$. Then, for all $\varphi \in \mathscr{B}$,

$$\langle \varphi_k - T, \varphi \rangle = \langle T_k, \rho_{1/j_k} * \varphi \rangle - \langle T, \varphi \rangle$$
$$= \langle T, \rho_{1/j_k} * \varphi \rangle - \langle T, \varphi \rangle$$
$$= \langle gT, \rho_{1/j_k} * \varphi \rangle - \langle gT, \varphi \rangle,$$

where g is an arbitrary function which belongs to $\mathscr{C}_c^\infty(\Omega)$ and is equal to one in Ω'. This shows that, for all $\varphi \in \mathscr{B}$, and k sufficiently large,

$$\langle \varphi_k - T, \varphi \rangle = \langle \rho_{1/j_k} * (gT) - (gT), \varphi \rangle.$$

But gT is a fixed distribution with compact support, and $\rho_{1/j_k} \to \delta$ in $\mathscr{E}'(\mathbf{R}^n)$; by applying Theorem 27.6, we conclude that $\rho_{1/j_k} * (gT) \to gT$ uniformly on $\mathscr{B}$, which is what we wanted to prove.

When the support of T is compact, it is easy to see that the φ_k converge to T in $\mathscr{E}'(\Omega)$. First of all, we may take $T_k = T$ for all k, and $\varphi_k = \rho_{1/k} * T$ for $1/k < d(\text{supp } T, \complement \Omega)$. As the $\rho_{1/k}$ converge to δ in $\mathscr{E}'(\Omega)$, a fortiori in $\mathscr{D}'(\Omega)$, it follows immediately from Theorem 27.6 that φ_k converges to T in $\mathscr{E}'(\Omega)$.

Property (ii) in Theorem 28.2 is obvious, by inspection of the definition of the functions φ_k.

COROLLARY. *Let Ω be an open subset of $\mathbf{R}^n$; $\mathscr{C}_c^\infty(\Omega)$ is sequentially dense in $\mathscr{E}'(\Omega)$ and in $\mathscr{D}'(\Omega)$.*

There is no need to underline the close relationship between Theorem 15.3 and Theorem 28.2; the properties stated in these theorems are often referred to as *"approximation by cutting and regularizing."* Convolution of a distribution with a $\mathscr{C}^\infty$ function is often called *regularization* (or *smoothing*). The word *cutting* refers to the multiplication of a distribution T by $\mathscr{C}^\infty$ functions which are equal to 1 in some relatively compact open set Ω' and equal to zero outside a neighborhood of the closure of Ω'.

Definition 28.1.   *A space of distributions in Ω, $\mathscr{A}$, is said to be normal if $\mathscr{C}_c^\infty(\Omega)$ is contained and dense in $\mathscr{A}$, and if the injection of $\mathscr{C}_c^\infty(\Omega)$ into $\mathscr{A}$ is continuous.*

We recall (Definition 23.1) that a space of distributions in Ω is a linear subspace of $\mathscr{D}'(\Omega)$ carrying a locally convex topology finer than the one induced by $\mathscr{D}'(\Omega)$.

PROPOSITION 28.1.   *If $\mathscr{A}$ is a normal space of distributions in Ω, the strong dual of $\mathscr{A}$ is canonically isomorphic to a space of distributions in Ω.*

We have already essentially proved this statement (in Chapter 23) and often used it: let $j : \mathscr{C}_c^\infty(\Omega) \to \mathscr{A}$ be the natural injection; since the image of j is dense, the transpose ${}^tj$ of j is a continuous one-to-one linear map of the strong dual $\mathscr{A}'$ of $\mathscr{A}$ into $\mathscr{D}'(\Omega)$.

Examples of normal spaces of distributions in Ω:

   (i) $\mathscr{C}^m(\Omega)$, $\mathscr{C}_c^m(\Omega)$ $(0 \leqslant m \leqslant +\infty)$ (Corollaries 1 and 2 of Theorem 15.3);

  (ii) $L^p(\Omega)$ $(1 \leqslant p < +\infty)$ (Corollary 3 of Theorem 15.3);

 (iii) $\mathscr{D}'(\Omega)$, $\mathscr{E}'(\Omega)$ (Corollary of Theorem 28.2.).

Example of spaces of distributions which are not normal:

$L^\infty(\Omega)$ (we suppose Ω nonempty!); $H(\mathbf{R}^n)$, space of functions in $\mathbf{R}^n$ which can be extended to $\mathbf{C}^n$ as entire functions (with the topology carried over from the space of entire functions in $\mathbf{C}^n$, $H(\mathbf{C}^n)$). Note that $L^\infty(\Omega)$ contains $\mathscr{C}_c^\infty(\Omega)$ whereas the intersection of $H(\mathbf{R}^n)$ with $\mathscr{C}_c^\infty(\Omega)$ is reduced to the zero function.

Remark 28.2.   The dual of a normal space of distributions is not necessarily a normal space of distributions (although it is a space of distributions, by Proposition 28.1), as shown by the example of $\mathscr{A} = L^1(\Omega)$, $\mathscr{A}' = L^\infty(\Omega)$.

Exercise 28.3.   Let $\{g_k\}$ be a sequence of $\mathscr{C}_c^\infty$ functions in $\mathbf{R}^n$ such that $g_k(x) = 1$ for $|x| < k$ $(k = 1, 2,...)$, and such that, to every n-tuple p, there is a constant $C_p > 0$ such that, for all $k = 1, 2,...$,

$$\sup_{x \in \mathbf{R}^n} |(\partial/\partial x)^p g_k(x)| \leqslant C_p .$$

Prove that, given any tempered distribution S in $\mathbf{R}^n$, $g_k S$ converges to S in $\mathscr{S}'(\mathbf{R}^n)$ as $k \to \infty$. (The student may simply prove that $g_k S$ converges weakly to S in $\mathscr{S}'$; we shall see that, for sequences in $\mathscr{S}'$, strong and weak convergences coincide.)
   Construct a sequence of functions g_k with the above properties.

Remark 28.3.   A corollary of Exercise 28.3 is that $\mathscr{S}'$ is a normal space of distributions in $\mathbf{R}^n$. We already knew that $\mathscr{S}$ is a normal space of distributions in $\mathbf{R}^n$ (Theorem 15.4).

Exercises

28.4. Prove the following lemma:

LEMMA 28.2. *The sequence of functions*

$$(k/\pi^{1/2})^n \exp(-k^2|x|^2), \qquad k = 1, 2, \ldots,$$

converge to the Dirac measure δ in $\mathscr{D}'(\mathbf{R}^n)$.
(Cf. Lemma 15.1 and Lemma 28.1).

28.5. Prove the following result:

THEOREM 28.3. *Let Ω be an open subset of $\mathbf{R}^n$. Any distribution in Ω is the limit of a sequence of polynomial functions.*
(Hint: Make use of Theorem 28.1 and Lemma 28.2; cf. Corollary 2 of Lemma 15.1 and Exercise 27.2).

29

Fourier Transforms of Distributions with Compact Support. The Paley-Wiener Theorem

Consider a continuous function f with compact support in $\mathbf{R}^n$. For $\zeta \in \mathbf{C}_n$, dual of $\mathbf{C}^n$, we may set

$$\hat{f}(\zeta) = \int f(x) \exp(-2i\pi \langle x, \zeta \rangle)\, dx.$$

Let us write $\zeta = \xi + i\eta(\xi, \eta \in \mathbf{R}_n$, dual of $\mathbf{R}^n$); the rule of differentiation under the integral sign shows immediately that $\hat{f}(\xi + i\eta)$ is a $\mathscr{C}^\infty$ function of (ξ, η) in $\mathbf{R}_{2n}$. We set, as usual, for $j = 1,\ldots, n$,

$$\partial/\partial\bar{\zeta}_j = \tfrac{1}{2}(\partial/\partial\xi_j + i\,\partial/\partial\eta_j).$$

As the integrand, in the definition of $\hat{f}(\zeta)$, is a solution of the system of n equations $\partial u/\partial\bar{\zeta}_j = 0$, so is $\hat{f}$. In other words, $\hat{f}$ is a holomorphic function of ζ in the whole complex space $\mathbf{C}_n$: $\hat{f}$ is an entire analytic function of ζ. Next, we show that $\hat{f}$ is of exponential type. We shall generalize the situation a little and consider a Radon measure μ with compact support, rather than the more special measure $f(x)\, dx$. We shall then use the following lemma:

LEMMA 29.1. *Let μ be a Radon measure in $\mathbf{R}^n$, with compact support, K. There is a constant $C > 0$ such that, for all functions $\phi \in \mathscr{C}^0(\mathbf{R}^n)$,*

(29.1) $$|\langle \mu, \phi \rangle| \leqslant C \sup_{x \in K} |\phi(x)|.$$

Proof. Let U_k be the set of points x such that $d(x, K) < 1/k$ $(k = 1, 2,\ldots;$ d is the Euclidean distance). Let g_k be a continuous function with compact support in U_k, equal to one in some neighborhood of supp μ, such that $0 \leqslant g_k(x) \leqslant 1$ for all x. We have $g_k\mu = \mu$. On the other hand, there is a constant $C > 0$ such that we have, for all $\psi \in \mathscr{C}^0$ with compact support contained in U_1,

$$|\langle \mu, \psi \rangle| \leqslant C \sup_x |\psi(x)|.$$

It suffices then, in order to get (29.1), to take $\psi = g_k\phi$, $\phi \in \mathscr{C}^0(\mathbf{R}^n)$ arbitrary. Indeed,

$$|\langle \mu, \phi \rangle| = |\langle \mu, g_k\phi \rangle| \leqslant C \sup_x |g_k(x)\phi(x)| \leqslant C \sup_{x \in U_k} |\phi(x)|.$$

By taking the limit when $k \to +\infty$, we obtain the desired inequality.

Remark 29.1.   When $\mu = f(x)\,dx$, we may take $C = \int |f(x)|\,dx$ in (29.1).

In (29.1), we take $\phi(x) = \exp(-2i\pi\langle x, \zeta \rangle)$, $\zeta \in \mathbf{C}_n$. We obtain immediately

$$(29.2) \quad |\langle \mu, \exp(-2i\pi \langle x, \zeta \rangle) \rangle| \leqslant C \sup_{x \in K} \exp(2\pi|\langle x, \eta \rangle|) \quad (\eta = \operatorname{Im} \zeta).$$

If we apply (29.2) to $\mu = f(x)\,dx$, we obtain (Remark 29.1)

$$(29.3) \qquad |\hat{f}(\zeta)| \leqslant \|f\|_{L^1} \sup_{x \in \operatorname{supp} f} \exp(2\pi|\langle x, \eta \rangle|).$$

This implies immediately our assertion: that f is an entire function of exponential type (Notation 22.2). For $\xi \in \mathbf{R}_n$, $\hat{f}(\xi)$ is the Fourier transform of f. We have therefore proved the following: *the Fourier transform of a continuous function with compact support can be extended to the complex space $\mathbf{C}_n$ as an entire analytic function of exponential type.*

In this chapter, we shall extend this result to arbitrary distributions with compact support. We shall prove that it then has a converse: every tempered distribution, whose Fourier transform can be extended to the complex space as an entire function of exponential type, has a compact support (in fact, we shall prove a slightly weaker implication: we shall assume that the growth of the Fourier transform of the given distribution is known on the manifolds parallel to the real space $\mathbf{R}_n$, generalizing (29.3)). The analogy between this result, known as the Paley–Wiener theorem, and the theorem on the Fourier–Borel transformation of the analytic functionals (Theorem 22.3), is obvious. Let j be the restriction of an entire function (in $\mathbf{C}^n$) to the real space $\mathbf{R}_n$; j is a one-to-one linear mapping of $H(\mathbf{C}_n)$ into $\mathscr{C}^\infty(\mathbf{R}_n)$. It is one-to-one since an entire function cannot vanish identically in $\mathbf{R}_n$ without also vanishing identically in $\mathbf{C}_n$.

Furthermore, the image of j is dense in view of Corollary 2 of Theorem 15.2. The transpose of j, ${}^t j : \mathscr{E}'(\mathbf{R}_n) \to H'(\mathbf{C}_n)$, is a continuous injection of the space of distributions with compact support into the space of analytic functionals (both spaces carrying the strong dual topology). Let, then, T be a distribution with compact support in $\mathbf{R}^n$, $\mathscr{F}T$ its Fourier transform, and $\mathscr{F}\mathscr{B}T$ its Fourier–Borel transform (Definition 22.3). Both transforms are entire functions in $\mathbf{C}_n$, respectively, defined (as will be shown in a moment for $\mathscr{F}T$) by

$$\mathscr{F}T(\zeta) = \langle T_x, \exp(-2i\pi \langle x, \zeta \rangle) \rangle; \qquad \mathscr{F}\mathscr{B}\,T(\zeta) = \langle ({}^t jT)_z, \exp(\langle z, \zeta \rangle) \rangle.$$

This shows right away that $\mathscr{F}T(\zeta) = \mathscr{F}\mathscr{B}T(-2i\pi\zeta)$.

PROPOSITION 29.1.   *The Fourier transform of a distribution T with compact support in $\mathbf{R}^n$ is the function, in $\mathbf{R}_n$,*

(29.4) $$\hat{T}(\xi) = \langle T_x, \quad \exp(-2i\pi \langle x, \xi \rangle) \rangle.$$

$\hat{T}$ can be extended to the complex space $\mathbf{C}_n$ as an entire analytic, given by

(29.5) $$\hat{T}(\zeta) = \langle T_x, \quad \exp(-2i\pi \langle x, \zeta \rangle) \rangle.$$

Proof. The validity of (29.4) can be established in a variety of ways, e.g. by noting that (29.4) is valid when $T \in \mathscr{C}_c^\infty$ and then, when $T \notin \mathscr{C}_c^\infty$, by going to the limit along a sequence of elements of $\mathscr{C}_c^\infty$ which converge to T in $\mathscr{E}'$ (Theorem 28.2). Or else, we may use the representation of T as a finite sum of derivatives of continuous functions with compact support. A third proof is obtained by reasoning directly on the definition of $\mathscr{F}T$.

Observe, next, that $\zeta \rightsquigarrow (x \rightsquigarrow \exp(-2i\pi \langle x, \zeta \rangle))$ is a $\mathscr{C}^\infty$ function of $\zeta \in \mathbf{C}_n \cong \mathbf{R}_{2n}$ with values in $\mathscr{C}^\infty(\mathbf{R}_x^n)$; in particular, (29.5) makes sense (and obviously extends (29.4)) and defines, in virtue of Theorem 27.1, a $\mathscr{C}^\infty$ function in $\mathbf{R}_{2n}$. We may apply to it the Cauchy-Riemann operators $\partial/\partial\bar{\zeta}_j$ $(j = 1,\dots, n)$ and by again using Theorem 27.1, this time part (b), we see that $\hat{T}$ is everywhere holomorphic.   Q.E.D.

Observe that Theorem 22.3 provides us with some information about the relation between the polydisks carrying an analytic functional and the growth of its Fourier–Borel transforms. Similarly, Estimate (29.3) points to a link between the growth of the Fourier transform of f on the parallels of the real space and the support of F. We shall obtain analog relations for a distribution with compact support.

The following definition will help us to achieve more precision in the statements and in the proofs:

Definition 29.1. Let A be a subset of $\mathbf{R}^n$. We shall call indicator of A the function, defined in $\mathbf{R}_n$,

$$\eta \rightsquigarrow I_A(\eta) = \sup_{x \in A} |\langle x, \eta \rangle|.$$

Remark 29.2. If A is a subset of a locally convex TVS E, we may consider the function I_A defined in the dual of E, E'. The set $\{x'; I_A(x') \leqslant 1\}$ is the polar A^0 of A. It is clear that $I_A = I_{\Gamma(A)}$, where $\Gamma(A)$ is the convex balanced closed hull of A. *When the scalars are the real numbers, balanced means symmetric and star-shaped with respect to the origin.* A convex set B is balanced, then, if $B = -B$.

Note that, with Definition 29.1, (29.2) can be rewritten as

(29.6) $|\langle \mu, \exp(-2i\pi \langle x, \zeta \rangle) \rangle| \leqslant C \exp(2\pi I_K(\eta))$ $(\zeta \in \mathbf{C}_n$, $\eta = \operatorname{Im} \zeta)$.

We begin by proving (following L. Hörmander) the section of the Paley–Wiener theorem which is relative to $\mathscr{C}^\infty$ functions.

THEOREM 29.1. *Let K be a convex balanced compact subset of $\mathbf{R}^n$. The following properties of a distribution $\phi \in \mathscr{S}'$ are equivalent:*

(a)   *ϕ is a $\mathscr{C}^\infty$ function with compact support contained in K.*

(b)   *The Fourier transform of ϕ can be extended to the complex space $\mathbf{C}_n$ as an entire analytic function $\zeta \rightsquigarrow \hat{\phi}(\zeta)$ such that, for every integer $m = 0, 1,...,$ there is a constant $C_m > 0$ such that, for all $\zeta = \xi + i\eta$,*

(29.7) $|\hat{\phi}(\zeta)| \leqslant C_m(1 + |\zeta|)^{-m} \exp\{2\pi I_K(\eta)\},$

where I_K is the indicator of K.

Proof. To see that (a) implies (b), it suffices to apply (29.6) with $\mu = \psi(x)\, dx$, where $\psi(x) = (1 - \Delta/4\pi^2)^k \phi(x)$ and $2k \geqslant m$. For then we have $\hat{\psi}(\zeta) = (1 + |\zeta|^2)^k \hat{\phi}(\zeta)$.

Next, we prove that (b) implies (a). In order to show that supp ϕ is contained in the convex balanced compact set K, it suffices (Corollary 1 of Proposition 18.2) to show that supp ϕ is contained in every polygon of the form

$$\Pi = \{x \in \mathbf{R}^n; |\langle x, N_j \rangle| \leqslant 1, \quad j = 1,..., n\},$$

which contains K and such that $N_1,..., N_n$ form a basis of $\mathbf{R}_n$. Let us therefore consider such a polygon. For simplicity, we take the functionals $x \rightsquigarrow \langle N_j, x \rangle$ as coordinates in $\mathbf{R}^n$ and we denote them respectively by x_j. We use the dual coordinates system in $\mathbf{R}_n$: the coordinates of $\xi \in \mathbf{R}_n$ are the numbers ξ_j such that $\xi = \sum_{j=1}^{n} \xi_j N_j$; we extend these coordinates to $\mathbf{C}_n$. With the coordinates x_j (resp. ξ_j) we associate the Euclidean norm $|x|$ (resp. $|\xi|$) and the Lebesgue measure dx (resp. $d\xi$); the latter is determined by the requirement that the measure of the unit cube be one. Now, if $\eta \in \mathbf{R}_n$, we see that the supremum of $|\langle x, \eta \rangle| = \sum_{j=1}^{n} \eta_j x_j$ is attained when $x_j = \operatorname{sgn} \eta_j$ for every $j = 1,..., n$, and that this supremum is then equal to

$$(29.8) \qquad\qquad I_{\Pi}(\eta) = \sum_{j=1}^{n} |\eta_j| .$$

On the other hand, since $K \subset \Pi$, we have $\Pi^0 \subset K^0$ and therefore, for all $\eta \in \mathbf{R}_n$, $I_K(\eta) \leqslant I_{\Pi}(\eta)$. Thus we see that it suffices to show that (b) implies (a) when $K = \Pi$; then $I_K(\eta)$ is given by (29.8).

First of all, a straightforward application of Cauchy's formulae (p. 90) shows that every derivative of ϕ, $\phi^{(p)}$ ($p \in \mathbf{N}^n$), satisfies also Inequality (29.7), possibly with a different constant C_m (indeed, this constant will depend on p). From this fact, by taking $\zeta = \xi$ real, i.e. $\eta = 0$, we derive that every function $|\hat{\phi}^{(p)}(\xi)|$ decreases at infinity faster than any power of $1/|\xi|$. In other words, ϕ belongs to $\mathscr{S}(\mathbf{R}_n)$, hence $\phi \in \mathscr{S}(\mathbf{R}^n)$ and we have the reciprocity formula (Theorem 25.1):

$$\phi(x) = \int \exp(2i\pi \langle x, \xi \rangle) \hat{\phi}(\xi) \, d\xi.$$

Consider, for fixed complex numbers $\zeta_2 ,..., \zeta_n$, the integral

$$(29.9) \qquad\qquad \int_{-M}^{M} \exp(2i\pi x_1 \xi_1) \hat{\phi}(\xi_1 , \zeta_2 ,..., \zeta_n) \, d\xi_1 .$$

Using the fact that ϕ is an entire analytic function, we can integrate $\hat{\phi}(\zeta)$, regarded as a function of ζ_1 alone, on the following path Γ of the complex plane.

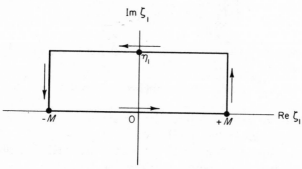

As Γ is a closed path, the integral of $\hat{\phi}$ along it is equal to zero, by Cauchy's theorem; the portion of this integral which is performed on the segment $(-M, +M)$ of the real line is equal to (29.9), which is therefore equal to

$$\int_{-M}^{+M} \exp(2i\pi(\xi_1 + i\eta_1)x_1)\, \hat{\phi}(\xi_1 + i\eta_1, \zeta_2, ..., \zeta_n)\, d\xi_1 + R(M),$$

where

$$|R(M)| \leqslant \int_0^{\eta_1} \exp(2\pi|x_1 t|) \{|\hat{\phi}(M + it, \zeta_2, ..., \zeta_n)|$$

$$+ |\hat{\phi}(-M + it, \zeta_2, ..., \zeta_n)|\}\, dt.$$

If we take into account (29.7), when $m = 1$, we see immediately that $R(M) \to 0$, when $M \to +\infty$. In other words, we have

$$\int_{-\infty}^{+\infty} \exp(2i\pi(\xi_1 + i\eta_1)x_1)\, \hat{\phi}(\xi_1 + i\eta_1, \zeta_2, ..., \zeta_n)\, d\xi_1$$

$$= \int_{-\infty}^{+\infty} \exp(2i\pi\xi_1 x_1)\, \hat{\phi}(\xi_1, \zeta_2, ..., \zeta_n)\, d\xi_1.$$

If we apply this argument, in turn, to each variable $\xi_1, ..., \xi_n$, we obtain

$$\int_{\mathbf{R}_n} \exp(2i\pi \langle \xi + i\eta, x \rangle)\, \hat{\phi}(\xi + i\eta)\, d\xi = \phi(x).$$

In this formula, η is an arbitrary vector of $\mathbf{R}_n$.

We again take into account (29.7), this time with $m = n + 1$. We obtain

$$|\phi(x)| \leqslant C_{n+1} \exp(2\pi(|\eta_1| + \cdots + |\eta_n| - \langle x, \eta \rangle))$$

$$\times \int (1 + |\xi|)^{-n-1}\, d\xi.$$

We then choose $\eta_j = \tau x_j$, $\tau > 0$, $\eta_k = 0$ for $k \neq j$ $(j = 1, ..., n)$. This gives us

$$|\phi(x)| \leqslant \text{const.} \exp(2\pi\tau(1 - |x_j|)|x_j|).$$

If $|x_j| > 1$, we take $\tau \to +\infty$; we obtain $\phi(x) = 0$. Therefore supp ϕ must be contained in the set Π.   Q.E.D.

Next we prove (following L. Schwartz) the extension of the Paley–Wiener theorem to tempered distributions.

THEOREM 29.2. *The following properties of a tempered distribution T in*
$\mathbf{R}^n$ *are equivalent:*

(a′) *The support of T is compact; its convex balanced hull is the set K.*
(b′) *The Fourier transform of T can be extended to the complex space $\mathbf{C}_n$*
      as an entire analytic function $\zeta \to \hat{T}(\zeta)$ such that there is an
      integer $m \geqslant 0$ and a constant $C > 0$ such that, for all $\zeta = \xi + i\eta$,

$$(29.10) \qquad |\hat{T}(\zeta)| \leqslant C(1 + |\zeta|)^m \exp\{2\pi I_K(\eta)\}.$$

Proof of (a′) $\Rightarrow$ (b′). We consider arbitrary coordinates $x_1,\ldots, x_n$ in $\mathbf{R}^n$
and the dual coordinates, $\xi_1,\ldots, \xi_n$ in $\mathbf{R}_n$; we use the associated Hilbert
space structures on $\mathbf{R}^n$ and $\mathbf{R}_n$ (inner product, e.g. in $\mathbf{R}_n$: $(\xi \mid \xi') =$
$\sum_{j=1}^n \xi_j \xi_j'$). Note that the notion of an orthogonal transformation then
makes sense: it is a linear map of the space into itself which preserves
the inner product (hence also the norm). Let us consider a polygon,
containing K,

$$\Pi = \{x \in \mathbf{R}^n; |\langle x, N_1\rangle| \leqslant B_1,\ldots, |\langle x, N_n\rangle| \leqslant B_n\},$$

(cf. the proof of Theorem 29.1); we require now that $N_1,\ldots, N_n$ be an
orthonormal basis of $\mathbf{R}_n$ ($B_1,\ldots, B_n$ are positive numbers). As the support
of T is compact, there is an integer $m \geqslant 0$ and a constant $A > 0$ such
that, for all $\phi \in \mathscr{C}_c^\infty(\mathbf{R}^n)$,

$$(29.11) \qquad |\langle T, \phi\rangle| \leqslant A \sup_{x \in \mathbf{R}^n} \sum_{|p| \leqslant m} |(\partial/\partial x)^p \phi(x)|.$$

Let, then, $g \in \mathscr{C}^\infty(\mathbf{R}^1)$ be equal to one on $(-\infty, \frac{1}{2})$ and to zero on $(1, +\infty)$.
The function

$$\phi(x) = \exp(-2i\pi \langle x, \zeta\rangle) \prod_{j=1}^n g(|\zeta|(|\langle x, N_j\rangle| - B_j))$$

is $\mathscr{C}^\infty$ in $\mathbf{R}^n$. Its support is contained in the set

$$\Pi_\zeta = \{x; |\langle x, N_j\rangle| < B_j + |\zeta|^{-1}, j = 1,\ldots, n\}.$$

It is equal to $\exp(-2i\pi \langle x, \zeta\rangle)$ in $\Pi_{2\zeta}$, which is a neighborhood of Π,
hence of K, hence of supp T. Consequently we have, in view of (29.11)
and of Proposition 29.1,

$$|\hat{T}(\zeta)| = |\langle T, \phi\rangle| \leqslant A_1(1 + |\zeta|)^m \sup_{x \in \Pi_\zeta} \exp(2\pi|\langle x, \eta\rangle|) \qquad (\eta = \text{Im } \zeta).$$

It is very important to observe that the constant A_1 depends only on the numbers m and A in (29.11) and not on the orthonormal basis $(N_1,\ldots, N_n)$ nor on the numbers $B_1,\ldots, B_n$. The independence from the B_j's is evident; and so is the independence from the N_j's if we perform an orthogonal linear change of variables in $\mathbf{R}_n$. Thus we see that

$$| \hat{T}(\zeta)| \leqslant A_1(1 + |\zeta|)^m \exp(2\pi I_\Pi(\eta) + 2\pi\sqrt{n})$$

since

$$\sup_{x\in\Pi_\zeta} |\langle x, \eta\rangle| \leqslant \sup_{x\in\Pi} |\langle x, \eta\rangle| + (|(N_1|\eta)| + \cdots + |(N_n|\eta)|)|\zeta|^{-1}$$
$$\leqslant I_\Pi(\eta) + \sqrt{n}.$$

Finally we obtain

$$| \hat{T}(\zeta)| \leqslant A_1 \exp(2\pi\sqrt{n})(1 + |\zeta|)^m \exp(2\pi I_\Pi(\eta)) \qquad \text{for all} \quad \zeta \in \mathbf{C}_n.$$

This is true for all polygons Π of the type considered. We may therefore put $\inf_\Pi$ in front of the right-hand side of the preceding inequality or, if one prefers, in front of $I_\Pi(\eta)$. But, as is easily proved,

$$\inf_\Pi I_\Pi(\eta) = I_K(\eta).$$

Proof of (b') $\Rightarrow$ (a'). The proof of this implication is based on a standard regularization and on application of Theorem 29.1. Let $\rho_\varepsilon(x) = \varepsilon^{-n} \rho(x/\varepsilon)$ be the usual mollifiers (cf. Lemma 28.1). It is immediately seen that the convolution $\rho_\varepsilon * T$ is also a tempered distribution for all $\varepsilon > 0$ and that its Fourier transform is equal to $\hat{\rho}_\varepsilon \hat{T}$ (cf. Theorem 30.4; a direct verification is easy). Also observe that

$$\hat{\rho}_\varepsilon(\zeta) = \hat{\rho}(\varepsilon\zeta).$$

Since the support of ρ is the ball $\{x; |x| \leqslant 1\}$, we have, by applying (29.7), for each $k = 0, 1,\ldots,$

$$| \hat{\rho}_\varepsilon(\zeta)| \leqslant C_k(1 + |\varepsilon\zeta|)^{-k} \exp(2\pi\varepsilon|\eta|).$$

If we now take into account (29.10), we obtain

$$| \hat{\rho}_\varepsilon(\zeta) \hat{T}(\zeta)| \leqslant CC_k\varepsilon^{-k}(1 + |\zeta|)^{-(k-m)} \exp(2\pi(I_K(\eta) + \varepsilon|\eta|)).$$

This shows that the Fourier transform of $\rho_\varepsilon * T$ satisfies Condition (b) in Theorem 29.1 with $I_K(\eta)$ replaced by $I_K(\eta) + \varepsilon|\eta|$, which is the indicator of the set $K_\varepsilon = \{x \in \mathbf{R}^n; d(x, K) \leqslant \varepsilon\}$ (d : Euclidean distance). We conclude that $\operatorname{supp}(\rho_\varepsilon * T) \subset K_\varepsilon$.

As $\varepsilon \to 0$, $\rho_\varepsilon * T \to T$ (proof of Theorem 28.2); for $\varepsilon < 1$, the support of $\rho_\varepsilon * T$ is contained in K_1; therefore this is also true of supp T, which implies that supp T is compact. Finally, $\mathrm{supp}(\rho_\varepsilon * T)$ converges to supp T (again by the proof of Theorem 28.2), hence

$$\mathrm{supp}\ T \subset \bigcap_{\varepsilon > 0} K_\varepsilon = K. \qquad\qquad \text{Q.E.D.}$$

Exercises

29.1. Give an example of a distribution T with compact support, which is not a Radon measure, whose Fourier transform $\hat{T}$ can be extended as an entire analytic function in $\mathbf{C}_n$ satisfying, for some C, $B > 0$ and all $\zeta \in \mathbf{C}_n$, $|\hat{T}(\zeta)| \leqslant C \exp(B \mid \zeta \mid)$. (Hint: Use Exercise 25.12 or 25.13 and multiplication by an element of $\mathscr{C}_c^\infty$.)

29.2. Characterize the distributions with compact support in $\mathbf{R}^n$ whose Fourier transform is an entire analytic function of exponential type $(\varepsilon,..., \varepsilon)$ (**Notation 22.2**) for all $\varepsilon > 0$. Give an example of an analytic functional on $\mathbf{C}^n$ whose Fourier–Borel transform is an entire function of exponential type $(\varepsilon,..., \varepsilon)$ for all $\varepsilon > 0$ and which is *not* a distribution with compact support.

29.3. By using the Paley–Wiener theorem, show that, if $P(\partial/\partial x)$ is a differential operator with constant coefficients (not all identically zero) in $\mathbf{R}^n$, the equation $P(\partial/\partial x)u = 0$ cannot have any solution u, other than $u = 0$, in the space $\mathscr{E}'$ of distributions with compact support.

29.4. Let A be a subset of $\mathbf{R}^n$. Set for $\eta \in \mathbf{R}_n$,

$$I'_A(\eta) = \sup_{x \in A} \langle \eta, x \rangle.$$

Set $A' = \{\eta \in \mathbf{R}_n \ ; \ I'_A(\eta) \leqslant 1\}$, $A'' = \{x \in R^n; \ \sup_{\eta \in A'} \langle \eta, x \rangle \leqslant 1\}$. Prove that A' is convex, that A'' is the convex hull of A and that $I'_{A''} = I'_A$.

29.5. Let K be a convex compact subset of $\mathbf{R}^n$. Prove that the following two properties of a tempered distribution T in $\mathbf{R}^n$ are equivalent:

(a)   the support of T is contained in the set K;

(b)   the Fourier transform of T can be extended to $\mathbf{C}^n$ as an entire function $\hat{T}(\zeta)$ satisfying, for some constants C, $m > 0$ and for all $\zeta \in \mathbf{C}^n$,

$$|\ \hat{T}(\zeta)\ | \leqslant C(1 + |\ \zeta\ |)^m \exp\{2\pi I'_K(\mathrm{Im}\ \zeta)\}.$$

30

Fourier Transforms of
Convolutions and Multiplications

If S and T are two distributions with compact support, the support of their convolution, $S * T$, is also compact, and, by Proposition 29.1, the Fourier transform of $S * T$ is the function equal to

$$x \rightsquigarrow \langle (S * T)_y , \exp(-2i\pi \langle x, y \rangle) \rangle.$$

By definition of the convolution of two distributions with compact support (Definition 27.3), we have

$$\langle (S * T)_y , \exp(-2i\pi\langle x, y \rangle) \rangle = \langle S_y , \langle \check{T}_\eta , \exp(-2i\pi \langle x, y - \eta \rangle) \rangle \rangle$$

$$= \langle S_y , \langle T_\eta , \exp(-2i\pi \langle x, y + \eta \rangle) \rangle \rangle$$

$$= \langle S_y , \exp(-2\pi \langle x, y \rangle) \langle T_\eta , \exp(-2i\pi \langle x, \eta \rangle) \rangle \rangle$$

$$= \langle S_y , \exp(-2\pi \langle x, y \rangle) \rangle \langle T_\eta , \exp(-2i\pi \langle x, \eta \rangle) \rangle,$$

which shows that

(30.1) $$\mathscr{F}(S*T) = \mathscr{F}S\mathscr{F}T.$$

Note that both sides are analytic functions (Theorem 29.2).

The same relation is valid when S and T are two L^1 functions in $\mathbf{R}^n$. We know then (Corollary 3 of Theorem 26.1) that $S * T$ is also an L^1 function. We also know (Theorem 25.3) that the Fourier transform of any L^1 function is a continuous function in $\mathbf{R}_n$, converging to zero at infinity, so that both sides of (30.1) make sense.

It is not difficult to extend the domain of validity of (30.1) in various directions. In the present chapter, we shall consider the case where one of the factors in the convolution, say S, is a tempered distribution. It is not difficult to see then that $S * T$ and $\mathscr{F}S \mathscr{F}T$ are also tempered distributions, whenever T is a distribution with compact support. But it is also easy to see that, now, the condition that T have compact support can be relaxed. As the growth of S is slow at infinity, it should be enough to require that the decay of T at infinity be faster than any power of $1/|x|$, in some reasonable sense. A similar remark applies to the right-hand side of (30.1): noting that $\mathscr{F}S$ is an arbitrary tempered distribution (since $\mathscr{F} : \mathscr{S}' \to \mathscr{S}'$ is an isomorphism onto), we must require, if we wish to have $\mathscr{F}S\,\mathscr{F}T \in \mathscr{S}'$, that $\mathscr{F}T$ be a $\mathscr{C}^\infty$ function slowly growing at infinity (this means that all the derivatives of $\mathscr{F}T$ should, as well, grow slowly at infinity); this is to say that $\mathscr{F}T$ should be an element of the space of $\mathscr{C}^\infty$ functions already introduced as the space $\mathscr{O}_M$ (Definition 25.3; see also Theorem 25.5). We are going to make all this more precise.

Definition 30.1. We denote by $\mathscr{O}'_C$ the space of distributions T having the following property:

(30.2) *Given any integer $h \geqslant 0$, there is a finite family of continuous functions in $\mathbf{R}^n$, $\{f_p\}$ $(p \in \mathbf{N}^n, |p| \leqslant m(h))$, such that*

$$T = \sum_{|p| \leqslant m(h)} (\partial/\partial x)^p f_p\,,$$

and such that, for all $p \in \mathbf{N}^n, |p| \leqslant m(h)$,

$$\lim_{|x|\to\infty} (1 + |x|)^h\,|f_p(x)| = 0.$$

The letter $\mathscr{O}$ stands for *operators*, the subscript C for convolution (the prime indicates that we are dealing with distributions, not functions). Obviously the elements of $\mathscr{O}'_C$ are tempered distributions (cf. Theorem 25.4). One often refers to the elements of $\mathscr{O}'_C$ as *the distributions rapidly decreasing at infinity.*

Examples

30.1. The continuous functions rapidly decreasing at infinity (i.e., decreasing at infinity faster than any power of $1/|x|$), in particular the functions φ belonging to $\mathscr{S}$.

30.2. The distributions with compact support.

30.3. The finite sums of derivatives of continuous functions rapidly decreasing at infinity. One should not think however that this type of distributions completely makes up $\mathscr{O}'_C$: Property (30.2) is *not* equivalent with saying that T is a finite sum of derivatives of continuous functions rapidly decreasing at infinity. Note, in Property (30.2), that, as the integer h increases, so will, in general, the integer $m(h)$. In connection with this, we propose the following exercise to the student:

Exercise 30.1. Prove that the distribution $e^{i|x|^2}$, that is to say the distribution

$$\mathscr{C}_c^\infty \ni \varphi \rightsquigarrow \int e^{i|x|^2} \varphi(x)\, dx,$$

belongs to $\mathscr{O}_C'$.

We shall not define any topology on $\mathscr{O}_C'$ (although this can be done in a natural way). The introduction of this space is justified by the following result:

THEOREM 30.1. *Let $T \in \mathscr{O}_C'$. Then the convolution mapping*

$$(30.3) \qquad \mathscr{C}_c^\infty(\mathbf{R}^n) \ni \varphi \rightsquigarrow (T * \varphi)(x) = \langle T_y\,, \varphi(x-y)\rangle$$

is a map of $\mathscr{C}_c^\infty(\mathbf{R}^n)$ into $\mathscr{S}(\mathbf{R}^n)$, which can be extended as a continuous linear map of $\mathscr{S}(\mathbf{R}^n)$ into itself.

Proof. The proof is very straightforward. Let k be a nonnegative integer, and $P(D)$ a differential operator with constant coefficients on $\mathbf{R}^n$. Let h be an integer $\geqslant k + n + 1$. We use a representation of T as in (30.2), with this choice of the integer h. We have

$$P(D)(T * \varphi) = T * (P(D)\varphi) = \sum_{p \leqslant m(h)} f_p * [(\partial/\partial x)^p\, P(D)\, \varphi].$$

Let now q be an arbitrary n-tuple such that $|q| \leqslant k$. We have (see Chapter 26, Exercise 26.4)

$$x^q\{f_p * [(\partial/\partial x)^p\, P(D)\, \varphi]\} = \sum_{r \leqslant q} \binom{q}{r}(x^{q-r}f_p) * [x^r(\partial/\partial x)^p\, P(D)\, \varphi],$$

where $r \leqslant q$ means, as usual, $r_1 \leqslant q_1, ..., r_n \leqslant q_n$. Because of our choice of $h \geqslant k + n + 1$, and since

$$|f_p(x)| \leqslant C_p(1 + |x|)^{-h},$$

we see that $x^{q-r}f_p \in L^1$; hence we have (Corollary 2 of Theorem 26.1)

$$\|(x^{q-r}f_p) * [x^r(\partial/\partial x)^p\, P(D)\, \varphi]\|_{L^\infty} \leqslant \|x^{q-r}f_p\|_{L^1}\, \|x^r(\partial/\partial x)^p\, P(D)\, \varphi\|_{L^\infty}.$$

Combining all these inequalities, we see that there is a constant $C > 0$ such that, for all functions $\varphi \in \mathscr{C}_c^\infty(R^n)$,

$$\sum_{|q| \leqslant k} \|x^q\, P(D)(T * \varphi)\|_{L^\infty} \leqslant C \sum_{|r| \leqslant k} \sum_{|p| \leqslant m(h)} \|x^r(\partial/\partial x)^p\, P(D)\, \varphi\|_{L^\infty},$$

which immediately proves our assertion.

Definition 30.2.   Let $T \in \mathcal{O}'_C$; we denote by $S * T$ or $T * S$ the image of a tempered distribution S under the transpose of the continuous linear map $\varphi \rightsquigarrow \check{T} * \varphi$ of $\mathscr{S}$ into itself.

We have denoted by $\check{T} * \varphi$ the image of $\varphi \in \mathscr{S}$ under the extension to $\mathscr{S}$ of Mapping (30.3). We recall that $\check{T}$ is the distribution defined by

$$\mathscr{C}^\infty_c \ni \varphi \rightsquigarrow \langle \check{T}, \varphi \rangle = \langle T, \check{\varphi} \rangle, \qquad \check{\varphi}(x) = \varphi(-x).$$

There are several consistency conditions to be verified; for instance, when both S and T belong to $\mathcal{O}'_C$, we must show that the image of S under the transpose of $\varphi \rightsquigarrow T * \varphi$ is equal to the image of T under the transpose of $\varphi \rightsquigarrow S * \varphi$. This is easily done by using Condition (30.2). By using the theorem on structure of tempered distributions, one proves easily the following theorem (regularization of tempered distributions):

THEOREM 30.2.   *Let* $\varphi \in \mathscr{S}(\mathbf{R}^n)$, $S \in \mathscr{S}'(\mathbf{R}^n)$. *The convolution* $S * \varphi$ *(Definition 30.2) is the* $\mathscr{C}^\infty$ *function slowly increasing at infinity (i.e.,* $\in \mathcal{O}_M$*):*

$$x \rightsquigarrow \langle S_y, \varphi(x - y) \rangle.$$

Again by using the structure of a tempered distribution, one may easily define the convolution of $S \in \mathscr{S}'$ with $T \in \mathcal{O}'_C$; suppose that

$$S = \sum_{|p| \leqslant m} (\partial/\partial x)^p g_p,$$

where the g_p are continuous functions such that, for a suitable integer $k \geqslant 0$, we have, for all $x \in \mathbf{R}^n$,

$$|g_p(x)| \leqslant C(1 + |x|)^k, \qquad p \in \mathbf{N}^n, \quad |p| \leqslant m.$$

Choose for T a representation as in Property (30.2) with $h \geqslant k + n + 1$. Then we may write

(30.4) $$S * T = \sum_{|p| \leqslant m} \sum_{|q| \leqslant m(h)} (\partial/\partial x)^{p+q}(g_p * f_q).$$

It is not difficult to see that

$$(g_p * f_q)(x) = \int g_p(x - y) f_q(y) \, dy,$$

that $g_p * f_q$ is a continuous function, and that

$$|(g_p * f_q)(x)| \leqslant C \int (1 + |x - y|)^k |f_q(y)| \, dy$$

$$\leqslant C(1 + |x|)^k \int (1 + |y|)^k |f_q(y)| \, dy$$

$$\leqslant C_1(1 + |x|)^k \int (1 + |y|)^{k-h} \, dy$$

$$\leqslant C_2(1 + |x|)^k \quad \text{since} \quad h \geqslant k + n + 1.$$

This proves that $g_p * f_q$ is slowly increasing at infinity and, therefore, that $S * T$, given by (30.4), is a tempered distribution.

We go back, next, to Fourier transformation. Observe that both $\mathcal{O}_M$ and $\mathcal{O}'_C$ are linear subspaces of $\mathcal{S}'$; the Fourier transformation is well defined for these sets of distributions. Furthermore, we have the following result:

THEOREM 30.3.   *The Fourier transformation is a one-to-one linear map of $\mathcal{O}'_C$ onto $\mathcal{O}_M$ and of $\mathcal{O}_M$ onto $\mathcal{O}'_C$.*

Proof.   It is enough to prove that $\mathscr{F}$ maps $\mathcal{O}'_C$ into $\mathcal{O}_M$ and $\mathcal{O}_M$ into $\mathcal{O}'_C$ since $\bar{\mathscr{F}}$, which is the inverse of $\mathscr{F}$ in $\mathcal{S}'$, will obviously have the same properties.

Let $T \in \mathcal{O}'_C$; we use a representation of T as in Property (30.2):

$$T = \sum_{|p| \leqslant m(h)} (\partial/\partial x)^p f_p.$$

A straightforward computation shows that the Fourier transform $\hat{T}$ of T is equal to

$$\sum_{|p| \leqslant m(h)} (2i\pi\xi)^p \hat{f}_p(\xi),$$

where

$$\hat{f}_p(\xi) = \int \exp(-2i\pi \langle x, \xi \rangle) f_p(x) \, dx.$$

Let k be an arbitrary integer; choose $h \geqslant k + n + 1$ and $q \in \mathbf{N}^n$, $|q| \leqslant k$. Then we may differentiate q times $\hat{f}_p$ under the integral sign. We obtain, in this way,

$$|(\partial/\partial\xi)^q \hat{f}_p(\xi)| \leqslant \int |(2\pi x)^q| \, |f_p(x)| \, dx$$

$$\leqslant C \int (2\pi |x|)^{|q|} (1 + |x|)^{-h} \, dx,$$

from which one derives easily that $\hat{f}_p$ belongs to $\mathscr{C}^k(\mathbf{R}_n)$; thus $\hat{T} \in \mathscr{C}^k(\mathbf{R}_n)$; since k is arbitrary, this means that $\hat{T}$ is a $\mathscr{C}^\infty$ function. Furthermore, since all the derivatives of order $\leqslant k$ of the f_p are bounded functions, we have, for every $q \in \mathbf{N}^n$, $| q | \leqslant h$,

$$|(\partial/\partial\xi)^q \, \hat{T}(\xi)| \leqslant C(1 + | \xi |)^{m(h)}$$

for all $\xi \in \mathbf{R}_n$ and a suitable constant $C > 0$, independent of ξ. This shows that $\hat{T} \in \mathcal{O}_M$.

Let, now, $\alpha \in \mathcal{O}_M$. If h is an arbitrary integer $\geqslant 0$, there is an integer $m = m(h) \geqslant 0$ and a constant $C > 0$ such that, for all x,

$$\sum_{|q|\leqslant h} |(\partial/\partial x)^q \, \alpha(x)| \leqslant C(1 + | x |^2)^m.$$

Set, then,

(30.5) $$\beta(x) = (1 + | x |^2)^{-m-n-1} \, \alpha(x);$$

an easy computation shows that there is a constant $C' > 0$ such that, for all $x \in \mathbf{R}^n$,

(30.6) $$\sum_{|q|\leqslant h} |(\partial/\partial x)^q \, \beta(x)| \leqslant C'(1 + | x |^2)^{-n-1}.$$

Let $\hat{\beta}$ be the Fourier transform of β; we have

$$(2i\pi\xi)^q \, \hat{\beta}(\xi) = \int \exp(-2i\pi \, \langle x, \xi \rangle) \, (\partial/\partial x)^q \, \beta(x) \, dx.$$

Combining this with (30.6), we obtain

$$(1 + | \xi |)^h \, | \hat{\beta}(\xi)| \leqslant C'' \int (1 + | x |^2)^{-n-1} \, dx = C'''.$$

But Relation (30.5) means that

$$\hat{\alpha} = \left(1 - \frac{\Delta}{4\pi^2}\right)^{m+n+1} \beta,$$

which shows that $\hat{\alpha} \in \mathcal{O}'_C$.   Q.E.D.

We prove now the announced extension of Eq. (30.1):

THEOREM 30.4.   *Let $S \in \mathscr{S}'$, $T \in \mathcal{O}'_C$, and $\alpha \in \mathcal{O}_M$. Then we have*

(30.7) $$\mathscr{F}(S * T) = \mathscr{F}S \, \mathscr{F}T;$$

(30.8) $$\mathscr{F}(\alpha S) = \mathscr{F}\alpha * \mathscr{F}S.$$

Proof. It suffices to prove (30.7). Indeed, the same formula is then true with $\bar{\mathscr{F}}$ replacing $\mathscr{F}$, from which we derive

$$S * T = \mathscr{F}(\bar{\mathscr{F}} S \, \bar{\mathscr{F}} T);$$

but $\bar{\mathscr{F}} S$ may be replaced by S and $\bar{\mathscr{F}} T$ by α; then S has to be replaced by $\mathscr{F} S$ and T by $\mathscr{F} \alpha$ (we have used Theorems 25.6 and 30.3).

In order to prove (30.7), we observe that we have

$$\langle \mathscr{F}(S * T), \varphi \rangle = \langle S * T, \mathscr{F}\varphi \rangle = \langle S, \check{T} * (\mathscr{F}\varphi) \rangle.$$

But on the other hand, we have (cf. Theorems 30.1 and 30.2)

$$[\check{T} * (\mathscr{F}\varphi)](x) = \langle T_y, (\mathscr{F}\varphi)(x + y) \rangle = \langle T_y, \int \exp(-2i\pi \langle x + y, \xi \rangle) \, \varphi(\xi) \, d\xi \rangle.$$

Let us use a representation of T as in Property (30.2) with $h \geqslant n + 1$. Then we see that

$$[\check{T} * (\mathscr{F}\varphi)](x) = \sum_{|p| \leqslant m(h)} \int f_p(y) \left\{ \int \exp(-2i\pi \langle x + y, \xi \rangle) \, (2i\pi\xi)^p \, \varphi(\xi) \, d\xi \right\} dy.$$

But we can interchange the integrations with respect to ξ and y, since (by our choice of $h \geqslant n + 1$) all the functions f_p are integrable. We obtain immediately

$$[\check{T} * (\mathscr{F}\varphi)](x) = \sum_{|p| \leqslant m(h)} \int \exp(-2i\pi \langle x, \xi \rangle) \, (2i\pi\xi)^p \, f_p(\xi) \, \varphi(\xi) \, d\xi$$

$$= [\mathscr{F}((\mathscr{F}T)\varphi)](x).$$

Thus we have obtained

$$\langle \mathscr{F}(S * T), \varphi \rangle = \langle \mathscr{F}S, (\mathscr{F}T)\varphi \rangle,$$

which is exactly what we wanted to prove.

From Eqs. (30.7) and (30.8), one easily derives the following formulas, valid for an arbitrary polynomial $P(X_1, ..., X_n)$ in n variables with complex coefficients, for any vector $a \in \mathbf{R}^n$, for any tempered distribution S in $\mathbf{R}^n$:

(30.9) $$\mathscr{F}(P(\partial/\partial x)S) = P(2i\pi\xi)(\mathscr{F}S);$$

(30.10) $$\mathscr{F}(P(x)S) = P\left(-\frac{1}{2i\pi} \, \partial/\partial\xi\right)(\mathscr{F}S);$$

(30.11) $$\mathscr{F}(\tau_a S) = \exp(-2i\pi \langle a, \xi \rangle)(\mathscr{F}S).$$

It suffices to observe that

$$P(\partial/\partial x)S = [P(\partial/\partial x)\,\delta] * S, \qquad \tau_a S = \delta_a * S.$$

Exercises

30.2. Compute the Fourier transform of the distribution $e^{i|x|^2}$. From the fact that $e^{i|x|^2}$ belongs obviously to $\mathcal{O}_M$, derive that $e^{i|x|^2}$ belongs to $\mathcal{O}'_C$ (cf. Exercise 30.1). Conclude that there is no natural way of establishing a duality between $\mathcal{O}_M$ and $\mathcal{O}'_C$.

30.3. Let H' be the space of analytic functionals in $\mathbf{C}^n$, Exp the space of entire functions of exponential type in $\mathbf{C}^n$, $\mathscr{F}$ the Fourier–Borel transformation $H' \to \text{Exp}$ (Definition 22.3). For all pairs $\mu, \nu \in H'$, let us set

$$\mu * \nu = \mathscr{F}^{-1}(\mathscr{F}\mu\,\mathscr{F}\nu).$$

Prove that the star composition law, which is called *convolution* of analytic functionals, turns H' into a commutative ring with a unit element, and that, if we embed $\mathscr{E}'(\mathbf{R}^n)$ into $H'(\mathbf{C}^n)$ (by transposing the restriction to the real space, $H(\mathbf{C}^n) \to \mathscr{C}^\infty(\mathbf{R}^n)$), the convolution in H' induces the convolution of distributions with compact support in $\mathbf{R}^n$.

Define and state the basic properties of the convolution $h * \mu$ of an entire analytic function $h \in H(\mathbf{C}^n)$ with an analytic functional $\mu \in H'(\mathbf{C}^n)$.

30.4. Let X be the set $\mathbf{N}^n$. Let E be the space of complex functions in X with compact support, and F the space of all the complex functions in X; E will carry the topology inductive limit of the spaces E_m (space of functions $p \rightsquigarrow f(p)$ which vanish identically for $|p| > m$; $m = 0, 1,...$); F will carry the topology of pointwise convergence in X. It is clear that

$$f \rightsquigarrow \sum_{p \in N^n} f(p)\, X^p$$

is a TVS isomorphism u of F onto $\mathscr{Q}_n$, the Fréchet space of formal power series with complex coefficients in n indeterminates $X = (X_1,..., X_n)$. The restriction of the same mapping to E is a TVS isomorphism v of E onto $\mathscr{P}_n$, the LF-space of polynomials in n letters.

We define the *convolution* in F

$$(f, g) \rightsquigarrow f * g : p \rightsquigarrow \sum_{q \leqslant p} f(p - q)\, g(q),$$

where $q \leqslant p$ means $q_j \leqslant p_j$ for every $j = 1,..., n$.

Describe the relations between u, v, their transposes, t_u, t_v, also between the transforms, under u and v, of multiplicative products and convolutions of functions on X, and the transforms, under t_u and t_v, of products of polynomials and formal power series.

30.5. Prove that there are no *zero divisors* in the following convolution algebras: $\mathscr{E}'$, $\mathcal{O}'_c$, $\mathscr{D}'_+$ (see Exercise 27.6).

31

The Sobolev Spaces

In the previous chapters, the main body of distribution theory has been described. Needless to say, the applications of this theory to analysis often require a much more detailed study of particular classes of distributions. The present chapter is devoted to the elementary theory of such a class, which has come to play an increasingly important role in the theory of partial differential equations. These distributions are grouped into a succession of spaces, the so-called Sobolev spaces and their variants. This succession, or ladder, of spaces is built on a notion of order of a distribution which is somehow different from the notion introduced in Definition 24.3 and turns out to be better adapted to the study of a large number of problems. Instead of looking at distributions which are sums of derivatives of order $\leqslant m$ of Radon measures (Theorem 24.4), we may look at distributions which are sums of derivatives of order $\leqslant m$ of functions belonging to L^p ($1 \leqslant p \leqslant +\infty$). The advantage lies in the fact that the spaces L^p are much easier to manipulate than the space of Radon measures; for one thing, the dual of L^p is as well known as L^p itself, provided that p be finite, for it is nothing more than $L^{p'}$, $p' = p/(p-1)$.

The distributions of order $\leqslant m$ (in the sense of Definition 24.3) form a space which is the dual of $\mathscr{C}_c^m$ (m-times continuously differentiable functions with compact support). One may say that the concept of a distribution which is a sum of derivatives of order $\leqslant m$ of Radon measures is the dual of the concept of a function (with compact support) whose derivatives of order $\leqslant m$ are continuous functions. A similar fact occurs in the theory of Sobolev spaces: for $p > 1$, the concept of a distribution which is a sum of derivatives of order $\leqslant m$ of L^p functions is, in a sense, the dual of the concept of a function whose derivatives of order $\leqslant m$ belong to $L^{p'}$ (again, $p' = p/(p-1)$) and furthermore, which is the limit (in a natural way) of $\mathscr{C}_c^\infty$ functions. We shall now proceed to give a precise form to these ideas.

In the forthcoming, Ω will be an arbitrary open subset of $\mathbf{R}^n$, m an integer $\geqslant 0$.

Definition 31.1.   *Let p be a number such that $1 \leqslant p \leqslant +\infty$. We denote by $H^{p,m}(\Omega)$ the set of distributions u in Ω such that all derivatives of order $\leqslant m$ of u belong to $L^p(\Omega)$.*

If $u \in H^{p,m}(\Omega)$, we have $u \in L^p(\Omega)$: thus u is a locally integrable function in Ω (Theorem 21.4). The space $H^{p,m}(\Omega)$ is equipped with the natural norm:

$$(31.1) \qquad u \rightsquigarrow \Big(\sum_{|\alpha| \leqslant m} \int_\Omega |(\partial/\partial x)^\alpha u(x)|^p \, dx \Big)^{1/p}$$

By α we have denoted an n-tuple $\alpha = (\alpha_1, ..., \alpha_n)$. The norm (31.1) will be denoted by $\| u \|_{p,m,\Omega}$, or simply by $\| u \|_{p,m}$ when there is no risk of confusion. As we shall soon see, the case $p = 2$ is particularly important; in this case, the norm (31.1) will usually be denoted by $\| u \|_m$.

PROPOSITION 31.1.   *The norm (31.1) turns $H^{p,m}(\Omega)$ into a Banach space.*

Proof.   Let $\{u_k\}$ $(k = 1, 2,...)$ be a Cauchy sequence in $H^{p,m}(\Omega)$; for each n-tuple α such that $| \alpha | \leqslant m$, $\{(\partial/\partial x)^\alpha u_k\}$ is a Cauchy sequence in $L^p(\Omega)$, and therefore converges in this space to a limit u^α. In particular, $\{u_k\}$ converges in $L^p(\Omega)$ to a limit $u = u^0$. Since the differential operators $(\partial/\partial x)^\alpha$ are continuous mappings of $\mathscr{D}'(\Omega)$ into itself, $(\partial/\partial x)^\alpha u_k$ converges in $\mathscr{D}'(\Omega)$ to $(\partial/\partial x)^\alpha u$; because of the uniqueness of the limit in a Hausdorff space, we must have $u^\alpha = (\partial/\partial x)^\alpha u$.   Q.E.D.

COROLLARY.   *The norm (31.1) turns $H^{2,m}(\Omega)$ into a Hilbert space.*

It suffices to combine Proposition 31.1 with the remark that, when $p = 2$, the norm (31.1) is associated with the sesquilinear form

$$(31.2) \qquad (u, v) \rightsquigarrow \sum_{|\alpha| \leqslant m} \int (\partial/\partial x)^\alpha u(x) \overline{(\partial/\partial x)^\alpha v(x)} \, dx.$$

Usually, we shall denote (31.2) by $(u, v)_{m,\Omega}$ or $(u, v)_m$.

The natural injection of $H^{p,m}(\Omega)$ into $L^p(\Omega)$ and, a fortiori, into $\mathscr{D}'(\Omega)$, is continuous. Also note that we have $H^{p,m_1}(\Omega) \subset H^{p,m_2}(\Omega)$ for $m_1 \geqslant m_2$ and that the norm of the natural injection of the first space into the second is $\leqslant 1$ (in fact, it is equal to one as is easily seen). An important feature of the Banach spaces $H^{p,m}(\Omega)$ is that, in general, they are *not* normal spaces of distributions; that is to say, $\mathscr{C}_c^\infty(\Omega)$ is not dense in $H^{p,m}(\Omega)$ (this is only true in general, of course, since $\mathscr{C}_c^\infty(\Omega)$ is obviously dense in $H^{p,m}(\Omega)$ when $p < +\infty$ and $m = 0$ or when $p < +\infty$ and $\Omega = \mathbf{R}^n$ as will soon be shown).

Definition 31.2. We denote by $H_0^{p,m}(\Omega)$ $(1 \leqslant p \leqslant +\infty, m \geqslant 1)$ the closure of $\mathscr{C}_c^\infty(\Omega)$ in $H^{p,m}(\Omega)$.

Notice that the norm of a function $\phi \in \mathscr{C}_c^\infty(\Omega)$ in $H^{p,m}(\Omega)$ is equal to its norm in $H^{p,m}(\Omega')$ for all open sets Ω' containing Ω. This means that the natural injection of $\mathscr{C}_c^\infty(\Omega)$ into $\mathscr{C}_c^\infty(\Omega')$ can be extended as an isometry of $H_0^{p,m}(\Omega)$ into $H_0^{p,m}(\Omega')$. We shall denote by $u \rightsquigarrow \| u \|_{p,m}$ the norm (31.1) on $H_0^{p,m}(\Omega)$.

From Proposition 31.1 and the fact that a closed subspace of a Banach space (resp. of a Hilbert space) is a Banach space (resp. a Hilbert space), we derive:

PROPOSITION 31.2. $H_0^{p,m}(\Omega)$ *is a Banach space* $(1 \leqslant p \leqslant +\infty)$; $H_0^{2,m}(\Omega)$ *is a Hilbert space.*

In order to see that $H_0^{p,m}(\Omega)$ is, in general, a proper subspace of $H^{p,m}(\Omega)$, we might look at the case of a *bounded* subset Ω of $\mathbf{R}^n$, when $p = 2$ and $m = 1$. In this case, the orthogonal of $\mathscr{C}_c^\infty(\Omega)$ in $H^{2,1}(\Omega)$ is made up by the elements $H^{2,1}(\Omega)$ satisfying $(u, \phi)_1 = 0$ for all $\phi \in \mathscr{C}_c^\infty(\Omega)$; but (cf. (31.2)):

$$(u, \phi)_1 = \int_\Omega u(x) \overline{\phi(x)}\, dx - \sum_{j=1}^m \int_\Omega u(x) \overline{(\partial/\partial x_j)^2 \phi(x)}\, dx,$$

which shows that these elements u are the solutions of the partial differential equation (to be understood in the sense of distributions in Ω)

$$(31.3) \qquad\qquad u - \Delta u = 0,$$

where $\Delta = (\partial/\partial x_1)^2 + \cdots + (\partial/\partial x_n)^2$ is the *Laplace operator*. It is easy to see that there are elements u of $H^{2,1}(\Omega)$ which satisfy (31.3): in fact, it suffices to take the restrictions to Ω of the solutions of (31.3) in the whole space $\mathbf{R}^n$. One can show that these solutions are $\mathscr{C}^\infty$ in $\mathbf{R}^n$; since Ω is bounded, their restrictions obviously belong to $H^{2,m}(\Omega)$ for all $m \geqslant 0$. If one does not want to use this result, one may take the restrictions to Ω of the functions $x \rightsquigarrow e^{\langle x, \zeta \rangle}$, where $\zeta \in \mathbf{C}^n$ satisfies the equation

$$\sum_{j=1}^n (\zeta_j)^2 = 1.$$

Since $H_0^{p,m}(\Omega)$ is a normal space of distributions in Ω, its dual may be canonically identified with a space of distributions in Ω, which we describe now, in the case where $p < +\infty$.

Definition 31.3. We denote by $H^{p,-m}(\Omega)$ $(1 \leqslant p \leqslant +\infty, m \geqslant 1)$ the *space of distributions in Ω which are equal to a finite sum of derivatives of order $\leqslant m$ of functions belonging to $L^p(\Omega)$.*

Suppose that $1 \leqslant p < +\infty$ and set $p' = p/(p-1)$. If $u \in H^{p',-m}(\Omega)$, there are functions g_α ($\alpha \in \mathbf{N}^n$, $|\alpha| \leqslant m$), belonging to $L^{p'}(\Omega)$, such that

$$u = \sum_{|\alpha| \leqslant m} (\partial/\partial x)^\alpha g_\alpha.$$

If, then, $\phi \in \mathscr{C}_c^\infty(\Omega)$, we have

$$\langle u, \phi \rangle = \sum_{|\alpha| \leqslant m} (-1)^{|\alpha|} \int g_\alpha(x)\,(\partial/\partial x)^\alpha \phi(x)\,dx,$$

whence, in view of Hölder's inequalities (Theorem 20.3),

$$|\langle u, \phi \rangle| \leqslant \sum_{|\alpha| \leqslant m} \|g_\alpha\|_{L^{p'}} \|(\partial/\partial x)^\alpha \phi\|_{L^p} \leqslant C \|\phi\|_{m,p},$$

This shows that $\phi \rightsquigarrow \langle u, \phi \rangle$ can be extended, in a unique way, as a continuous linear form on $H_0^{p,m}(\Omega)$; let us call this extension *the canonical extension* of u to $H_0^{p,m}(\Omega)$.

PROPOSITION 31.3. *Let $p \geqslant 1$ be finite, and m an integer $\geqslant 1$. The canonical extension is a one-to-one linear map of the space $H^{p',-m}(\Omega)$ onto the dual of $H_0^{p,m}(\Omega)$.*

We recall that $p' = p/(p-1)$.

Proof of Proposition 31.3. We have seen that the canonical extension is "into"; it is of course linear, and it is one-to-one since $\mathscr{C}_c^\infty(\Omega)$ is dense in $H_0^{p,m}(\Omega)$. We must therefore only show that it is onto. This is very easily done in the following way. Let $N = N(m, n)$ be the number of n-tuples $\alpha = (\alpha_1, ..., \alpha_n)$ such that $|\alpha| = \alpha_1 + \cdots + \alpha_n \leqslant m$; let E be the product space of N copies of the space $L^p(\Omega)$, equipped with the product topology: an element f on E is an N-tuple (f_α) whose components are elements of $L^p(\Omega)$; the topology of E can be defined by the norm $(f_\alpha) \rightsquigarrow (\sum_{|\alpha| \leqslant m} \|f_\alpha\|_{L^p(\Omega)}^p)^{1/p}$ which turns E into a Banach space (a Hilbert space if $p = 2$). The dual of E is canonically isomorphic to the product of N copies of the dual of $L^p(\Omega)$, which is canonically isomorphic with $L^{p'}(\Omega)$ (Theorem 20.3). On the other hand, there is a canonical isometry of $H^{p,m}(\Omega)$, and therefore of $H_0^{p,m}(\Omega)$, into E, namely the mapping

(31.4) $v \rightsquigarrow ((\partial/\partial x)^\alpha v)_{|\alpha| \leqslant m}.$

If we transpose this mapping, we obtain a continuous linear map of E' onto the dual of $H_0^{p,m}(\Omega)$: that the last map is onto follows immediately from the Hahn–Banach theorem. Indeed, any continuous linear form on $H_0^{p,m}(\Omega)$ can be transferred as a continuous linear form on the image of $H_0^{p,m}(\Omega)$ under (31.4) and then extended to the whole of E as a continuous linear form; the image of the latter under the transpose of (31.4) gives back the form on $H_0^{p,m}(\Omega)$ we started from.

Let, now, L be a continuous linear form on $H_0^{p,m}(\Omega)$; it is the image, by the transpose of (31.4), of an element of the dual of E, that is to say of an N-tuple (g_α) of functions $g_\alpha \in L^{p'}(\Omega)$; if we set $u = \sum_{|\alpha| \leqslant m} (-1)^{|\alpha|} (\partial/\partial x)^\alpha g_\alpha$, we see immediately that L is the canonical extension of u: it suffices to compare $L(\phi)$ and $\langle u, \phi \rangle$ for all $\phi \in \mathscr{C}_c^\infty(\Omega)$.

Remark 31.1.   We may define on $H^{p',-m}(\Omega)$ the following norm:

$$\| u \|_{p',-m} = \inf \left(\sum_{|\alpha| \leqslant m} \| g_\alpha \|_{L^{p'}(\Omega)}^{p'} \right)^{1/p'}$$

where the infimum is computed over all the representations of u of the form

$$u = \sum_{|\alpha| \leqslant m} (\partial/\partial x)^\alpha g_\alpha, \qquad g_\alpha \in L^{p'}(\Omega).$$

Then, inspection of the proof of Proposition 31.3 shows that the canonical extension is an *isometry* of $H^{p',-m}(\Omega)$ onto the dual of $H_0^{p,m}(\Omega)$ (provided with the dual norm).

Again, the case $p = 2$ is worth examining more closely. Let J be the canonical isometry of the Hilbert space $H_0^{2,m}(\Omega)$ onto its anti-dual (Theorem 12.2), and let K be the inverse mapping of the canonical extension, $K : (H_0^{2,m}(\Omega))' \to H^{2,-m}(\Omega)$. If L is a continuous antilinear functional on $H_0^{2,m}(\Omega)$, $\bar{L}$ is a continuous linear functional on the same space; we then define $\bar{K}(L) = \overline{K(\bar{L})}$ (the complex conjugate of any distribution T is defined by $\langle \bar{T}, \phi \rangle = \overline{\langle T, \bar{\phi} \rangle}$, $\phi \in \mathscr{C}_c^\infty$); thus $\bar{K}$ is a *linear* isometry of the anti-dual of $H_0^{2,m}(\Omega)$ onto $H^{2,-m}(\Omega)$ (cf. Remark 31.1). Finally, we have obtained a linear isometry of $H_0^{2,m}(\Omega)$ onto $H^{2,-m}(\Omega)$, namely the compose $\bar{K} \circ J$. We may refer to this linear isometry as *the canonical isometry* of $H_0^{2,m}(\Omega)$ onto $H^{2,-m}(\Omega)$.

PROPOSITION 31.4.   *The canonical isometry of $H_0^{2,m}(\Omega)$ onto $H^{2,-m}(\Omega)$ is the map*

$$u \rightsquigarrow \sum_{|\alpha| \leqslant m} (-1)^{|\alpha|} (\partial/\partial x)^{2\alpha} u.$$

We have used the notation $2\alpha = (2\alpha_1, ..., 2\alpha_n)$. For example, we see that, if $m = 1$, the above canonical isometry is the mapping

$$u \rightsquigarrow u - \Delta u, \qquad \Delta: \text{Laplace operator.}$$

The proof of Proposition 31.4 consists essentially of the remark that, for $u \in H_0^{2,m}(\Omega)$ and $\phi \in \mathscr{C}_c^\infty(\Omega)$, we have

$$(u, \phi)_m = \sum_{|\alpha| \leqslant m} (-1)^{|\alpha|} \int u(x) \, (\partial/\partial x)^{2\alpha} \overline{\phi(x)} \, dx.$$

We leave the details to the student.

COROLLARY. *For all* $m = 1, 2,...,$ $H^{2,-m}(\Omega)$ *is a normal space of distributions.*

Indeed, the canonical isometry of $H_0^{2,m}(\Omega)$ onto $H^{2,-m}(\Omega)$ maps $\mathscr{C}_c^\infty(\Omega)$ into itself; but the image of a dense subset under an isometry onto is a dense subset.

We have defined $H_0^{p,m}(\Omega)$ and $H^{p,-m}(\Omega)$ only for $m \geqslant 1$; it is clear that we could have defined these spaces also for $m = 0$, taking them to be equal to $H^{p,0}(\Omega) = L^p(\Omega)$.

We now center our attention on the case $\Omega = \mathbf{R}^n$.

PROPOSITION 31.5. *Let us assume* $1 \leqslant p < +\infty$, $m \geqslant 1$. *Then* $\mathscr{C}_c^\infty(\mathbf{R}^n)$ *is dense in* $H^{p,m}(\mathbf{R}^n)$.

Proof. By cutting and regularizing (cf. Chapter 28). Let $g \in \mathscr{C}_c^\infty$ be equal to one in the ball $\{x; \, |x| < 1\}$ and to zero for $|x| > 2$. Set

$$g_k(x) = g(x/k), \qquad k = 1, 2, ...$$

The function g_k is equal to one for $|x| < k$ and to zero for $|x| > 2k$. Let u be an arbitrary function belonging to $H^{p,m}(\mathbf{R}^n)$, and D any one of the differential operators $(\partial/\partial x)^\alpha$, $|\alpha| \leqslant m$. We have, by Leibniz' formula,

$$D(g_k u) - g_k D u = \sum_{0 \neq \beta \leqslant \alpha} \binom{\alpha}{\beta} [(\partial/\partial x)^\beta g_k](\partial/\partial x)^{\alpha - \beta} u,$$

whence

$$\int |D(g_k u) - g_k D u|^p \, dx \leqslant C \sum_{|\beta| < m} \int_{|x| > k} |(\partial/\partial x)^\beta u|^p \, dx,$$

Here, C is a positive constant independent of $k = 1, 2,...$; if we choose suitably, we may also write

$$\int | g_k Du - Du |^p \, dx \leqslant C \int_{|x| > k} | Du |^p \, dx,$$

Since $(\partial/\partial x)^\gamma u \in L^p(\mathbf{R}^n)$ for $| \gamma | \leqslant m$, we see that, as $k \to +\infty$,

$$\| D(g_k u) - Du \|_{L^p(\mathbf{R}^n)} \leqslant \| D(g_k u) - g_k Du \|_{L^p(\mathbf{R}^n)}$$
$$+ \| g_k Du - Du \|_{L^p(\mathbf{R}^n)} \to 0.$$

This proves that $g_k u$ converges to u in $H^{p,m}(\mathbf{R}^n)$. In view of this fact, it suffices to prove that every distribution $u \in H^{p,m}(\mathbf{R}^n)$ *with compact support* is the limit of a sequence of test functions. In order to prove this, we need not assume $\Omega = \mathbf{R}^n$.

LEMMA 31.1. *Let $p \geqslant 1$ be finite. Every $u \in H^{p,m}(\Omega)$ having compact support is the limit of a sequence of functions belonging to $\mathscr{C}_c^\infty(\Omega)$.*

Proof of Lemma 31.1. Let ρ be the usual function employed to define mollifiers (e.g., see Lemma 28.1); set $\rho_\varepsilon(x) = \varepsilon^{-n} \rho(x/\varepsilon)$ for $\varepsilon > 0$. For ε sufficiently small, the support of $\rho_\varepsilon * u$ is contained in Ω; of course, $\rho_\varepsilon * u$ is a $\mathscr{C}_c^\infty$ function. It is enough to prove that, for all n-tuples α such that $| \alpha | \leqslant m$, when $k \to +\infty$, $(\partial/\partial x)^\alpha(\rho_{1/k} * u)$ converges to $(\partial/\partial x)^\alpha u$ in $L^p(\Omega)$. Since

$$(\partial/\partial x)^\alpha(\rho_\varepsilon * u) = \rho_\varepsilon * (\partial/\partial x)^\alpha u,$$

it suffices to prove that $\rho_{1/k} * v \to v$ in $L^p(\mathbf{R}^n)$, as $k \to +\infty$, where now v is an arbitrary function in $L^p(\mathbf{R}^n)$, not necessarily with compact support. The shortest proof of this fact is based on a consequence of the Banach–Steinhaus theorem, which will be stated and proved later on (Theorem 33.1). Anticipating a little, the consequence relevant in the present situation is the following one: *let E be a Banach space, and $\{u_k\}$ ($k = 1, 2,...$) a sequence of continuous linear mappings of E into itself with the following two properties:* (1) *there is a constant $C > 0$ such that the norm of u_k is $\leqslant C$ for all k;* (2) *there is a dense subset A of E such that $u_k(x)$ converges to x in E, as $k \to +\infty$. Under these circumstances, $u_k(x)$ converges to x as $k \to +\infty$ for every $x \in E$* (and not only for every $x \in A$). We apply this to the sequence of mappings $u \rightsquigarrow \rho_{1/k} * u$ of $L^p(\mathbf{R}^n)$ into itself. The hypotheses of the preceding statement are satisfied. Indeed, the norms of these mappings are $\leqslant \| \rho_{1/k} \|_{L^1} = \| \rho \|_{L^1}$ (Corollary 2 of Theorem 26.1); when $k \to +\infty$, $\rho_{1/k} * u$ converges uniformly to u in $\mathbf{R}^n$

provided that u be continuous and have compact support (Lemma 15.2); we know that $\mathscr{C}_c^0$ is dense in $L^p(\mathbf{R}^n)$ (Theorem 11.3). We conclude that $u = \lim_{k \to +\infty}(\rho_{1/k} * u)$ for all $u \in L^p(\mathbf{R}^n)$. This proves the lemma and consequently also Proposition 31.5.

Remark 31.2. Lemma 31.1 proves that the elements of $H^{p,m}(\Omega)$ $(1 \leqslant p < +\infty)$ which have compact support (contained in Ω!) belong to $H_0^{p,m}(\Omega)$.

Since $H^{p,m}(\mathbf{R}^n) \subset L^p(\mathbf{R}^n)$, we see that all the distributions which belong to the spaces $H^{p,m}(\mathbf{R}^n)$ are tempered, and so also are the distributions belonging to $H^{p,-m}(\mathbf{R}^n)$, as follows immediately from their definition. This fact enables us to perform the Fourier transformation on the distributions belonging to $H^{p,k}(\mathbf{R}^n)$, where now k is an integer, $\geqslant 0$ or < 0. We shall restrict ourselves to the case $p = 2$, which is the most simple to study, in view of the fact that the Fourier transformation is an isometry of $L^2(\mathbf{R}^n)$ onto itself (Plancherel formula, Corollary 1 of Theorem 25.2).

PROPOSITION 31.6. *Let k be an integer $\geqslant 0$ or < 0. A distribution u belongs to $H^{2,k}(\mathbf{R}^n)$ if and only if u is tempered and if its Fourier transform $\hat{u}$ is a function which is square integrable with respect to the measure $(1 + |\xi|^2)^k \, d\xi$.*

Proof. Suppose first that k is $\geqslant 0$. For all n-tuples α whose length $|\alpha|$ is $\leqslant k$, we have $(\partial/\partial x)^\alpha u \in L^2(\mathbf{R}^n)$; by Fourier transformation, this means that $\xi^\alpha \hat{u} \in L^2(\mathbf{R}_n)$, which implies immediately our assertion in this case.

Suppose now $k < 0$. By Proposition 31.4, every element of $H^{2,k}(\mathbf{R}^n)$ is of the form $\sum_{|\alpha| \leqslant |k|} (-1)^{|\alpha|}(\partial/\partial x)^{2\alpha}u$, with $u \in H^{2,|k|}(\mathbf{R}^n)$. Performing a Fourier transformation and applying the result when $k \geqslant 0$, we obtain the desired conclusion easily.

Proposition 31.6 points to the natural way of interpolating the spaces $H^{2,k}(\mathbf{R}^n)$, that is to say of incorporating them in a one-parameter family of Hilbert spaces. The interpolation is achieved by the spaces H^s (s real arbitrary) defined in the following way:

Definition 31.4. *Let s be an arbitrary real number: H^s, or $H^s(\mathbf{R}^n)$, is the space of tempered distributions u whose Fourier transform $\hat{u}$ is a square-integrable function with respect to the measure $(1 + |\xi|^2)^s \, d\xi$.*

Thus, $u \in H^s$ means that $(1 + |\xi|^2)^{s/2} u \in L^2$. Proposition 31.6 says that $H^s = H^{2,s}(\mathbf{R}^n)$ when s is an integer $\geqslant 0$ or < 0. On H^s, we consider the Hermitian product

$$(u, v)_s = \int \hat{u}(\xi) \, \overline{\hat{v}(\xi)} \, (1 + |\xi|^2)^s \, d\xi,$$

and the associated norm

$$\| u \|_s = \left(\int | \hat{u}(\xi)|^2 (1 + | \xi |^2)^s \, d\xi \right)^{1/2}$$

PROPOSITION 31.7.   *The Hermitian product* $(\, , \,)_s$ *turns* H^s *into a Hilbert space.*

This is nothing else but the well-known statement (Fischer-Riesz theorem) about the space L^2 relative to a positive Radon measure. For $s \geqslant s'$, there is a natural injection, continuous with norm $\leqslant 1$, $H^s \to H^{s'}$. In view of the Plancherel theorem, H^0 is identical with $L^2(\mathbf{R}^n)$ as a Hilbert space (which is to say that the identity applies to Hermitian products and norms).

Note that $\xi \leadsto (1 + | \xi |^2)^{s/2}$ is a $\mathscr{C}^\infty$ function in $\mathbf{R}_n$, belonging to $\mathscr{O}_M$; its Fourier transform is a distribution belonging to $\mathscr{O}'_C$ (Theorem 30.3) which we denote by U_s. We may consider the convolution mapping $S \leadsto U_s * S$, which is a continuous linear map of the space $\mathscr{S}'$ of tempered distributions into itself (Definition 30.1). The next result follows immediately, by Fourier transformation (see Theorem 30.4):

PROPOSITION 31.8.   *Let* s, a *be two arbitrary real numbers. The convolution*

$$u \leadsto U_s * u$$

is an isometry of H^{s+a} *onto* H^a.

In particular, $u \leadsto U_s * u$ is an open isometry of H^s onto $H^0 = L^2(\mathbf{R}^n)$. Thus, we see that the normed spaces H^s are all copies of $L^2(\mathbf{R}^n)$. This fact implies that H^s is a Hilbert space for all real s; on the other hand, as $\mathscr{S}(\mathbf{R}^n) = \mathscr{S}$ is a dense subspace of $L^2(\mathbf{R}^n)$ and as $\phi \leadsto U_s * \phi$ is obviously a continuous linear map of $\mathscr{S}$ into itself, and because the image under an isometry onto of a dense subset is a dense subset, we obtain the following result:

PROPOSITION 31.9.   *The injection* $\mathscr{S} \to H^s$ *is continuous and has a dense image. In particular,* H^s *is a normal space of distributions in* $\mathbf{R}^n$.

On the product $H^s \times H^{-s}$ we may consider the bilinear form

(31.5)
$$(u, v) \leadsto (U_s * u, U_{-s} * \bar{v})_0 = \int (1 + | \xi |^2)^{s/2} \hat{u}(\xi) (1 + | \xi |^2)^{-s/2} \hat{v}(-\xi) \, d\xi,$$

where $\bar{v}$ is the complex conjugate of v. If v belongs to $\mathscr{S}$, we have

$$(U_s * u, U_{-s} * \bar{v})_0 = \int \hat{u}(\xi) \hat{v}(-\xi) \, d\xi = \langle \mathscr{F}u, \mathscr{F}v \rangle = \langle u, v \rangle,$$

where the bracket is the bracket of the duality between $\mathscr{S}$ and $\mathscr{S}'$ and where we have used the definition of the Fourier transform of a tempered distribution. The above remark justifies the use of the notation $(u, v) \rightsquigarrow \langle u, v \rangle$ for the bilinear form (31.5). Observing that $u \in H^s$ implies $U_{2s} * u \in H^{-s}$, we obtain the following:

$$\langle u, U_{2s} * \bar{u} \rangle = (U_s * u, U_s * u)_0 = \| u \|_s^2.$$

If v is a second element of H^s, we have

$$\langle u, U_{2s} * \bar{v} \rangle = (u, v)_s.$$

Using these facts, we see easily that the following is true:

PROPOSITION 31.10.   *The bilinear form* (31.5) *turns H^s and H^{-s} into the dual of each other; the mapping $u \rightsquigarrow U_{2s} * \bar{u}$ is the canonical antilinear isometry of the Hilbert space H^s onto its dual, H^{-s}.*

When s is a nonnegative integer, the mapping $u \rightsquigarrow U_{2s} * u$ is nothing else but the differential operator $u \rightsquigarrow (1 - \Delta/4\pi^2)^s u$, where Δ is the Laplace operator (cf. Proposition 31.4).

The one-parameter family of spaces H^s is often used to measure the regularity of distributions. The next result throws some light on this role of the spaces H^s:

PROPOSITION 31.11.   *If $s > n/2$, the elements of H^s are continuous functions.*

Proof.   Let s be $> n/2$, $\phi \in \mathscr{S}$. We have, by Schwarz's inequality,

$$\int | \hat{\phi}(\xi) | \, d\xi \leqslant \int | \hat{\phi}(\xi) | \, (1 + | \xi |^2)^{s/2} (1 + | \xi |^2)^{-s/2} \, d\xi$$

$$\leqslant \| \phi \|_s \left(\int (1 + | \xi |^2)^{-s} \, d\xi \right)^{1/2}.$$

By extending this inequality from $\mathscr{S}$ to H^s, we see that the Fourier transformation maps H^s into $L^1(\mathbf{R}_n)$. In other words, every function $u \in H^s$ is the inverse Fourier transform of an integrable function. Proposition 31.11 then follows immediately from the Lebesgue theorem (Theorem 25.3).

COROLLARY.   *Let m be an integer $\geqslant 0$. If $s > m + n/2$, the elements of H^s are m times continuously differentiable functions in $\mathbf{R}^n$.*

Indeed, if $u \in H^s$, $(\partial/\partial x)^\alpha u \in H^{s-m}$ for all n-tuples α, $|\alpha| \leqslant m$, as one can check easily. If $s - m > n/2$, $(\partial/\partial x)^\alpha u$ is a continuous function.

Finally, we must mention the spaces of distributions with compact support, on one hand, and the local spaces of distributions, on the other, that one builds out of the Banach spaces $H^{p,m}$. Let Ω, as before, be an open subset of $\mathbf{R}^n$, and K an arbitrary compact subset of Ω. We shall denote by $H^{p,m}(K)$ $(1 \leqslant p \leqslant +\infty$, m: integer $\geqslant 0$ or $< 0)$ the space of distributions in $\mathbf{R}^n$ which belong to $H^{p,m}(\mathbf{R}^n)$ and have their support contained in K; we equip $H^{p,m}(K)$ with the norm induced by $H^{p,m}(\mathbf{R}^n)$ with which it obviously becomes a Banach space (a Hilbert space if $p = 2$). Then, by $H_c^{p,m}(\Omega)$ we denote the inductive limit of the spaces $H^{p,m}(K)$ as K ranges over the family of all compact subsets of Ω. Obviously, $H_c^{p,m}(\Omega)$ is a space LF.

If Ω' is an open subset of Ω, let us denote by $T \mid \Omega'$ the restriction of a distribution T in Ω to Ω'. We then denote by $H_{loc}^{p,m}(\Omega)$ $(1 \leqslant p \leqslant +\infty$, m: integer $\geqslant 0$ or $< 0)$ the space of distributions T in Ω such that $T \mid \Omega'$ belongs to $H^{p,m}(\Omega')$ for all *relatively compact* open subsets Ω' of Ω. We equip $H_{loc}^{p,m}(\Omega)$ with the least-fine locally convex topology such that all the mappings $T \rightsquigarrow T \mid \Omega'$ from $H_{loc}^{p,m}(\Omega)$ into $H^{p,m}(\Omega')$ will be continuous $(\bar{\Omega}' \subset \Omega$ and compact). By taking a sequence of relatively compact open sets Ω' whose union is equal to Ω, one sees that $H_{loc}^{p,m}(\Omega)$ is a Fréchet space.

For s real, not necessarily an integer, the space $H_c^s(\Omega)$ can be defined exactly in the same manner as we have defined the $H_c^{p,m}(\Omega)$ above: as the inductive limit of the Hilbert spaces $H^s(K) = \{u \in H^s;$ supp $u \subset K\}$, equipped with the Hilbert space structure induced by H^s, as K ranges over the compact subsets of Ω. However, we cannot define the spaces $H_{loc}^s(\Omega)$ in exactly the same manner as the spaces $H_{loc}^{p,m}(\Omega)$, for we have made use of the spaces $H^{p,m}(\Omega)$, which have no equivalence here (at least in the framework to which we have limited ourselves). But the difficulty is easily turned by using cutting-off functions. Thus $H_{loc}^s(\Omega)$ is the space of distributions T in Ω such that, for every $\phi \in \mathscr{C}_c^\infty(\Omega)$, ϕT belongs to H^s. We equip $H_{loc}^s(\Omega)$ with the least-fine topology such that all the mappings $T \rightsquigarrow \phi T$ from $H_{loc}^s(\Omega)$ into H^s will be continuous (as ϕ ranges over $\mathscr{C}_c^\infty(\Omega)$). As we could have limited ourselves to a sequence of functions ϕ equal to one on (relatively compact) open subsets whose union is equal to Ω, we see that $H_{loc}^s(\Omega)$ is a Fréchet space. Needless to say, we could have defined $H_{loc}^{p,m}(\Omega)$ in the same fashion and the two definitions agree when $p = 2$ and s is an integer, equal to m.

Many of the basic properties of the spaces $H_c^{p,m}(\Omega)$ and $H_{loc}^{p,m}(\Omega)$, as well as $H_c^s(\Omega)$ and $H_{loc}^s(\Omega)$, will now be stated in the exercises, to be proved by the student. But one property, which is a trivial consequence of

Proposition 31.11 and of its corollary, should be mentioned right now, namely that

$$\mathscr{C}^\infty(\Omega) = \bigcap_{s\in\mathbf{R}} H^s_{\mathrm{loc}}(\Omega), \qquad \mathscr{C}^\infty_c(\Omega) = \bigcap_{s\in\mathbf{R}} H^s_c(\Omega).$$

But the student should be very careful not to think that the last one of these equalities extends to the topologies: it is not true that a subset of $\mathscr{C}^\infty_c(\Omega)$ is a neighborhood of zero in this space only if it is the intersection, with $\mathscr{C}^\infty_c(\Omega)$, of a neighborhood of zero in some space $H^s_c(\Omega)$ (this is true however, when the subscript c is replaced by loc, so that the first equality indeed extends to the topologies). These facts are connected with the "dual" (see Exercise 31.4) equalities

$$(31.6) \qquad\qquad \mathscr{E}'(\Omega) = \bigcup_{s\in\mathbf{R}} H^s_c(\Omega),$$

$$(31.7) \qquad\qquad \mathscr{D}'^F(\Omega) = \bigcup_{s\in\mathbf{R}} H^s_{\mathrm{loc}}(\Omega),$$

where $\mathscr{D}'^F(\Omega)$ denotes the space of distributions of *finite* order in Ω. The validity of (31.6) and (31.7) follows from the fact (Corollary 2 of Theorem 24.5) that a distribution of finite order is a finite sum of derivatives of locally L^2 functions and that multiplication by a test-function does not alter the value of the maximum order of these derivatives. By Fourier transformation, one checks immediately that every distribution with compact support of order $\leqslant m$ belongs to some space $H^s_c(\Omega)$ with s depending only on m and on n (dimension of the surrounding space, $\mathbf{R}^n$).

Exercises

31.1. Let Δ be the Laplace operator in n variables, and λ a number > 0. Show that to every $f \in H^{2,-1}(\Omega)$ there is a unique $u \in H^{2,1}_0(\Omega)$ such that

$$(\Delta - \lambda)u = f$$

and that

$$H^{2,1}(\Omega) = H^{2,1}_0(\Omega) \oplus N_\lambda,$$

where $\oplus$ means Hilbert sum and N_λ is the subspace of $H^{2,1}(\Omega)$ consisting of the solutions of the homogeneous equation $(\Delta - \lambda)u = 0$ (the student must therefore also prove that N_λ is closed).

31.2. Let Δ and λ be as in Exercise 31.1. Prove that to every $f \in H^{2,-1}(\Omega)$ and to every $g \in H^{2,1}(\Omega)$ there is $u \in H^{2,1}(\Omega)$ such that

$$(\Delta - \lambda)u = f, \qquad u - g \in H^{2,1}_0(\Omega).$$

31.3. Prove that $H^{p,m}_{\mathrm{loc}}(\Omega), H^{p,m}_c(\Omega)$ $(1 \leqslant p < +\infty, m = 0, \pm 1,...)$ and $H^s_{\mathrm{loc}}(\Omega)$, $H^s_c(\Omega)$ (s real) are *normal* spaces of distributions.

31.4. Let $1 \leqslant p < +\infty$ and $1 < p' \leqslant +\infty$ be such that $1/p + 1/p' = 1$. Prove that the transpose of the canonical injection $\mathscr{C}^\infty_c(\Omega) \to H^{p,m}_{\mathrm{loc}}(\Omega)$ is a TVS isomorphism of the strong dual of $H^{p,m}_{\mathrm{loc}}(\Omega)$ onto $H^{p',-m}_c(\Omega) \subset \mathscr{D}'(\Omega)$ (the latter H space carries the inductive limit topology of the $H^{p',-m}(K)$, K : compact subset of Ω). Prove that $H^s_c(\Omega)$ and $H^{-s}_{\mathrm{loc}}(\Omega)$ are (in a natural way—to be made explicit) the dual of each other.

31.5. Motivate the assertion that the topology of $\mathscr{C}^\infty_c(\Omega)$ is *not* the intersection of the topologies of the spaces $H^s_c(\Omega)$, $s \in \mathbf{R}$.

31.6. Prove that the topology of $\mathscr{D}'(\Omega)$ is the least-fine locally convex topology such that all the restriction mappings $T \rightsquigarrow T \mid \Omega'$ from $\mathscr{D}'(\Omega)$ into $H^{-\infty}_{\mathrm{loc}}(\Omega')$ are continuous as Ω' ranges over the family of all relatively compact open subsets of Ω ($H^{-\infty}_{\mathrm{loc}}(\Omega')$ is the vector space $\bigcup_{s \in \mathbf{R}} H^s_{\mathrm{loc}}(\Omega)$ equipped with the finest locally convex topology such that the injections $H^s_{\mathrm{loc}}(\Omega') \to H^{-\infty}_{\mathrm{loc}}(\Omega')$ are continuous).

31.7. Let $\phi \in \mathscr{C}^\infty_c(\Omega)$. Prove that $u \rightsquigarrow \phi u$ is a continuous linear map of $H^{p,m}(\Omega)$ into $H^{p,m}_0(\Omega)$ $(1 \leqslant p \leqslant +\infty; m = 0, \pm 1, \pm 2,...)$.

31.8. Prove that $(\phi, u) \rightsquigarrow \phi u$ is a continuous bilinear map of $\mathscr{S} \times H^s$ into H^s (Hint: use the corollary of Theorem 34.1).

32

Equicontinuous Sets of Linear Mappings

In the previous chapters, we have presented the theory of distributions in the form given to it by L. Schwartz. Most of the spaces of distributions have been introduced as duals of suitable spaces of $\mathscr{C}^\infty$ functions. The basic operations on distributions (differentiation, multiplication by functions, Fourier transformation, convolution) were systematically introduced as transposes of similar operations defined in classes of $\mathscr{C}^\infty$ functions. What we were required to know about topological vector spaces is remarkably little. Practically the only theorem we have made use of is the Hahn–Banach theorem (Chapter 18). The only concepts we have been manipulating are the dual of a locally convex Hausdorff space and the transpose of a continuous linear map. But now the time has come to investigate in greater detail the structure of the spaces of functions and distributions which have been introduced, and, in order to do that, we must go back to abstract functional analysis. We shall now direct our efforts toward one of the most important results on topological vector spaces, the Banach–Steinhaus theorem. This theorem generalizes a property of Hilbert spaces, which goes back to the early part of the century and has long been known as *Osgood's theorem* or, sometimes, as *the principle of uniform boundedness*, and which states that every weakly converging sequence in a Hilbert space is strongly bounded, i.e., is bounded in norm. The generalization which we shall present is due to N. Bourbaki; it has an optimal range of application, as far as our purposes are concerned. It deals with the space $L(E; F)$ of all continuous linear mappings of a Hausdorff TVS E into another one, F (eventually, E will have to be *barreled* and F locally convex). We shall first provide $L(E; F)$ with various locally convex topologies, in straight generalization of what we have done in Chapter 19, when we have defined the various dual topologies. As we did there, we begin by considering a family $\mathfrak{S}$ of *bounded* subsets of E, satisfying the following two conditions:

$(\mathfrak{S}_\mathrm{I})$ *If $A, B \in \mathfrak{S}$, there is $C \in \mathfrak{S}$ such that $A \cup B \subset C$.*

$(\mathfrak{S}_\mathrm{II})$ *If $\lambda \in \mathbf{C}$ and $A \in \mathfrak{S}$, there is $B \in \mathfrak{S}$ such that $\lambda A \subset B$.*

Consider, then, a bounded subset B of E and a neighborhood of zero, V, in F; this pair of sets defines the following subset of $L(E; F)$:

$$\mathscr{U}(B; V) = \{u \in L(E; F); u(B) \subset V\}.$$

PROPOSITION 32.1. *The subset* $\mathscr{U}(B; V)$ *of* $L(E; F)$ *is absorbing; it is convex (resp. balanced) if* V *is convex (resp. balanced).*

Proof. The part of the statement about convex and balanced sets is obvious. Let u be an arbitrary continuous linear map of E into F; we know that the image of B under u, $u(B)$, is bounded in F (Proposition 14.2), therefore there is a number $\lambda \neq 0$ such that $u(B) \subset \lambda V$, which means that

$$\lambda^{-1}u \in \mathscr{U}(B; V) \quad \text{or} \quad u \in \lambda\, \mathscr{U}(B; V). \qquad \text{Q.E.D.}$$

When B varies over a family $\mathfrak{S}$ of bounded subsets of E satisfying $(\mathfrak{S}_\mathrm{I})$ and $(\mathfrak{S}_\mathrm{II})$, while V varies over a basis of neighborhoods of zero in F, the sets $\mathscr{U}(B; V)$ form a basis of filter in $L(E; F)$. The filter they generate is the filter of neighborhoods of zero in a topology on $L(E; F)$ which is compatible with the linear structure of $L(E; F)$. In order to see this, it suffices to apply Theorem 3.1 and Proposition 32.1, combined with the remark that, if $V \subset W \subset F$, then $\mathscr{U}(B; V) \subset \mathscr{U}(B; W)$.

Definition 32.1. *Given a family of bounded subsets of* E, $\mathfrak{S}$, *satisfying Conditions* $(\mathfrak{S}_\mathrm{I})$, $(\mathfrak{S}_\mathrm{II})$, *we shall call* $\mathfrak{S}$-*topology on* $L(E; F)$ *the topology on this vector space defined by the basis of neighborhoods of zero consisting of the sets*

$$\mathscr{U}(B; V) = \{u \in L(E; F); u(B) \subset V\},$$

when B *runs over* $\mathfrak{S}$ *and* V *over an arbitrary basis of neighborhoods of* F. *When carrying the* $\mathfrak{S}$-*topology, the space* $L(E; F)$ *will be denoted by*

$$L_\mathfrak{S}(E; F).$$

The next statement generalizes results in Chapter 19 (in particular, Proposition 19.1):

PROPOSITION 32.2. *If* F *is a locally convex TVS, so is* $L_\mathfrak{S}(E; F)$. *If* F *is Hausdorff and if the union of the sets belonging to* $\mathfrak{S}$ *is dense in* E, *then* $L_\mathfrak{S}(E; F)$ *is Hausdorff.*

Proof. The first part follows immediately from Proposition 32.1. As for the last part, let $u \in L(E; F)$ be nonzero. This means that there is an element x in some set B belonging to $\mathfrak{S}$ such that $u(x) \neq 0$.[†] Since F is Hausdorff, there is a neighborhood V of zero in F such that $u(x) \notin V$. Hence we have $u(B) \not\subset V$, i.e., $u \notin \mathscr{U}(B; V)$.

If one prefers to visualize a topology in terms of convergence (rather than in terms of neighborhoods of zero), the following can be said: To any filter $\mathscr{F}$ on $L(E; F)$ and to any point x of E, we can associate the filter $\mathscr{F}(x)$ in F generated by the basis of filter consisting of the sets

$$M(x) = \{u(x) \in F; u \in M\},$$

where M varies over $\mathscr{F}$. Then the filter $\mathscr{F}$ converges in the $\mathfrak{S}$-topology on $L(E; F)$ if the filters $\mathscr{F}(x)$ converge in F, uniformly on every subset of E belonging to the family $\mathfrak{S}$. For instance, suppose that both E and F are normed spaces and that $\mathfrak{S}$ is the family of all balls of finite radius centered at the origin in E; then, a sequence of continuous linear maps $u_n : E \to F$ converges to zero in this $\mathfrak{S}$-topology on $L(E; F)$ if and only if the norms of the u_n (cf. Part I, Chapter 11, p. 107) converge to zero. Another important example is the case of $\mathfrak{S} =$ the family of all finite subsets of E; then $\mathscr{F}$ converges to zero in the $\mathfrak{S}$-topology if and only if $\mathscr{F}(x)$ converges to zero in F for each single point x of E.

The families of bounded subsets of E in which we are interested are the same as in Chapter 19:

(1) The family of all *finite* subsets of E; the corresponding $\mathfrak{S}$-topology is called *the topology of pointwise convergence*; when carrying it, the space $L(E; F)$ will be denoted by $L_\sigma(E; F)$ (many authors write $L_s(E; F)$ instead, but we shall make the notation consistent with the one used in the duality case, when F is the complex field $\mathbf{C}$: then, of course, $L(E; F) = E'$ and E'_σ is precisely the weak dual of E).

(2) The family of all *convex compact* subsets of E (the $\mathfrak{S}$-topology is then the *topology of convex compact convergence*); equipped with it, $L(E; F)$ will be denoted by $L_\gamma(E; F)$.

(3) The family of all *compact* subsets of E, which leads to *the topology of compact convergence*; equipped with it, $L(E; F)$ will be denoted by $L_c(E; F)$.

(4) The family of all *bounded* subsets of E, leading to *the topology of bounded convergence*; thus topologized, $L(E; F)$ will be denoted by $L_b(E; F)$. In agreement with the definitions of Chapter 19, $L_b(E; \mathbf{C}) = E'_b$, strong dual of E.

[†] Right at this point, one sees that we needed not assume the union of the sets in $\mathfrak{S}$ to be dense but only *total*, i.e., to span a dense linear subspace.

If we call σ, γ, c, b the topologies above, we see that each one is locally convex Hausdorff whenever this is true of F; we have the following comparison relations:

$$\sigma \leqslant \gamma \leqslant c \leqslant b.$$

Observe that this implies that a subset of $L(E; F)$ which is bounded for one of these topologies is also bounded for the weaker ones. In particular, any b-bounded set is σ-bounded. For the time being, we shall center our attention on the two extreme topologies, σ and b. The result which we are seeking concerns σ-bounded and b-bounded subsets of $L(E; F)$, as well as equicontinuous sets of linear maps, which we must now introduce. We have already defined equicontinuous sets of mappings of a topological space X into a TVS F (Definition 14.3). In the particular case of *linear* mappings, the definition is simplified (cf. the fact that a linear map is continuous everywhere if and only if it is continuous at the origin):

Definition 32.2. Let E, F be two TVS. *A set H of linear maps of E into F is said to be equicontinuous if, to every neighborhood of zero V in F, there is a neighborhood of zero U in E such that, for all mappings $u \in H$,*

$$x \in U \quad implies \quad u(x) \in V.$$

The condition in Definition 32.2 can be rewritten in a variety of ways: H is equicontinuous if, to every neighborhood of zero V in F, there is a neighborhood of zero U in E such that

$$(32.1) \qquad H(U) = \bigcup_{u \in H} u(U) \subset V,$$

or, equivalently, if, given any neighborhood of zero V in F,

$$(32.2) \qquad H^{-1}(V) = \bigcap_{u \in H} u^{-1}(V)$$

is a neighborhood of zero in E.

We are now going to state and prove a few simple properties of equicontinuous sets of linear maps. It is obvious that a subset of an equicontinuous set is equicontinuous. Moreover:

PROPOSITION 32.3. *Let E, F be two* TVS; *suppose that F is locally convex. Then the balanced convex hull of an equicontinuous subset of $L(E; F)$ is equicontinuous.*

Proof. Let $H \subset L(E; F)$ be equicontinuous, and K the balanced convex hull of H. Let V be an arbitrary neighborhood of zero in F; let W be another neighborhood of zero in F which is balanced, convex, and contained in V. By hypothesis, there is a neighborhood U of zero in E such that $H(U) \subset W$; hence the balanced convex hull of $H(U)$, which is obviously $K(U)$, is contained in the balanced convex hull of W, which is equal to W and therefore contained in V.   Q.E.D.

Let us denote by F^E the product set $\prod_{x \in E} F_x$, where F_x is a copy of F for every $x \in E$; F^E is then equipped with the structure of product topological vector space. A basis of neighborhoods of zero in F^E consists of the sets

$$\prod_{x \in E} V_x,$$

where V_x is a neighborhood of zero in F for every x and where $V_x = F$ for all $x \in E$ except possibly a finite number of them. The vector space F^E can be identified with the vector space of all functions (linear or nonlinear) defined in E and taking their values in F, $\mathscr{F}(E; F)$. The canonical isomorphism is the mapping

$$\mathscr{F}(E; F) \ni f \rightsquigarrow (f(x))_{x \in E} \in \prod_{x \in E} F_x.$$

In other words, this isomorphism assigns to a function $f \in \mathscr{F}(E; F)$ the element of F^E whose projection on the "axis" F_x is the value $f(x)$ of f at x, this for every $x \in E$.

The canonical isomorphism $\mathscr{F}(E; F) \cong F^E$ extends to the topologies (i.e., becomes an isomorphism for the structures of topological vector spaces) if we provide $\mathscr{F}(E; F)$ with the topology of pointwise convergence and F^E with the product topology. In particular, $L_\sigma(E; F)$ can be regarded as a linear subspace of F^E: the σ-topology on $L(E; F)$ is exactly the topology induced by F^E.

It is not difficult to see that $L(E; F)$ is *not* closed in $\mathscr{F}(E; F)$, in general. In connection with this, we propose the following exercises to the student:

Exercises

32.1. Let us denote by $\mathscr{L}(E; F)$ the linear subspace of $\mathscr{F}(E; F)$ consisting of all linear maps of E into F, continuous or not. Prove that $\mathscr{L}(E; F)$ is closed in $\mathscr{F}(E; F)$ (for the topology of pointwise convergence).

32.2. Let E, F be locally convex TVS and E be Hausdorff. Prove that $L(E; F)$ is dense in $\mathscr{L}(E; F)$ for the topology of pointwise convergence.

32.3. Prove that $L(E; F) = \mathscr{L}(E; F)$ if and only if E is finite dimensional and Hausdorff (cf. Theorem 9.1).

32.4. What is the dual of $\mathscr{L}(E; F)$?
Using this dual and the Hahn–Banach theorem, give a second proof that $L(E; F)$ is dense in $\mathscr{L}(E; F)$ when E and F are locally convex and E is Hausdorff.

In general, the closure in $\mathscr{F}(E; F)$ of a subset A of $L(E; F)$ will not be contained in $L(E; F)$, i.e., will contain mappings which are linear but not continuous. This is not, however, the case, when we deal with an equicontinuous set of linear maps.

PROPOSITION 32.4.   *The closure of an equicontinuous subset H of $L(E; F)$ in the space F^E is an equicontinuous set of linear maps.*

Proof.   Let $\bar{H}$ be the closure of H in F^E; $\bar{H}$ is a set of linear maps (Exercise 32.1). Let V be an arbitrary neighborhood of zero in F, W another neighborhood of zero in F such that $W + W \subset V$. There is a neighborhood of zero, U, in E, such that $H(U) \subset W$. Select now an arbitrary point x of U; to every $v \in \bar{H}$ there is an element $u \in H$ such that

$$u(x) \in v(x) - W;$$

this follows simply from the fact that every neighborhood of v in F^E intersects H. It can be rewritten as $v(x) \in u(x) + W$. But $u(U) \subset W$, hence $v(x) \in W + W \subset V$. As x is an arbitrary point of U, we have $v(U) \subset V$; as v is an arbitrary element of $\bar{H}$, this proves the result.

PROPOSITION 32.5.   *On an equicontinuous set H of linear maps of E into F the following topologies coincide:*

  the topology of pointwise convergence in a dense subset A of E;
  the topology of pointwise convergence in E;
  the topology of compact convergence.

Proof.   It suffices to prove the identity, on H, of the first topology with the third one. Let K be an arbitrary compact subset of E, and V an arbitrary neighborhood of zero in F. Let W be another neighborhood of zero in F such that $W = -W$ and that $W + W + W \subset V$. Let us select a neighborhood of zero U in E such that $U = -U$ and that $H(U) \subset W$.

Since A is dense in E, the sets $U + y$ form a covering of E as y runs over A. Therefore there is a finite number of points $y_1, ..., y_r$ in A such that

$$K \subset (U + y_1) \cup \cdots \cup (U + y_r).$$

Let u_0 be an arbitrary element of H, and u an arbitrary element of

(32.3) $u_0 + \mathscr{U}(\{y_1, \ldots, y_r\}; W)$. .

If x is an arbitrary point of K, there is an index j such that $x \in U + y_j$ ($1 \leqslant j \leqslant r$). Then

$$u(x) - u_0(x) = [u(x) - u(y_j)] + [u(y_j) - u_0(y_j)] + [u_0(y_j) - u_0(x)].$$

In the right-hand side, the first and the last brackets belong to W since $x - y_j \in -U = U$ and $H(U) \subset W$; the middle bracket belongs to W since u is in (32.3), hence $u(x) - u_0(x) \in W + W + W \subset V$. As x is arbitrary in K, this means that $u \in u_0 + \mathscr{U}(K; V)$; as u is arbitrary in (32.3), this proves the desired result.

Remark 32.1. One notices, in the preceding proof, that the set K need not be compact, but only *precompact* (Definition 6.3): this means that the topology of compact convergence could have been replaced, in the statement of Proposition 32.5, by *the topology of precompact convergence*.

PROPOSITION 32.6.   *An equicontinuous set H of linear maps of E into F is bounded for the topology of bounded convergence.*

Proof. We must show that H can be absorbed (or swallowed) by an arbitrary neighborhood of zero in $L_b(E; F)$. We may assume that this neighborhood of zero is of the form $\mathscr{U}(B; V)$, where B is an arbitrary bounded subset of E and V, an arbitrary neighborhood of zero in F. Since H is equicontinuous, there is a neighborhood of zero, U, in E, such that $H(U) \subset V$. On the other hand, there is a number $\lambda \neq 0$ such that $B \subset \lambda U$, whence

$$\lambda^{-1} H(B) \subset H(U) \subset V,$$

i.e.,

$$\lambda^{-1} H \subset \mathscr{U}(B, V) \quad \text{or} \quad H \subset \lambda \mathscr{U}(B, V). \qquad \text{Q.E.D.}$$

In the remainder of this chapter, we center our attention on equicontinuous sets of linear forms. In other words, F will now be the complex field, $\mathbf{C}$.

PROPOSITION 32.7.   *A set of continuous linear functionals on a TVS E is equicontinuous if and only if it is contained in the polar of some neighborhood of zero in E.*

For the notion of polar, see Definition 19.1.

Proof. Let H be an equicontinuous set of linear forms on E. Let us denote by D_ρ the closed disk of the complex plane centered at the origin and having a radius equal to ρ. There is a neighborhood of zero, U, in E, such that $H(U) \subset D_1$, which means exactly that $H \subset U^0$.

Let U, now, be an arbitrary neighborhood of zero in E; let us set $H = U^0$. For every $\rho > 0$, we have $H(\rho U) \subset D_\rho$. Therefore H, and a fortiori any subset of H, is equicontinuous.

PROPOSITION 32.8.   *The weak closure of an equicontinuous set of linear forms on E is a compact subset of E'_σ.*

Proof. Let H be an equicontinuous subset of E'. Note that the closure of H in E'_σ is identical to its closure in the product space

$$\mathbf{C}^E = \prod_{x \in E} \mathbf{C}_x \qquad (\mathbf{C}_x \cong \mathbf{C}; \text{ see Proposition 32.4}).$$

We may assume that H is equal to this closure. For each $x \in E$, let us denote by $H(x)$ the set of complex numbers $\langle x', x \rangle$ as x' runs over H; this set is closed. Moreover, it is canonically isomorphic to the coordinate projection of H into the "axis" $\mathbf{C}_x$, in the product space $\mathbf{C}^E$. On the other hand, $H(x)$ is a bounded subset of $\mathbf{C}$ for all x. Indeed, this would mean that H is weakly bounded; but we know that H is more than that: H is strongly bounded, i.e., bounded in E'_b (Proposition 32.6). Thus we see that $H(x)$ is a *compact* subset of the complex plane $\mathbf{C}$ for all $x \in E$. At this stage, we use *Tychonoff's theorem* (see Exercise 6.1), which asserts the following: *let $\{X_\alpha\}(\alpha \in A)$ be an arbitrary family of compact topological spaces; then the product topological space $\prod_{\alpha \in A} X_\alpha$ is also compact.* Applying this in our situation, we see that

$$\prod_{x \in E} H(x)$$

is a compact subset of $\mathbf{C}^E$. As H is a closed subset of it, we conclude that H is compact.   Q.E.D.

Exercises

32.5.   Let E be a normed space. Prove that, in the dual E' of E, there is identity between equicontinuous sets and strongly bounded sets.

32.6.   Let us denote by $\tau_a(a \in \mathbf{R}^n)$ the translation mapping $f(x) \rightsquigarrow f(x - a)$. By using Proposition 32.5, prove the following result:

THEOREM 32.1.   *Let p be a real number such that $1 \leqslant p < +\infty$. When the vector $a \in \mathbf{R}^n$*

converges to zero in $\mathbf{R}^n$, *the translation mappings* τ_a *converge to the identity in*

$$L_\sigma(L^p(\mathbf{R}^n); L^p(\mathbf{R}^n)).$$

What can you say when $p = +\infty$?

32.7. Let $\{\mu_k\}$ ($k = 0, 1,\ldots$) be a sequence of Radon measures in $\mathbf{R}^n$ having their support contained in a fixed compact subset K of $\mathbf{R}^n$. Suppose furthermore that there is a constant $C > 0$ such that, for all $k \geqslant 0$ and all bounded continuous functions f in $\mathbf{R}^n$,

$$|\langle \mu_k, f\rangle| \leqslant C \sup_{x \in \mathbf{R}^n} |f(x)|.$$

By using Proposition 32.5, prove the following result; if the sequence $\{\mu_k\}$ converges weakly to zero in $\mathscr{D}'(\mathbf{R}^n)$, the Fourier transforms $\hat{\mu}_k$ converge to zero uniformly on every compact subset of $\mathbf{R}^n$.

32.8. Let E be a separable TVS (Exercise 12.7), A a dense countable subset of E, F a metrizable TVS. Prove that the topology of pointwise convergence in A turns $L(E;F)$ into a metrizable TVS (Chapter 8). Derive from this that, on every equicontinuous set of linear mappings $E \to F$, the topology of pointwise convergence in E can be defined by a metric.

32.9. Let E be a separable TVS. Prove that every weakly closed equicontinuous set of linear functionals in E is, when equipped with the weak topology, a metrizable compact space (use Exercise 32.8). Prove that every equicontinuous infinite sequence of linear functionals in E contains a subsequence which converges weakly in E' (use Exercise 8.7).

32.10. Let E be a separable normed space. Prove that every infinite sequence $\{x'_k\}$ ($k = 1, 2,\ldots$) in E' such that the norms $\{x'_k\}$ are bounded independently of k, contains a subsequence which converges weakly in E' (use Exercise 32.9).

32.11. Let E be a Hilbert space. Prove that every infinite sequence $\{x_k\}$ ($k = 1, 2,\ldots$) in E such that the norms $\| x_k \|$ are bounded independently of k, contains a subsequence $\{x_{k_j}\}$ with the following property: there is $x^0 \in E$ such that, for all $x \in E$, $(x^0 \mid x) = \lim_{j \to +\infty} (x_{k_j} \mid x)$. (*Hint*: Show that E may be assumed separable and use Exercise 32.10.)

For want of a better place, we present, as an appendix to Chapter 32, a result on completeness of spaces $L_\mathfrak{S}(E; F)$. This result is extremely simple to prove; it is also very important in so far as it implies the completeness of many spaces of continuous linear mappings in distribution theory, and in particular the completeness of the most important duals occurring in that theory.

THEOREM 32.2. *Let E, F be two locally convex Hausdorff spaces. Suppose that F is complete. Let $\mathfrak{S}$ be a family of bounded subsets of E, satisfying* $(\mathfrak{S}_\mathrm{I})$ *and* $(\mathfrak{S}_\mathrm{II})$ *and forming a covering of E. Suppose that the following condition holds:*

(*) *If the restriction of a linear map $u : E \to F$ to every set $A \in \mathfrak{S}$ is continuous, then u is continuous.*

Under these circumstances, the TVS $L_\mathfrak{S}(E; F)$ is complete.

Proof. Let $\mathscr{F}$ be a Cauchy filter in $L_{\mathfrak{S}}(E; F)$; $\mathscr{F}$ is a fortiori a Cauchy filter for the topology of pointwise convergence. As F is complete, $\mathscr{F}$ converges pointwise to a function $u : E \to F$. The function u is linear because the conditions defining linearity involve only a finite number of points. Now the filter $\mathscr{F}$ converges to u *uniformly* on every set $A \in \mathfrak{S}$. Indeed, let V, W be two neighborhoods of zero in F such that $V + V \subset W$. As $\mathscr{F}$ is a Cauchy filter in the $\mathfrak{S}$-topology there is $M \in \mathscr{F}$ such that, for all $f, g \in M$ and all $x \in A$,

$$f(x) - g(x) \in V.$$

On the other hand, given any point x_0 of A, there is $M_1 \in \mathscr{F}$ such that $h \in M_1$ implies $h_1(x_0) \in V + u(x_0)$. But $M \cap M_1 \neq \varnothing$; taking $g \in M \cap M_1$, we see that, for all $f \in M$,

$$f(x_0) \in V + V + u(x_0) \subset U + u(x_0).$$

As $x_0 \in A$ is arbitrary, this means that $M - u \in \mathscr{U}(A; U)$, whence our assertion. As u is the uniform limit of a filter of continuous functions on A, u, restricted to A, is continuous. By (*), u is continuous. We have already shown that $\mathscr{F}$ converges to u in the $\mathfrak{S}$-topology.   Q.E.D.

COROLLARY 1. *Let E be a locally convex Hausdorff space such that a linear mapping of E into a locally convex space which is bounded on every bounded set is continuous. Then for all complete locally convex Hausdorff spaces $F, L_b(E; F)$ is complete; in particular, E'_b is complete.*

Proof. It suffices to show that Property (*) holds when $\mathfrak{S}$ is the family of all bounded subsets of E. Let u be a linear map whose restriction to every bounded set is continuous. Then u transforms bounded sets into bounded sets (we leave the verification of this statement to the student), hence is continuous in view of our hypothesis.

The class of locally convex spaces having the property assumed in Corollary 1 is very important (it is called *bornological*). We know (Proposition 14.8 and Corollary) that metrizable spaces and *LF*-spaces are bornological. Thus:

COROLLARY 2. *Let E be either a metrizable space or a space LF, and F any complete locally convex Hausdorff space. Then $L_b(E; F)$ is complete. In particular, E'_b is complete.*

COROLLARY 3. *The spaces $\mathscr{D}'$, $\mathscr{E}'$, $\mathscr{S}'$ (with their strong dual topology) are complete.*

COROLLARY 4. *Let E be either a Fréchet space or an LF-space. For any complete locally convex Hausdorff space F, $L_c(E; F)$ is complete.*

Proof of Corollary 4. We will show that Condition (*) is satisfied. It suffices to show that, if the restriction of a linear map u to every convex compact subset K of E is continuous, then u is sequentially continuous (for then we may apply Propositions 8.5 and 14.7). But if S is a sequence in E, which converges to an element x_0, the set $S \cup \{x_0\}$ is compact, its closed convex hull K is compact (Corollary of Proposition 7.11); as the restriction of u to it is continuous, $u(S)$ converges to $u(x_0)$.

Q.E.D.

An immediate corollary of Theorem 32.2 is Theorem 11.5, which we have already proved directly. Theorem 11.5 states that, if E is a normed space and F a Banach space, $L(E; F)$ is a Banach space when equipped with the operator's norm

$$L(E; F) \ni u \rightsquigarrow \| u \| = \sup_{x \in E, \, \| u \| = 1} \| u(x) \|,$$

where we have denoted by $\| \; \|$ the norm both in E and F. In particular, the strong dual of a normed space is a Banach space (Corollary of Theorem 11.5).

33

Barreled Spaces.
The Banach-Steinhaus Theorem

We recall (Definition 7.1) that a subset T of a topological vector space E is called a *barrel* if T is absorbing (Definition 3.1), convex, balanced (Definition 3.2), and closed.

Definition 33.1. *A topological vector space E is said to be barreled if every barrel in E is a neighborhood of zero in E.*

A barreled space need not be locally convex; it will be locally convex if and only if it has a basis of neighborhoods of zero consisting of barrels. Many authors include local convexity in the definition of barreled spaces.

PROPOSITION 33.1. *Let E be a barreled space, and M a linear subspace of E. The quotient space E/M is barreled.*

Proof. Let $\dot{T}$ be a barrel in E/M; its preimage under the canonical homomorphism π of E onto E/M is a barrel, T, hence a neighborhood of zero in E. But the image under π of a neighborhood of zero in E is a neighborhood of zero in E/M, and $\dot{T} = \pi(T)$.

A linear subspace of a barreled space need *not* be barreled. One can show that *a product of barreled spaces is barreled.*

PROPOSITION 33.2. *A TVS which is a Baire space is barreled.*

We recall (cf. Proposition 8.3) that a Baire space is a space having the property:

(B) *The union of any countable family of closed sets, none of which has interior points, has no interior points.*

Proof of Proposition 33.2. Let E be a Baire TVS, and T a barrel in E. Since T is absorbing and balanced, we have $E = \bigcup_{k=0}^{+\infty} kT$. Since every set kT is closed, at least one of them must have a nonempty interior;

since $x \rightsquigarrow k^{-1}x$ is a homeomorphism, T itself must have at least one interior point, x_0. If $x_0 = 0$, T is a neighborhood of zero. If $x_0 \neq 0$, $-x_0$ must also be an interior point of T. But the interior of a convex set, in any TVS, is a convex set (Proposition 7.1), therefore all the points of the segment joining $-x_0$ to x_0, in particular the origin, must be interior points. Q.E.D.

COROLLARY 1. *Fréchet spaces are barreled.*

COROLLARY 2. *Banach spaces and Hilbert spaces are barreled.*

COROLLARY 3. *LF-spaces are barreled.*

It suffices to prove Corollary 3, the first two being evident. Let E be a strict inductive limit of a sequence $\{F_k\}$ ($k = 0, 1,...$) of Fréchet spaces (see Chapter 13), and T a barrel in E. The intersection $T \cap F_k$ is obviously a barrel in F_k for every $k = 0, 1,...$; hence $T \cap F_k$ is a neighborhood of zero in F_k. Since T is convex, this implies that T is a neighborhood of zero in E.

The importance of barreled spaces stems mainly from the following result:

THEOREM 33.1. *Let E be a barreled TVS, and F a locally convex space. The following properties of a subset H of the space $L(E;F)$ of continuous linear maps of E into F are equivalent:*

(a) *H is bounded for the topology of pointwise convergence;*
(b) *H is bounded for the topology of bounded convergence;*
(c) *H is equicontinuous.*

Proof. That (b) $\Rightarrow$ (a) is trivial; that (c) $\Rightarrow$ (b) has been proved already (Proposition 32.6). Note that both these implications are true in general, whether E is barreled or not. Theorem 33.1 will be proved if we prove that (a) $\Rightarrow$ (c). This is where we use the fact that E is barreled. Let H be a bounded subset of $L_\sigma(E;F)$. We must show that, if V is an arbitrary neighborhood of zero in F, $H^{-1}(V)$ is a neighborhood of zero in E. As F is locally convex, we may take V to be a barrel (Proposition 7.2). Then, if u is any continuous linear map of E into F, $u^{-1}(V)$ is obviously a barrel. As we have

$$H^{-1}(V) = \bigcap_{u \in H} u^{-1}(V),$$

we see immediately that $H^{-1}(V)$ is convex, balanced, and closed. The fact that H is bounded for the topology of pointwise convergence will imply that $H^{-1}(V)$ is absorbing, hence that $H^{-1}(V)$ is a barrel and therefore a neighborhood of zero in E, which will complete the proof.

Let x be an arbitrary element of E: the set $H(x)$ consisting of the elements $u(x)$ of F as u varies over H is bounded. Therefore, there is a number $\lambda > 0$ such that $H(x) \subset \lambda V$. This means that $x \in \lambda H^{-1}(V)$.

Q.E.D.

The corollary which follows is often referred to as *the Banach–Steinhaus theorem;* we shall, however, use this name for the Theorem 33.1 itself.

COROLLARY. *Let E be a barreled space, F a locally convex Hausdorff space, and $\mathscr{F}$ a filter on $L(E; F)$ which converges pointwise in E to a linear map u_0 of E into F. Suppose that $\mathscr{F}$ has either one of the following two properties:*

(33.1)   *There is a set H, belonging to $\mathscr{F}$, which is bounded for the topology of pointwise convergence.*

(33.2)   *$\mathscr{F}$ has a countable basis.*

Then u_0 is a continuous linear map of E into F and $\mathscr{F}$ converges to u_0 in $L_c(E; F)$ (i.e., uniformly on every compact subset of E).

Proof. Suppose that (33.1) holds. Then H is an equicontinuous set and u_0 belongs to the closure of H in F^E, $\bar{H}$. But $\bar{H}$ is an equicontinuous set of linear maps of E into F (Proposition 32.4), hence u_0 is continuous and $\mathscr{F}$ converges to u_0 in $L_\sigma(E; F)$. In view of Proposition 32.5, $\mathscr{F}$ converges to u_0 in $L_c(E; F)$ (as $H \in \mathscr{F}$, to say that $\mathscr{F}$ converges to u_0 in $L_c(E; F)$ or that the filter induced by $\mathscr{F}$ on $\bar{H}$ converges to u_0 in $\bar{H}$ when this set carries the topology of compact convergence, is one and the same thing).

Next we suppose that (33.2) holds. Let $\{M_1, M_2, ...\}$ be a countable basis of $\mathscr{F}$. For each $k = 1, 2, ...$, we select an element u_k of M_k. By hypothesis, for each $x \in E$, the sequence $\{u_k(x)\}$ converges in F (to $u_0(x)$, of course). This implies that the set of continuous mappings $\{u_k\}$ is bounded in $L_\sigma(E; F)$.

Therefore, the filter associated with that sequence has Property (33.1). From the first part of the proof, it follows that u_0 is continuous and that the u_k converge to u_0 in $L_c(E; F)$. Let, then, $\mathscr{U}$ be a neighborhood of u_0 in $L_c(E; F)$: suppose that none of the sets M_k is contained in $\mathscr{U}$. Then we could find, for each k, an element u_k of M_k which is not contained in $\mathscr{U}$. But this would contradict the fact that any such sequence $\{u_k\}$ converges to u_0 in $L_c(E; F)$. Therefore, some set M_k must be contained in $\mathscr{U}$.   Q.E.D.

When F is the complex field, Theorem 33.1 can be given a stronger version:

THEOREM 33.2. *Let E be a barreled TVS. The following properties of a subset H of the dual E' of E are equivalent:*

(a) *H is weakly bounded;*
(b) *H is strongly bounded;*
(c) *H is equicontinuous;*
(d) *H is relatively compact in the weak dual topology.*

Proof of Theorem 33.2. The equivalence of (a), (b), and (c) follows from Theorem 33.1. On the other hand, (d) implies trivially (a). It suffices to show that (c) implies (d): but this has already been stated and proved (Proposition 32.8).

As a first application of Theorems 33.1 and 33.2, we shall give an example of a normed space which is not barreled.

Example 33.1. Let $\mathscr{C}_c^0$ be the space of continuous functions with compact support in the real line $\mathbf{R}^1$. We equip it with the maximum norm: $\|f\| = \sup_{t \in \mathbf{R}^1} |f(t)|$. Note that the normed space thus obtained, which we shall denote by E, is not a Banach space: indeed, it is not complete: the completion of E can be identified with the space of continuous functions decaying at infinity, in $\mathbf{R}^1$. Consider then the sequence of continuous linear functionals on E,

$$x_k' : f \rightsquigarrow k f(k), \qquad k = 1, 2, \ldots.$$

Since every $f \in E$ is a function with compact support, $f(k) = 0$ for sufficiently large k. This means that the sequence $\{x_k'\}$ converges weakly to zero. But it is obvious that this set of continuous linear forms is *not* strongly bounded; in fact, we have $\|x_k'\| = k$ (here, $\|\ \|$ is the norm in the strong dual of E). Thus E cannot be barreled, otherwise we would have found a fact contradicting Theorem 33.2.

Remark 33.1. The normed space E in Example 33.1 is not a Baire space, in view of Proposition 33.2.

Exercises

33.1. Give an example of a normed space of sequences which is not barreled.

33.2. Let E be the Banach space of continuous functions (with complex values) in the closed interval $[0, 1]$, provided with the norm $\|f\| = \sup_{0 \leqslant t \leqslant 1} |f(t)|$. Let $\sigma = (\sigma_n)$ be a sequence of complex numbers such that, for *all* $f \in E$,

$$\sum_{n=0}^{+\infty} |\sigma_n c_n(f)| < +\infty,$$

where $c_n(f)$ is the nth *Fourier coefficient of f*:

$$c_n(f) = \int_0^1 e^{2i\pi nt} f(t) \, dt.$$

By applying the Banach–Steinhaus theorem, prove that $f \leadsto \sum_{n=0}^{+\infty} \sigma_n c_n(f)$ is a continuous linear functional on E.

33.3.   Let E be a Baire TVS, and F an arbitrary TVS. Prove that every subset of $L(E;F)$ which is bounded for the topology of pointwise convergence is equicontinuous.

33.4.   Let E be a locally convex Hausdorff space, and E' its dual. Prove that, if every weakly bounded subset of E' is equicontinuous, then E is barreled.

33.5.   Let E be a Fréchet space, and E' its dual. Show that there is a countable basis of bounded sets in E' (weakly or strongly bounded, as they are identical).

34

Applications of
the Banach-Steinhaus Theorem

34.1. Application to Hilbert Spaces

Let E be a Hilbert space, $\bar{E}'$ its anti-dual (see Chapter 12, p. 116), and J the canonical isometry of E onto $\bar{E}'$. The complex conjugate of an antilinear form is a linear form, and complex conjugation is therefore a one-to-one antilinear mapping of $\bar{E}'$ onto the dual of E, E'. Composing this mapping with the isometry J, we obtain the canonical antilinear isometry of E onto its dual; we denote it by $\bar{J}$. The weak topology on E is then the topology carried over from E'_σ via the mapping $\bar{J}$. For instance, a sequence $\{x_k\}$ of elements of E converges to x weakly in E if, for every $y \in E$, the inner products $(x_k \mid y)$ converge to $(x \mid y)$ (we have denoted by $(\mid)$ the sesquilinear form on E defining the Hilbert structure of E; the norm in E will be denoted by $\|\ \|$; we have $\|x\|^2 = (x \mid x)$).

PROPOSITION 34.1.   *The unit ball of the Hilbert space E, $\{x \in E; \|x\| \leqslant 1\}$, is weakly compact.*

Proof.   We begin by proving that the unit ball of E is *weakly closed*. Let $x \in E$ belong to its closure, and let $\{x_k\}$ ($k = 1, 2,...$) be a sequence of elements of norm $\leqslant 1$ converging to x weakly. In particular, the numbers $(x_k \mid x)$ converge to $\|x\|^2$. For every k, we have $\mid (x_k \mid x) \mid \leqslant \|x\|$. Therefore we must have $\|x\|^2 \leqslant \|x\| \leqslant 1$.

By the Banach–Steinhaus theorem applied to duals (Theorem 33.2), we know that the unit ball of E' is relatively compact for the weak topology on E'; its image under $\bar{J}^{-1}$ is the unit ball of E, which must therefore be weakly relatively compact. As we have just seen that it is weakly closed, this proves Proposition 34.1.

We recall the following result, already stated (Exercise 32.11):

Let $\{x_k\}$ be a sequence in E such that $\|x_k\|$ is bounded independently of k; then there is a subsequence $\{x_{k_j}\}$ which converges weakly.

Remark 34.1.   Suppose that the Hilbert space E is infinite dimensional. Let $\{e_k\}$ be an orthonormal sequence in E (p. 121); *the whole sequence* $\{e_k\}$ *converges weakly to zero.* Indeed, Bessel's inequality states that, for all $x \in E$, we have

$$\sum_{k=0}^{+\infty} |(x \mid e_k)|^2 \leqslant \| x \|^2,$$

which implies that the numbers $| (x \mid e_k) | \geqslant 0$ must converge to zero.

In particular, we see that *in an infinite dimensional Hilbert space, the origin belongs to the weak closure of the unit sphere* $\{x; \| x \| = 1\}$.

We can also derive from Theorem 33.2 the theorem of Osgood:

PROPOSITION 34.2.   *If a sequence* $\{x_k\}$ *converges weakly in the Hilbert space* E, *there is a constant* $C > 0$ *such that* $\| x_k \| \leqslant C$ *for all* k.

Indeed, the sequence $\{x_k\}$ is weakly relatively compact, therefore strongly bounded.

34.2. Application to Separately Continuous Functions on Products

A topological space T is *metrizable* if its topology can be defined by a metric.

THEOREM 34.1.   *Let* T *be a metrizable topological space,* E *a TVS which is metrizable and barreled,* F *a locally convex space, and* M *a set of mappings of* $E \times T$ *into* F. *We make the following hypotheses:*

(34.1)   *For every* $t_0 \in T$, *the set of mappings* $x \rightsquigarrow f(x, t_0), f \in M$, *is an equicontinuous set of linear maps of* E *into* F.

(34.2)   *For every* $x_0 \in E$, *the set of mappings* $t \rightsquigarrow f(x_0, t), f \in M$, *is an equicontinuous set of maps of* T *into* F.

Under these circumstances, the set M *is equicontinuous.*

We should underline the fact that T carries no linear structure: T is just a topological space whose topology can be defined by a metric. The proof of Theorem 34.1 will make use of the following elementary lemma:

LEMMA 34.1.   *Let* A *be a metrizable topological space,* F *a TVS, and* H *a set of mappings of* A *into* F. *Suppose that, for every sequence* $\{a_n\}$

converging to a limit a in A, the sequence $\{f(a_n)\}$ converges to $f(a)$ in F uniformly with respect to $f \in H$. Then H is equicontinuous.

Lemma 34.1 is a straightforward generalization of Proposition 8.5.

Proof of Lemma 34.1.   Let a be an arbitrary point of A. We must prove that, to every neighborhood V of zero in F, there is a neighborhood U of a in A such that, for all $f \in H$ and all $x \in U$,

$$f(x) - f(a) \in V.$$

Suppose this were not the case. Let $\{U_k\}$ be a basis of neighborhoods of a in A such that $U_k \subset U_{k-1}$ $(k = 1, 2,...)$. For some V and for each k there would be a point $a_k \in U_k$ and a function $f_k \in H$ such that $f_k(a_k) - f_k(a) \notin V$. The sequence $\{a_k\}$ converges to a in A. By our hypothesis, given an arbitrary neighborhood of zero W in F, there is an integer $k(W) > 0$ such that, for all $f \in H$ and all $k \geqslant k(W)$, $f(a_k) - f(a) \in W$. Taking $W = V$, $k \geqslant k(V)$, and $f = f_k$, we see that we have reached a contradiction.

Proof of Theorem 34.1.   We notice that the product space $E \times T$ in the statement of Theorem 34.1 is metrizable. In virtue of Lemma 34.1, it will suffice to show that, for every sequence (x_k, t_k) $(k = 1, 2,...)$ converging to a limit (x_0, t_0) in $E \times T$, $f(x_k, t_k)$ converges to $f(x_0, t_0)$ in F, uniformly for all $f \in M$.

Let us denote by f_{t_k} the mapping $x \rightsquigarrow f(x, t_k)$ from E into F. Hypothesis (34.1) says that, for fixed k, when f varies over M, this is an equicontinuous set of linear maps. We contend that, when $k = 0, 1,...$ varies, and when f runs over M, the set of mappings $f_{t_k} - f_{t_0}$ is bounded in $L_\sigma(E; F)$ (σ: topology of pointwise convergence).

Indeed, let x be an arbitrary point of E. We must show that the set N_x of elements $f(x, t_k) - f(x, t_0)$ $(k = 1, 2,...; f \in M)$ is bounded in F. Let V be an arbitrary balanced neighborhood of zero in F. We must prove that there is a number $\lambda > 0$ such that $N_x \subset \lambda V$. On one hand, in view of (34.2), we may find an integer $k(x) \geqslant 1$ such that $k \geqslant k(x)$ implies $f_{t_k}(x) - f_{t_0}(x) \in V$. On the other hand, when $t \in T$ is fixed and f varies over M, the set $f_t(x)$ is bounded in F: this follows from (34.1) and from the fact that an equicontinuous subset of $L(E; F)$ is bounded in $L_\sigma(E; F)$ (cf. Proposition 32.7). Therefore the union of the finite number of sets $\{f_{t_k}(x) - f_{t_0}(x)\}$ $(f \in M)$, $k = 1, 2,..., k(x)$, is bounded in F. From there, our contention follows immediately.

We have not yet used the fact that E is barreled. We use it now: the set of mappings $f_{t_k} - f_{t_0}$ $(f \in M, k = 1, 2,...)$ is σ-bounded, therefore

it is equicontinuous (Theorem 33.1). This means that, to every neighborhood of zero V in F, there is a neighborhood U of zero in E such that

$$(f_{t_k} - f_{t_0})(U) \subset V \quad \text{for all} \quad f \in M \quad \text{and all} \quad k = 1, 2,\ldots.$$

Observe then that we may write

$$\begin{aligned}
f(x_k, t_k) - f(x_0, t_0) &= [f(x_k, t_k) - f(x_k, t_0)] \\
&\quad + [f(x_k, t_0) - f(x_0, t_0)] \\
&= [f(x_k - x_0, t_k) - f(x_k - x_0, t_0)] \\
&\quad + [f(x_0, t_k) - f(x_0, t_0)] + [f(x_k, t_0) - f(x_0, t_0)].
\end{aligned}$$

It suffices to choose $k_0 \geqslant 1$ so large as to have, for all $k \geqslant k_0$,

$$x_k - x_0 \in U,$$

and, for all $f \in M$,

$$\begin{aligned}
f(x_k, t_0) - f(x_0, t_0) &\in V \quad \text{(by using (34.1))}, \\
f(x_0, t_k) - f(x_0, t_0) &\in V \quad \text{(by using (34.2))}.
\end{aligned}$$

For those $k \geqslant k_0$ and all $f \in M$,

$$f(x_0, t_k) - f(x_0, t_0) \in V + V + V,$$

which obviously proves what we wanted.

One of the very important consequences of Theorem 34.1 is the
following one:

COROLLARY. *Let E be a Fréchet space, E_1 a metrizable TVS, F a locally
convex space, and $(x, y) \rightsquigarrow f(x, y)$ a separately continuous bilinear map
of $E \times E_1$ into F. Then the mapping f is continuous.*

That f is separately continuous means that, for every fixed $x_0 \in E$,
the linear map $y \rightsquigarrow f(x_0, y)$ of E_1 into F is continuous, and that, for
every fixed $y_0 \in E_1$, the linear map $x \rightsquigarrow f(x, y_0)$ of E into F is continuous.

34.3. Complete Subsets of $L_{\mathfrak{S}}(E; F)$

Definition 34.1. *A TVS E is said to be quasi-complete if every bounded
closed subset of E is complete.*

Obviously, complete implies quasi-complete. But there are TVS
which are quasi-complete without being complete. However, we have:

PROPOSITION 34.3. *If a metrizable space E is quasi-complete, it is complete.*

Proof. It suffices to show that every Cauchy sequence in E converges (Proposition 8.2). But the closure of a Cauchy sequence is a bounded set (Corollary 3 of Proposition 14.1), whence the result.

THEOREM 34.2.   *Let E, F be two TVS, and $\mathfrak{S}$ a family of bounded subsets of E satisfying Conditions $(\mathfrak{S}_{\mathrm{I}})$, $(\mathfrak{S}_{\mathrm{II}})$ (p. 335). Suppose that F is Hausdorff and quasi-complete and that $\mathfrak{S}$ is a covering of E.*

Under these circumstances, every closed equicontinuous set H of $L_{\mathfrak{S}}(E; F)$ is complete.

Proof. Let Φ be a Cauchy filter on H, and x an arbitrary point of E. We denote by $H(x)$ the image of H under the mapping $u \rightsquigarrow u(x)$ of $L(E; F)$ into F and by $\Phi(x)$ the family of sets $M(x)$ as M varies over Φ. We contend that $\Phi(x)$ is the basis of a Cauchy filter on $H(x)$. That it is a basis of filter is evident. Since $\mathfrak{S}$ is a covering of E, we may find a set A belonging to $\mathfrak{S}$ which contains x; let V be an arbitrary neighborhood of zero in F, $\mathscr{U}(A; V)$ the set of mappings $u \in L(E; F)$ such that $u(A) \subset V$. There is a set $M \in \Phi$ such that $M - M \subset \mathscr{U}(A; V)$; hence $M(x) - M(x) \subset V$, which proves our contention.

Since H is equicontinuous, H is bounded for the topology of pointwise convergence (Proposition 37.2); in other words, for all $x \in E$, $H(x)$ is bounded in F and so is its closure, $\overline{H(x)}$, which is therefore complete in view of our hypothesis. The filter on $\overline{H(x)}$ generated by $\Phi(x)$ converges to a unique element $u_0(x)$ of $\overline{H(x)}$ (unique, since F is Hausdorff). This defines a mapping u_0 of E into F; and we have proved that Φ converges pointwise in E to u_0. In particular, u_0 belongs to the closure of H in F^E which is an equicontinuous set of linear maps of E into F (Proposition 32.4); therefore $u_0 \in L(E; F)$.

It remains to show that Φ converges to u_0 in $L_{\mathfrak{S}}(E; F)$. This follows from the fact that there is a basis of neighborhoods of zero for the $\mathfrak{S}$-topology on $L(E; F)$ which are closed for the topology of pointwise convergence (namely the sets $\mathscr{U}(A; V)$ with $A \in \mathfrak{S}$ and V, a closed neighborhood of zero in F), and from the following general lemma:

LEMMA 34.2.   *Let $\mathscr{T}$, $\mathscr{T}'$ be two topologies on the same vector space G, both compatible with the linear structure of G. Suppose that $\mathscr{T}'$ is finer than $\mathscr{T}$ and that there is a basis of neighborhoods of zero for $\mathscr{T}'$ which are closed for $\mathscr{T}$.*

Let, then, A be a subset of G and Φ a filter on A which is a Cauchy filter for $\mathscr{T}'$ and which converges to $x \in A$ for $\mathscr{T}$. Then Φ converges to x for $\mathscr{T}'$.

Proof of Lemma 34.2.   Let $\mathscr{B}$ be a basis of neighborhood of zero in G, for $\mathscr{T}'$, which are closed for $\mathscr{T}$; we may assume that the neighborhoods belonging to $\mathscr{B}$ are symmetric (V is symmetric if $V = -V$). To every $V \in \mathscr{B}$ there is $M \in \Phi$ such that $M - M \subset V$. The latter means that, given any point y of M, $M \subset y + V$. On the other hand, since x is a limit point of Φ for $\mathscr{T}$, x belongs to the closure (for $\mathscr{T}$) of any set belonging to Φ, in particular to the closure of M and therefore to $y + V$, which is closed. But $x \in y + V$ means $y \in x + V$; as y is arbitrary in M, we conclude that $M \subset x + V$. As V is arbitrary in $\mathscr{B}$, this proves that Φ converges to x for $\mathscr{T}'$.

We now derive some important consequences of Theorem 34.2.

COROLLARY 1. *Let E, F be two TVS, F being Hausdorff and quasi-complete. Let A be a dense subset of E, and H an equicontinuous subset of $L(E; F)$. If a filter Φ on H converges pointwise in A to a mapping u_0 of A into F, then u_0 has a unique extension which is a continuous linear map of E into F, $\tilde{u}_0$, and Φ converges to $\tilde{u}_0$ in $L_c(E; F)$.*

We apply Proposition 32.5 twice: first to see that Φ, which is a Cauchy filter for the topology of pointwise convergence in A, must also be a Cauchy filter for the topology of pointwise convergence in E, and therefore (by Theorem 34.2) converges to $\tilde{u}_0 \in \bar{H}$ (closure of H in F^E; Proposition 32.4). As $\bar{H}$ is equicontinuous, Φ converges to $\tilde{u}_0$ for the topology of compact convergence, again by Proposition 32.5.

In the proof of the next corollary, we apply the Banach–Steinhaus theorem:

COROLLARY 2. *Let E be a barreled TVS, and F a locally convex Hausdorff TVS. If $\mathfrak{S}$ is a covering of E, $L_{\mathfrak{S}}(E; F)$ is Hausdorff and quasi-complete.*

Needless to say, $\mathfrak{S}$ is a family of bounded sets of E satisfying ($\mathfrak{S}_I$) and ($\mathfrak{S}_{II}$).

The $\mathfrak{S}$-topology on $L(E; F)$ is finer than the topology of pointwise convergence, since $\mathfrak{S}$ is a covering of E; hence every subset of $L(E; F)$ which is bounded for the $\mathfrak{S}$-topology is bounded for the topology of pointwise convergence. Since E is barreled, it follows from Theorem 33.1 that such a subset is equicontinuous. If it is closed, it must be complete by Theorem 34.2.

34.4. Duals of Montel Spaces

We now introduce the following definition:

Definition 34.2.   A TVS E is called a Montel space if E is locally convex Hausdorff and barreled and if every closed bounded subset of E is compact.

Every Montel space is quasi-complete (Definition 34.1) since a compact subset of a Hausdorff TVS is complete (Corollary 2 of Proposition 6.8). A normed space is a Montel space if and only if it is finite dimensional: indeed, in a normed space which is a Montel space the closed unit ball must be compact, hence the space must be locally compact; but a locally compact TVS is necessarily finite dimensional (Theorem 9.2).

There are Fréchet spaces which are Montel spaces, as the next statement shows:

PROPOSITION 34.4. *The spaces* $\mathscr{C}^\infty(\Omega)$, $\mathscr{C}^\infty_c(\Omega)$ (Ω: *open subset of* $\mathbf{R}^n$), $\mathscr{S}(\mathbf{R}^n)$, *and* $H(\mathcal{O})$ ($\mathcal{O}$: *open subset of* $\mathbf{C}^n$) *are Montel spaces.*

For the definition of $\mathscr{S}$ and H, see Chapter 10, Examples IV and II, respectively. Proposition 34.4 is a direct consequence of Theorem 14.4 (for $\mathscr{C}^\infty$), of Corollary 2 of Theorem 14.4 (for $\mathscr{C}^\infty_c$), of Theorem 14.5 (for $\mathscr{S}$), of Theorem 14.6 (for H).

There are Fréchet spaces which are neither Banach spaces nor Montel spaces (see Exercise 34.4).

A first straightforward consequence of Definition 34.2 is the following one:

PROPOSITION 34.5. *Let E be a Montel space, and F a TVS. On* $L(E; F)$, *the topology of compact convergence in E and the topology of bounded convergence in E are identical.*

Indeed, let B be an arbitrary bounded subset of E, and V a closed neighborhood of zero in F; we have (with the notation of p. 336) $\mathscr{U}(B; V) = \mathscr{U}(\bar{B}; V)$ since $u(B) \subset V$ implies obviously $u(\bar{B}) \subset \overline{u(B)} \subset \bar{V} = V$ for any mapping $u \in L(E; F)$. But since E is a Montel space, $\bar{B}$ is compact.   Q.E.D.

PROPOSITION 34.6. *Let E be a Montel space. Every closed bounded subset of its strong dual,* E'_b, *is compact in* E'_b. *Furthermore, on the bounded subsets of* E'_b, *strong and weak topologies coincide.*

Proof. Let B' be a bounded subset of E'_b. By Theorem 33.2, B' is equicontinuous. By Proposition 32.5, the weak topology and the topology of compact convergence coincide on B'; therefore they coincide with the strong dual topology by Proposition 34.5. This proves the last part of the statement.

Suppose now that B' is strongly closed, in addition to being strongly bounded. Let $\bar{B}'$ be its weak closure; since B' is equicontinuous, its weak closure is equicontinuous and weakly compact (Proposition 32.8).

By the first part of the proof (and the last part of the statement), strong and weak topologies coincide on $\bar{B}'$. Since B' is weakly dense in $\bar{B}'$, it is also strongly dense, therefore $\bar{B}' = B'$. But then B' must be weakly compact, therefore strongly compact.   Q.E.D.

COROLLARY 1. *In the dual E' of a Montel space E, every weakly convergent sequence is strongly convergent.*

Let S' be the union of a convergent sequence S'_0 in E' and of its limit point x'_0. The set S' is weakly compact, hence strongly bounded. Therefore, by Proposition 34.6, the strong topology is identical with the weak one, on S', and therefore S'_0 converges strongly to x'_0.

COROLLARY 2. *In the strong dual spaces $\mathscr{E}'(\Omega)$, $\mathscr{D}'(\Omega)$ (Ω: open subset of $\mathbf{R}^n$), $\mathscr{S}'$, and $\mathscr{H}'(\mathcal{O})$ ($\mathcal{O}$: open subset of $\mathbf{C}^n$), every weakly converging sequence is strongly converging.*

It suffices to combine Corollary 1 with Proposition 34.4.

Note that Corollary 2 stands in complete contrast with what happens in the dual of an infinite dimensional normed space, for instance, in an infinite dimensional Hilbert space: in such a space, an infinite orthonormal set converges weakly to zero (Remark 34.1) but, obviously, does not converge strongly (that is to say, in the sense of the norm) since, when two elements x, y of norm one are orthogonal, we have

$$\| x - y \|^2 = \| x \|^2 + \| y \|^2 = 2.$$

Exercises

34.1. Give an example of a sequence of continuous linear forms $\{x'_n\}$ ($n = 0, 1,...$) on a Banach space E which converge to zero for the topology of pointwise convergence in a dense subset A of E but which do not converge to zero for the topology of pointwise convergence in E (hint: use Example 33.1).

34.2. Let Θ be a locally compact metrizable topological space (this means that every point of Θ has a compact neighborhood and that the topology of Θ can be defined by a metric). Let E be a barreled TVS, and F a locally convex Hausdorff TVS. Let $t \rightsquigarrow x(t)$ and $t \rightsquigarrow A(t)$ be continuous functions defined in Θ, and valued in E and $L_\sigma(E; F)$, respectively. Prove that $t \rightsquigarrow A(t) x(t)$ is a continuous map of Θ into F.

34.3. Let Θ be a metrizable topological space, t_0 a point of Θ, and $t \rightsquigarrow T(t)$ a mapping of Θ into the space $\mathscr{D}'(\Omega)$ of distributions in an open subset Ω of $\mathbf{R}^n$. Suppose that, for each test function $\varphi \in \mathscr{C}_c^\infty(\Omega)$, the complex numbers $\langle T(t), \varphi \rangle$ converge as $t \rightsquigarrow t_0$. Prove that this implies that, as $t \rightsquigarrow t_0$, the distributions $T(t)$ converge strongly to a distribution T_0 in Ω.

34.4. Let Ω be an open subset of $\mathbf{R}^n$. Prove that, if $k \geqslant 0$ is finite, $\mathscr{C}^k(\Omega)$ is not a Montel space.

34.5. Let $\mathcal{L}_n$ be the Fréchet space of formal power series in n indeterminates with complex coefficients (with the topology of simple convergence of the coefficients). Prove that $\mathcal{L}_n$ is a Montel space.

34.6. Prove the following extensions of Theorem 24.3 and of its corollary:

THEOREM 34.3. *Let U be a relatively compact open subset of an open subset Ω of $\mathbf{R}^n$, and $\mathfrak{B}$ a bounded set of distributions in Ω. Let us denote by $\mathfrak{B} \mid U$ the set of restrictions to U of distributions belonging to $\mathfrak{B}$. There exists an integer $m \geqslant 0$ such that $\mathfrak{B} \mid U$ is contained and is bounded in $\mathcal{D}'^m(U)$, the dual of $\mathscr{C}^m_c(U)$.*

THEOREM 34.4. *Let $\mathfrak{B}$ be a bounded subset of $\mathscr{E}'(\Omega)$. There is a compact subset K of Ω and an integer $m \geqslant 0$ such that $\mathfrak{B}$ is contained and bounded in $\mathscr{E}'^m(\Omega)$, the dual of $\mathscr{C}^m(\Omega)$, and such that all the distributions belonging to $\mathfrak{B}$ have their support contained in K.*

Can one replace, in the above statements, *bounded set* by *convergent sequence* and *bounded* by *convergent*?

34.7. Prove the following extension of Theorem 24.4:

THEOREM 34.5. *Let $\mathfrak{B}$ be a bounded set (resp. a convergent sequence) in $\mathcal{D}'^m(\Omega)$ $(0 \leqslant m < +\infty)$. For every $T \in \mathfrak{B}$, let U_T be a neighborhood of supp T in Ω.*

Then, for every $T \in \mathfrak{B}$ and every $p \in \mathbf{N}^n$, $\mid p \mid \leqslant m$, there is a Radon measure $\mu_{p,T}$ in Ω such that the following will be true:

(i) *for every $T \in \mathfrak{B}$,*

$$T = \sum_{|p| \leqslant m} (\partial/\partial x)^p \mu_{p,T} \, ;$$

(ii) *for every $T \in \mathfrak{B}$ and all $p \in \mathbf{N}^n$, $\mid p \mid \leqslant m$, supp $\mu_{p,T} \subset U_T$;*

(iii) *for every $p \in \mathbf{N}^n$, when T ranges over $\mathfrak{B}$, $\mu_{p,T}$ ranges over a bounded subset (resp. a convergent sequence) in the space $\mathcal{D}'^0(\Omega)$ of Radon measures in Ω.*

34.8. Let E be a Fréchet space. Prove that the dual E'_c of E, equipped with the topology of compact convergence, is complete.

Derive from Lemma 34.2 that $E'_{\mathfrak{S}}$ is complete for any family $\mathfrak{S}$ of bounded subsets of E, satisfying $(\mathfrak{S}_\mathrm{I})$ and $(\mathfrak{S}_\mathrm{II})$ and containing all the compact subsets of E.

35

Further Study of the Weak Topology

In this chapter, E will always denote a locally convex Hausdorff TVS, and E' its dual.

Let M be a finite dimensional linear subspace of E, and M^0 its orthogonal (see Chapter 19, p. 196). Suppose that $d = \dim M$ and let $e_1, \ldots, e_d$ be a basis of the vector space M. Let us apply Corollary 3 of the Hahn–Banach theorem (Theorem 18.1): there is a continuous linear form f_j on M such that $f_j(e_k) = 0$ if $k \neq j$ ($1 \leqslant j, k \leqslant d$), without f_j being identically zero in M; the latter implies that $f_j(e_j) \neq 0$; possibly by multiplying f_j by a complex number, we assume that $f_j(e_j) = 1$. Again by the Hahn–Banach theorem, we may extend f_j as a continuous linear form e'_j in the whole of E. Let M' be the linear subspace of E' spanned by $e'_1, \ldots, e'_d$ and let x' be an arbitrary element of E'; to x' corresponds the following element of M',

$$p(x') = \sum_{j=1}^{d} \langle x', e_j \rangle e'_j.$$

It is clear that the forms x' and $p(x')$ take the same values on every e_j, therefore on M. In other words, $x' - p(x')$ belongs to M^0; conversely, if $x' \in M^0$, we obviously have $p(x') = 0$; finally, observe that if $x' \in M'$, we have

$$x' = \sum_{j=1}^{d} a_j e'_j \qquad (a_j: \text{complex numbers}),$$

whence

$$\langle x', e_k \rangle = a_k \qquad \text{for each} \quad k = 1, \ldots, d,$$

therefore $x' = p(x')$. All this shows that p is a linear map of E' onto M' whose kernel is exactly M^0 and which is equal to the identity in M'. We see thus that E'/M^0 is isomorphic (as a vector space) to M' and therefore also to M (in particular, $\dim E'/M^0 = d$).

Let now L be a continuous linear form on E' when the latter carries the weak dual topology, σ. By definition of this topology, there is a finite subset S of E and a constant $C > 0$ such that, for all $x' \in E'$,

$$(35.1) \qquad |L(x')| \leqslant C \sup_{x \in S} |\langle x', x \rangle|.$$

We shall apply the preceding considerations to the linear subspace M of E spanned by S (it is obvious that M is finite dimensional); we construct M' as we have said. Set, then,

$$x_L = \sum_{j=1}^{d} L(e'_j) e_j .$$

We have, for all $x' \in E'$, with the above definition of the mapping p,

$$\langle x', x_L \rangle = \sum_{j=1}^{d} L(e'_j) \langle x', e_j \rangle = L(p(x')).$$

On the other hand, observe that $x' - p(x')$ is orthogonal to M, hence $\langle x' - p(x'), x \rangle = 0$ for all $x \in S$, which implies, by (35.1), that

$$L(x' - p(x')) = 0, \qquad \text{i.e.,} \quad L(p(x')) = L(x').$$

We reach the conclusion that, for all $x' \in E'$,

$$\langle x', x_L \rangle = L(x').$$

Now, we know that every element x of E defines a linear form on E', namely the "value" at x, $x' \rightsquigarrow \langle x', x \rangle$. By Proposition 19.6, we know that the value at x is a linear map of E into the dual of E'_σ; and by Proposition 19.7 we know that this mapping is one-to-one. As we have just proved that it is also onto, we may state:

PROPOSITION 35.1. *The mapping $x \rightsquigarrow$ value at x is a linear isomorphism of E onto the dual of its weak dual.*

COROLLARY. *The dual of E, equipped with $\sigma(E, E')$, is identical to E'.*

Indeed, by Proposition 35.1, E'_σ is canonically isomorphic to the weak dual of E_σ. It suffices then to apply Proposition 35.1 with E and E' exchanged.

We shall identify E with the dual of E'_σ through $x \rightsquigarrow$ value at x. Note that the *set* E' depends on the topology initially given on E, but that the weak topology on E' does *not*. We may now consider the weak topology on E

when we regard E as the dual of E'_σ. In order not to mix up the two weak topologies which we have now on E and E' we shall denote them by $\sigma(E, E')$ and $\sigma(E', E)$, respectively. *In the notation $\sigma(F, G)$ the topology is carried by the first space, F.*

If we provide E with the topology $\sigma(E, E')$ and E' with $\sigma(E', E)$, we find ourselves in a perfectly symmetric situation; we are now going to exploit this symmetry. First of all, observe that we may write indifferently $\langle x', x \rangle$ or $\langle x, x' \rangle$ for the value of x' at x (or the value of x at x').

We may combine the corollary of Proposition 35.1 with Proposition 18.3. We obtain:

PROPOSITION 35.2.   *Let A be a convex subset of the locally convex Hausdorff space E. The closure of A for the initial topology of E is identical to the closure of A for the weak topology $\sigma(E, E')$.*

This proposition enables us to simplify considerably many forthcoming statements. The student should however beware of the following possible mistake: Let E' be the dual of E, and A' a convex subset of E'; it is not true that the closure of A' in the weak topology on E' and the closure of A' in the strong dual topology on E' are the same (they will be the same whenever the dual of E'_b can be identified with E, but not otherwise—in general).

Let A' be a subset of E'. By A'^0 we mean the polar of A' which is a subset of E when E is regarded as the dual of E' (see Definition 19.1). The set A' itself could have been the polar of some subset A of E, in which case $A'^0 = (A^0)^0$. We introduce the following definition:

Definition 35.1.   Let A be a subset of E, and A^0 the polar of A. The polar of A^0 (when we identify E with the dual of E') is called the bipolar of A and is denoted by A^{00}.

The next proposition gives a simple description of the bipolar of a set.

PROPOSITION 35.3.   *Let A be a subset of E. The bipolar A^{00} of A is the closed convex balanced hull of A.*

Proof.   Let $\hat{A}$ be the closure, for the initial topology of E as well as for the topology $\sigma(E, E')$, of the convex balanced hull of A (i.e., the smallest closed convex balanced set containing A). Since A^{00} is obviously balanced and convex (cf. p. 195), and also weakly closed (the polar of a set is always weakly closed, as immediately seen), we see that $\hat{A} \subset A^{00}$. We must prove the inclusion in the opposite direction: we must prove that if $x^0 \notin \hat{A}$ then $x^0 \notin A^{00}$.

As $\hat{A}$ is weakly closed, there is a neighborhood of x^0 in the topology $\sigma(E, E')$ which does not intersect $\hat{A}$; in other words, there is a continuous linear form x' on E such that

$$(35.2) \qquad d = \inf_{x \in \hat{A}} |\langle x', x - x^0 \rangle| > 0.$$

Let D be the image of $\hat{A}$ under the mapping $x \rightsquigarrow \langle x', x \rangle$; D is convex balanced and closed (in the complex plane), therefore D is a closed disk with radius $\rho \geqslant 0$ centered at the origin. In view of (35.2), $|\langle x', x^0 \rangle| \geqslant d + \rho$. Since $|\langle x', x \rangle| \leqslant \rho$ for all $x \in \hat{A}$, in particular for all $x \in A$, we see that $(\frac{1}{2}d + \rho)^{-1}x' \in A^0$ but that

$$|\langle (\tfrac{1}{2}d + \rho)^{-1}x', x^0 \rangle| > 1,$$

which proves that $x^0 \notin A^{00}$.

COROLLARY 1. $(A^{00})^0 = A^0$.

Indeed, $(A^{00})^0 = (A^0)^{00}$ is the closed (for $\sigma(E', E)$) convex balanced hull of A^0; but A^0 is weakly closed convex and balanced.

COROLLARY 2.   *Let M be a linear subspace of E. The bipolar M^{00} of M is the closure of M.*

When M is a linear subspace of E, M^0 is a linear subspace of E', hence M^{00} is a linear subspace of E.

Let, now, F be another locally convex Hausdorff TVS, and u a continuous linear map of E into F. As usual, we set

$$\text{Im } u = \{y \in F; \text{ there exists } x \in E \text{ such that } u(x) = y\};$$

similarly,

$$\text{Ker } {}^t u = \{y' \in F'; {}^t u(y') = 0\}.$$

We recall (Proposition 23.2) that we have

$$(35.3) \qquad \text{Ker } {}^t u = (\text{Im } u)^0.$$

Now, of course, we may also consider the transpose of ${}^t u : F' \to E'$, when we regard E (resp. F) as the dual of E'_σ (resp. F'_σ). It is immediately seen that ${}^{tt} u = u$. Substituting then ${}^t u$ for u in (35.3), we obtain

$$\text{Ker } u = (\text{Im } {}^t u)^0.$$

We may then take the polars of both sides. Combining with the Corollary 2 of Proposition 35.3, we see that we have:

PROPOSITION 35.4. *Let u be a continuous linear map of E into F. The orthogonal of* Ker *u is the weak closure of the image of* $^t u$.

We shall apply Eq. (35.3) and Proposition 35.4 when u is either the canonical injection of a subspace into the space or the canonical homomorphism of the space onto a quotient. We shall use them to study the dual of a linear subspace and the dual of quotient spaces.

First, let N be a linear subspace of E; the injection of N into E can be transposed into a linear mapping of E' *onto* N' (onto, by virtue of the Hahn–Banach theorem). The kernel of the transpose is N^0, because of (35.3). We obtain thus a canonical linear mapping of E'/N^0 onto N'.

Let, now, φ be the canonical homomorphism of E onto $E/\bar{N}$. We suppose that E carries the initial topology (or the weak one) and $E/\bar{N}$, the quotient modulo $\bar{N}$ of the topology carried by E. The transpose $^t\varphi$ is an injection of the dual of $E/\bar{N}$ into E'. By Proposition 35.4, the weak closure of the image of $^t\varphi$ is the orthogonal of Ker φ, that is to say of $\bar{N}$. In other words, the weak closure of Im $^t\varphi$ is N^0. This implies Im $^t\varphi \subset N^0$. But, conversely, let $x' \in N^0$; the continuous linear form x' vanishes on $\bar{N}$ and therefore is constant along the equivalence classes modulo $\bar{N}$. In other words, x' defines a continuous linear form $\tilde{x}'$ on $E/\bar{N}$. It is immediately seen that $^t\varphi(\tilde{x}') = x'$. This proves that Im $^t\varphi = N^0$. We obtain thus a canonical isomorphism of the dual of $E/\bar{N}$ ($E/\bar{N}$ carries the quotient modulo $\bar{N}$ of the topology of E) onto N^0.

Let us summarize:

PROPOSITION 35.5. *Let E be a locally convex Hausdorff space, and N a linear subspace of E. The following is true:*

(a)   *The transpose of the natural injection* $N \to E$ *is a one-to-one linear map of* E'/N^0 *onto the dual* N' *of* N.

(b)   *The transpose of the canonical homomorphism* $E_0 \to E/\bar{N}$ *is a one-to-one linear map of the dual of* $E/\bar{N}$ *onto* N^0.

It should be underlined that the isomorphisms in (a) and (b) are algebraic, that is to say, they do not involve the topologies on the intervening spaces. Observe that E'/N^0 can now carry two apparently distinct topologies: (1) the quotient modulo N^0 of the weak topologies $\sigma(E', E)$ on E'; let us denote this topology by $\sigma(E', E)/N^0$; (2) the weak dual topology $\sigma(E'/N^0, N)$ when we regard E'/N^0 as the dual of N.

PROPOSITION 35.6. *Let E and N be as in Proposition 35.5. The following two properties are equivalent:*

(35.4)   *N is closed;*

(35.5)   *the topologies* $\sigma(E', E)/N^0$ *and* $\sigma(E'/N^0, N)$ *on* E'/N^0 *are identical.*

Proof. We begin by some preliminary considerations. Let us denote by $\mathcal{W}(x_1,...,x_r;\varepsilon)$ $(x_1,...,x_r$, an arbitrary finite subset of E, ε an arbitrary number >0) the set of $x' \in E'$ such that $|\langle x',x_j\rangle| \leqslant \varepsilon$ for every $j = 1,...,r$. Let φ' be the canonical homomorphism of E' onto E'/N^0. A basis of neighborhoods of zero in the topology $\sigma(E',E)/N^0$ consists of the images under φ' of the sets $\mathcal{W}(x_1,...,x_r;\varepsilon)$.

Let now $y_1,...,y_s$ be a finite subset of N. In the duality between E'/N^0 and N, we have $\langle \varphi'(x'),y_j\rangle = \langle x',y_j\rangle$, regardless of what is the representative x' of the class $\varphi'(x')$ modulo N^0. This means that we obtain a basis of neighborhoods of 0 in the topology $\sigma(E'/N^0,N)$ by taking the images under φ' of the sets $\mathcal{W}(y_1,...,y_s;\varepsilon)$, when $\varepsilon > 0$ and the finite subset of N, $\{y_1,...,y_s\}$, vary in all the possible ways. This proves, in particular, that the topology $\sigma(E',E)/N^0$ is always finer than the topology $\sigma(E'/N^0,N)$. We are going to show that, when N is closed, to every finite subset of E, $\{x_1,...,x_r\}$, there is a finite subset of N, $\{y_1,...,y_r\}$ (having the same number of elements), such that, for all $\varepsilon > 0$,

$$(35.6) \qquad \varphi'(W(y_1,...,y_r;\varepsilon)) \subset \varphi'(W(x_1,...,x_r;\varepsilon)).$$

This will obviously imply that $\sigma(E'/N^0,N)$ is finer than, therefore equal to, $\sigma(E',E)/N^0$.

We now give the "construction" of the y_j's. Let L be the linear subspace of E generated by N together with $x_1,...,x_r$; N is of finite codimension in L, hence N has a (finite dimensional) supplementary P in L. Let $z_1,...,z_d$ be a basis of P. The restrictions to N^0 of the linear forms

$$(35.7) \qquad x' \rightsquigarrow \langle z_j,x'\rangle, \qquad j = 1,...,d,$$

are linearly independent. Indeed, consider a linear combination $\sum_{j=1}^d \lambda_j z_j$ vanishing on N^0, in other words belonging to N^{00}. Since N is closed, we have $N = N^{00}$; and $\sum \lambda_j z_j \in N$ only if all the coefficients λ_j are equal to zero, from the very choice of the z_j (it is in this argument that we use the hypothesis N closed; cf. Proposition 35.3). Because the restrictions to N^0 of the forms (35.7) are linearly independent, for each $j = 1,...,d$, we may find $z_j' \in N^0$ such that $\langle z_j',z_j\rangle = 1$ and such that $\langle z_j',z_k\rangle = 0$ for $j \neq k$. Now let x' be an arbitrary element of E'; for each $j = 1,...,d$, $\langle x',z_j\rangle$ is a complex number; set

$$p(x') = -\sum_{j=1}^d \langle x',z_j\rangle z_j'.$$

The element $p(x')$ belongs to N^0 and we have

$$\langle x' + p(x'), z_j \rangle = 0 \qquad \text{for all} \quad j = 1,..., d.$$

Let us go back to x_1 , ..., x_r ; for each $l = 1,..., r$, we have

$$x_l = y_l + w_l ,$$

where $y_l \in N$ and $w_l \in P$, i.e., is a linear combination of the z_j . For every $x' \in E'$, we have

$$\langle x', y_l \rangle = \langle x' + p(x'), y_l \rangle = \langle x' + p(x'), x_l \rangle.$$

This implies the following: if $x' \in W(y_1 ,..., y_r ; \varepsilon)$, then there is an element of E', congruent to x' modulo N^0, which belongs to $W(x_1 , ..., x_r ; \varepsilon)$. From there, (35.6) follows immediately.

We prove now that, if $\sigma(E'/N^0, N) = \sigma(E', E)/N^0$, then N must be closed. In view of the first part of the proof, we know that

(35.8) $\sigma(E', E)/N^0 = \sigma(E'/N^0, \bar{N}).$

(Indeed, observe that N^0 is also the orthogonal of $\bar{N}$.) We apply twice Proposition 35.1: (35.5) implies that the dual of E'/N^0 (when this space carries the topology $\sigma(E', E)/N^0$) is canonically isomorphic to N. But (35.8) implies that this dual is canonically isomorphic to $\bar{N}$. Hence we have $N = \bar{N}$.   Q.E.D.

PROPOSITION 35.7. *Let E, F be two locally convex Hausdorff spaces, and u a continuous linear map of E into F. The following two properties are equivalent:*

(a) $u(E)$ *is closed in F;*

(b) ${}^t u$ *is a homomorphism* (for the TVS structures) *of F'_σ onto ${}^t u(F') \subset E'_\sigma$.*

For the notion of homomorphism, see Chapter 17, p. 166. If we set $M = u(E)$, the image of u, we know by (35.1) that Ker ${}^t u = M^0$; let φ be the canonical projection $F' \to F'/M^0$; there is a linear map $v : F'/M^0 \to E'$ such that the diagram below is commutative:

$$
\begin{array}{ccc}
F' & \xrightarrow{\;{}^t u\;} & E' \\
{\scriptstyle\varphi}\big\downarrow & \nearrow{\scriptstyle v} & \\
F'/M^0 & &
\end{array}
$$

Property (b), in Proposition 35.7, states that v is an isomorphism onto ${}^t u(F')$ for the structures of TVS, when F'/M^0 carries the quotient topology $\sigma(F', F)/M^0$ and ${}^t u(F')$ carries the topology induced by $\sigma(E', E)$.

Proof of Proposition 35.7. For $x \in E$ and $y' \in F'$, we have

$$\langle v(\varphi(y')), x \rangle = \langle y', u(x) \rangle.$$

The right-hand side is equal to $\langle \varphi(y'), u(x) \rangle$, where we use now the bracket of the duality between F'/M^0 and M (Proposition 35.5). Thus we see that $\varphi(y') \to 0$ for $\sigma(F'/M^0, M)$ if and only if $v(\varphi(y')) \to 0$ for $\sigma(E', E)$. This means that v is an isomorphism of F'/M^0, equipped with $\sigma(F'/M^0, M)$, onto ${}^t u(F') \subset E'$, equipped with $\sigma(E', E)$. But $\sigma(F'/M^0, M) = \sigma(F', F)/M^0$ if and only if M is closed (Proposition 35.6).   Q.E.D.

Concerning weak and strong topologies on a TVS, in relation to linear mappings, we have the following result:

PROPOSITION 35.8.   *Let* E, F *be two locally convex Hausdorff spaces, and* u *a continuous linear map of* E *into* F. *Then* u *is continuous from* E, *equipped with the weak topology* $\sigma(E, E')$, *into* F, *equipped with* $\sigma(F, F')$.

Proof. If $u : E \to F$ is continuous, the transpose ${}^t u$ of u is a continuous linear mapping of the weak dual of E, E'_σ, into the weak dual of F, F'_σ (corollary of Proposition 19.5). By iterating this result, we see that the transpose of ${}^t u$, which is equal to u, is a continuous linear mapping of the weak dual of E'_σ into the weak dual of F'_σ. It suffices then to apply Proposition 35.1.

We shall see later (Lemma 37.6) that the converse of Proposition 35.8 is true when the space E is metrizable: then continuity for $\sigma(E, E')$ and $\sigma(F, F')$ implies continuity for the initial topologies.

Exercises

35.1.   Give a necessary and sufficient condition on a subset A of a locally convex Hausdorff E in order that the polar A^0 of A be equal to $\{0\}$.

35.2.   Let E be a locally convex Hausdorff space, and $\hat{E}$ its completion. Prove that, if $E \neq \hat{E}$, then $\sigma(E', \hat{E})$ is strictly finer than $\sigma(E', E)$.

35.3.   Let E be a locally convex Hausdorff TVS, and E' its dual. Prove that, if M is a linear subspace of dimension $n < +\infty$, its orthogonal $M^0 \subset E'$ has codimension equal to n. Is it true that, if M is of codimension n, then M^0 is necessarily of dimension n?

35.4.   Let E and E' be as in Exercise 35.3. We suppose that E, provided with the weak topology $\sigma(E, E')$, is the topological direct sum of two linear subspaces M and N, i.e., the mapping $(x, y) \leadsto x + y$ of the product TVS $M \times N$ into E is an isomorphism onto. Prove then that E', provided with $\sigma(E', E)$, is the topological direct sum of M^0 and N^0.

36

Topologies Compatible with a Duality. The Theorem of Mackey. Reflexivity

In this chapter, we shall consider a locally convex Hausdorff space E. It will be convenient to distinguish the topology originally given on E from other topologies that are going to be defined on E; for this reason, we shall refer to it as *the initial topology on E*. The dual of E is denoted E', as usual. We might ask the following question: which locally convex Hausdorff topologies on E have the property that, when E carries them, its dual is identical to E'? In Chapter 35 we have encountered such a topology: the weak topology $\sigma(E, E')$ on E. The initial topology on E is another one. We shall see, now, that one can characterize all of them. In order to do this, we identify E to the dual of E'_σ (Proposition 35.1). We may then equip E with a $\mathfrak{S}$-topology, where $\mathfrak{S}$ is a family of bounded subsets of E'_σ (i.e., of weakly bounded subsets of E') satisfying Conditions $(\mathfrak{S}_I)$ and $(\mathfrak{S}_{II})$ (cf. Chapter 19, p. 196, or Chapter 32, p. 335).

THEOREM 36.1. *Let $\mathscr{T}$ be a locally convex Hausdorff topology on E. The following two properties are equivalent:*

(a) *$\mathscr{T}$ is identical to a $\mathfrak{S}$-topology on E, where $\mathfrak{S}$ is a covering of E' consisting of convex balanced weakly compact sets;*

(b) *the dual of E, equipped with the topology $\mathscr{T}$, is identical to E'.*

Proof. (a) *implies* (b). The $\mathfrak{S}$-topology on E is not modified if we add to $\mathfrak{S}$ all the subsets of the sets which belong to $\mathfrak{S}$. Since $\mathfrak{S}$ is a covering of E, we see that now all the finite subsets of E' belong to $\mathfrak{S}$, so that the $\mathfrak{S}$-topology is certainly finer than the weak topology $\sigma(E, E')$. This has the following immediate consequence: E', dual of E_σ, is contained in the dual of E when this space carries the $\mathfrak{S}$-topology; we shall denote by E'_1 the latter dual. We must prove that $E'_1 \subset E'$. On E', the weak topology $\sigma(E', E)$ is obviously identical to the topology induced by $\sigma(E'_1, E)$. This implies that all the sets $K \in \mathfrak{S}$, which are compact for $\sigma(E', E)$, are also compact, hence closed, for $\sigma(E'_1, E)$. As they are also convex and balanced, we derive from Proposition 35.3 that each one of them is equal to its bipolar. Let, then, $f \in E'_1$. By definition of E'_1, there exists $K \in \mathfrak{S}$ such that

$$|f(x)| \leqslant 1 \qquad \text{for all} \quad x \in K^0.$$

This means that $f \in K^{00} \subset E'$. Thus $f \in E'$ and therefore $E'_1 = E'$.

(b) *implies* (a). Let U be a neighborhood of zero in E for the topology $\mathscr{T}$. The polar U^0 of U is a weakly compact subset of E' (Propositions 32.7 and 32.8). Let us denote by $\mathfrak{S}$ the family of all these subsets U^0 as U varies over the filter of neighborhoods of zero; the elements of $\mathfrak{S}$ are convex, balanced, weakly compact. Moreover, they form a covering of E'. Indeed, if $x' \in E'$, there is a neighborhood U of zero in E such that $|\langle x', x \rangle| \leqslant 1$ for all $x \in U$, i.e., $x' \in U^0$. It remains to check that $\mathscr{T}$ is identical to the $\mathfrak{S}$-topology on E. But observe that there is a basis of neighborhoods of zero for $\mathscr{T}$ consisting of convex balanced closed neighborhoods of zero. If V is such a neighborhood, V is equal to its bipolar V^{00} (Proposition 35.3), hence V is the polar of $V^0 \in \mathfrak{S}$. On the other hand, there is a basis of neighborhoods of zero in the $\mathfrak{S}$-topology consisting of the polars of the sets belonging to $\mathfrak{S}$, that is to say of the bipolars U^{00} of the neighborhoods of zero U in E (for the topology $\mathscr{T}$); as $U \subset U^{00}$, the bipolars are neighborhoods of zero for $\mathscr{T}$. We have thus proved that the $\mathfrak{S}$-topology is finer and coarser than $\mathscr{T}$. Q.E.D.

A topology $\mathscr{T}$ on E which is locally convex Hausdorff and is such that the dual of E, equipped with $\mathscr{T}$, is identical to E', will be said to be *compatible with the duality between E and E'*. Theorem 36.1 characterizes these topologies; it shows that they form a partially ordered set having a minimum element, the weak topology $\sigma(E, E')$ (the family of finite subsets of E' being the smallest $\mathfrak{S}$ family of weakly compact subsets of E'), and a maximum element, the topology of uniform convergence on every weakly compact convex balanced subset of E'. The latter topology is called *the Mackey topology on E* and is denoted by $\tau(E, E')$. Since the initial topology on E is compatible with the duality between E and E', it is coarser than $\tau(E, E')$; this is also evident by inspection of Proof 2 of Theorem 36.1. In connection with this, we have:

PROPOSITION 36.1. *The topology of a locally convex Hausdorff space E is identical to the topology of uniform convergence on every equicontinuous subset of E'.*

Indeed, we have seen in the proof of Theorem 36.1 that a topology $\mathscr{T}$ compatible with the duality between E and E' is identical to the $\mathfrak{S}$-

topology on E, where $\mathfrak{S}$ consists of the polars U^0 of the neighborhoods of zero U for $\mathcal{T}$. It suffices then to apply Proposition 32.7.

We have seen (Proposition 35.2) that the closure of a convex subset of E is the same for the initial topology or for the weak topology $\sigma(E, E')$. More generally we have:

PROPOSITION 36.2.   *The closure of a convex subset A of E is the same for all the locally convex Hausdorff topologies on E compatible with the duality between E and E'.*

Proof.   The continuous **R**-linear forms on E are in one-to-one correspondence with the continuous **C**-linear forms on E, that is to say the elements of E' (see proof of Proposition 35.2), therefore they are the same for all the topologies compatible with the duality between E and E'; the closed hyperplanes and the closed half-spaces are therefore also the same for these topologies. But then it suffices to apply Proposition 18.3.

A deeper result is the Mackey theorem, which states that the *bounded* sets are the same for all topologies compatible with the duality between E and E'. In order to prove it, we shall perform a construction which will also be used later on, in a different context.

The Normed Space E_B

Let E be a locally convex Hausdorff TVS, and B a bounded convex balanced subset of E. Let E_B be the vector subspace of E spanned by B; note that B is an absorbing subset of E_B (but, in general, not of E). For $x \in E$, set

$$p_B(x) = \inf_{x \in \lambda B} |\lambda|;$$

since B is convex, balanced, and absorbing, p_B is a seminorm on E_B. In fact, it is a norm: For let U be an arbitrary neighborhood of zero in E. There exists a number $\rho > 0$ such that $B \subset \rho U$, i.e., $\rho^{-1}B \subset U \cap E_B$. This means that the topology defined by the seminorm p_B on E_B, i.e., the topology defined by taking the multiples of B, εB, $\varepsilon > 0$, as a basis of neighborhoods of zero, is finer than the topology induced by E. As the latter is Hausdorff, so is the former. Thus we have obtained a normed space E_B with a continuous natural injection $E_B \to E$.

LEMMA 36.1.   *Suppose that the bounded convex balanced subset B of E is complete. Then the normed space E_B is a Banach space.*

Proof. Let $\{x_k\}$ $(k = 1, 2,...)$ be a Cauchy sequence in E_B ; it is a bounded subset of E_B (Corollary 3 of Proposition 14.1), hence it is contained in some multiple ρB of B. But ρB is complete for the topology induced by E, hence the sequence $\{x_k\}$ converges to a limit $x \in \rho B$ in the sense of the induced topology. Observe that the multiples ρB are *closed* for the induced topology. We are in the right conditions to apply Lemma 34.2. Lemma 36.1 follows immediately.

COROLLARY. *Let K be a compact convex balanced subset of E; the normed space E_K is a Banach space.*

In the situation of the corollary, observe that the natural injection $E_K \to E$ transforms the closed unit ball of E_K into a compact set.

From Lemma 36.1 we derive the following useful consequence:

LEMMA 36.2. *Let T be a barrel in the locally convex Hausdorff TVS E, and B a complete bounded convex balanced subset of E. There is a number $\rho > 0$ such that*

$$B \subset \rho T.$$

Lemma 36.2 is of interest when E itself is not barreled; otherwise T would be a neighborhood of zero in E and would therefore absorb every bounded set.

Proof. $T \cap E_B$ is obviously a barrel in E_B : it is convex balanced, absorbing; as the topology induced by E is coarser than the topology of the norm on E_B and since T is closed in E, $T \cap E_B$ is also closed in the normed space E_B . In view of Lemma 36.1, E_B is a Banach space, hence is barreled. Therefore $T \cap E_B$ contains a multiple of the closed unit ball (which is B). Q.E.D.

We may now easily prove Mackey's theorem:

THEOREM 36.2. *The bounded subsets of E are the same for all locally convex Hausdorff topologies on E compatible with the duality between E and E'.*

Proof. A set which is bounded in some topology is also bounded in every coarser topology. Therefore, it suffices to prove that the weakly bounded subsets of E are bounded for the Mackey topology $\tau(E, E')$.

Let K be a convex balanced weakly compact subset of E', and B a weakly bounded subset of E. Let us denote by D_ε the closed disk in the complex plane with center at zero and radius $\varepsilon > 0$, and by $x^{-1}(D_\varepsilon)$ the set of elements $x' \in E'$ such that $\langle x, x' \rangle \in D_\varepsilon$. The subset of E',

$$T = \bigcap_{x \in B} x^{-1}(D_\varepsilon),$$

is a barrel. Indeed, it is the intersection of closed convex balanced subsets of E'_σ. It is absorbing. Indeed, if x' is an arbitrary element of E', there is a number $\tau > 0$ such that $\tau B \subset W(\{x'\}; \varepsilon) = \{x \in E; \, | \, \langle x', x \rangle \, | \leqslant \varepsilon\}$ since B is weakly bounded. But this means precisely that $\tau x' \in T$. Then, from Lemma 36.2, we derive that there is a number $\rho > 0$ such that $K \subset \rho T$. But this means that

$$B \subset \rho \, W(K; \varepsilon), \qquad W(K; \varepsilon) = \{x \in E; \sup_{x' \in K} |\langle x', x \rangle| \leqslant \varepsilon\}. \qquad \text{Q.E.D.}$$

Given a locally convex Hausdorff TVS E, we may now talk of its bounded sets without specifying for which one, among the topologies on E compatible with the duality between E and E', they are bounded. In particular, there is no need to distinguish between bounded and weakly bounded sets.

PROPOSITION 36.3. *If the locally convex Hausdorff space E is either metrizable or barreled, the initial topology of E is identical to the Mackey topology $\tau(E, E')$.*

Proof. 1. Suppose E metrizable. Let U be a neighborhood of zero for the Mackey topology, and $\{U_n\}$ a basis of neighborhoods of zero for the initial topology. If U were not a neighborhood of zero for the initial topology, we would be able to find, for each $n = 1, 2,\ldots$, a point $x_n \in (1/n) \, U_n$ such that $x_n \notin U$. Note that the points nx_n converge to zero, hence form a bounded subset of E. In view of Theorem 36.2, there should be a number $\rho > 0$ such that $nx_n \in \rho U$ for all n. But this contradicts the fact that $x_n \notin U$ for every n.

2. Suppose E barreled. Then every weakly compact subset K of E' is equicontinuous (Theorem 33.2). In this case, Proposition 36.3 follows from Proposition 36.1.

We proceed now to study the reflexivity of locally convex Hausdorff spaces.

Definition 36.1. The dual of the strong dual E'_b of E is called the bidual of E and is denoted by E''.

The strong dual topology being finer than the weak dual topology, the dual of E'_σ can be regarded as a linear subspace of E''. By Proposition 35.1, we see that the mapping $x \rightsquigarrow$ value at x is a one-to-one linear mapping of E into E''.

Definition 36.2. The space E is said to be semireflexive if the mapping $x \rightsquigarrow$ value at x maps E onto E''. The space E is said to be reflexive if this *mapping is an isomorphism* (for the TVS structures) *of E onto E_b'' , the strong dual of its strong dual E_b' .*

In other words, E is reflexive if it is semireflexive and if its initial topology is equal to the strong dual topology when we regard E as the dual of E_b' .

PROPOSITION 36.4.   *The strong dual of a semireflexive space is barreled.*

Proof.   Let E be a semireflexive space, and E_b' its strong dual. Let T' be a barrel in E_b' . Since both the strong dual topology and the weak dual one, on E', are compatible with the duality between E' and E, Proposition 36.2 implies that T' is weakly closed; hence $T' = T'^{00}$. Therefore, we will have proved that T' is a neighborhood of zero in E_b' if we prove that T'^0 is a bounded subset of E (since T' is the polar of T'^0). But in view of Mackey's theorem (Theorem 36.2), it will suffice to prove that T'^0 is bounded in E for the topology $\sigma(E, E')$.

As T' is absorbing, to every $x' \in E'$ there is a number $c > 0$ such that $cx' \in T'$. This implies immediately, for every $\varepsilon > 0$,

$$c\varepsilon T'^0 \subset W(\{x'\}; \varepsilon) = \{x \in E; |\langle x', x\rangle| \leqslant \varepsilon\}.$$

The finite intersections of sets $W(\{x'\}; \varepsilon)$, as $x' \in E'$ and $\varepsilon > 0$ vary, form a basis of neighborhoods of zero in $\sigma(E, E')$, whence the result.

THEOREM 36.3.   *Let E be a locally convex Hausdorff space. The following two properties are equivalent:*

(a)   *E is semireflexive;*

(b)   *for $\sigma(E, E')$, every closed bounded subset of E is compact.*

Proof.   (a) implies (b), for if E is semireflexive, E is the dual of a barreled space, its strong dual E_b' , according to Proposition 36.4. It suffices then to apply Theorem 33.2.

(b) implies (a). It suffices to show that the strong dual topology on E' is compatible with the duality between E' and E. According to Theorem 36.1, we must show that the strong dual topology is identical to a $\mathfrak{S}$-topology, where $\mathfrak{S}$ is a covering of E consisting of convex balanced weakly compact (i.e., compact for $\sigma(E, E')$) subsets of E. But the strong dual topology on E' is obviously identical to the topology of uniform convergence on the weakly closed convex balanced and bounded subsets of E. By hypothesis, these sets are weakly compact.   Q.E.D.

PROPOSITION 36.5.   *If a barreled locally convex Hausdorff space E is semireflexive, E is reflexive.*

Proof. As E is barreled, there is identity between bounded subsets of E'_b and equicontinuous subsets of E'; the initial topology of E is the topology of uniform convergence on the equicontinuous subsets of E'; the strong dual topology on E (regarded as dual of E'_b if we use the semi-reflexivity of E) is the topology of uniform convergence on the bounded subsets of E'_b, whence the result.

Before proving the last general result in the matter of reflexivity, let us observe that the strong dual of a reflexive space is trivially reflexive.

THEOREM 36.4.   *A locally convex Hausdorff space E is reflexive if and only if E is barreled and if, for the topology $\sigma(E, E')$, every closed bounded set is compact.*

Proof. The second of the two conditions is equivalent with the fact that E is semireflexive, by Theorem 36.3. If E is reflexive, then it is the dual of the reflexive space E'_b and therefore E is barreled, by virtue of Proposition 36.4. Conversely, if E is semireflexive and barreled, E is reflexive, by virtue of Proposition 36.5.

Examples of Semireflexive and Reflexive Spaces

Example 36.1.   The finite dimensional Hausdorff spaces. On these spaces, all the locally convex Hausdorff topologies are identical; every closed bounded set is compact; every vector basis defines an isomorphism of E onto its dual.

Example 36.2.   The Hilbert spaces. Let J be the canonical isometry of a Hilbert space E onto its anti-dual, $\bar{E}'$: let K be the canonical isometry of the Hilbert space $\bar{E}'$ onto its own anti-dual, which is easily seen to be the strong dual of the strong dual of E, E''_b ; $K \circ J$ is an isometry of E onto E''_b which is nothing else but the mapping $x \rightsquigarrow$ value at x.

Example 36.3.   In order that a Banach space be reflexive, it suffices that it be semireflexive (by Proposition 36.5). If Ω is an open subset of $\mathbf{R}^n$, the spaces $L^p(\Omega)$ are reflexive when $1 < p < +\infty$ (Theorem 20.3); similarly, the spaces of sequences l^p, for $1 < p < +\infty$, are reflexive (Theorem 20.1.). These spaces $L^p(\Omega)$ and l^p for $p = 1$ or $p = +\infty$ are not reflexive (cf. Exercises 36.1 and 36.2; also Corollary 2 of Lemma 44.2).

In relation with biduals and reflexivity, a certain number of results concerning normed spaces are of interest. Let E be a normed space,

$x \rightsquigarrow \| x \|$ its norm, E' its dual, and $x' \rightsquigarrow \| x' \|$ the dual norm on E', that is to say the norm

$$(36.1) \qquad \qquad \| x' \| = \sup_{x \in E, \, \| x' \| = 1} |\langle x', x \rangle|.$$

On the bidual E'' of E, we may consider the dual norm of (36.1), which we denote by $x'' \rightsquigarrow \| x'' \|$.

PROPOSITION 36.6.   *The mapping $x \rightsquigarrow$ value at x is an isometry of the normed space E into its bidual E'' (equipped with the bidual norm).*

Proof.   Let B (resp. B', resp. B'') be the *closed* unit ball in E (resp. E', resp. E''). In view of Propositions 35.2 and 35.3, B is equal to its own bipolar, B^{00}; $B' = B^0$ and $B'' = B'^0$ in the duality between E and E' and E' and E'', respectively. Therefore $B = E \cap B''$, where we have identified E with its image in E'', under the mapping $x \rightsquigarrow$ value at x.

COROLLARY.   *If E is a Banach space, $x \rightsquigarrow$ value at x is an isometry of E onto a closed linear subspace of its strong bidual, E''_b.*

Proposition 36.6 can be expressed by the relation

$$\| x \| = \sup_{x' \in E', \, \| x' \| \leqslant 1} |\langle x, x' \rangle|.$$

If E is normed but is not a Banach space, its closure $\bar{E}$ in E''_b is a Banach space, since E''_b is a Banach space (the strong dual of a normed space is a Banach space: corollary of Theorem 11.5). It is clear that $\bar{E}$ is canonically isomorphic to the completion $\hat{E}$ of E and can be identified with $\hat{E}$.

PROPOSITION 36.7.   *A normed space E which is semireflexive is a Banach space and therefore is reflexive.*

Proof.   To say that E is semireflexive is equivalent with saying that the mapping $x \rightsquigarrow$ value at x is a linear isometry of E onto E''_b; the latter being a Banach space, E must also be one. But then semireflexivity of E implies reflexivity of E, by Proposition 36.5.

Proposition 36.6 should be contrasted with the next one:

PROPOSITION 36.8.   *Let E be a normed space, and E'' its bidual equipped with the bidual norm. The unit ball of E, $\{x \in E; \| x \| \leqslant 1\}$, is dense in the unit ball of E'', $\{x'' \in E''; \| x'' \| < 1\}$, for the topology $\sigma(E'', E')$.*

Proof. We identify E with a linear subspace of E'' through the mapping $x \rightsquigarrow$ value at x. The unit ball B of E is a convex subset of E''; its bipolar, in the sense of the duality between E'' and E', is equal to its weakly closed convex balanced hull (here, weakly must be taken in the sense of $\sigma(E'', E')$). But this bipolar is the set $\{x'' \in E''; \| x'' \| \leqslant 1\}$, by definition of the bidual norm.

COROLLARY. *A normed space E is reflexive if and only if its closed unit ball is compact for the weak topology $\sigma(E, E')$.*

Indeed, the topology $\sigma(E, E')$ is obviously identical to the topology induced on E by $\sigma(E'', E')$; the closed unit ball of E is dense in the one of E'' for $\sigma(E'', E')$. If the closed unit ball of E is compact for $\sigma(E, E')$, it is equal to the one of E''; hence $E = E''$ (as normed spaces).

Example 36.4.   The Montel spaces (Definition 34.2). This is quite a different class of locally convex Hausdorff spaces from the Banach spaces: the only spaces which are both Banach spaces and Montel spaces are the finite dimensional Hausdorff TVS!

PROPOSITION 36.9.   *On a bounded subset B of a Montel space E, the initial topology and the weak topology $\sigma(E, E')$ coincide.*

Proof. The closure of B, $\bar{B}$, is compact (for the initial topology); $\sigma(E, E')$ is Hausdorff and coarser than the initial topology, whence the result by a well-known property of compact sets (Proposition 6.4).

COROLLARY.   *A Montel space is reflexive.*

According to Proposition 36.9, for $\sigma(E, E')$, the closed bounded subsets of E (supposed to be a Montel space) are compact. Hence E is semireflexive, by Theorem 36.3. But as E is barreled (Definition 34.2), E is reflexive, in view of Proposition 36.5.

PROPOSITION 36.10.   *The strong dual of a Montel space is a Montel space.*

Proof. Let E be a Montel space, and E'_b its strong dual. Since E is reflexive, E'_b is barreled (Proposition 36.4). Every closed bounded subset of E'_b is strongly compact, as has already been proved (Proposition 34.6).

In addition to the Montel spaces $\mathscr{C}^\infty(\Omega)$, $\mathscr{C}^\infty_c(\Omega)$ (Ω: open subset of $\mathbf{R}^n$), $\mathscr{S}$, and $H(\mathcal{O})$ ($\mathcal{O}$: open subset of $\mathbf{C}^n$), we have now the Montel spaces $\mathscr{E}'(\Omega)$, $\mathscr{D}'(\Omega)$, $\mathscr{S}'$, $H'(\mathcal{O})$, their strong duals (see Proposition 34.4). All these spaces are reflexive.

Exercises

36.1. Prove that a Banach space E is reflexive if and only if E_b' is reflexive.

36.2. Let l^1 be the Banach space of complex sequences $\sigma = (\sigma_k)$ $(k = 0, 1,...)$ such that $|\sigma|_1 = \Sigma_{k=0}^{+\infty} |\sigma_k| < +\infty$, and l^∞ the Banach space of bounded complex sequences $\tau = (\tau_k)$ $(k = 0, 1,...)$, with its natural norm $|\tau|_\infty = \sup_{k=0,1,...} |\tau_k|$ (Chapter 11, Example IV). Let us denote by l_∞ the linear subspace of l^∞ consisting of the sequences $\tau = (\tau_k)$ such that $\tau_k \to 0$ as $k \to +\infty$. Prove that l_∞ is a closed linear subspace of l^∞ (hence is a Banach space for the norm $|\ |_\infty$) and that there is a canonical isometry of l^1 onto the dual of l_∞. Derive from this that l^1 and l^∞ are not reflexive (use also Exercise 36.1).

36.3. Using the Hahn–Banach theorem and the reflexivity of the spaces involved, prove that $\mathscr{C}_c^\infty(\Omega)$ is dense in $\mathscr{E}'(\Omega)$ and in $\mathscr{D}'(\Omega)$ (Ω: open subset of $\mathbf{R}^n$), that the (finite) linear combinations of Dirac measures at the points of Ω are dense in $\mathscr{E}'(\Omega)$ and $\mathscr{D}'(\Omega)$, and that the finite linear combinations of Dirac measures at the points of $\mathbf{R}^n$ are dense in $\mathscr{S}'$.

36.4. Let T be a topological space, E a TVS, E' its dual, and $t \rightsquigarrow x(t)$ a mapping of T into E. Let us say that this mapping is *scalarly continuous* if for every $x' \in E'$ the mapping $t \rightsquigarrow \langle x', x(t) \rangle$ of T into $\mathbf{C}$ is continuous. When T is an open subset of $\mathbf{R}^n$ (resp. of $\mathbf{C}^n$), we may say that $t \rightsquigarrow x(t)$ is *scalarly k-times continuously differentiable* $(0 \leqslant k \leqslant +\infty)$ (resp. *scalarly holomorphic*) if this is true of the mapping $t \rightsquigarrow \langle x', x(t) \rangle$ for all $x' \in E'$, etc. Prove the following result:

PROPOSITION 36.11. *If E is a Montel space, and if the mapping $t \rightsquigarrow x(t)$ is scalarly continuous (resp. scalarly k-times continuously differentiable, resp. scalarly holomorphic), then this mapping is continuous (resp. $\mathscr{C}^k$, resp. holomorphic).*

36.5. Prove the following result by using Lemma 36.1:

Let E be a locally convex Hausdorff space, and F a locally convex space. A subset H of $L(E; F)$, space of continuous linear mappings of E into F, which is bounded for the topology of pointwise convergence, is also bounded for the topology of uniform convergence on the convex balanced *complete* bounded subsets of E.

36.6. Prove that the spaces $H_{\mathrm{loc}}^s(\Omega)$ and $H_c^s(\Omega)$ are reflexive (see p. 332).

37

Surjections of Fréchet Spaces

In this chapter, we shall state and prove a very important theorem due to S. Banach. Let E, F be two Fréchet spaces, and u a continuous linear map of E into F. The theorem gives necessary and sufficient conditions, bearing on the transpose ${}^t u$ of u, in order that u be a surjection, i.e., u be a mapping of E *onto* F (i.e., $u(E) = F$). From Eq. (35.3) we derive immediately that Im u is dense if and only if Ker ${}^t u = \{0\}$, in other words if and only if ${}^t u$ is one-to-one. Note that the fact that u is onto means that Im u is dense and closed. But in general, the fact that ${}^t u$ is one-to-one will not be enough to ensure that Im u is closed. It is of course so when F is finite dimensional, since every linear subspace of a finite dimensional Hausdorff TVS is closed; but it is not so, in general, when F is infinite dimensional. Examples are easy to exhibit: it suffices to consider two Hilbert spaces H_1, H_2 such that there is a continuous injection with dense image of H_1 into H_2 which is not an isomorphism (in which case, the injection cannot be onto, in view of the open mapping theorem). Take for instance $H_1 = H^1, H_2 = H^0 = L^2(\mathbf{R}^n)$ (Definition 31.4; cf. Proposition 31.9): the injection of H^1 into H^0 is the natural one, expressing that $H^1 \subset H^0$ when we regard both spaces as subsets of the space of distributions in $\mathbf{R}^n$.

Thus, we see that, if $u(E)$ is to be equal to F, we must have some further condition, in addition to the fact that ${}^t u : F' \to E'$ be one-to-one. As we shall see, the additional condition on ${}^t u$ will simply be that ${}^t u$ itself have its image closed—provided that *closed* be understood in the sense of the weak dual topology $\sigma(E', E)$ on E'. When E is reflexive, for instance when E is a Hilbert space, this is equivalent with saying that ${}^t u(F')$ is closed for the strong dual topology.

The introduction of the condition that Im ${}^t u$ be weakly closed raises the following question: is there a way of characterizing the weakly closed linear subspaces of the dual of a Fréchet space? We shall begin by stating and proving such a characterization; it is due to S. Banach and is quite simple, as will be seen. We shall soon apply it to the proof of an important complement (Theorem 37.3) to the main theorem (Theorem 37.2).

THEOREM 37.1. *Let E be a Fréchet space. A linear subspace M' of the dual E' of E is weakly closed if and only if the following property holds*:

(37.1) *There is a basis of neighborhoods of the origin in E, $\mathscr{B}$, such that, for every $U \in \mathscr{B}$, the intersection of M' with the polar U^0 of U is weakly closed.*

Remark 37.1. We recall that the subsets U^0 of E are weakly compact (Propositions 32.7 and 32.8). It should also be pointed out that (37.1) is equivalent to the following property:

(37.2) *The intersection of M' with every equicontinuous subset H' of E' is relatively closed in H' (for the topology induced by $\sigma(E', E)$).*

In particular (cf. Proposition 32.7), (37.1) is equivalent to the fact that the intersection of M' with the polar of every neighborhood of zero in E is weakly closed. Since U^0 is weakly compact and equicontinuous (Proposition 32.7), (37.2) implies (37.1). Conversely, every equicontinuous subset H' of E' is contained in some U^0 for $U \in \mathscr{B}$ (again by Proposition 32.7). If $M' \cap U^0$ is closed in E', $M' \cap H' = M' \cap U^0 \cap H'$ must be closed in H'. Q.E.D.

Proof of Theorem 37.1

As U^0 is weakly compact, it is obvious that $M' \cap U^0$ will be weakly closed whenever M' is weakly closed. Therefore, it will suffice to prove that (37.1) or, equivalently, (37.2) implies that M' is weakly closed.

We shall begin by showing that, under the hypotheses of the theorem, it is equivalent to say that M' is weakly closed or that M' is closed for the topology of compact convergence. This will follow from the fact that E'_σ and E'_c have the same dual, namely E. It will then suffice to apply Proposition 36.2. Our statement will be a consequence of the following two lemmas:

LEMMA 37.1. *Let E be a locally convex Hausdorff TVS. The topology γ (on the dual E' of E) of compact convex convergence in E is compatible with the duality between E' and E (see Chapter 36, p. 369).*

In other words, $x \rightsquigarrow$ value at x is an isomorphism (for the vector spaces structures) of E onto the dual of E'_γ.

LEMMA 37.2. *On the dual E' of a Fréchet space E, the topology γ of compact convex convergence is identical to the topology c of compact convergence.*

Proof of Lemma 37.1. The topology γ is identical to the topology of uniform convergence on every compact convex balanced subset of E. Such a set is obviously compact also for the weak topology $\sigma(E, E')$ on E, since $\sigma(E, E')$ is Hausdorff and weaker than the initial topology of E (Proposition 6.8). It then suffices to apply Theorem 36.1.

Proof of Lemma 37.2. Whether E is a Fréchet space or not, the topology of compact convergence on E' is equivalent to the topology of uniform convergence over the closed convex hulls of the compact subsets of E. It will therefore suffice to show that the closed convex hull $\hat{K}$ of a compact set K of a Fréchet space E is compact. As $\hat{K}$ is closed, hence complete, it will suffice to prove that $\hat{K}$ is precompact (Proposition 8.4). But this is stated in Proposition 7.11.

At this stage, we have reduced the problem to showing that, under the hypothesis that E is a Fréchet space, (37.2) implies that M' is closed, i.e., that the complement of M' is open, for the topology of compact convergence $c(E', E)$. In order to prove this last implication, we only need that E be metrizable, as stated in the following lemma:

LEMMA 37.3. *Let W' be a subset of E' whose intersection with every equicontinuous subset of E' is weakly open. If E is metrizable, W is open for the topology of compact convergence.*

Proof of Lemma 37.3. Let $x' \in W'$; we must show that, under our hypotheses, there is a compact subset K of E such that $x' + K^0 \subset W'$; by performing a translation, if necessary, we may assume that $x' = 0$. Let $\{U_k\}$ $(k = 1, 2,...)$ be a basis of neighborhoods of zero in E such that $U_{k+1} \subset U_k$ for all k; the polars U_k^0 are equicontinuous sets.

Let us set $U_0 = E$. We shall construct a sequence of finite sets $B_n \subset U_n$ $(n = 0, 1,...)$ such that, for every $n = 1, 2,...$,

$$U_n^0 \cap A_n^0 \subset W', \qquad A_n^0 = B_0 \cup \cdots \cup B_{n-1}.$$

We do this by induction on n. No problem in selecting A_1 since $U_0^0 = \{0\}$ and since the origin belongs to W'. Suppose that we have selected A_n. Then we contend that there is a finite subset B_n of U_n such that

$$U_{n+1}^0 \cap (A_n \cup B_n)^0 \subset W'.$$

If we prove this assertion, we will have established the existence of the

B_k's for all k. We reason by contradiction. Suppose that a set like B_n did not exist. Let us denote by C' the complement of W' and set $C'_n = U^0_{n+1} \cap A^0_n \cap C'$. Since, by hypothesis, W' is weakly relatively open in U^0_{n+1}, C'_n is a closed subset of $U^0_{n+1} \cap A^0_n$, hence is weakly compact. According to our line of argument, given any finite subset B of U_n, B^0 would intersect C'_n; as any finite intersection of sets B^0 is of the same form, we see that, because of the compactness of C'_n, the intersection of all the sets $B^0 \cap C'_n$ should contain at least one point x'^0. This x'^0 belongs to the polar of every finite subset of U_n, hence to the polar of U_n, i.e.,

$$x' \in U^0_n \cap A^0_n \cap C'.$$

But in view of the induction on n, $U^0_n \cap A^0_n \subset W'$, the complement of C'. We have thus reached a contradiction and therefore proved the existence of the sequence of sets B_n. Let S be their union. We have

$$\bigcup_{n=1}^{\infty} U^0_n \cap S^0 = \bigcup_{n=1}^{\infty} (U^0_n \cap A^0_n) \subset W'.$$

Now, the union of the sets U^0_n ($n = 1, 2,...$) is obviously the whole dual space E', whence $S^0 \subset W'$. But S is a sequence which converges to zero in E, therefore $S \cup \{0\}$ is a compact set. This means that S^0 is a neighborhood of the origin for the topology of compact convergence.

This proves Lemma 37.3 and therefore Theorem 37.1.

Before stating and proving the main theorem of this chapter, we shall introduce the following notation:

Notation 37.1. Let E be a locally convex Hausdorff TVS, and p a continuous seminorm on E. We denote by E'_p the vector space of the linear functionals on E which are continuous for the topology defined by the seminorm p.

Let us denote by E_p the vector space E equipped with the topology defined by p. Since p is continuous, the canonical mapping

$$E \to E_p/\text{Ker } p$$

is continuous (when E carries its initial topology and $E_p/\text{Ker } p$, the quotient of the topology defined by p); its transpose is a continuous injection of the dual of $E_p/\text{Ker } p$ into E', whose image is exactly E'_p as one checks immediately. Let U_p be the closed unit semiball of the seminorm p, $U_p = \{x \in E \mid p(x) \leqslant 1\}$; let U^0_p be the polar of U_p. Then E'_p is the linear subspace of E' spanned by U^0_p. As E'_p is algebraically isomorphic to the dual of the normed space $E_p/\text{Ker } p$, it can be canonically equipped with a structure of Banach space; the norm, in this structure, is nothing else but the "gauge" of the weakly compact convex balanced set U_p^0, which is absorbing in E_p',

$$E_p' \ni x' \rightsquigarrow p'(x') = \inf_{\lambda > 0, x' \in \lambda U_p^0} \lambda.$$

The fact that p' turns E_p' into a Banach space can also be derived from the corollary of Lemma 36.1.

Exercises

37.1. Show that $E_p' \subset (\mathrm{Ker}\ p)^0$ and that E_p' is dense in $(\mathrm{Ker}\ p)^0$ for the weak topology $\sigma(E', E)$.

37.2. Give an example where $E_p' \neq (\mathrm{Ker}\ p)^0$.

37.3. Let $E = \mathscr{C}^0(\Omega)$, the space of continuous complex functions in an open subset Ω of $\mathbf{R}^n$, equipped with the topology of uniform convergence on every compact subset K of Ω. Let p be the seminorm $f \rightsquigarrow \sup_{x \in K} |f(x)|$. Prove that, in this case, $E_p' = (\mathrm{Ker}\ p)^0 =$ set of Radon measures in Ω with support in K.

37.4. Prove the following result (which is going to be used later on):

PROPOSITION 37.1.   *Let E, F be two locally Hausdorff TVS. If u is a homomorphism of E onto F, to every continuous seminorm p on E there is a continuous seminorm q on F, such that, for all $y' \in F'$,*

$$^t u(y') \in E_p' \qquad implies \qquad y' \in F_q'.$$

Now, we state and prove the main theorem of this chapter:

THEOREM 37.2.   *Let E, F be two Fréchet spaces, and u a continuous linear map of E into F. Then u maps E onto F if and only if the following two conditions are satisfied:*

(a)   *the transpose of u, $^t u : F' \to E'$, is one-to-one;*

(b)   *the image of $^t u$, $^t u(F')$, is weakly closed in E'.*

We state also the announced complement to this result; this complement will be proved and applied in the next chapter.

THEOREM 37.3.   *Let E, F, and u be as in Theorem 37.2. Then the following facts are equivalent:*

(I)   *u maps E onto F;*

(II)   *to every continuous seminorm p on E there is a continuous seminorm q on F such that the following is true:*

(II_1) to every $y \in F$ there is $x \in E$ such that $q(u(x) - y) = 0$;

(II_2) for all $y' \in F'$, $^t u(y') \in E'_p$ implies $y' \in F'_q$;

(III)   to every continuous seminorm p on E there is a linear subspace N of F such that the following is true:

(III_1) to every $y \in F$ there is $x \in E$ such that $u(x) - y \in N$;

(III_2) for all $y' \in F'$, $^t u(y') \in E'_p$ implies $y' \in N^0$;

(IV)   there is a nonincreasing sequence $N_1 \supset N_2 \supset \cdots \supset N_k \supset \cdots$ of closed linear subspaces of F, whose intersection is equal to $\{0\}$, and such that the following is true:

(IV_1) to every $k = 1, 2,...$ and to every $y \in F$, there is $x \in E$ such that $u(x) - y \in N_k$;

(IV_2) to every continuous seminorm p on E there is an integer $k \geqslant 1$ such that every $x \in E$ satisfying

$$u(x) \in N_k$$

is the limit, in the sense of the seminorm p, of a sequence of elements x_ν ($\nu = 1, 2,...$) of E satisfying, for all ν,

$$u(x_\nu) = 0.$$

By saying that x is the limit, in the sense of p, of the x_ν's, we mean that $p(x - x_\nu) \to 0$ as $\nu \to + \infty$.

Proof of Theorem 37.2

Since Ker $^t u = (\text{Im } u)^0$, we see that (a) is equivalent with the fact that Im $u = u(E)$ is dense in F. The theorem will be proved if we show that (b) is equivalent with the fact that $u(E)$ is closed in F. This will be done through the application of Proposition 35.7, where we substitute $^t u$ for u, F'_σ for E, E'_σ for F, thus obtaining:

LEMMA 37.4.   Let E, F be two locally convex Hausdorff TVS, and u a continuous linear map of E into F. The following two properties are equivalent:

(a) $^t u(F')$ is weakly closed in E';

(b) u is a homomorphism of E, equipped with the weak topology $\sigma(E, E')$, onto $u(E)$, equipped with the topology induced by $\sigma(F, F')$.

The crux of the proof of Theorem 37.2 lies in the next lemma:

LEMMA 37.5. *If E and F are both metrizable, the following properties are equivalent:*

(a) *u is a homomorphism of E, equipped with $\sigma(E, E')$, onto $u(E)$, equipped with the topology induced by $\sigma(F, F')$;*

(b) *u is a homomorphism of E onto $u(E) \subset F$ (for the initial topologies).*

Before presenting the proof of Lemma 37.5, we shall show how it implies the equivalence of Property (a) of Lemma 37.4 with the fact that $u(E)$ is closed in F. Consider the usual diagram associated with the linear mapping u,

(37.3)

$$
\begin{array}{ccc}
E & \xrightarrow{\quad u \quad} & u(E) \subset F \\
{\scriptstyle \varphi}\downarrow & \nearrow_{v} & \\
E/\mathrm{Ker}\, u, &
\end{array}
$$

where φ is the canonical homomorphism of E onto the quotient space $E/\mathrm{Ker}\, u$ and v is the unique one-to-one linear map of $E/\mathrm{Ker}\, u$ onto $u(E)$ which makes the triangle (37.3) commutative. Since u is continuous, so is v. By definition, u is a homomorphism if v is an isomorphism (for the TVS structures). But when both E and F are Fréchet spaces, v is an isomorphism if and only if $u(E)$ is closed. Indeed, if v is an isomorphism, $u(E) = \mathrm{Im}\, v$ is isomorphic to the Fréchet space $E/\mathrm{Ker}\, u$, hence is closed. Conversely, if $u(E)$ is closed in F, $u(E)$ is a Fréchet space for the induced topology and therefore v is a one-to-one continuous linear map of a Fréchet space, $E/\mathrm{Ker}\, u$, onto another one, $u(E)$; v must be an isomorphism in view of the open mapping theorem (Theorem 17.1).

Thus, we are left with the proof of Lemma 37.5:

PROOF OF LEMMA 37.5.   We go back to Diagram (37.3). What we ought to show is the following; v is an isomorphism (for the initial topologies on E and F and the quotient topology on $E/\mathrm{Ker}\, u$) if and only if v is an isomorphism when E carries $\sigma(E, E')$, F carries $\sigma(F, F')$, and $E/\mathrm{Ker}\, u$ carries $\sigma(E, E')/\mathrm{Ker}\, u$.

We shall apply to v and to its inverse v^{-1} (defined on $u(E)$) the following result:

LEMMA 37.6. *Let E be a metrizable locally convex TVS, F a locally convex Hausdorff TVS, and u a linear map of E into F. If u is continuous when E and F carry their respective weak topologies $\sigma(E, E')$ and $\sigma(F, F')$, then u is continuous (for the initial topologies).*

PROOF OF LEMMA 37.6.   Let $\{U_k\}$ $(k = 1, 2, ...)$ be a basis of neighborhoods of zero in E such that $U_k \subset U_{k-1}$ ($k \geqslant 2$). If u were not continuous, there would be a balanced neighborhood of zero V in F such that $u^{-1}(V)$ would not be a neighborhood of zero in E, hence would not contain any one of the sets $(1/k)\, U_k$; for each k, we would be able to select an element x_k of E such that $kx_k \in U_k$ and such that $u(x_k) \notin V$. Since the sequence kx_k converges to zero in E, and a fortiori converges to zero for $\sigma(E, E')$, we see that the sequence $k\, u(x_k)$ is relatively compact in F for $\sigma(F, F')$. In view of Mackey's theorem (Theorem 36.2), this sequence is also bounded in F (for the initial topology). Therefore there is a number $\lambda > 0$ such that $k\, u(x_k) \in \lambda V$ for all k; this implies that, for k sufficiently large, $u(x_k) \in (\lambda/k)\, V \subset V$, contrary to our assumption about the vectors x_k .   Q.E.D.

If we combine Lemma 37.6 with Proposition 35.8, we see that a linear mapping of a metrizable locally convex TVS into a locally convex Hausdorff TVS is continuous if and only if it is continuous when both spaces carry their respective weak topologies. Keeping this in mind, we proceed with the proof of Lemma 37.5.

Recalling that $\mathrm{Ker}\, u$ is closed in E (since u is continuous), we see that we are dealing with a one-to-one linear map v of a metrizable space $G = E/\mathrm{Ker}\, u$, onto the metrizable space $H = u(E) \subset F$. According to what has just been said, v is continuous as a mapping of G, equipped with $\sigma(G, G')$, onto H, equipped with $\sigma(H, H')$; similarly, $v^{-1} : H \to G$ is continuous if and only if it is continuous when H carries $\sigma(H, H')$ and G carries $\sigma(G, G')$. Lemma 37.5 will therefore be proved if we prove the following two facts:

(37.4) $\sigma(G, G') = \sigma(E, E')/\mathrm{Ker}\, u;$

(37.5) $\sigma(H, H')$ is identical to the topology induced on H by $\sigma(F, F')$.

Recalling that H' is canonically isomorphic to F'/H^0, the statement (37.5) is absolutely obvious. As for (37.4), it will follow immediately from Propositions 35.5 and 35.6. Indeed, Part (a) of Proposition 35.5 says that the dual G' of $G = E/\mathrm{Ker}\, u$ is canonically isomorphic to $(\mathrm{Ker}\, u)^0$. On the other hand, if we apply Proposition 35.6 with E and E' exchanged and with $N = (\mathrm{Ker}\, u)^0$ (observing that this is a *closed* linear subspace of E'_σ), we see that

$$\sigma(E/\mathrm{Ker}\, u, (\mathrm{Ker}\, u)^0) = \sigma(E, E')/\mathrm{Ker}\, u.$$

Combining these facts, we obtain (37.4).

The proof of Theorem 37.2 is complete. The proof of Theorem 37.3 will be given in the next chapter.

37.5. Prove that Theorem 37.2 remains valid if we replace the assumption that E and F be Fréchet spaces by the one that E and F be duals of reflexive Fréchet spaces.

37.6. Let $P(\partial/\partial X) = P(\partial/\partial X_1, ..., \partial/\partial X_n)$, and $P \in \mathscr{P}_n$, i.e., a polynomial in n indeterminates with complex coefficients. We let $P(\partial/\partial X)$ operate on $\mathscr{Q}_n$, the space of formal power series in $X = (X_1, ..., X_n)$. Prove that $P(\partial/\partial X)$ maps $\mathscr{Q}_n$ *onto* itself, unless $P = 0$ (use Theorem 37.2).

37.7. Prove that every ideal $u\mathscr{Q}_n$, $u \in \mathscr{Q}_n$, is closed in $\mathscr{Q}_n$ (study the transpose of the mapping $v \to uv$ of $\mathscr{Q}_n$ into itself; show that it is surjective and apply Exercise 37.5).

37.8. Prove the following result:

LEMMA 37.7. *Let E, F be two locally convex Hausdorff TVS, and $u : E \to F$ a continuous linear map. The following two properties are equivalent:*

(a) *u is a homomorphism of E onto $u(E) \subset F$;*
(b) *${}^t u(F')$ is weakly closed in E' and every equicontinuous set $A' \subset {}^t u(F')$ is the image, under ${}^t u$, of an equicontinuous subset of F'.*

38

Surjections of Fréchet Spaces (continued).
Applications.

We proceed to give the proof of Theorem 37.3:

Proof of Theorem 37.3

(I) implies (II) by virtue of Proposition 37.1. Indeed, if $u(E) = F$, u is a homomorphism of E onto F, by the open mapping theorem. Trivially, (II) implies (III): it suffices to take $N = \text{Ker } q$.

We prove, now, that (III) implies (IV). For this, we prove first that (III) implies Property (b) in Theorem 37.2, that is to say that ${}^t u(F')$ is weakly closed in E'. We apply Theorem 37.1, in the following manner. As a basis of neighborhoods of zero in E, we take the closed unit semiballs U_p of the continuous seminorms p on E, and we prove that ${}^t u(F') \cap U_p^0$ is weakly closed. Let H' be the preimage of this set under the mapping ${}^t u$. Let N be a linear subspace of F associated with the seminorm p as in (III). If $y' \in H'$, then ${}^t u(y') \in U_p^0 \subset E_p'$ (Notation 37.1), therefore $y' \in N^0$. Let y be an arbitrary element of F, $x \in E$ such that $y - u(x) \in N$. We have (for $y' \in H'$):

$$\langle y', y \rangle = \langle y', u(x) \rangle = \langle {}^t u(y'), x \rangle.$$

This equality proves, first of all, that $y' \leadsto {}^t u(y')$ is a homeomorphism of H' onto its image, ${}^t u(F') \cap U_p^0$, for the topologies induced by $\sigma(F', F)$ and $\sigma(E', E)$, respectively, and, second, that H' is weakly bounded, since ${}^t u(y')$ varies in the weakly bounded set U_p^0 as y' runs over H'. But since F is barreled (Corollary 1 of Proposition 33.2), we derive from the Banach–Steinhaus theorem (Theorem 33.2) that H' is equicontinuous and its weak closure $\bar{H}'$ is a weakly compact subset of F' (Proposition 32.8). The restriction of ${}^t u$ to $\bar{H}'$ is then a homeomorphism of $\bar{H}'$ onto the weak closure of ${}^t u(F') \cap U_p^0$. The latter weak closure is contained in the weakly compact set U_p^0 ; since it is equal to ${}^t u(\bar{H}')$, it is also contained in ${}^t u(F')$. Consequently, it is equal to ${}^t u(F') \cap U_p^0$.

Thus (III) implies that ${}^t u(F')$ is weakly closed. As we have

$$\operatorname{Ker} {}^t u = (\operatorname{Im} u)^0, \qquad \text{hence} \quad \operatorname{Ker} u = (\operatorname{Im} {}^t u)^0,$$

hence

$$(\operatorname{Ker} u)^0 = (\operatorname{Im} {}^t u)^{00} = \text{weak closure of} \quad \operatorname{Im} {}^t u,$$

we have, in the present situation,

(38.1) $\operatorname{Im} {}^t u = (\operatorname{Ker} u)^0.$

This identity will soon be used.

We now complete the proof of the implication (III) $\Rightarrow$ (IV). Let $p_1 \leqslant p_2 \leqslant \cdots \leqslant p_k \leqslant \cdots$ be a sequence of continuous seminorms on E such that, for every continuous seminorm p on E, there is an integer $k \geqslant 1$ such that $p \leqslant p_k$. That such a sequence exists is evident: it suffices to take the seminorms associated with the sets belonging to a decreasing countable basis of neighborhoods of zero, consisting of convex closed balanced neighborhoods of zero. For each $k = 1, 2,...$, let N^k be a linear subspace of F associated with p_k as in (III). We take as linear subspace N_k the closure of the algebraic sum of the subspaces N^l for $l \geqslant k$ (an element of this algebraic sum is the sum of a finite number of vectors belonging to the $N^l, l \geqslant k$). We start by proving that the intersection of the N_k is equal to $\{0\}$. Let $y' \in F'$ be arbitrary. Since ${}^t u(y')$ is a continuous linear form on E, there is an integer $k \geqslant 1$ such that ${}^t u(y') \in E'_{p_k}$ (Notation 37.1), therefore, in view of (III$_2$), $y' \in (N^k)^0$. As a matter of fact, since $p_k \leqslant p_l$ for $l \geqslant k$, we have $E'_{p_k} \subset E'_{p_l}$ and therefore $y' \in (N^l)^0$ for those l. Since y' belongs to the orthogonal of every $N^l, l \geqslant k$, y' belongs to the orthogonal of their algebraic sum and therefore also to the orthogonal of the closure of their algebraic sum, hence to $(N_k)^0$. In particular, y' belongs to the orthogonal of the intersection of all the $N_j, j = 1, 2,....$ Since y' is arbitrary, it follows from the Hahn–Banach theorem that this intersection must be equal to $\{0\}$.

Since $N^k \subset N_k$, (IV$_1$) is a trivial consequence of (III$_1$).

We derive now (IV$_2$). Let p be an arbitrary continuous seminorm on E, and $k \geqslant 1$ be such that $p \leqslant p_k$. We show that Ker u is dense in $H_k = u^{-1}(N_k)$ for the topology defined by the seminorm p. In view of Corollary 1 of the Hahn–Banach theorem (Theorem 18.1), it suffices to show that every linear form on E, continuous for the seminorm p, which vanishes on Ker u also vanishes on H_k . Let x' be such a form. Since $x' \in E'$ is orthogonal to Ker u, we derive from (38.1) that there is $y' \in F'$ such that $x' = {}^t u(y')$. Since $x' \in E'_{p_k}$, we derive from (III_2) that $y' \in (N^k)^0$. As a matter of fact, $x' \in E'_{p_l}$ for all $l \geqslant k$, therefore, by the same argument used above, $y' \in (N_k)^0$. But then, if $x \in H_k$,

$$\langle x', x \rangle = \langle {}^t u(y'), x \rangle = \langle y, u(x) \rangle = 0 \qquad \text{since} \quad u(x) \in N_k.$$

The implication (III) $\Rightarrow$ (IV) is completely proved; it remains to show that (IV) implies (I).

Let us use a basis of continuous seminorms p_k on E as before. By possibly renaming the N_k's, we may assume that, for each k, N_k is associated to p_k as in (IV_2): every $x \in u^{-1}(N_k)$ is the limit for p_k of a sequence in Ker u. Let y be an arbitrary element of F. By applying (IV_1), we may find $x_1 \in E$ such that $y_1 = y - u(x_1) \in N_1$ and then, by induction on $k = 2, 3, \ldots$, a sequence of elements x_k in E, y_k in F such that

$$y_k = y_{k-1} - u(x_k) \in N_k.$$

Observing that $u(x_k) = y_{k-1} - y_k \ (k \geqslant 2)$, we see that there is $h_k \in \text{Ker } u$ such that $p_{k-1}(x_k - h_k) \leqslant 2^{-k}$. The series

$$x_1 + \sum_{k=2}^{\infty} (x_k - h_k)$$

converges absolutely in E, defining there an element x (since E is complete). Let us set, for $r > 1$,

$$z_r = x_1 + \sum_{k=2}^{r} (x_k - h_k).$$

We have

$$u(z_r) = \sum_{k=1}^{r} u(x_k) = y - y_r.$$

Therefore

$$u(z_r) - y \in N_r.$$

But since $N_s \subset N_r$ for all $s \geqslant r$, we also have

$$u(z_s) - y \in N_r.$$

By going to the limit as $s \to +\infty$ and recalling that N_r is closed, we obtain

$$u(x) - y \in N_r.$$

As the integer r is arbitrarily large, $y - u(x)$ belongs to the intersection of all the N_r's, therefore is equal to zero.

The proof of Theorem 37.3 is complete.

An Application of Theorem 37.2: A Theorem of E. Borel

We shall now give a very simple application of Theorem 37.2. We shall show how it enables us to prove the following classical theorem of E. Borel:

THEOREM 38.1.   *Let Φ be an arbitrary formal power series in n indeterminates, with complex coefficients. There is a $\mathscr{C}^\infty$ function φ in $\mathbf{R}^n$ whose Taylor expansion at the origin is identical to Φ.*

In other words if, for every n-tuple $p = (p_1, \ldots, p_n)$ of integers $p_j \geqslant 0$, we give ourselves arbitrarily some complex number a_p, there is a $\mathscr{C}^\infty$ function φ in $\mathbf{R}^n$ such that $(\partial/\partial x)^p \varphi \mid_{x=0} = a_p$ for every p. Of course, the origin in $\mathbf{R}^n$ can be replaced by any other point.

Proof of Theorem 38.1.   Let us denote by u the mapping which assigns to every function $\varphi \in \mathscr{C}^\infty(\mathbf{R}^n)$ its Taylor expansion at the origin; we regard the latter as an element of the space $\mathscr{Q}_n$ of formal power series in n letters with complex coefficients. We must show that u is a surjection. We provide $\mathscr{C}^\infty(\mathbf{R}^n)$ with the natural $\mathscr{C}^\infty$ topology and $\mathscr{Q}_n$ with the topology of simple convergence of the coefficients (Chapter 10, Examples I and III). The dual of $\mathscr{C}^\infty(\mathbf{R}^n)$ is the space of distributions with compact support in $\mathbf{R}^n$; the dual of $\mathscr{Q}_n$ is the space $\mathscr{P}_n$ of polynomials in n letters with complex coefficients (Chapter 22). What is then the transpose ${}^t u$ of u?

Observe that the mapping u is the mapping

$$\varphi \rightsquigarrow \sum_{p \in \mathbf{N}^n} \frac{1}{p!} [(\partial/\partial x)^p \varphi(0)] X^p.$$

If $\langle \, , \, \rangle$ denotes the bracket of the duality between $\mathscr{C}^\infty(\mathbf{R}^n)$ and $\mathscr{E}'$ on one hand, and between $\mathscr{P}_n$ and $\mathscr{Q}_n$ on the other, we see that we have, for any polynomial

$$P(X) = \sum_{p \in \mathbf{N}^n} P_p X^p,$$

$$\langle P, u(\varphi) \rangle = \sum_{p \in \mathbf{N}^n} \frac{1}{p!} P_p [(\partial/\partial x)^p \varphi(0)] = \langle \check{P}(-\partial/\partial x)\delta, \varphi \rangle,$$

where δ is the Dirac measure at the origin and where we have set

$$\tilde{P}(-\partial/\partial x) = \sum_{p \in N^n} (-1)^{|p|} \frac{1}{p!} P_p(\partial/\partial x)^p.$$

This means that the transpose $^t u$ of u is the mapping $P \rightsquigarrow \tilde{P}(-\partial/\partial x)\delta$ of $\mathscr{P}_n$ into $\mathscr{E}'$; it is clear what the image of $^t u$ is: the space of all linear combinations of derivatives of the Dirac measure at 0. But we know (Theorem 24.6) that this space is identical to the space of distributions having the origin $\{0\}$ as support (plus the zero distribution!); the latter space is trivially closed, i.e., weakly closed. As $^t u$ is obviously one-to-one (apply for instance the Fourier transformation to $\tilde{P}(-\partial/\partial x)\delta$), we see that Conditions (a) and (b) of Theorem 37.2 are satisfied, whence the result.   Q.E.D.

An Application of Theorem 37.3: A Theorem of Existence of $\mathscr{C}^\infty$ Solutions of a Linear Partial Differential Equation

As usual, Ω will denote an open subset of $\mathbf{R}^n$. We consider a linear partial differential operator D, with $\mathscr{C}^\infty$ coefficients, defined in Ω (see Chapter 23, Example III). We are going to prove necessary and sufficient conditions, bearing on the pair of objects Ω, D, in order that the equation

$$Du = f$$

have a solution $u \in \mathscr{C}^\infty(\Omega)$ for every $f \in \mathscr{C}^\infty(\Omega)$. This property of the equation $Du = f$ can be rephrased by saying that D, which is a continuous linear operator of $\mathscr{C}^\infty(\Omega)$ into itself, is in fact a surjection of that Fréchet space onto itself, i.e.,

$$D\mathscr{C}^\infty(\Omega) = \mathscr{C}^\infty(\Omega).$$

In order to state the announced necessary and sufficient conditions for this to be true, we shall make use of two definitions, which we now state.

Definition 38.1.   *We say that the open set Ω is D-convex if to every compact subset K of Ω and to every integer $k \geqslant 0$ there is a compact subset $\hat{K}(k)$ of Ω such that, for every distribution μ with compact support in Ω, the following is true:*

(38.2)   *If $^t D\mu$ is of order $\leqslant k$ and if* supp $^t D\mu \subset K$, *then* supp $\mu \subset \hat{K}(k)$.

We explain the notation used: if T is a distribution, supp T is the support of T. We have viewed the differential operator D as a continuous linear mapping of $\mathscr{C}_c^\infty(\Omega)$ (or $\mathscr{C}^\infty(\Omega)$) into itself (Proposition 23.4); then ^tD is the differential operator defined as the transpose of that mapping; ^tD is a continuous linear map of $\mathscr{D}'(\Omega)$ (or $\mathscr{E}'(\Omega)$) into itself. If the expression of D in the coordinates $(x_1,\ldots,x_n)$ is

$$P(x, \partial/\partial x) = \sum_{|p|\leqslant m} a_p(x)\,(\partial/\partial x)^p, \qquad m\text{: order of } D, \quad a_p \in \mathscr{C}^\infty(\Omega),$$

the expression of ^tD in the same coordinate system is given by the "formal transpose" of $P(x, \partial/\partial x)$ (Definition 23.3),

$$^tP(x, \partial/\partial x) = \sum_{|p|\leqslant m} (-1)^{|p|}(\partial/\partial x)^p\, a_p(x).$$

For the concept of the order of a distribution, see Definition 24.3.

We shall also need the following definition:

Definition 38.2.   We say that the differential operator D is semiglobally solvable in Ω if, for every relatively compact open subset Ω' of Ω, the following property holds:

(38.3)   *To every function $\phi \in \mathscr{C}^\infty(\Omega)$ there is $\psi \in \mathscr{C}^\infty(\Omega)$ such that $D\psi = \phi$ in Ω'.*

We may now state the announced result:

THEOREM 38.2.   *Let D be a linear partial differential operator with $\mathscr{C}^\infty$ coefficients in the open set $\Omega \subset \mathbf{R}^n$. The following two properties are equivalent:*

(38.4)   *To every $f \in \mathscr{C}^\infty(\Omega)$, there is $u \in \mathscr{C}^\infty(\Omega)$ such that $Du = f$.*

(38.5)   *The open set Ω is D-convex and D is semiglobally solvable in Ω.*

Proof of Theorem 38.2.   We shall apply Theorem 37.3 with $E = F = \mathscr{C}^\infty(\Omega)$, $u = D$, and show that (38.5) is equivalent with Conditions (II) and (III) there. Let us show that (38.5) implies (III). Let p be some continuous seminorm on $\mathscr{C}^\infty(\Omega)$. By definition of the $\mathscr{C}^\infty$ topology, there is a compact subset K of Ω, an integer $m \geqslant 0$, and a constant $C > 0$ such that, for all $\phi \in \mathscr{C}^\infty(\Omega)$,

$$p(\phi) \leqslant C \sup_{x\in K} \sum_{|p|\leqslant m} |(\partial/\partial x)^p\, \phi(x)|.$$

This fact implies immediately that all the distributions $\mu \in \mathscr{E}'(\Omega)$ which are continuous in the sense of the seminorm p must have their support in K and must be of order $\leqslant m$. We now use the D-convexity of Ω. There is a compact subset K' of Ω such that, for all distributions $\mu \in \mathscr{E}'(\Omega)$ such that the order of ${}^tD\mu$ is $\leqslant m$ and that supp ${}^tD\mu \subset K$, we have supp $\mu \subset K'$. Let then Ω' be a relatively compact open subset of Ω containing K'. Since D is semiglobally solvable, to every $f \in \mathscr{C}^\infty(\Omega)$ there is $u \in \mathscr{C}^\infty(\Omega)$ such that $Du - f \in N$, the subspace of $\mathscr{C}^\infty(\Omega)$ consisting of the functions which vanish in Ω'. We have just seen moreover that, if $\mu \in \mathscr{E}'(\Omega)$, ${}^tD\mu \in E'_p$ implies $\mu \in N^0$. Whence (III).

Finally, let us show that (II) implies (38.5). Let m be a nonnegative integer, and K a compact subset of Ω. Let K_1 be another compact subset of Ω containing K in its interior. Every distribution of order $\leqslant m$ in Ω with support in K defines a linear form on $\mathscr{C}^\infty(\Omega)$ continuous for the seminorm

$$p(\phi) = \sup_{x \in K_1} \sum_{|p| \leqslant m} |(\partial/\partial x)^p \phi(x)|.$$

In view of (II), there is another continuous seminorm q on $\mathscr{C}^\infty(\Omega)$ such that, for all $\mu \in \mathscr{E}'(\Omega)$, if ${}^tD\mu$ is continuous in the sense of p, μ itself must be continuous in the sense of q. The argument already used in the first part of the proof shows that to every continuous seminorm q on $\mathscr{C}^\infty(\Omega)$ there is a compact subset K' of Ω such that the distributions $\mu \in \mathscr{E}'(\Omega)$ which are continuous in the sense of q must all have their support contained in K'.   Q.E.D.

Although we have not used it, Condition (IV) in Theorem 37.3 has an important interpretation when we apply it to the situation of Theorem 38.2. Let us select an increasing basis of continuous seminorms in $\mathscr{C}^\infty(\Omega)$. We may select the sequence of seminorms

$$\phi \rightsquigarrow p_k(\phi) = k \sup_{x \in K_k} \sum_{|p| \leqslant k} |(\partial/\partial x)^p \phi(x)|, \qquad k = 1, 2, \dots.$$

Here $\{K_k\}$ $(k = 1, 2, \dots)$ is a sequence of compact subsets of Ω such that $K_k \subset K_{k+1}$ and whose interiors cover Ω. As Ω is D-convex, to every $k = 1, 2, \dots$, there is a compact subset K'_k of Ω such that, if a distribution $\mu \in \mathscr{E}'(\Omega)$ is continuous on $\mathscr{C}^\infty(\Omega)$ for the topology defined by the seminorm p_k, then supp $\mu \subset K'_k$. Let, then, Ω'_k be a relatively compact open neighborhood of K'_k in Ω; let us choose these open sets Ω'_k such that $\Omega'_k \subset \Omega'_{k+1}$, and let us denote by N_k the (closed) linear subspace of $\mathscr{C}^\infty(\Omega)$ consisting of the functions ϕ which vanish identically in Ω'_k. By inspection of the proof of the implications (II) $\Rightarrow$ (III) $\Rightarrow$ (IV) in Theorem 37.3, we see immediately that the subspaces N_k have all the properties listed in (IV). In particular, with respect to (IV$_2$), we see that, for every $k \geqslant 1$, every function $\phi \in \mathscr{C}^\infty(\Omega)$ which satisfies the linear partial differential equation $D\phi = 0$ in Ω_k' is the limit, for the uniform convergence of functions and of their derivatives of order $\leqslant k$ on the compact set K_k, of a sequence of functions $\phi_\nu \in \mathscr{C}^\infty(\Omega)$ which satisfy $D\phi_\nu = 0$ in the whole of Ω (for all ν). (Observe that $K_k \subset K_k'$ for all k; see Exercise 38.4.)

Remark 38.1.   One can give examples of differential operators D which are not semiglobally solvable in any open subset of $\mathbf{R}^n$, although some of these subsets are D-convex. On the other hand, differential operators with *constant* coefficients are always semiglobally solvable (see Exercise 38.1 below) but, in general, given such an operator D, one can find open subsets of $\mathbf{R}^n$ which are not D-convex. This shows the independence of the two properties: D-convexity of Ω, semiglobal solvability of D in Ω.

Exercises

38.1.   Suppose that the differential operator D has constant coefficients (with respect to some coordinate system $x_1, ..., x_n$ in $\mathbf{R}^n$) and that D is not identically zero. Then, it can be proved that there exists a distribution E in $\mathbf{R}^n$ such that $DE = \delta$. Admitting this, derive the fact that D is semiglobally solvable in every open subset of $\mathbf{R}^n$.

38.2.   Let the differential operator D have constant coefficients (not all zero). Prove that an open set $\Omega \subset \mathbf{R}^n$ is D-convex if and only if, to every compact subset K of Ω, there is another compact subset K' of Ω such that, for all functions $\phi \in \mathscr{C}_c^\infty(\Omega)$,

$$\text{supp } P(-D)\phi \subset K \qquad \text{implies} \quad \text{supp } \phi \subset K'.$$

38.3.   Let D be a differential operator with $\mathscr{C}^\infty$ coefficients in an open subset Ω of $\mathbf{R}^n$, having the following property:

(AHE)   For every open subset Ω' of Ω and every distribution T in Ω', the fact that DT is an analytic function in Ω' implies that T is an analytic function in Ω'.

Prove then that Ω is D-convex.

38.4.   Let D be a linear partial differential operator with $\mathscr{C}^\infty$ coefficients defined in some open subset Ω of $\mathbf{R}^n$. Suppose that D is semiglobally solvable in Ω. Let then K be a compact subset of Ω, and m an integer $\geqslant 0$. Prove the following: if K' is a compact subset of Ω such that, for all distributions $\mu \in \mathscr{E}'(\Omega)$ such that $^tD\mu$ be of order $\leqslant m$ and have its support in K, we have supp $\mu \subset K'$, then necessarily $K \subset K'$.

Give an example of an isomorphism J of $\mathscr{C}^\infty(\mathbf{R}^n)$ onto itself (for the TVS structure), which induces an isomorphism of $\mathscr{C}_c^\infty(\mathbf{R}^n)$ onto itself, having the following property: there are compact subsets K of $\mathbf{R}^n$ such that, for all distributions $\mu \in \mathscr{E}'(\mathbf{R}^n)$ with the property that supp $^tJ\mu \subset K$, we have $K \cap$ supp $\mu = \varnothing$.

PART III

Tensor Products. Kernels

The topics discussed in Part III are tensor products, mainly of spaces of functions and distributions, the topologies that such tensor products carry naturally, the locally convex spaces which arise by completion of the tensor products so topologized. The elements of these completions are often referred to as "kernels," whence the title of Part III.

This section of the book begins with the definition of the tensor product of two vector spaces (Chapter 39). We have departed slightly from the now generally adopted definition by the "universal property" (which we state as a theorem, 39.1). In practice, one needs to know if a space already given, M, is the tensor product of two others, E and F. This is so if there exists a bilinear map $\phi : E \times F \to M$ whose image spans M and such that, for all pairs of linearly independent sets of vectors (x_α) and (y_β) in E and F, respectively, the vectors $\phi(x_\alpha, y_\beta)$ are linearly independent in M (then it is natural to refer to the pair (M, ϕ) rather than to M as a tensor product of E and F). In the same chapter, a few examples are presented, essentially of spaces of functions. Examples among spaces of distributions are given in Chapter 40. Prior to this, we introduce the functions which take their values in a locally convex space E and which are differentiable. This enables us to define the tensor products $\mathscr{C}^k \otimes E$ ($0 \leqslant k \leqslant +\infty$): their elements are those $\mathscr{C}^k$ functions with values in E whose images span a finite dimensional subspace of E. The tensor product of two distributions is defined; its basic properties (among which the Fubini type theorem, 40.4) are proved. The following chapter, Chapter 41, is devoted to bilinear mappings and the important notion of hypocontinuity. The student uncertain as to the advisability of advancing further but who has gone as far as Chapter 37 or 38 should definitely try to assimilate the contents of Chapters 39, 40, and 41 and to gain some familiarity with tensor products and bilinear mappings.

In any representation of a vector space as a tensor product, the first feature that strikes the eye is that of a certain splitting. Splitting of the tensor product type are common in algebra. In the problems that concern us, they usually originate in a splitting of the variables in the "base space." We could be dealing, for instance, with functions defined in a product space $\mathbf{R}^m \times \mathbf{R}^n$ where the variable is denoted by (x, y). It might then be convenient to regard those functions as functions with respect to the first variable, x, taking their values in a suitable space of functions (or distributions) with respect to the second one, y. In this way, for example, $\mathscr{C}^\infty_{x,y}$ can be viewed as the space of $\mathscr{C}^\infty$ functions of x with values in $\mathscr{C}^\infty_y$. It is only natural to try to study this situation in more comprehensive, i.e., general, terms, and to deal with functions and distributions valued in arbitrary locally convex spaces. Among those, the simplest are the ones whose image spans a finite dimensional subspace of the "values space" E. They form a linear space which can generally be viewed as the tensor product of E with a suitable space of complex-valued functions; we have encountered this in dealing with $\mathscr{C}^k(E)$ (see above). However, in analysis, where one is forced to go beyond the limitations of finite dimensionality, other operations, of a topological nature—definition of a notion of convergence, adjunction of limit points, which is to say completion—must follow up the formation of tensor products. The space $\mathscr{C}^\infty_{x,y}$ can certainly not be equated to the tensor product $\mathscr{C}^\infty_x \otimes \mathscr{C}^\infty_y$ in any reasonable manner, for the elements of the latter must be recognizable as the finite sums $\sum_j u_j(x)\, v_j(y)$, $u_j \in \mathscr{C}^\infty_x$, $v_j \in \mathscr{C}^\infty_y$. But one cannot fail to notice that, although $\mathscr{C}^\infty_x \otimes \mathscr{C}^\infty_y$ is not the whole of $\mathscr{C}^\infty_{x,y}$, it is a dense linear subspace of the latter (by Corollary 4 of Theorem 15.3), so that $\mathscr{C}^\infty_{x,y}$ can indeed be regarded as a completion of $\mathscr{C}^\infty_x \otimes \mathscr{C}^\infty_y$. In particular, $\mathscr{C}^\infty_{x,y}$ induces on $\mathscr{C}^\infty_x \otimes \mathscr{C}^\infty_y$ a certain topology, which is to be considered as "natural" for many good reasons. In general, when we wish to topologize and form the completion of a tensor product $E \otimes F$, we are forced to look for an intrinsic definition of the topologies, either relying directly on the seminorms on E and F, or else using an embedding of $E \otimes F$ in some space related to E and F over which a "natural" topology already exists. The first method leads to the so-called projective or π topology. The second method may lead to a variety of topologies, the most important of which is the ε topology (ε stands for equicontinuous). This is the only one of the latter class which we study here, although the other ones may be of considerable importance in special problems.

Let us sketch how the projective topology is defined. The aim is to build, out of any two continuous seminorms p and q on E and F, respectively, a seminorm, denoted by $p \otimes q$, on $E \otimes F$, and to define the topology π by these "tensor products of seminorms," $p \otimes q$, as p and q vary in all possible ways. Consider an arbitrary tensor $\theta \in E \otimes F$; θ has, in general, many representations $\sum_k x_k \otimes y_k$ (finite sum; $x_k \in E$, $y_k \in F$). To each one of them, we associate the nonnegative number $\sum_k p(x_k)\, q(y_k)$. The value $(p \otimes q)(\theta)$ is the infimum of all these numbers. This definition is due to R. Schatten.

The definition of the topology ε is based on the relationship between tensor products and linear mappings—or bilinear forms. This relationship is evident in the finite dimensional case. Suppose that dim E and dim F are finite and let θ be an element of $E \otimes F$. We may associate with θ the following bilinear form on $E' \times F'$:

(III.1) $$(x', y') \rightsquigarrow \sum_k \langle x', x_k \rangle \langle y', y_k \rangle,$$

where $\sum_k x_k \otimes y_k$ is an arbitrary representation of θ (that the above bilinear form does not depend on the chosen representation of θ is inherent to the definition of tensor product). We may also associate with θ a linear map of E' into F, namely

(III.2) $$x' \rightsquigarrow \sum_k \langle x', x_k \rangle y_k .$$

It is easy to see that these correspondences establish an isomorphism between $E \otimes F$, $B(E', F')$ (the space of bilinear forms on $E' \times F'$), and $L(E'; F)$ (the space of linear mappings $E' \to F$). But it is clear that these convenient isomorphisms will not subsist if we give up the finite dimensionality of E and F. Indeed, one notices that any linear map of the type (III.2) must have a finite dimensional image (the image of (III.2) is contained in the linear subspace of F spanned by the vectors y_k). In addition to this fact, problems concerning continuity arise. Of course, we shall make use of the "topological" dual E', that is to say the space of *continuous* linear functionals of E. Then (III.2) establishes a one-to-one correspondence between $E \otimes F$ and the mappings of E' into F which are continuous when both spaces carry their weak topology and which have a finite dimensional image. It is easy to see that (III.1) establishes a one-to-one correspondence between $E \otimes F$ and the space $B(E'_\sigma, F'_\sigma)$ of *all* continuous bilinear forms on $E'_\sigma \times F'_\sigma$. We may embed the latter in the space $\mathscr{B}(E'_\sigma, F'_\sigma)$ of *separately* continuous bilinear functionals on $E'_\sigma \times F'_\sigma$. The advantages of such an embedding are two-fold: the space $\mathscr{B}(E'_\sigma, F'_\sigma)$ carries a natural topology: the topology of uniform convergence on the products $A' \times B'$ of equicontinuous sets; equipped with this topology, $\mathscr{B}(E'_\sigma, F'_\sigma)$ is complete if and only if both E and F are complete. Taking all this into account, we identify $E \otimes F$ to $B(E'_\sigma, F'_\sigma)$ (space of continuous bilinear functionals on $E'_\sigma \times F'_\sigma$); we regard it as a linear subspace of $\mathscr{B}(E'_\sigma, F'_\sigma)$. The topology induced by this embedding is the ε topology. The completion of $E \otimes F$, provided with it, will be identified with its closure in $\mathscr{B}(E'_\sigma, F'_\sigma)$ (assuming that both E and F are complete). Now, if we regard the tensor product $\mathscr{C}^k \otimes E$ as a linear subspace of $\mathscr{C}^k(E)$, the space of $\mathscr{C}^k$ functions valued in E (and defined, say, in some open subset of $\mathbf{R}^n$), we see easily that the topology of uniform convergence on compact sets of the functions and of all their derivatives of order $\leqslant k$

induces on $\mathscr{C}^k \otimes E$ precisely the topology ε; as furthermore $\mathscr{C}^k \otimes E$ is dense in $\mathscr{C}^k(E)$, it makes sense to write $\mathscr{C}^k(E) \cong \mathscr{C}^k \widehat{\otimes}_\varepsilon E$. Chapters 42 and 43 are devoted to the definitions and elementary properties of the topologies π and ε (Chapter 42 studies the spaces $B(E'_\sigma, F'_\sigma)$ and $\mathscr{B}(E'_\sigma, F'_\sigma)$, and related spaces of continuous linear maps, alluded to above). Chapter 44 presents two important examples of completed ε-tensor products: the space $\mathscr{C}(K; E) \cong \mathscr{C}(K) \widehat{\otimes}_\varepsilon E$ of continuous functions, defined in a compact set K and valued in a locally convex space E, Hausdorff and complete; $l^1(E) \cong l^1 \widehat{\otimes}_\varepsilon E$, the space of sequences $\{x_k\}$ in E such that the series $\sum_k x_k$ converges (here again, E is Hausdorff and complete). Chapter 45 is devoted to one of the most important results of the theory, the representation of the elements of the completed π-product $E \widehat{\otimes}_\pi F$ of two Fréchet spaces E, F as series

$$(\text{III.3}) \qquad \sum_k \lambda_k x_k \otimes y_k$$

absolutely convergent in $E \widehat{\otimes}_\pi F$, with $\sum_k |\lambda_k| < +\infty$, $\{x_k\}$ and $\{y_k\}$, sequences converging to 0 in E and F, respectively. This representation is constantly used in the sequel. Chapter 46 presents one more example of completed π-tensor product: the space $L^1(E) \cong L^1 \widehat{\otimes}_\pi E$ of integrable functions valued in the complete Hausdorff locally convex space E. When E is a Banach space and $L^1 \widehat{\otimes}_\pi E$ carries the tensor product norm, the above isomorphism is an isometry.

Chapter 47 introduces and studies nuclear mappings, Chapter 49 does the same for integral mappings. Suppose that E and F are Banach spaces and let the series (III.3) represent an element θ of $E \widehat{\otimes}_\pi F$; it is clear that θ defines a continuous linear map of E' into F, namely

$$x' \rightsquigarrow \sum_k \lambda_k \langle x', x_k \rangle y_k \,.$$

This is a typical nuclear map of E' into F (if we want to deal with nuclear mappings of E into F, we must exchange E and E'). A nuclear map is compact (that is to say, transforms the unit ball of E into a relatively compact subset of F); the compose of nuclear maps, whether on the right or on the left, with continuous mappings is nuclear. In the case of Hilbert spaces, nuclear mappings are exactly those compact operators u such that the sequence of eigenvalues $\{\lambda_k\}$ of their absolute value $(u * u)^{1/2}$ is summable, i.e., $\sum_k \lambda_k < +\infty$ (Theorem 48.2). As a matter of fact, in the case of Hilbert spaces E and F, $E \widehat{\otimes}_\varepsilon F$ can be identified with the space of *compact* operators of E into F (we are identifying E and E') and the dual of $E \widehat{\otimes}_\varepsilon F$ can be identified with the space of nuclear operators (with the so-called trace-norm, which is equal to the sum of the eigenvalues of the absolute value of the operator); the dual of the latter is none other than the space of all bounded operators of E into F (with the operators norm; Theorem 48.5'). When E and F are locally convex spaces, not necessarily Banach spaces, one defines the nuclear operators by means of Banach spaces E_p and F_B naturally associated with the continuous seminorms p on E and the bounded closed convex balanced subsets of F, B, such that the space E_B is complete.

The topology π is finer than the topology ε. Thus, the identity mapping of $E \otimes F$ onto itself can be extended as a continuous linear map of $E \widehat{\otimes}_\pi F$ into $E \widehat{\otimes}_\varepsilon F$. That this mapping may not be an isomorphism, or even simply surjective, is shown by the case of Hilbert spaces. But its image is dense. Therefore, its transpose is an injection of the dual of $E \widehat{\otimes}_\varepsilon F$, denoted by $J(E,F)$, into the one of $E \widehat{\otimes}_\pi F$, which is identifiable with the space of *continuous* bilinear functionals on $E \times F$, $B(E,F)$. Thus the elements of $J(E,F)$ are continuous bilinear forms on $E \times F$, of a special type, called *integral* forms. An operator $u : E \to F$ is said to be integral if the associated bilinear form on $E \times F'$,

$$(x, y') \rightsquigarrow \langle y', u(x) \rangle,$$

is integral. All the nuclear operators are integral but the converse is not generally true unless E and F are Hilbert spaces. However the compose of three integral operators (as a matter of fact, of two only—but this will not be proved here) is nuclear. A typical integral form is provided by the bilinear form on $\mathscr{C}(X) \times \mathscr{C}(X)$,

$$(f, g) \rightsquigarrow \int_X f(x)\, g(x)\, dx,$$

where X is a compact space and dx a positive Radon measure on it (this form is so typical that any integral form on any product $E \times F$ possesses such a representation—for a suitable choice of X and dx). What is important, for the subsequent chapters, is that any integral operator $E \to F$ can be decomposed into two continuous linear mappings $E \to H \to F$, with H a Hilbert space.

With Chapter 50, we get to nuclear spaces. These are the locally convex Hausdorff spaces E such that, given any other space F, the topologies π and ε on $E \otimes F$ are identical. Nuclear spaces do exist: the main spaces occurring in distribution theory are nuclear: $\mathscr{C}^\infty$, $\mathscr{C}_c^\infty$, $\mathscr{E}'$, $\mathscr{D}'$, $\mathscr{S}$, and $\mathscr{S}'$; and so also are the space of holomorphic functions in an open set of $\mathbf{C}^n$ (and its dual, the space of analytic functionals in that open set). Banach spaces are not nuclear, unless they are finite dimensional. Complete nuclear spaces, when they are barreled, are

Montel spaces, hence they are reflexive (see Chapter 36). Nuclear spaces have beautiful "stability properties": linear subspaces, quotient spaces, products, and projective limits of nuclear spaces are nuclear spaces; so also are countable topological direct sums and countable inductive limits of nuclear spaces. A Fréchet space is nuclear if and only if its strong dual is nuclear. Most important, the space of continuous linear mappings of E into F (with its topology of bounded convergence) can be identified, under reasonable conditions, when the strong dual E' of E is nuclear, to the completed tensor product $E' \otimes F$. This fact leads easily to the *kernels theorem* of L. Schwartz, which states that there is a one-to-one correspondence between distributions $K(x, y)$ in two sets of variables x and y and the continuous linear mappings of $(\mathscr{C}_c^\infty)_y$ into $\mathscr{D}'_x$. The correspondence is the natural one, given by the formula

$$u \rightsquigarrow Ku, \qquad Ku(x) = \int K(x, y)\, u(y)\, dy.$$

This important theorem is proved in Chapter 51, where the nuclearity of the main spaces of distribution theory is established. The importance of kernels $K(x, y)$ in the field of partial differential equations has been recognized long before the advent of distributions or topological tensor products! Very deep and intensive study has been made of kernels in relation to operators in the spaces L^p and particularly in L^2. In Chapter 52, the last in this book, a few applications are presented to linear partial differential equations, involving nuclear spaces (therefore, not Banach spaces) and based on some of the main theorems of the previous chapters.

The theory of topological tensor products and nuclear spaces is due to A. Grothendieck. We have followed very closely the work (13) of this author, as well as the exposition of L. Schwartz (14). We have omitted many of the questions discussed in these two books, to which we refer the reader for further information.

39

Tensor Product of Vector Spaces

As before, we consider only vector spaces over the field $\mathbf{C}$ of complex numbers. Let E, F be two vector spaces. Let ϕ be a bilinear map of $E \times F$ into a third vector space M.

Definition 39.1. *We say that E and F are ϕ-linearly disjoint if the following holds:*

(LD) *Let $\{x_1,\ldots, x_r\}$ be a finite subset of E, and $\{y_1,\ldots, y_r\}$ a finite subset of F, consisting of the same number of elements and satisfying the relation*

$$\sum_{j=1}^{r} \phi(x_j, y_j) = 0.$$

Then, if $x_1,\ldots, x_r$ are linearly independent, $y_1 = \cdots = y_r = 0$, and if $y_1,\ldots, y_r$ are linearly independent, $x_1 = \cdots = x_r = 0$.

The reason for introducing this definition lies in the next one:

Definition 39.2. *A tensor product of E and F is a pair (M, ϕ) consisting of a vector space M and of a bilinear mapping ϕ of $E \times F$ into M such that the following conditions be satisfied:*

(TP 1) *The image of $E \times F$ spans the whole space M.*

(TP 2) *E and F are ϕ-linearly disjoint.*

We shall now prove the existence of a tensor product of any two vector spaces, its uniqueness up to isomorphisms and the well-known "universal property." But before doing this, we shall give an equivalent definition of ϕ-linear disjointness:

PROPOSITION 39.1. *Let E, F, and M be three vector spaces, and ϕ a bilinear map of $E \times F$ into M. Then E and F are ϕ-linearly disjoint if and only if the following is true:*

(LD′) *Let* $\{x_j\}$ *and* $\{y_k\}$ $(1 \leqslant j \leqslant r, 1 \leqslant k \leqslant s)$ *be arbitrary linearly independent sets of vectors in E and F, respectively. Then the set of vectors of M,* $\{\phi(x_j, y_k)\}$, *are linearly independent.*

Proof. (LD) *implies* (LD′). Let $\{x_j\}$ and $\{y_k\}$ be as in (LD′) and suppose that $\sum_{j,k} \lambda_{jk} \phi(x_j, y_k) = 0$. Set $z_j = \sum_k \lambda_{jk} y_k$; we have $\sum_j \phi(x_j, z_j) = 0$. From (LD) we derive that every z_j must be equal to zero. As the y_k are linearly independent, this implies that the coefficients λ_{jk} be all equal to zero.

(LD′) *implies* (LD). Let $\{x_1, ..., x_r\}$ and $(y_1, ..., y_r)$ be as in (LD); suppose that the x_j's are linearly independent and let $z_1, ..., z_s$ be a basis of the linear subspace spanned by the y_j's. Let us set $y_j = \sum_k \lambda_{jk} z_k$; from $\sum_j \phi(x_j, y_j) = 0$ we derive $\sum_{j,k} \lambda_{jk} \phi(x_j, z_k) = 0$, whence $\lambda_{jk} = 0$ for every pair (j, k), in view of (LD′). Thus all the y_j's are equal to zero.   Q.E.D.

THEOREM 39.1. *Let E, F be two vector spaces.*

(a) *There exists a tensor product of E and F.*

(b) *Let* (M, ϕ) *be a tensor product of E and F. Let G be any vector space, and b any bilinear mapping of* $E \times F$ *into G. There exists any unique linear map* $\tilde{b}$ *of M into G such that the diagram*

(39.1)

$$E \times F \xrightarrow{\ b\ } G$$
$$\phi \downarrow \quad \nearrow \tilde{b}$$
$$M$$

is commutative.

(c) *If* (M_1, ϕ_1) *and* (M_2, ϕ_2) *are two tensor products of E and F, there is a one-to-one linear map u of* M_1 *onto* M_2 *such that the diagram*

(39.2)

$$E \times F \xrightarrow{\ \phi_2\ } M_2$$
$$\phi_1 \downarrow \quad \nearrow u$$
$$M_1$$

is commutative.

Property (b) is sometimes referred to as the "universal property"; (c) states the uniqueness of tensor product up to isomorphisms; (a) states its existence.

Proof of (a). Let $\mathscr{X}$ be the vector space of all complex-valued functions on $E \times F$ which vanish outside a finite set. Let us denote by $e_{(x,y)}$ the function equal to 1 at the point (x, y) and to zero everywhere else; as (x, y) varies over $E \times F$, the functions $e_{(x,y)}$ form a basis of $\mathscr{X}$. Let N be the linear subspace of $\mathscr{X}$ spanned by the functions

$$(39.3) \quad e_{(\alpha x' + \beta x'', \gamma y' + \delta y'')} - \alpha\gamma e_{(x',y')} - \alpha\delta e_{(x',y'')} - \beta\gamma e_{(x'',y')} - \beta\delta e_{(x'',y'')} ,$$

where α, β, γ, and δ vary in all possible ways in the complex field $\mathbf{C}$ whereas (x', y') and (x'', y'') do the same in $E \times F$. We then denote by M the quotient vector space $\mathscr{X}/N$, by π the canonical mapping of $\mathscr{X}$ onto M, and by ϕ the mapping of $E \times F$ into M defined by

$$\phi(x, y) = \pi(e_{(x,y)}).$$

It is obvious, in view of our definition of N, that ϕ is bilinear. In order to conclude that (M, ϕ) is a tensor product of E and F, it remains to show that E and F are ϕ-linearly disjoint.

Let $(x_1, ..., x_r)$ and $(y_1, ..., y_s)$ be linearly independent sets of vectors in E and F, respectively. We assume that there are complex numbers λ_{jk} $(1 \leqslant j \leqslant r, 1 \leqslant k \leqslant s)$ such that

$$\sum_{j,k} \lambda_{jk} \phi(x_j, y_k) = 0.$$

This is equivalent with saying that the function

$$f = \sum_{j,k} \lambda_{j,k} e_{(x_j, y_k)}$$

belongs to the subspace N of $\mathscr{X}$. Let $\{x_\alpha\}$ and $\{y_\beta\}$ $(\alpha \in A, \beta \in B)$ be bases of E and F, respectively, containing the sets $\{x_1, ..., x_r\}$ and $\{y_1, ..., y_s\}$. It is immediately seen that every function (39.3), and consequently every function belonging to N, is a linear combination of elements of the form

$$(39.4) \quad e_{(\sum_\alpha a_l^\alpha x_\alpha, \sum_\beta b_l^\beta y_\beta)} - \sum_{\alpha,\beta} a_l^\alpha b_l^\beta e_{(x_\alpha, y_\beta)} ,$$

where, needless to say, all the linear combinations are finite. If we denote by g_l the element (39.4), we see that there is a finite family of indices l and, for each one of them, a constant c_l such that

$$f = \sum_l c_l g_l .$$

We may assume that no pair $(\sum_\alpha a_l^\alpha x_\alpha, \sum_\beta b_l^\beta y_\beta)$ is equal to a pair $(x_{\alpha_0}, y_{\beta_0})$

for $\alpha_0 \in A$, $\beta_0 \in B$. For this would imply $a_i^\alpha = 1$ if $\alpha = \alpha_0$, $= 0$ otherwise, $b_l^\beta = 1$ if $\beta = \beta_0$, $= 0$ otherwise, and therefore g_l would be identically zero. We may also assume that the pairs $(\sum_\alpha a_i^\alpha x_\alpha, \sum_\beta b_l^\beta y_\beta)$ are pairwise distinct; for if two of them were equal, the corresponding functions g_l would be equal, and we could reorder the sum expressing f. But if these pairs are pairwise distinct and different from every pair (x_α, y_β), the linear independence of the functions $e_{(x,y)}$ implies immediately that all the coefficients c_l must be zero, i.e., $f = 0$; but then, for the same reason, all the coefficients $\lambda_{j,k}$ must also be equal to zero. Q.E.D.: Property (LD′) is satisfied (see Proposition 39.1).

Proof of (b). Let G, b be as in (b). Let (x_α), (y_β) $(\alpha \in A, \beta \in B)$ be bases of E and F, respectively. We know that $\phi(x_\alpha, y_\beta)$ form a basis of M as (α, β) vary over $A \times B$. The linear mapping $\tilde{b}$ will therefore be the (unique) linear map of M into G such that, for all α, β,

$$\tilde{b}(\phi(x_\alpha, y_\beta)) = b(x_\alpha, y_\beta).$$

Proof of (c). Let (M_i, ϕ_i) $(i = 1, 2)$ be two tensor products of E and F. Apply Diagram (39.1) with $M = M_1$, $G = M_2$, $\phi = \phi_1$, and $b = \phi_2$; this yields a linear map $u : M_1 \to M_2$ (the $\tilde{b}$ in (39.1)). Then do this once more with (M_1, ϕ_1), (M_2, ϕ_2) interchanged. It yields a linear map $v : M_2 \to M_1$. It is easy to see that u and v are inverse of each other, hence isomorphisms.   Q.E.D.

We shall never use the tensor product constructed in Part (a) of the proof of Theorem 39.1; whenever we shall need the tensor product of two spaces, we will have a "concrete realization" of it. In accordance with a well-established custom, we shall denote by $E \otimes F$ the tensor product of E and F which we shall happen to be using. The canonical mapping ϕ of $E \times F$ into $E \otimes F$ will be denoted by

$$(x, y) \rightsquigarrow x \otimes y$$

rather than by ϕ.

PROPOSITION 39.2. *Let E, F, E_1, and F_1 be four vector spaces over the complex numbers. Let $u : E \to E_1$ and $v : F \to F_1$ be linear mappings. There is a unique linear map of $E \otimes F$ into $E_1 \otimes F_1$, called the tensor product of u and v and denoted by $u \otimes v$, such that*

$$u \otimes v(x \otimes y) = u(x) \otimes v(y) \qquad \text{for all} \quad x \in E, \quad y \in F.$$

Proof. $(x, y) \rightsquigarrow u(x) \otimes v(y)$ is a bilinear map of $E \times F$ into $E_1 \otimes F_1$

and $u \otimes v$ is the linear map of $E \otimes F$ into $E_1 \otimes F_1$ associated with it by (b), Theorem 39.1.

Example I.   Finite dimensional vector spaces

Let $E = \mathbf{C}^m$, $F = \mathbf{C}^n$ (m, n: positive integers). Then $\mathbf{C}^{mn}$ is a tensor product of E and F, the canonical bilinear map of $E \times F$ into $\mathbf{C}^{mn}$ being given by

$$((x_j), (y_k))_{1 \leqslant j \leqslant m, 1 \leqslant k \leqslant n} \rightsquigarrow (x_j y_k)_{1 \leqslant j \leqslant m, 1 \leqslant k \leqslant n} .$$

Example II.   Tensor product of functions

Let X, Y be two *sets*, and f (resp. g) a complex-valued function defined in X (resp. Y). We shall denote by $f \otimes g$ the function defined in $X \times Y$

$$(x, y) \rightsquigarrow f(x) g(y).$$

Let, now, E (resp. F) be an arbitrary linear space of complex-valued functions defined in X (resp. Y). We shall denote by $E \otimes F$ the linear subspace of the space of all complex functions defined in $X \times Y$ spanned by the elements of the form $f \otimes g$ where f varies over E and g over F. It is immediately seen that $E \otimes F$ is a tensor product of E and F.

Suppose that both X and Y carry a topology. We recall that the support of a function is the closure of the set of points at which the function is $\neq 0$. It is immediately checked that

$$\operatorname{supp}(f \otimes g) = (\operatorname{supp} f) \times (\operatorname{supp} g).$$

We shall take a quick look at a few particular cases of Example II.

Example IIa. Functions with finite support

Suppose that E (resp. F) is the vector space of complex functions in X (resp. Y) which vanish outside a finite set. Then it is immediately seen that $E \otimes F$ is the space of complex functions in $X \times Y$ which vanish outside a finite subset of this product.

Let $\mathbf{N}$ be the set of nonnegative integers; let us take $X = \mathbf{N}^m$, $Y = \mathbf{N}^n$ (m, n: positive integers). The space E (resp. F) of complex functions with finite support in X (resp. in Y) can be identified with the space $\mathscr{P}_m$ (resp. $\mathscr{P}_n$) of polynomials in m indeterminates (resp. in n indeterminates): if $f \in E$, we assign to f the polynomial $P_f(X) = \sum_{p \in \mathbf{N}^m} f(p) X^p$; similarly with $g \in F$. The tensor product $\mathscr{P}_m \otimes \mathscr{P}_n$ is therefore canonically isomorphic to $\mathscr{P}_{m+n}$, according to what we have said at the beginning of this discussion.

Example IIb. Formal power series, entire functions, analytic functionals

Let $\mathscr{Q}_m$ and $\mathscr{Q}_n$ be the vector spaces of formal power series in m and n indeterminates, respectively (with complex coefficients). To an arbitrary power series $u(X) = \sum_{p \in \mathbf{N}^m} u_p X^p$ we assign the function $p \rightsquigarrow u_p$ defined in $\mathbf{N}^m$ (with complex values). This is an isomorphism of $\mathscr{Q}_m$ onto the space of all complex-valued functions in $\mathbf{N}^m$ or, if one prefers, of all complex sequences in m indices. The tensor product $\mathscr{Q}_m \otimes \mathscr{Q}_n$ is canonically isomorphic to a linear subspace of $\mathscr{Q}_{m+n}$; this linear subspace is always distinct from $\mathscr{Q}_{m+n}$. This simply means that a formal power series $u(X, Y)$ in $m + n$ indeterminates $(X_1, ..., X_m, Y_1, ..., Y_n)$ cannot be written, in general, as a finite sum

$$\sum_{j=1}^{J} u_j(X)\, v_j(Y)$$

of products of formal power series in the X_j's and formal power series in the Y_k's. Observe however that, if we provide $\mathscr{Q}_{m+n}$ with the topology of simple convergence of the coefficients (Chapter 10, Example III), then $\mathscr{Q}_m \otimes \mathscr{Q}_n$ is dense in $\mathscr{Q}_{m+n}$. Indeed, $\mathscr{P}_{m+n}$ is dense in $\mathscr{Q}_{m+n}$; this is evident. On the other hand, we have just seen that $\mathscr{P}_{m+n} = \mathscr{P}_m \otimes \mathscr{P}_n$; the latter is obviously contained in $\mathscr{Q}_m \otimes \mathscr{Q}_n$.

The space $H(\mathbf{C}^m)$ of entire analytic functions in the m-dimensional complex space $\mathbf{C}^m$ can be identified to a linear subspace of $\mathscr{Q}_m$, precisely the subspace consisting of the convergent power series whose radius of convergence is infinite. Then $H(\mathbf{C}^m) \otimes H(\mathbf{C}^n)$, viewed as a space of functions on $\mathbf{C}^m \times \mathbf{C}^n$, can be canonically identified to the subspace of $\mathscr{Q}_m \otimes \mathscr{Q}_n$ consisting of the series with infinite radius of convergence; of course, $H(\mathbf{C}^m) \otimes H(\mathbf{C}^n) \subset H(\mathbf{C}^{m+n})$. Indeed, $H(\mathbf{C}^m) \otimes H(\mathbf{C}^n)$ is a proper dense subspace of $H(\mathbf{C}^{m+n})$, as it contains the polynomials in $m + n$ (complex) variables and as these are dense in the space of entire functions in $\mathbf{C}^{m+n}$ (Theorem 15.1).

We might also consider the space $Exp(\mathbf{C}^m)$ of entire functions of exponential type in $\mathbf{C}^m$ (Notation 22.2). It is immediately seen that

$$Exp(\mathbf{C}^m) \otimes Exp(\mathbf{C}^n)$$

is a linear subspace of $Exp(\mathbf{C}^{m+n})$. If μ (resp. ν) is an analytic functional in $\mathbf{C}^m$ (resp. $\mathbf{C}^n$), its Fourier–Borel transform $\hat{\mu}$ (resp. $\hat{\nu}$) belongs to $Exp(\mathbf{C}^m)$ (resp. to $Exp(\mathbf{C}^n)$; Definition 22.3, Theorem 22.3). By Theorem 22.3, the tensor product $\hat{\mu} \otimes \hat{\nu}$ is the Fourier–Borel transform of an analytic functional in $\mathbf{C}^{m+n}$ which we shall denote by $\mu \otimes \nu$ and call the tensor product of μ and ν. Later on, we shall see that $\mu \otimes \nu$ can be defined without making use of the Fourier–Borel transformation.

Example IIc.   Functions in open subsets of Euclidean spaces

Let X and Y be open subsets of $\mathbf{R}^m$ and $\mathbf{R}^n$, respectively. Let k, l be two nonnegative integers, possibly infinite. We may form the tensor products

$$\mathscr{C}^k(X) \otimes \mathscr{C}^l(Y), \qquad \mathscr{C}^k_c(X) \otimes \mathscr{C}^l_c(Y), \qquad \text{etc.}$$

They are spaces of functions defined in the product set $X \times Y$, regarded as an open subset of $\mathbf{R}^{m+n}$. As a matter of fact, they are linear subspaces of $\mathscr{C}^{k,l}(X \times Y)$, the latter notation having an obvious meaning. The functions belonging to $\mathscr{C}^k_c(X) \otimes \mathscr{C}^l(Y)$, for instance, have supports whose projection into X is compact, etc.

The approximation results in Chapter 15 imply easily the following:

THEOREM 39.2.   *Let X (resp. Y) be an open subset of $\mathbf{R}^m$ (resp. $\mathbf{R}^n$). Then*

$$\mathscr{C}^\infty_c(X) \otimes \mathscr{C}^\infty_c(Y)$$

is sequentially dense in $\mathscr{C}^\infty_c(X \times Y)$.

Proof.   Let $\phi \in \mathscr{C}^\infty_c(X \times Y)$. By Corollary 2 of Lemma 15.1, ϕ is the limit in $\mathscr{C}^\infty(X \times Y)$ of a sequence of polynomials $\{P_k(x, y)\}$ ($k = 1, 2,...$). Let $K = \operatorname{supp} \phi$, K_1 (resp. K_2) be the projection of K into X (resp. into Y). Both K_1 and K_2 are compact sets. Let $g \in \mathscr{C}^\infty_c(X)$, $h \in \mathscr{C}^\infty_c(Y)$ be equal to one in a neighborhood of K_1 and K_2, respectively. Then $g \otimes h$ is identically equal to one in a neighborhood of K, and belongs to $\mathscr{C}^\infty_c(X) \otimes \mathscr{C}^\infty_c(Y)$. This tensor product then contains the sequence of functions

$$(g \otimes h)P_k, \qquad k = 1,...,$$

which converges in $\mathscr{C}^\infty(X \times Y)$, therefore also in $\mathscr{C}^\infty_c(X \times Y)$ (as the elements of the sequence have their support in a fixed compact subset of $X \times Y$) to the function $(g \otimes h)\phi = \phi$.   Q.E.D.

COROLLARY 1. $C^\infty_c(X) \otimes \mathscr{C}^\infty_c(Y)$ *is sequentially dense in $\mathscr{C}^{k,l}_c(X \times Y)$, in $\mathscr{C}^{k,l}(X \times Y)$, and in $L^p(X \times Y)$ $(1 \leqslant p < +\infty)$.*

It suffices to combine Theorem 39.2 with Corollaries 1, 2, and 3 of Theorem 15.3 (where $\Omega = X \times Y$).

COROLLARY 2. $\mathscr{C}^k_c(X) \otimes \mathscr{C}^l_c(Y)$ *is sequentially dense in $\mathscr{C}^{k,l}_c(X \times Y)$ and in $\mathscr{C}^{k,l}(X \times Y)$.*

COROLLARY 3. $L^p(X) \otimes L^p(Y)$ *is dense in $L^p(X \times Y)$ $(1 \leqslant p < +\infty)$.*
Many more results of a similar nature can easily be stated and proved.

Exercises

39.1. Let (M, ϕ) be a tensor product of E and F, and (N, ψ) a tensor product of F and E. Prove that there is a canonical isomorphism J (for the tensor product structures) of (M, ϕ) onto (N, ψ). What is the mapping S that then makes the following diagram commutative?

$$
\begin{array}{ccc}
E \times F & \xrightarrow{\ S\ } & F \times E \\
\phi \downarrow & & \downarrow \psi \\
M & \xrightarrow{\ J\ } & N
\end{array}
$$

39.2. Let E and F be two vector spaces over $\mathbf{C}$, and $\{e_\alpha\}$, $\{f_\beta\}$ ($\alpha \in A$, $\beta \in B$) bases in E and F, respectively. Let $E \otimes F$ be a tensor product of E and F. Prove that $\{e_\alpha \otimes f_\beta\}$ ($\alpha \in A$, $\beta \in \beta$) form a basis of $E \otimes F$.

39.3. Let E_j, F_j ($j = 1, 2$) be four vector spaces, $u_j : E_j \to F_j$ two linear mappings ($j = 1, 2$), and $\{e_{\alpha,j}\}$, $\{f_{\beta,j}\}$ ($\alpha \in A_j$, $\beta = B_j$) bases in E_j and F_j, respectively ($j = 1, 2$). Let $(c_{\alpha,j}^\beta)$ be the matrix of u_j with respect to those bases. What is the matrix of the tensor product $u_1 \otimes u_2$ with respect to the bases $(e_{\alpha,1} \otimes e_{\alpha',2})$ of $E_1 \otimes E_2$ and $(f_{\beta,1} \otimes f_{\beta',2})$ of $F_1 \otimes F_2$?

39.4. Let E_j, F_j, and u_j ($j = 1, 2$) be as in Exercise 39.3. What is the kernel (resp. the image) of $u_1 \otimes u_2$? Derive that $u_1 \otimes u_2$ is one-to-one (resp. onto) if this is true of both u_1 and u_2 .

39.5. Let E, F, and M be three vector spaces, and ϕ a bilinear mapping of $E \times F$ into M. Prove that the following properties are equivalent:

(a) (M, ϕ) is a tensor product of E and F;
(b) the mapping

$$x^* \rightsquigarrow x^* \circ \phi$$

is an isomorphism of the algebraic dual M^* of M onto the vector space $\mathfrak{B}(E, F)$ of bilinear forms on $E \times F$.

40

Differentiable Functions with Values in Topological Vector Spaces. Tensor Product of Distributions

Let X and Y be open subsets of $\mathbf{R}^m$ and $\mathbf{R}^n$, respectively (m, n: integers $\geqslant 1$). It is convenient, in many a situation, to regard a function $\phi(x, y)$ of the pair of variables $x \in X$, $y \in Y$ as a function of one of them, say y, with values in a space of functions with respect to the other one, x. More generally, one might be interested in dealing with functions defined in Y and taking their values in some topological vector space E. In certain circumstances, E could be a space of distributions (and not merely a space of functions) in X. This is why it is reasonable to introduce the concepts of differentiable functions with values in a TVS E and, having done this, to study the spaces of these functions, and their duals. Eventually, one may also need the theory of distributions with values in E.

In the present chapter, we limit ourselves to recalling the definition of a differentiable function, defined in the open set Y, with values in a TVS E, and to introducing the spaces $\mathscr{C}^k(Y; E)$ and $\mathscr{C}_c^k(Y; E)$ of k-times differentiable functions (with arbitrary support and with compact support, respectively), defined in Y and valued in E. These spaces can be made to carry a natural $\mathscr{C}^k$ topology, whose definition is a straightforward generalization of the scalar case.

We use the concepts and facts thus introduced to define the tensor product of a distribution S in X with a distribution T in Y. The approach through functions valued in a TVS has the considerable advantage of revealing the general facts underlying the definition of the tensor product $S \otimes T$ (and, in particular, the so-called Fubini theorem for distributions: see Theorem 40.3).

Let f be a mapping of the open set $Y \in \mathbf{R}^n$ into the TVS E. We recall the meaning of "f is differentiable at a point y^0 of Y" (cf. the remarks following the statement of Theorem 27.1):

411

Definition 40.1.   The function f is said to be differentiable at $y^0 \in Y$ if there are n vectors in E, $\mathbf{e}_1 ,..., \mathbf{e}_n$, such that

$$| y - y^0 |^{-1} \left\{ f(y) - f(y^0) - \sum_{j=1}^n (y_j - y_j^0) \mathbf{e}_j \right\}$$

converges to zero in E as the number $| y - y^0 | > 0$ converges to zero. The vectors $\mathbf{e}_j$ are then called the first partial derivatives of f at the point y^0: one sets

$$\mathbf{e}_j = \frac{\partial f}{\partial y_j}(y^0), \qquad j = 1,..., n.$$

If f is differentiable at y^0, it is obviously continuous at that point. The traditional terminology is extended to functions valued in a TVS: f is said to be differentiable in a set $A \subset Y$ if it is differentiable at every point of A; f is said to be continuously differentiable if it is differentiable at every point and if its first partial derivatives are continuous functions; f is said to be k-times continuously differentiable (or $\mathscr{C}^k$) if f is differentiable at every point and if its first partial derivatives are $\mathscr{C}^{k-1}$; f is said to be infinitely differentiable if it is $\mathscr{C}^k$ for all $k = 0, 1,...,$ etc.

Notation 40.1.   We shall denote by $\mathscr{C}^k(Y; E)$ the vector space of $\mathscr{C}^k$ mappings of Y into E $(0 \leqslant k \leqslant + \infty)$. We shall denote by $\mathscr{C}_c^k(Y; E)$ the subspace of $\mathscr{C}^k(Y; E)$ consisting of the functions with compact support.

The support of a vector-valued function is the closure of the set of points at which the function is nonzero.

Definition 40.2.   The $\mathscr{C}^k$ topology on $\mathscr{C}^k(Y; E)$ is the topology of uniform convergence of the functions together with their derivatives of order $< k + 1$ on every compact subset of Y.

Consider a sequence $\Omega_1 \subset \Omega_2 \subset \cdots \subset \Omega_j \subset \cdots$ of relatively compact open subsets of Y whose union is equal to Y, an arbitrary integer $l < k + 1$, a basis of neighborhoods of zero in E, $\{U_\alpha\}$. As j, l, and α vary in all possible ways, the subsets of $\mathscr{C}^k(Y; E)$,

$$\mathscr{U}_{j,l,\alpha} = \{ f; (\partial/\partial y)^q f(y) \in U_\alpha \text{ for all } y \in \Omega_j \text{ and all } q \in \mathbf{N}^n, | q | \leqslant l \},$$

form a basis of neighborhoods of zero for the $\mathscr{C}^k$ topology. If E is metrizable, so is $\mathscr{C}^k(Y; E)$; if E is metrizable and complete, so is $\mathscr{C}^k(Y;E)$. Noting that $\mathscr{U}_{j,l,\alpha}$ is a convex set whenever U is a convex set, we see also that $\mathscr{C}^k(Y; E)$ is locally convex whenever this is true of E. Needless to say, $\mathscr{C}^k(Y; E)$ is Hausdorff if and only if this is true of E.

When E is locally convex, it is easy to obtain a basis of continuous seminorms on $\mathscr{C}^k(Y; E)$. It suffices to select a basis of continuous seminorms on E, $\{\mathfrak{p}_\kappa\}$, and to form the seminorms

$$f \rightsquigarrow \mathscr{P}_{j,l,\kappa}(f) = \sup_{y \in \Omega_j} \Big(\sum_{|q| \leqslant l} \mathfrak{p}_\kappa((\partial/\partial y)^q f(y)) \Big).$$

When j, l, and κ vary in all possible ways, the $\mathscr{P}_{j,l,\kappa}$ form a basis of continuous seminorms for the topology $\mathscr{C}^k$.

Given an arbitrary compact subset K of Y, we denote by $\mathscr{C}^k_c(K; E)$ the subspace of $\mathscr{C}^k(Y; E)$ consisting of the functions with support contained in K. We provide $\mathscr{C}^k_c(K; E)$ with the topology induced by $\mathscr{C}^k(Y; E)$. Let us suppose, as we shall always do from now on, that E is locally convex. Then we provide $\mathscr{C}^k_c(Y; E)$ with the topology inductive limit of the topologies of the spaces $\mathscr{C}^k_c(K; E)$ as K varies over the family of all compact subsets of Y. A convex subset of $\mathscr{C}^k_c(Y; E)$ is a neighborhood of zero if its intersection with every subspace $\mathscr{C}^k_c(K; E)$ is a neighborhood of zero in $\mathscr{C}^k_c(K; E)$. The definition of the topology of $\mathscr{C}^k_c(Y; E)$ duplicates the definition of the topology of a space LF, except that the subspaces $\mathscr{C}^k_c(K; E)$, which serve as building blocks, are not (in general) Fréchet spaces. But notice that, if $K \subset K'$ are two compact subsets of Y, the topology induced on $\mathscr{C}^k_c(K; E)$ by $\mathscr{C}^k_c(K'; E)$ is identical to the initially given topology on $\mathscr{C}^k_c(K; E)$, which is the topology induced by $\mathscr{C}^k(Y; E)$. Note also that $\mathscr{C}^k_c(Y; E)$ can be defined as the inductive limit of a *sequence* of subspaces $\mathscr{C}^k_c(K_j ; E)$: it suffices to take an arbitrary sequence of compact subsets K_j of Y whose union is equal to Y (for instance, $K_j = \bar{\Omega}_j$, where the Ω_j are the relatively compact open subsets of Y considered after Definition 40.2). We see then easily that $\mathscr{C}^k_c(Y; E)$ induces on every $\mathscr{C}^k_c(K; E)$ its original topology (cf. Lemma 13.1). In analogy with the properties of LF-spaces proved in Chapter 13, we have:

PROPOSITION 40.1. *Let Y be an open subset of $\mathbf{R}^n$, E a locally convex Hausdorff TVS, and k an integer $\geqslant 0$, possibly infinite.*

A linear map u of $\mathscr{C}^k_c(Y; E)$ into a locally convex TVS F is continuous if and only if the restriction of u to every subspace $\mathscr{C}^k_c(K; E)$ (K: compact subset of Y) is continuous.

COROLLARY. *A linear functional on $\mathscr{C}^k_c(Y; E)$ is continuous if and only if its restriction to every subspace $\mathscr{C}^k_c(K; E)$ is continuous.*

Proof. It suffices to observe that the proof of Proposition 13.1 and of its corollary never makes use of the fact that the sequence of definition $\{E_n\}$ ($n = 1, 2,...$) consists of Fréchet spaces (cf. Exercise 13.6)!

Example 40.1.   *The space of values E is finite dimensional (and Hausdorff)*

Let $d = \dim E$ be finite, $\mathbf{e}_1, ..., \mathbf{e}_d$ a basis of E, and $\mathbf{e}'_1, ..., \mathbf{e}'_d$ the dual basis in E' (this means that $\langle \mathbf{e}'_i, \mathbf{e}_j \rangle = 0$ if $i \neq j$, $= 1$ if $i = j$). Consider a function $f \in \mathscr{C}^k(Y; E)$. For each $y \in Y$, we may write

$$f(y) = \sum_{j=1}^{n} f_j(y)\mathbf{e}_j;$$

we have

$$f_j(y) = \langle \mathbf{e}'_j, f(y) \rangle.$$

It is immediately seen (cf. Theorem 27.1) that f_j is a complex function belonging to $\mathscr{C}^k(Y)$. Conversely, let f be such a function, and $\mathbf{e}$ a vector in E. Let us denote by $f \otimes \mathbf{e}$ the function, valued in E, $y \rightsquigarrow f(y)\mathbf{e}$. What we have just said means that the functions of the form $f \otimes \mathbf{e}$ span $\mathscr{C}^k(Y; E)$ when f varies over $\mathscr{C}^k(Y)$ and $\mathbf{e}$ over E; this is true if and only if E is finite dimensional. Then the bilinear map $(f, \mathbf{e}) \rightsquigarrow f \otimes \mathbf{e}$ of $\mathscr{C}^k(Y) \times E$ into $\mathscr{C}^k(Y; E)$ turns the latter into a tensor product of $\mathscr{C}^k(Y)$ and E.

We go back to the general case (in which $\dim E$ is not necessarily finite).

Notation 40.2.   *Let E be a vector space over the field of complex numbers, f a complex-valued function defined in $Y \subset \mathbf{R}^n$, and $\mathbf{e}$ a vector belonging to E. We denote by $f \otimes \mathbf{e}$ the function, defined in Y and valued in E, $y \rightsquigarrow f(y)\,\mathbf{e}$.*

PROPOSITION 40.2.   *Let E be a Hausdorff TVS. The bilinear mapping*

$$(f, \mathbf{e}) \rightsquigarrow f \otimes \mathbf{e}$$

of $\mathscr{C}^k(Y) \times E$ into the subspace of $\mathscr{C}^k(Y; E)$, consisting of the functions whose image is contained in a finite dimensional subspace of E, turns this subspace into a tensor product of $\mathscr{C}^k(Y)$ and E (which we shall denote by $\mathscr{C}^k(Y) \otimes E$).

Proof.   That the functions $f \otimes \mathbf{e}$ have their image contained in a finite dimensional (in fact, a one-dimensional) linear subspace of E is trivial. Conversely, let f be a $\mathscr{C}^k$ mapping of Y into E whose image is contained in some linear subspace E_0 of E such that $d = \dim E_0$ is finite. If $\mathbf{e}_1, ..., \mathbf{e}_d$ is a basis of E_0, we may write $f = f_1 \otimes \mathbf{e}_1 + \cdots + f_d \otimes \mathbf{e}_d$ with $f_j \in \mathscr{C}^k(Y)$ (see Example 40.1 above). This shows that f belongs to the linear subspace of $\mathscr{C}^k(Y; E)$ spanned by the functions of the form $f \otimes \mathbf{e}$. From there the statement follows easily.

We shall use the notation $\mathscr{C}_c^k(Y) \otimes E$ to denote the subspace of $\mathscr{C}^k(Y) \otimes E$ consisting of the functions with compact support. Of course, $\mathscr{C}_c^k(Y) \otimes E$ is a tensor product of $\mathscr{C}_c^k(Y)$ and E.

THEOREM 40.1.   *Let X, Y be open subsets of $\mathbf{R}^m, \mathbf{R}^n$, respectively. The mapping*

$$(40.1) \qquad \phi \rightsquigarrow (y \rightsquigarrow (x \rightsquigarrow \phi(x, y)))$$

is an isomorphism, for the structures of topological vector spaces, of

$$\mathscr{C}^\infty(X \times Y) \quad onto \quad \mathscr{C}^\infty(Y; C^\infty(X)).$$

Proof. That (40.1) is a one-to-one continuous linear map into is a straightforward consequence of the definitions. If, on the other hand, f is an element of $\mathscr{C}^\infty(Y; E)$, with $E = \mathscr{C}^\infty(X)$, we observe that, for every $y \in Y$, $f(y)$ is a function in X. If we denote by $\phi(x, y)$ its value at $x \in X$, it is immediately seen that $\phi \in \mathscr{C}^\infty(X \times Y)$ and that f is the image of ϕ by (40.1). The continuity of the mapping $f \rightsquigarrow \phi$ is evident by virtue of the definitions (or by application of the open mapping theorem: see Corollary 1 of Theorem 17.1, since both $\mathscr{C}^\infty(X \times Y)$ and $\mathscr{C}^\infty(Y; \mathscr{C}^\infty(X))$ are Fréchet spaces).

COROLLARY 1.   *The restriction of (40.1) to $\mathscr{C}_c^\infty(X \times Y)$ is an isomorphism of this space onto $\mathscr{C}_c^\infty(Y; \mathscr{C}_c^\infty(X))$.*

Proof. That the restriction of (40.1) is a one-to-one continuous linear map into follows immediately from Theorem 40.1, from the definition of the topology on $\mathscr{C}_c^\infty(Y; \mathscr{C}_c^\infty(X))$ and from Proposition 40.1. Let $f \in \mathscr{C}_c^\infty(Y; \mathscr{C}_c^\infty(X))$; the support of f is a compact subset of Y, K. The image of K by f is therefore a compact subset of $\mathscr{C}_c^\infty(X)$; such a subset is necessarily contained in some subspace of the form $\mathscr{C}_c^\infty(H)$, with H a compact subset of X (Proposition 14.6). This means that the preimage of f under (40.1) is a $\mathscr{C}^\infty$ function of (x, y) with support contained in $H \times K$. This shows that the restriction of (40.1) to $\mathscr{C}_c^\infty(X \times Y)$ maps it *onto* $\mathscr{C}_c^\infty(Y; \mathscr{C}_c^\infty(X))$. In order to prove that the inverse mapping is continuous, the shortest way is probably by observing that $\mathscr{C}_c^\infty(Y; \mathscr{C}_c^\infty(X))$ is the inductive limit of the Fréchet spaces $\mathscr{C}_c^\infty(K; \mathscr{C}_c^\infty(H))$ as H (resp. K) runs over the family of all the compact subsets of X (resp. Y). Then we may either use the fact that a one-to-one continuous linear map of a space LF onto another space LF is an isomorphism (i.e., is bicontinuous) or that (40.1) induces an isomorphism of $\mathscr{C}_c^\infty(H \times K)$ onto $\mathscr{C}_c^\infty(K; \mathscr{C}_c^\infty(H))$.

Going back now to the general case, we state a strengthened version of Theorem 27.1:

THEOREM 40.2. *Let E be a locally convex Hausdorff space, and Y an open subset of* $\mathbf{R}^n$. *For every continuous linear form e' on E,*

$$f \rightsquigarrow (y \rightsquigarrow \langle e', f(y) \rangle)$$

is a continuous linear map of $\mathscr{C}^\infty(Y; E)$ *(resp.* $\mathscr{C}^\infty_c(Y; E)$*) into* $\mathscr{C}^\infty(Y)$ *(resp.* $\mathscr{C}^\infty_c(Y)$*).*

The proof is absolutely standard; it is essentially done by inspection of the definitions and will not be given here. Instead, we present a rather important application of the preceding theory. As before, X and Y are two open subsets of $\mathbf{R}^m$ and $\mathbf{R}^n$, respectively.

THEOREM 40.3. *Let S be a distribution in X; then*

$$\phi \rightsquigarrow (y \rightsquigarrow \langle S_x, \phi(x, y) \rangle)$$

is a continuous linear map of $\mathscr{C}^\infty_c(X \times Y)$ *into* $\mathscr{C}^\infty_c(Y)$; *if the support of S is compact, it is a continuous linear map of* $\mathscr{C}^\infty(X \times Y)$ *into* $\mathscr{C}^\infty(Y)$.

The notation we have used has an obvious meaning: $\langle S_x, \phi(x, y) \rangle$ is the value of the distribution S in X on the test function $x \rightsquigarrow \phi(x, y)$, with y playing the role of a parameter.

Proof of Theorem 40.3. It suffices to combine Theorem 40.1, or its corollary, with Theorem 40.2.

Let now T be a distribution in Y. In virtue of Theorem 40.3,

$$\mathscr{C}^\infty_c(X \times Y) \ni \phi \rightsquigarrow \langle T_y, \langle S_x, \phi(x, y) \rangle \rangle$$

defines a distribution in $X \times Y$. Similarly,

$$\mathscr{C}^\infty_c(X \times Y) \ni \phi \rightsquigarrow \langle S_y, \langle T_x, \phi(x, y) \rangle \rangle$$

is a distribution in $X \times Y$. The next result states that these two distributions are equal. It can be viewed as a kind of rule of interchanging integrations with respect to x and y. In analogy with integration theory, it is often referred to as *Fubini's theorem for distributions*.

THEOREM 40.4. *Let S be a distribution in X, and T a distribution in Y. For every test function* $\phi \in \mathscr{C}^\infty_c(X \times Y)$, *we have*

$$(40.2) \qquad \langle S_x, \langle T_y, \phi(x, y) \rangle \rangle = \langle T_y, \langle S_x, \phi(x, y) \rangle \rangle.$$

Proof. The equality (40.2) is evident if $\phi(x, y) = u(x)\, v(y)$, with $u \in \mathscr{C}_c^\infty(X)$ and $v \in \mathscr{C}_c^\infty(Y)$. It is therefore also true if ϕ is a finite sum of products $u(x)\, v(y)$, in other words if $\phi \in \mathscr{C}_c^\infty(X) \otimes \mathscr{C}_c^\infty(Y)$ (Chapter 39, Example IIc). But $\mathscr{C}_c^\infty(X) \otimes \mathscr{C}_c^\infty(Y)$ is a dense subspace of $\mathscr{C}_c^\infty(X \times Y)$ (Theorem 39.2) and both sides of (40.2) are continuous with respect to ϕ, whence the result.

Definition 40.3.   *Let S be a distribution in X, and T a distribution in Y. The distribution in $X \times Y$,*

$$\mathscr{C}_c^\infty(X \times Y) \ni \phi \rightsquigarrow \langle S_x, \langle T_y, \phi(x, y) \rangle \rangle = \langle T_y, \langle S_x, \phi(x, y) \rangle \rangle,$$

is called the tensor product of S and T (or of T and S) and denoted by

$$S \otimes T \quad or \quad T \otimes S \ (or \quad S_x \otimes T_y, \quad etc.).$$

We now state a few of the basic properties of the tensor product of distributions (without proving them):

PROPOSITION 40.3.   (a) $(S, T) \rightsquigarrow S \otimes T$ *is a bilinear map of $\mathscr{D}'(X) \times \mathscr{D}'(Y)$ into $\mathscr{D}'(X \times Y)$.*

(b) $\mathrm{supp}(S \otimes T) = (\mathrm{supp}\ S) \times (\mathrm{supp}\ T)$;

(c) $(S, T) \rightsquigarrow S \otimes T$ *is a bilinear map of $\mathscr{E}'(X) \times \mathscr{E}'(Y)$ into $\mathscr{E}'(X \times Y)$.*

(d)   *If $P(x, D_x)$ (resp. $Q(y, D_y)$) is a differential operator (with $\mathscr{C}^\infty$ coefficients, see Chapter 23, Example III) in X (resp. in Y).*

$$P(x, D_x)\, Q(y, D_y)\, (S_x \otimes T_y) = (P(x, D_x)\, S_x) \otimes (Q(y, D_y)\, T_y).$$

(e)   *If both S and T are locally L^1 functions, $S \otimes T$ is equal to the tensor product of S and T in the functions sense (Chapter 39, Example II).*

Definition 40.4.   *We shall denote by $\mathscr{D}'(X) \otimes \mathscr{D}'(Y)$ the linear subspace of $\mathscr{D}'(X \times Y)$ spanned by the distributions of the form $S \otimes T$, $S \in \mathscr{D}'(X)$, $T \in \mathscr{D}'(Y)$.*

$\mathscr{D}'(X) \otimes \mathscr{D}'(Y)$ is obviously a tensor product of $\mathscr{D}'(X)$ and $\mathscr{D}'(Y)$. We shall also use the notation $\mathscr{D}'(X) \otimes \mathscr{E}'(Y)$, $\mathscr{E}'(X) \otimes \mathscr{E}'(Y)$, etc., with obvious meanings.

PROPOSITION 40.4. $\mathscr{D}'(X) \otimes \mathscr{D}'(Y)$ (resp. $\mathscr{E}'(X) \otimes \mathscr{E}'(Y)$) *is a dense subspace of $\mathscr{D}'(X \times Y)$ (resp. of $\mathscr{E}'(X \times Y)$).*

Indeed, both tensor products contain $\mathscr{C}_c^\infty(X) \otimes \mathscr{C}_c^\infty(Y)$, which is dense in $\mathscr{C}_c^\infty(X \times Y)$ (by Theorem 39.2); the latter, in turn, is dense in $\mathscr{E}'(X \times Y)$ and in $\mathscr{D}'(X \times Y)$ (by the corollary of Theorem 28.2).

As a conclusion to the present chapter, let us indicate how the convolution of distributions can be linked to their tensor product.

THEOREM 40.5. *Let S, T be two distributions with compact support in $\mathbf{R}^n$. For every test function $\phi \in \mathscr{C}_c^\infty(\mathbf{R}^n)$, we have*

$$(40.3) \qquad \langle S * T, \phi \rangle = \langle S_x \otimes T_y , \phi(x + y) \rangle.$$

If only the support of S is compact, supp T *being arbitrary, let $g \in \mathscr{C}_c^\infty(\mathbf{R}^n)$ be equal to one in a neighborhood of* supp S. *Then we have*

$$(40.4) \qquad \langle S * T, \phi \rangle = \langle S_x \otimes T_y , g(x) \phi(x + y) \rangle.$$

The right-hand side of the equation (40.4) makes sense because $(x, y) \rightsquigarrow g(x) \phi(x + y)$ is a function with compact support in $\mathbf{R}^n \times \mathbf{R}^n$; obviously (40.3) and (40.4) coincide when the support of T is also compact.

The proof is left to the student.

Exercises

40.1. Let $X \subset \mathbf{R}^m$, $Y \subset \mathbf{R}^n$ be open sets, S a distribution in X, and T a distribution in Y. Prove that $S \otimes T$ is a Radon measure in $X \times Y$ if and only if both S and T are Radon measures.

40.2. Let E be a locally convex Hausdorff space, and Y an open subset of the complex space $\mathbf{C}^n$. Compare the following concepts:

(1) *Analytic function $f : Y \to E$*: here, $f \in \mathscr{C}^\infty(Y; E)$ and, for every point $y^0 \in Y$, there is a family $\{\mathbf{f}_p(y^0)\}$ ($p \in \mathbf{N}^n$) of elements of E such that:

   (i)   the series

$$\sum_{p \in \mathbf{N}^n} \frac{1}{p!} (y - y^0)^p \, \mathbf{f}_p(y^0)$$

   converges absolutely in some neighborhood of y^0;

   (ii) the sum of the preceding series is equal to $f(y)$ for all y in some neighborhood of y^0.

(2) *Complex differentiable function in Y*: one duplicates Definition 40.1 for complex y, y^0.

(3) *Solution of the Cauchy–Riemann equations in Y*: here, $f : Y \to E$ is once continuously differentiable (in the sense of Definition 40.1 after we have identified $\mathbf{C}^n$ to $\mathbf{R}^{2n}$) and, for every $y^0 \in Y$, satisfies the n conditions

$$\left(\frac{\partial f}{\partial \xi_j} + (-1)^{1/2} \frac{\partial f}{\partial \eta_j} \right) (y^0) = 0, \qquad j = 1,\dots, n,$$

where we set $y_j = \xi_j + (-1)^{1/2}\eta_j$ $(1 \leqslant j \leqslant n,\ \xi,\ \eta$ real). Could you state a reasonable condition on E sufficient in order that the three concepts above coincide?

40.3.   Let $X \subset \mathbf{C}^m,\ Y \subset \mathbf{C}^n$ be open sets. Give the definition and the basic properties of the tensor product $\mu_x \otimes \nu_y$ of an analytic functional μ on X with an analytic functional ν on Y.

40.4.   Let $\mathscr{F}$ be the Fourier–Borel transformation in $\mathbf{C}^n$, and $\mu,\ \nu$ two analytic functionals in $\mathbf{C}^n$. Prove that we have, for all $h \in H$ (entire functions in $\mathbf{C}^n$),

$$\langle \mu * \nu, h \rangle = \langle \mathscr{F}^{-1}(\mathscr{F}\mu\ \mathscr{F}\nu), h \rangle = \langle \mu_x \otimes \nu_y,\, h(x + y) \rangle$$

(cf. Exercise 30.3 and 40.3).

41

Bilinear Mappings. Hypocontinuity

Let E, F, and G be three topological vector spaces,

$$\Phi : (x, y) \rightsquigarrow \Phi(x, y)$$

a bilinear mapping of $E \times F$ into G. This means that, for every $x_0 \in E$ (resp. every $y_0 \in F$), the mappings

$$\Phi_{x_0} : y \rightsquigarrow \Phi(x_0, y)$$
$$(\text{resp. } \Phi_{y_0} : x \rightsquigarrow \Phi(x, y_0))$$

from F (resp. E) into G are linear. The bilinear map Φ is said to be *separately continuous* if, for all x_0, y_0, these two linear mappings are continuous. Practically all bilinear mappings considered in analysis are separately continuous. But many of them are not *continuous*. Let us make more explicit what the latter means. It means that to every neighborhood of zero W in G, there are neighborhoods of zero U and V in E and F, respectively, such that

$$x \in U, \quad y \in V \quad \text{implies} \quad \Phi(x, y) \in W.$$

When E, F, and G are all three locally convex, this condition can be rephrased as follows: to every continuous seminorm $\mathfrak{r}$ on G, there are continuous seminorms $\mathfrak{p}$ and $\mathfrak{q}$ on E and F, respectively, such that, for all $x \in E$, $y \in F$,

$$\mathfrak{r}(\Phi(x, y)) \leqslant \mathfrak{p}(x)\, \mathfrak{q}(y).$$

Indeed, it suffices to take for W the closed unit semiball of $\mathfrak{r}$ and select then $\mathfrak{p}$ and $\mathfrak{q}$ so that their closed unit semiballs are contained in U and V, respectively. Observe then that, for all $\varepsilon > 0$, $[\mathfrak{p}(x) + \varepsilon]^{-1}x \in U$ and $[\mathfrak{q}(y) + \varepsilon]^{-1}y \in V$, and that, for all λ, $\mu > 0$,

$$\mathfrak{r}(\Phi(\lambda x, \mu y)) = \lambda\mu\mathfrak{r}(\Phi(x, y)).$$

We have already seen (Corollary of Theorem 34.1) that, when E is a Fréchet space, F a metrizable TVS, and G a locally convex space, then every separately continuous bilinear map of $E \times F$ into G is continuous. An interesting result of a similar kind is the following one:

THEOREM 41.1. *Let F, G be strong duals of reflexive Fréchet spaces, and E either a normed space or the strong dual of a reflexive Fréchet space. Then every separately continuous bilinear map of $E \times F$ into G is continuous.*

Proof of Theorem 41.1

1. *E is normed.* Let $\Phi : E \times F \to G$ be a separately continuous bilinear map. For $x \in E$ and $z' \in G'$, we set

$$\Phi'(x, z') = {}^t\Phi_x(z'),$$

where ${}^t\Phi_x : G' \to F'$ is the transpose of $\Phi_x : y \to \Phi(x, y)$. We claim that $\Phi' : E \times G' \to F'_\sigma$ is separately continuous (which will imply that Φ' is continuous since E is metrizable and G', a Fréchet space). Indeed, we have, for $x \in E$, $y \in F$, $z' \in G'$,

$$(41.1) \qquad \langle \Phi'(x, z'), y \rangle = \langle z', \Phi(x, y) \rangle.$$

If we fix x, and if $z' \to 0$ for $\sigma(G', G)$, we see that $\Phi'(x, z') \to 0$ for $\sigma(F', F)$. Recalling that F is the dual of F' and G the dual of G' (when F' and G' carry their Fréchet space structure), it follows from Lemma 37.6 that $z' \rightsquigarrow \Phi'(x, z')$ is a continuous linear map of G' into F'. Now, let us fix z'. We derive from (41.1) that, if $x \to 0$ in E, $\Phi'(x, z') \to 0$ weakly in F'. Our claim is proved.

Let, now, W be an arbitrary neighborhood of zero in G; W contains .a neighborhood of zero of the form C^0, where C is some bounded subset of G'. Let U be the closed unit ball of E; U is a bounded subset of E (this is where we really use the fact that E is normed). Therefore, as immediately seen, $\Phi'(U, C)$ is bounded in F'_σ. But since the Fréchet topology of F' is compatible with the duality between F' and F, it follows from Theorem 36.2 (Mackey's theorem) that $\Phi'(U, C)$ is a bounded subset of F', which we denote by B. Suppose now that $x \in E$ belongs to U, $y \in F$ belongs to B^0, and $z' \in G'$ to C. We derive from (41.1) that

$$|\langle z', \Phi(x, y) \rangle| \leqslant 1,$$

in other words, that $\Phi(x, y) \in C^0$. Thus we have $\Phi(U, B^0) \subset W$, which proves the continuity of Φ in this case.

2. *E is the strong dual of a reflexive Fréchet space.* Let $U_1 \supset U_2 \supset \cdots \supset$ $U_n \supset \cdots$ be a countable decreasing basis of neighborhoods of zero in E'; we assume the U_n's closed convex and balanced. Let us denote by E_n the linear subspace of E spanned by U_n^0 ; E_n is canonically equipped with a structure of Banach space for which the closed unit ball is U_n^0 (see Notation 37.1 and the remarks following it; cf. also Lemma 36.1). The topology of E_n is finer than the topology induced by E, since U_n^0 is bounded in E (U_n^0 is bounded for $\sigma(E, E')$ but E is the strong dual of the barreled space E'). From there it follows immediately that the restriction of Φ to $E_n \times F$, as a bilinear mapping of this product into G, is separately continuous. Since E_n is normed, it follows from the first part of the proof that $\Phi : E_n \times F \to G$ is continuous. Let, now, W be an arbitrary closed convex balanced neighborhood of zero in G. For each $n = 1, 2,...$, there is a bounded subset B_n of F' such that

$$(41.2) \qquad\qquad \Phi(U_n^0, B_n^0) \subset W.$$

At this stage, we use the following easy lemma:

LEMMA 41.1. *If $\{B_k\}$ ($k = 1, 2,...$) is a sequence of bounded subsets of a metrizable TVS M, there is a sequence of numbers $\varepsilon_k > 0$ such that the union $\bigcup_{k=1}^{\infty} \varepsilon_k B_k$ is bounded.*

Proof of Lemma 41.1. Select a countable decreasing basis of neighborhoods of zero in M, $V_1 \supset V_2 \supset \cdots \supset V_n \supset \cdots$. For each $k = 1, 2.,..$, select $\varepsilon_k > 0$ such that $\varepsilon_k B_k \subset V_k$; the sequence $\{\varepsilon_k\}$ fulfills our requirement. Indeed, let $n = 1, 2,...$ be arbitrary. There is $\eta_n > 0$ such that $\eta_n \varepsilon_k B_k \subset V_n$ for $k < n$; for $k \geqslant n$, $\varepsilon_k B_k \subset V_k \subset V_n$.

Let us return to the proof of Theorem 41.1. Since F' is a Fréchet space, we may apply Lemma 41.1 to the sequence $\{B_n\}$ and select a sequence of positive numbers ε_n such that $B = \bigcup_{n=1}^{\infty} \varepsilon_n B_n$ is bounded in F'. Recalling that $(\varepsilon_n B_n)^0 = \varepsilon_n^{-1} B_n^0$, we derive from (41.2),

$$\Phi(\varepsilon_n U_n^0, (\varepsilon_n B_n)^0) \subset W.$$

But the polar of B is contained in the polar of $\varepsilon_n B_n$, hence

$$\Phi(\varepsilon_n U_n^0, B^0) \subset W.$$

On the other hand, since W is closed convex and balanced, we have

$$\Phi(U, B^0) \subset W,$$

where U is the closed convex balanced hull of $\bigcup_{n=1}^{\infty} \varepsilon_n U_n^0 \subset E$. The proof of Theorem 41.1 will be complete if we prove that U is a neighborhood of zero in E; of course, $U = U^{00}$. It suffices therefore to show that U^0 is a bounded subset of E'; but, for every n, we have $\varepsilon_n U_n^0 \subset U$, which implies $\varepsilon_n^{-1} U_n = (\varepsilon_n U_n^0)^0 \supset U^0$.   Q.E.D.

COROLLARY 1.   *The convolution mapping*

$$(S, T) \rightsquigarrow S*T$$

is a continuous bilinear mapping of $\mathscr{E}' \times \mathscr{E}'$ into $\mathscr{E}'$.

COROLLARY 2.   *The tensor product of distributions*

$$(S, T) \rightsquigarrow S \otimes T$$

is a continuous bilinear mapping of $\mathscr{E}'(R^m) \times \mathscr{E}'(R^n)$ into $\mathscr{E}'(R^{m+n})$. *($\mathscr{E}'$ can be replaced by $\mathscr{S}'$.)*

Even if we exploit fully Theorems 34.1 and 41.1, there remains quite a stock of important bilinear mappings which are not continuous. Let us mention two:

(1)   the multiplication mappings $(\phi, \psi) \rightsquigarrow \phi\psi$ from $\mathscr{C}^{\infty} \times \mathscr{C}_c^{\infty}$ into $\mathscr{C}_c^{\infty}$ and $(\phi, T) \rightsquigarrow \phi T$ from $\mathscr{C}^{\infty} \times \mathscr{D}'$ into $\mathscr{D}'$ (see Exercises 41.1 and 41.2);

(2)   the convolution mappings $(\phi, T) \rightsquigarrow \phi * T$ from $\mathscr{C}_c^{\infty} \times \mathscr{D}'$ into $\mathscr{C}^{\infty}$ (or into $\mathscr{D}'$) and $(S, T) \rightsquigarrow S * T$ from $\mathscr{E}' \times \mathscr{D}'$ into $\mathscr{D}'$.

Many more examples could be given. But these bilinear mappings, while they are not continuous, have a property which is stronger than separate continuity and which palliates, in many a situation, the disadvantages resulting from the absence of continuity. This property is called hypocontinuity; here is its definition:

Definition 41.1.   *A bilinear mapping $\Phi : E \times F \rightsquigarrow G$ is said to be hypocontinuous if the following holds:*

(HC 1)   *For every bounded subset A of E, the mappings $\Phi_x : F \to G$ form, when x varies over A, an equicontinuous set of linear mappings;*

(HC 2)   *for every bounded subset B of E, the mappings $\Phi_y : E \to G$ form, when y varies over B, an equicontinuous set of linear mappings.*

This definition can be rephrased in various ways. We recall that

$$\Phi_x : y \rightsquigarrow \Phi(x, y), \qquad \Phi_y : x \rightsquigarrow \Phi(x, y).$$

Then, to say that Φ is hypocontinuous is saying that $\Phi(x, y)$ converges to zero in G when x (resp. y) converges to zero in E (resp. in F) while the other variable remains in a bounded set, and that the convergence of $\Phi(x, y)$ in G is then uniform on this bounded set. In other words, to every neighborhood of zero W in G and, to every bounded subset A (resp. B) of E (resp. F), there is a neighborhood of zero V (resp. U) in F (resp. E) such that

$$\Phi(A, V) \subset W, \qquad \Phi(U, B) \subset W.$$

For us, the most useful criterion of hypocontinuity will be the following one:

THEOREM 41.2. *Let E, F be barreled spaces* (Definition 33.1), *and G a locally convex space. Every separately continuous bilinear map of $E \times F$ into G is hypocontinuous.*

Proof of Theorem 41.2. Let A be a bounded subset of E. Let y be an arbitrary point of F. Since Φ is separately continuous, to every neighborhood of zero W in G there is a neighborhood of zero U in E such that $\Phi(U, y) \subset W$. Let, then, $\rho > 0$ be such that $A \subset \rho U$. We have $\Phi(A, y) \subset \rho \Phi(U, y) \subset \rho W$. This shows that $\Phi(A, y)$ is bounded in G. In other words, when x varies over A, the mappings Φ_x form a set of continuous linear mappings of F into G which is bounded for the topology of pointwise convergence. Our hypotheses allow us to apply the Banach–Steinhaus theorem (Theorem 33.1): the Φ_x ($x \in A$) form an equicontinuous set of linear mappings. Same argument after exchange of E and F.   Q.E.D.

As all the bilinear mappings which we have encountered (or which we are liable to encounter) are separately continuous, and as most of the spaces on which they are defined are barreled, these mappings will mostly be hypocontinuous. This applies in particular to the examples mentioned on p. 423, and to many other similar ones. We should also say that the notion of hypocontinuity which we have introduced here, although very important, is in a way rather particular. The bounded subsets of E and F are given a dominant role; this need not be so. For instance, this role might be played by the compact subsets (or the convex compact subsets) of E or F or, for instance, if E and F are duals of other spaces, one might be interested in bilinear mappings defined on $E \times F$

which are hypocontinuous with respect to the equicontinuous subsets of E and F; etc. It is clear that one may define the notion of $\mathfrak{S}$-$\mathfrak{H}$-hypocontinuity, where $\mathfrak{S}$ and $\mathfrak{H}$ are suitable families of bounded subsets of E and F, respectively. For instance, when $\mathfrak{S}$ and $\mathfrak{H}$ are the families of finite sets, the notion of $\mathfrak{S}$-$\mathfrak{H}$-hypocontinuity simply reduces to the one of separate continuity.

Exercises

41.1 Prove that if the locally convex Hausdorff space E is barreled, the bilinear form on $E \times E'$,

$$(x, x') \rightsquigarrow \langle x', x \rangle,$$

is hypocontinuous.

41.2. Do Exercise 19.5.

41.3. Let $\mathscr{P}_n$ and $\mathscr{Q}_n$ be the space of polynomials and formal power series in n letters (with complex coefficients), equipped with their Fréchet and LF topologies. Prove that the duality bracket on $\mathscr{P}_n \times \mathscr{Q}_n$,

$$(P, u) \rightsquigarrow \langle P, u \rangle,$$

is hypocontinuous but not continuous.

41.4. We use the same notation as in Exercise 41.3, but we assume now that $\mathscr{P}_n$ carries the topology induced by $\mathscr{Q}_n$. Prove that, now, the duality bracket between $\mathscr{P}_n$ and $\mathscr{Q}_n$ is not separately continuous.

41.5. Let E, F, and G be three TVS. Suppose that E is a Baire space and that F is metrizable. Prove that every separately continuous bilinear map of $E \times F$ into G is continuous.

41.6. Let E, F be two Fréchet spaces, G a locally convex Hausdorff space, E', F', and G' the duals of E, F, and G, respectively, and u a separately continuous bilinear map of $E'_\sigma \times F'_\sigma$ into G'_σ.
For all $z \in G$, $x' \in E'$, $y' \in F'$, set

$$\langle z, u(x', y') \rangle = \langle v_z(x'), y' \rangle.$$

This defines a linear map $v_z : E' \to F''$. Prove the following facts:

(1) v_z is a continuous linear map of E'_σ into F_σ ;

(2) when z varies in a bounded subset C of G, v_z varies in an equicontinuous subset of $L(E'_b ; F)$;

(3) for every bounded set C of G there is a neighborhood of zero, U', in E'_b, such that the set of points $v_z(x')$, $z \in C$, $x' \in U'$, is bounded in F.

Derive from this that the bilinear map

$$u : E'_b \times F'_b \to G'_b$$

is continuous.

41.7. Derive from the preceding result that any separately continuous bilinear maps of the product of two duals of reflexive Fréchet spaces into a third one is continuous (cf. Theorem 41.1).

41.8. Give the example of a Fréchet space E, an LF-space F, both reflexive, and of bilinear forms u on $E \times F$ and v on $E' \times F'$ with the following properties:

(a) u is hypocontinuous but not continuous;

(b) v is separately continuous on $E'_\sigma \times F'_\sigma$ and on $E'_b \times F'_b$ but not continuous on any of these two products.

41.9. Let E, F, and G be three locally convex Hausdorff spaces, and E_0 (resp. F_0) a dense linear subspace of E (resp. F). Let $u : E \times F \to G$ be a separately continuous bilinear map. Prove the following facts:

(i)   if $u = 0$ in $E_0 \times F_0$, $u = 0$ in $E \times F$;

(ii)   if the restriction of u to $E_0 \times F_0$ is hypocontinuous, then, for every bounded subset A_0 (resp. B_0) of E_0 (resp. F_0), the set of mappings

$$y \rightsquigarrow u(x, y), \qquad x \in A_0 \,,$$
$$(\text{resp. } x \rightsquigarrow u(x, y), \qquad y \in B_0)$$

is equicontinuous.

41.10. Let E, F, G, E_0, and F_0 be as in Exercise 41.9. We suppose furthermore that every point of E (resp. F) belongs to the closure of some *bounded* subset of E_0 (resp. F_0). Let $u_0 : E_0 \times F_0 \to G$ be a hypocontinuous bilinear map. There is a unique separately continuous bilinear map $u : E \times F \to G$ extending u_0. Moreover, u has Property (ii) of Exercise 41.9. Prove these assertions.

41.11. Let E, F, and G be three locally convex Hausdorff TVS. We suppose that the three spaces $L(E; F)$, $L(F; G)$, and $L(E; G)$ carry the topology of bounded convergence (or else that all three of them carry the topology of compact convergence, or that all three carry the topology of pointwise convergence). Prove the following facts:

(a)   for every equicontinuous subset H of $L(F; G)$, the composition mapping

$$(u, v) \rightsquigarrow v \circ u$$

from $L(E; F) \times H$ into $L(E; G)$ is continuous;

(b)   if F is barreled, for every sequence $\{u_k\}$ converging to u in $L(E; F)$ and every sequence $\{v_k\}$ converging to v in $L(F; G)$, the sequence $\{v_k \circ u_k\}$ converges to $v \circ u$ in $L(E; G)$.

42

Spaces of Bilinear Forms.
Relation with Spaces of Linear Mappings
and with Tensor Products

All the topological vector spaces considered in this chapter will be locally convex and Hausdorff. Let E, F, and G be three such spaces.

Notation 42.1. We denote by $\mathscr{B}(E, F; G)$ the space of separately continuous bilinear maps of $E \times F$ into G, and by $B(E, F; G)$ the space of continuous bilinear maps of $E \times F$ into G. When G is the scalar field (**R** or **C**), we write $\mathscr{B}(E, F)$ and $B(E, F)$, respectively.

That $\mathscr{B}(E, F; G)$ is a linear space (for the natural addition and scalar multiplication) is obvious; $B(E, F; G)$ is a linear subspace of it.

Let $\mathfrak{S}$ (resp. $\mathfrak{H}$) be a family of bounded subsets of E (resp. F). We may consider on $B(E, F; G)$ the $\mathfrak{S}$-$\mathfrak{H}$-topology, or topology of uniform convergence on subsets of the form $A \times B$ with $A \in \mathfrak{S}$, $B \in \mathfrak{H}$. We obtain a basis of neighborhoods of zero in this topology by taking the sets

$$\mathscr{U}(A, B; W) = \{\Phi \in B(E, F; G); \Phi(A, B) \subset W\},$$

where A (resp. B) varies over $\mathfrak{S}$ (resp. $\mathfrak{H}$) and W over a basis of neighborhoods of zero in G. Of course, one must check that the sets $\mathfrak{U}(A, B; W)$ fulfill the requirements on neighborhoods of zero in a TVS, in particular they must be absorbing. This is easily checked to be so, keeping in mind that we are dealing there with *continuous* bilinear mappings. However, if we were dealing with separately continuous bilinear mappings, this need not be: indeed, for a bilinear mapping Φ to be absorbed by a set $\mathfrak{U}(A, B; W)$, it is necessary and sufficient that $\Phi(A, B)$ be absorbed by W. If we wish this to be true for all W, we see that $\Phi(A, B)$ must be a *bounded* subset of G. But, in general, if A (resp. B) is a bounded subset of E (resp. F), and if Φ is only separately continuous, it is not true that $\Phi(A, B)$ will be bounded in G. Of course, as is readily

427

seen, this is true whenever Φ is hypocontinuous. Thus we may define the $\mathfrak{S}$-$\mathfrak{H}$-topology on $\mathscr{B}(E, F; G)$ (and not only on $B(E, F; G)$) whenever E and F are barreled and G is locally convex. There are other cases where this definition is possible, as shown by the next result:

PROPOSITION 42.1. *Let E, F, and G be three locally convex spaces, E_b', F_b' the strong duals of E, F, respectively, A' (resp. B') an equicontinuous subset of E' (resp. F'), and Φ a separately continuous bilinear mapping of $E_b' \times F_b'$ into G. Then $\Phi(A', B')$ is a bounded subset of G.*

Proof of Proposition 42.1. We may assume that A' and B' are closed convex and balanced, hence weakly compact convex and balanced (Propositions 32.3 and 32.8). When x' varies over A', the mappings $\Phi_{x'} : F_b' \to G$ form a set of continuous linear mappings which is bounded for the topology of pointwise convergence in F_b'. Therefore (Exercise 36.5) it is also bounded for the topology of uniform convergence on the convex balanced complete bounded subsets of F_b' ; in particular,

$$\bigcup_{x' \in A'} \Phi_{x'}(B') = \Phi(A', B')$$

must be bounded in G.   Q.E.D.

It is easily checked that the topology of uniform convergence on the products $A' \times B'$, A' (resp. B') an equicontinuous subset of E' (resp. F'), turns $\mathscr{B}(E_b', F_b'; G)$ into a locally convex TVS, which we shall denote by $\mathscr{B}_\varepsilon(E_b', F_b'; G)$. Later on, we shall be interested in the subspace $\mathscr{B}(E_\sigma', F_\sigma'; G)$ of $\mathscr{B}(E_b', F_b'; G)$ consisting of the bilinear mappings $E' \times F' \to G$ which are separately continuous when E' and F' carry their weak topologies. We shall denote by

$$\mathscr{B}_\varepsilon(E_\sigma', F_\sigma'; G)$$

the space in question provided with the topology induced by $\mathscr{B}_\varepsilon(E_b', F_b'; G)$; this topology will often be referred to as the ε-topology. Finally, let us remark that all these topologies are Hausdorff as soon as this is true of G.

We now focus our attention on spaces of bilinear forms and, first of all, for reasons that will be clear later on, on the space $\mathscr{B}(E_\sigma', F_\sigma')$ of separately continuous bilinear forms on the product of the weak duals E_σ' and F_σ' of E and F. We wish to show that there is a canonical isomorphism (for the vector space structures) of $\mathscr{B}(E_\sigma', F_\sigma')$ onto $L(E_\sigma'; F_\sigma)$, space of continuous linear mappings of E_σ' into F equipped with its weak topology $\sigma(F, F')$. We then want this isomorphism to extend to the topological structures, once we have provided $\mathscr{B}(E_\sigma', F_\sigma')$ with the

ε-topology. At this point, we discover that the topology to be put on $L(E'_\sigma ; F_\sigma)$ is the topology of uniform convergence on the equicontinuous subsets of E' when F carries its initial topology (and not its weak topology). This would appear as rather mysterious if one did not observe that continuous linear maps from E'_σ into F_σ are one and the same thing as continuous linear mappings of E'_τ into F, where the index τ means Mackey's topology on E', i.e., the topology of uniform convergence (of continuous linear forms) on the convex balanced weakly compact subsets of E. We now state and prove these various facts:

PROPOSITION 42.2.   *Let E, F be two locally convex Hausdorff TVS. Then:*

(1) $L(E'_\sigma ; F_\sigma) = L(E'_\tau ; F)$;

(2)   *the mapping $\Phi \rightsquigarrow \tilde{\Phi}$, where*

$$\tilde{\Phi} : x' \rightsquigarrow \Phi_{x'}, \qquad \Phi_{x'} : y' \rightsquigarrow \Phi(x', y'),$$

is an isomorphism (for the vector space structures) of $\mathscr{B}(E'_\sigma, F'_\sigma)$ onto $L(E'_\sigma ; F_\sigma)$;

(3)   *the mapping $\Phi \rightsquigarrow \tilde{\Phi}$ is an isomorphism (for the TVS structure) of $\mathscr{B}_\varepsilon(E'_\sigma, F'_\sigma)$ onto $L_\varepsilon(E'_\tau ; F)$, the space $L(E'_\tau ; F)$ equipped with the topology of uniform convergence on the equicontinuous subsets of E'.*

Proof of (1). If $u : E'_\tau \to F$ is continuous, its transpose ${}^t u : F'_\sigma \to E$ is continuous (as E is the dual of E'_τ), and the transpose of ${}^t u$, which is nothing else but $u : E'_\sigma \to F_\sigma$, is also continuous. Conversely, let u be a continuous linear map of E'_σ into F''_σ. Let ${}^t u : F'_\sigma \to E''_\sigma$ be its transpose, which is continuous. Let V be an arbitrary closed convex balanced neighborhood of zero in F; the polar of V, V^0, is a convex balanced compact subset of F'_σ (Propositions 32.7 and 32.8), hence ${}^t u(V^0)$ is a compact subset of E_σ, say K. The polar K^0 of K is a neighborhood of zero in E'_τ and it is readily seen (using the fact that V is equal to its weakly closed convex balanced hull) that $u(K^0) \subset V$.

Proof of (2).   As the dual of F'_σ is F, we identify $\Phi_{x'}$ with some element of F, which we denote by $\tilde{\Phi}(x')$. We have, for all $x' \in E'$, $y' \in F'$,

(42.1) $$\Phi(x', y') = \langle y', \tilde{\Phi}(x')\rangle.$$

On this equality, the fact that $\tilde{\Phi} : E'_\sigma \to F_\sigma$ is continuous is evident. If $\tilde{\Phi}$ is given, (42.1) defines Φ, which is obviously separately continuous on $E'_\sigma \times F'_\sigma$.

Proof of (3). A basis of neighborhoods of zero in $\mathcal{B}_\varepsilon(E'_\sigma, F'_\sigma)$ is obtained by taking the sets $\mathcal{W}(A', B')$ consisting of the forms Φ such that $|\Phi(x', y')| \leqslant 1$ for all $x' \in A'$, $y' \in B'$, where A' (resp. B') is an equicontinuous subset of E' (resp. F'). The mapping $\Phi \rightsquigarrow \tilde{\Phi}$ transforms $\mathcal{W}(A', B')$ into the set $\mathcal{U}(A', B'^0)$ of continuous linear mappings $E'_\tau \to F$ which map A' into B'^0. By Proposition 36.1, there is a basis of neighborhoods of zero in F consisting of sets of the form B'^0, B': equicontinuous subset of F'. Conversely, if V is a closed convex balanced neighborhood of zero in F, its polar V^0 is an equicontinuous subset of F' and the inverse of the mapping $\Phi \rightsquigarrow \tilde{\Phi}$ transforms $\mathcal{U}(A', V)$ into $\mathcal{W}(A', V^0)$.　　　　　　　　　　　　　　　　　　Q.E.D.

PROPOSITION 42.3. *Let E, F be locally convex Hausdorff TVS. Then $L_\varepsilon(E'_\tau; F)$ (and consequently $\mathcal{B}_\varepsilon(E'_\sigma; F'_\sigma)$) is complete if and only if both E and F are complete.*

Proof

Necessity of the condition. Let $y \in F$ be arbitrarily chosen except that y must be $\neq 0$. To every $x \in E$, we associate the mapping $u_x : E' \to F$ defined by $x' \rightsquigarrow \langle x', x \rangle y$. This yields a linear mapping $x \rightsquigarrow u_x$ of E into $L(E'_\tau; F)$. Indeed, if $x' \to 0$ in E'_τ, $x' \to 0$ in E'_σ a fortiori and therefore $\langle x', x \rangle \to 0$. We contend that $x \rightsquigarrow u_x$ is an isomorphism (for the TVS structures) of E into $L_\varepsilon(E'_\tau; F)$ and that the image of E under this mapping is a *closed* linear subspace.

The following facts are clear: for each $x \in E$, u_x is continuous from E'_τ (and in fact from E'_σ) into F; $x \rightsquigarrow u_x$ is linear and one-to-one; u_x converges to zero in $L_\varepsilon(E'_\tau; F)$ if and only if x converges to zero uniformly on every equicontinuous subset of E', i.e., (Proposition 36.1) converges to zero in E. Thus $x \rightsquigarrow u_x$ is an isomorphism into. The image of E under the mapping $x \rightsquigarrow u_x$ is the set of all continuous linear mappings of E'_τ into the one-dimensional subspace generated by y, $\mathbf{C}y$ (if $\mathbf{C}$ is the scalar field, otherwise $\mathbf{R}y$). Indeed, if u is such a mapping, we have, for all $x' \in E'$, $u(x') = f(x') y$, where f is a continuous linear form on E'_τ. But then there is $x \in E$ such that $f(x') = \langle x', x \rangle$ for all x'. It is clear that the set of all continuous linear mappings of E'_τ into $\mathbf{C}y$ is closed in $L_\varepsilon(E'_\tau; F)$ (and even in $L_s(E'_\tau; F)$). This implies that E is complete if this is true of $L_\varepsilon(E'_\tau; F)$. Since the latter TVS is isomorphic with $\mathcal{B}_\varepsilon(E'_\sigma, F'_\sigma)$, we see that the situation is perfectly symmetric in E and F and therefore the completeness of $L_\varepsilon(E'_\tau; F)$ also implies that of F.

Sufficiency of the condition. If F is complete, the vector space $\mathcal{L}(E'; F)$ of all linear mappings (whether continuous or not) of E' into F is complete when we provide it with the topology of uniform convergence on the equicontinuous subsets of E' (this topology is not compatible, in general, with the vector space structure of $\mathscr{L}(E'; F)$, but it is clear that the notions of Cauchy filter and of completeness make sense for it). It will suffice to show that $L_\epsilon(E'_\tau; F)$ is a closed subspace of $\mathscr{L}(E'; F)$ for this topology.

Let u be a linear mapping of E' into F which is the limit, uniformly on every equicontinuous subset of E', of continuous linear mappings of E'_τ into F. It suffices to show that u is continuous from E'_σ into F_σ (Proposition 42.2, Part (1)). Let, therefore, $y' \in F'$ be arbitrary; the linear form $y' \circ u$ on E' is the limit, uniform on the equicontinuous subsets of E', of linear forms on E' which are continuous for $\tau(E', E)$. The latter are of the form $x' \rightsquigarrow \langle x', x \rangle$ with $x \in E$, since E is the dual of E'_τ. To say that they converge uniformly on the equicontinuous subsets of E' is equivalent to saying that the corresponding x converge in the completion of E, in view of Proposition 36.1. Therefore, if E is complete, these x have a limit $x_0 \in E$ and we have, for all $x' \in E'$,

$$\langle y' \circ u, x' \rangle = \langle x', x_0 \rangle$$

This proves the continuity of $y' \circ u$.   Q.E.D.

We now turn our attention to tensor products. There is a canonical bilinear mapping ϕ of $E \times F$ into $B(E'_\sigma, F'_\sigma)$, space of continuous bilinear forms on $E'_\sigma \times F'_\sigma$:

(42.2) $(x, y) \rightsquigarrow \phi(x, y) : (x', y') \rightsquigarrow \langle x', x \rangle \langle y', y \rangle.$

We observe first that E and F are ϕ-linearly disjoint (Definition 39.1). Indeed, consider two linearly independent finite sets of vectors $\{x_j\}$, $\{y_k\}$ in E and F, respectively; select two sets $\{x'_j\}$, $\{y'_k\}$ in E' and F', respectively, having the same number of elements as $\{x_j\}$ and $\{y_k\}$, respectively, and such that $\langle x'_{j'}, x_j \rangle = \delta_{jj'}$ and $\langle y'_{k'}, y_k \rangle = \delta_{kk'}$ ($\delta_{jj'}$, $\delta_{kk'}$: Kronecker symbols). The value of the bilinear form $\phi(x_j, y_k)$ is equal to one on (x'_j, y'_k), and to zero on $(x'_{j'}, y'_{k'})$ as soon as either $j' \neq j$ or $k \neq k'$. This implies that the forms $\phi(x_j, y_k)$ are linearly independent. Our next observation is that the forms $\phi(x, y)$ span $B(E'_\sigma, F'_\sigma)$ as (x, y) varies over $E \times F$. Let us prove this statement. Let Φ be a continuous bilinear map on $E'_\sigma \times F'_\sigma$. There is a finite subset A of E and a finite subset B of F such that

$$x' \in A^0, \quad y' \in B^0 \quad \text{implies} \quad |\Phi(x', y')| \leqslant 1.$$

Let E_A (resp. F_B) be the linear subspace of E (resp. F) spanned by

A (resp. B); of course, E_A and F_B are finite dimensional. Their orthogonals, $(E_A)^0$ and $(F_B)^0$ have finite codimension, and we may write

$$E' = M' \oplus (E_A)^0, \qquad F' = N' \oplus (F_B)^0 \qquad (\oplus: \text{direct sum}).$$

Obviously, Φ vanishes on the subspace of $E' \times F'$,

$$((E_A)^0 \times F') \oplus (E' \times (F_B)^0),$$

which is a supplementary of $M' \times N'$ in $E' \times F'$; in other words, Φ is completely determined by its restriction to the finite dimensional subspace $M' \times N'$. Obviously, one can find a finite set of vectors in E_A, $x_1, ..., x_r$, a finite set in F_B, $y_1, ..., y_s$, such that this restriction to $M' \times N'$ (and therefore Φ in the whole of $E' \times F'$) is given by

$$(x', y') \rightsquigarrow \sum_{j=1}^{r} \sum_{k=1}^{s} \langle x', x_j \rangle \langle y', y_k \rangle.$$

Summarizing, we may state:

PROPOSITION 42.4. $B(E'_\sigma, F'_\sigma)$ *is a tensor product of E and F.*

In the forthcoming chapters, we shall often write $E \otimes F = B(E'_\sigma, F'_\sigma)$; the mapping (42.2) will be denoted by $(x, y) \rightsquigarrow x \otimes y$. Let us show rapidly that this is in agreement with our definition of the tensor product of two linear mappings (Proposition 39.2). In view of the general properties of tensor products, there is a one-to-one correspondence between bilinear forms on $E' \times F'$ and linear forms on $E' \otimes F'$ (Theorem 39.1). If we identify every element x (resp. y) of E (resp. F) with the linear form $x' \rightsquigarrow \langle x', x \rangle$ (resp. $y' \rightsquigarrow \langle y', y \rangle$) it defines on E' (resp. F'), we see that the bilinear form $\phi(x, y)$ of (42.2) is associated with the linear form $x \otimes y$ (Notation of Proposition 39.2).

Let us go back to $B(E'_\sigma, F'_\sigma)$; we may regard it as a linear subspace of $\mathscr{B}(E'_\sigma, F'_\sigma)$, space of separately continuous bilinear forms on $E'_\sigma \times F'_\sigma$. Then, by using the canonical isomorphism of $\mathscr{B}(E'_\sigma, F'_\sigma)$ onto $L(E'_\sigma; F_\sigma)$ (Proposition 42.2), we may regard $B(E'_\sigma, F'_\sigma)$ as a linear subspace of the latter. We leave the proof of the characterization below to the student:

PROPOSITION 42.5.   *The canonical image of $B(E'_\sigma, F'_\sigma)$ into $L(E'_\sigma; F_\sigma)$ is equal to the space of continuous linear mappings of E'_σ into F whose image is finite dimensional.*

For these mappings, it is irrelevant to specify the topology on F, as long as it is Hausdorff: indeed, on the image of the mapping all Hausdorff topologies (compatible with the linear structure) coincide.

It is also clear that, in the finite dimensional cases, all the above iso-morphisms become *onto*, and most statements are trivial. In the infinite dimensional case, not only is $B(E'_\sigma, F'_\sigma)$ distinct from $\mathscr{B}(E'_\sigma, F'_\sigma)$ (equi-valently, $E \otimes F$ is distinct from $L(E'_\sigma; F_\sigma)$), but it is not closed in $\mathscr{B}_\varepsilon(E'_\sigma, F'_\sigma)$ (i.e., in $L_\varepsilon(E'_\tau; F)$). We shall see, later on, that there are important classes of spaces for which $E \otimes F$ is dense in $L_\varepsilon(E'_\tau; F)$, as well as others for which this is not true.

We close this chapter with a few words about the *normed* case. By Mackey's theorem, we know that every convex balanced weakly compact subset of E is bounded (for the initial topology); therefore Mackey's topology $\tau(E', E)$ on E' is always *weaker* than the strong dual topology on E', in contrast with the fact that its Mackey topology $\tau(E', E'')$ is always *stronger* than the strong dual topology. From this it follows that a continuous linear map of E'_τ into F is a fortiori a continuous linear map of E'_b into F, i.e., $L(E'_\tau; F) \subset L(E'_b; F)$. Now, when E is normed, E'_b is a Banach space, and the equicontinuous subsets of E'_b are the subsets of its balls centered at the origin (with finite radius!). Thus $L_\varepsilon(E'_\tau; F)$ can be regarded as a subspace of the space $L_b(E'_b; F)$. By applying Propositions 42.2 and 42.3, we may state:

PROPOSITION 42.6.   *Assume that the spaces E and F are normed. Then $\mathscr{B}_\varepsilon(E'_\sigma, F'_\sigma)$ is a normed space, canonically isomorphic to the subspace $L_\varepsilon(E'_\tau; F)$ of $L_b(E'_b; F)$; $\mathscr{B}_\varepsilon(E'_\sigma, F'_\sigma)$ is a Banach space if and only if both E and F are Banach spaces.*

Note that if F is a Banach space so is $L_b(E'_b; F)$, even when E is not complete. In this case, we see that $L_\varepsilon(E'_\tau; F)$ is a subspace of $L_b(E'_b; F)$ which is *not* closed.

Exercises

42.1. Let E, F, and G be three locally convex Hausdorff spaces, and $\mathscr{H}$ the vector space of hypocontinuous bilinear mappings of $E \times F$ into G. Prove that the topology of uniform convergence on the product sets $A \times B$, where A and B are bounded subsets of E and F, respectively, is compatible with the linear structure of $\mathscr{H}$. Let us suppose then that $\mathscr{H}$ carries this topology. Prove that

$$\mathscr{H} \ni u \rightsquigarrow (x \rightsquigarrow (y \rightsquigarrow u(x, y))) \in L(E; L(F; G))$$

is a TVS isomorphism of $\mathscr{H}$ onto $L(E; L(F; G))$.

42.2. Same notation as in Exercise 42.1. Prove that if E and F are barreled and G quasi-complete (every closed bounded set is complete) the TVS $\mathscr{H}$ is quasi-complete.

42.3. Let E, F be two locally convex Hausdorff TVS. Let $\mathfrak{S}$ be a family of bounded subsets of E, covering E and having the usual properties $(\mathfrak{S}_I)$ and $(\mathfrak{S}_{II})$. We suppose that the initial topology of E is identical to its Mackey topology $\tau(E, E')$ (for instance, E is metrizable or barreled). Prove then that *the space $L_\mathfrak{S}(E; F)$* (Definition 32.1) *is complete if and only if $E'_\mathfrak{S}$* (Definition 19.2) *and F are both complete* (cf. Proposition 42.3).

43

The Two Main Topologies on Tensor Products. Completion of Topological Tensor Products

In this chapter, E and F will be two locally convex TVS, and $E \otimes F$ a tensor product of E and F.

We recall that $E \otimes F$ is isomorphic to $B(E'_\sigma, F'_\sigma)$ (Proposition 42.4).

Definition 43.1. We call ε-topology on $E \otimes F$ the topology carried over from $B(E'_\sigma, F'_\sigma)$ when we regard the latter as a vector subspace of $\mathscr{B}_\varepsilon(E'_\sigma, F'_\sigma)$, the space of separately continuous bilinear forms on $E'_\sigma \times F'_\sigma$ provided with the topology of uniform convergence on the products of an equicontinuous subset of E' and an equicontinuous subset of F'. Equipped with the ε-topology, the space $E \otimes F$ will be denoted by $E \otimes_\varepsilon F$.

The canonical mapping $(x, y) \rightsquigarrow x \otimes y$ of $E \times F$ into $E \otimes_\varepsilon F$ is continuous; this is obvious. The ε-topology on $E \otimes F$ is locally convex; it is Hausdorff if and only if both E and F are Hausdorff. When E and F are normed spaces, this is also true of $E \otimes_\varepsilon F$ (Proposition 42.6).

We proceed now to give the definition of the second main topology on tensor products.

Definition 43.2. We call π-topology (or projective topology) on $E \otimes F$ the strongest locally convex topology on this vector space for which the canonical bilinear mapping $(x, y) \rightsquigarrow x \otimes y$ of $E \times F$ into $E \otimes F$ is continuous. Provided with it, the space $E \otimes F$ will be denoted by $E \otimes_\pi F$.

A convex subset of $E \otimes F$ is a neighborhood of zero for the π-topology if and only if its preimage under $(x, y) \rightsquigarrow x \otimes y$ contains a neighborhood of zero in $E \times F$, i.e., if it contains a set of the form

$$U \otimes V = \{x \otimes y \in E \otimes F; x \in U, y \in V\},$$

where U (resp. V) is a neighborhood of zero in E (resp. F). In other

434

words, we obtain a basis of neighborhoods of O in $E \otimes_\pi F$ by taking the convex balanced hulls of sets $U_\alpha \otimes V_\beta$, where U_α (resp. V_β) runs over a basis of neighborhoods of zero in E (resp. F).

If we wish to describe the projective topology by means of semi-norms, the best way is to introduce the notion of tensor product of two seminorms. Let $\mathfrak{p}$ (resp. $\mathfrak{q}$) be a seminorm on E (resp. F), $U_\mathfrak{p}$ (resp. $V_\mathfrak{q}$) its closed unit semiball, and W the balanced convex hull of $U_\mathfrak{p} \otimes V_\mathfrak{q}$. Observe that W is absorbing. Let us then set

(43.1) $$(\mathfrak{p} \otimes \mathfrak{q})(\theta) = \inf_{\theta \in \rho W, \rho > 0} \rho, \qquad \theta \in E \otimes F.$$

Definition 43.3. *The seminorm* $\mathfrak{p} \otimes \mathfrak{q}$ *is called the tensor product of the seminorms* $\mathfrak{p}$ *and* $\mathfrak{q}$.

PROPOSITION 43.1. *For all* $\theta \in E \otimes F$,

$$(\mathfrak{p} \otimes \mathfrak{q})(\theta) = \inf \sum_j \mathfrak{p}(x_j)\, \mathfrak{q}(y_j),$$

where the infimum is taken over all finite sets of pairs (x_j, y_j) *such that*

$$\theta = \sum_j x_j \otimes y_j.$$

Furthermore, for all $x \in E$ *and* $y \in F$,

(43.2) $$(\mathfrak{p} \otimes \mathfrak{q})(x \otimes y) = \mathfrak{p}(x)\, \mathfrak{q}(y).$$

Proposition 43.1 can be interpreted in the following way. Let us introduce the space $\mathscr{X}$ of complex-valued functions on $E \times F$ which have a finite support, as in the proof of Theorem 39.1. The tensor product $E \otimes F$ is isomorphic to a quotient $\mathscr{X}/N$ of $\mathscr{X}$. Here, the isomorphism is purely algebraic. Consider on $\mathscr{X}$ the seminorm

$$f \rightsquigarrow \sum_{(x,y) \in E \times F} |f(x, y)|\, \mathfrak{p}(x)\, \mathfrak{q}(y).$$

Its quotient modulo N of this seminorm (cf. Proposition 7.9) "is" the seminorm $\mathfrak{p} \otimes \mathfrak{q}$.

Proof of Proposition 43.1. Let U (resp. V) be the closed unit semiball of $\mathfrak{p}$ (resp. $\mathfrak{q}$); let W be the closed convex balanced hull of $U \otimes V$. To say that $\theta \in \rho W, \rho > 0$, is equivalent to saying that

$$\theta = \sum_{k=1}^N t_k\, x_k \otimes y_k, \qquad \sum_{k=1}^N |t_k| \leqslant \rho, \qquad \mathfrak{p}(x_k) \leqslant 1, \quad \mathfrak{q}(x_k) \leqslant 1.$$

Let us set $\xi_k = t_k x_k$, $\eta_k = y_k$. We see that

$$\theta = \sum_{k=1}^{N} \xi_k \otimes \eta_k \qquad \text{with} \quad \sum_k \mathfrak{p}(\xi_k)\,\mathfrak{q}(\eta_k) \leqslant \rho.$$

Conversely, let us start from such a representation of θ. Let ε be an arbitrary number > 0. Let us set

$$x_k = \mathfrak{p}(\xi_k)^{-1}\,\xi_k\,, \qquad y_k = \mathfrak{q}(\eta_k)^{-1}\,\eta_k\,, \qquad t_k = \mathfrak{p}(\xi_k)\,\mathfrak{q}(\eta_k)$$

when $\mathfrak{p}(\xi_k)\,\mathfrak{q}(\eta_k) \neq 0$ and, otherwise,

$$x_k = \mathfrak{p}(\xi_k)^{-1}\,\xi_k\,, \qquad y_k = [\mathfrak{q}(\eta_k)N/\varepsilon]\eta_k\,, \qquad t_k = \varepsilon/N$$

if $\mathfrak{p}(\xi_k) \neq 0$ and $\mathfrak{q}(\eta_k) = 0$, the analog in the symmetric case, and lastly

$$x_k = (N/\varepsilon)\xi_k\,, \qquad y_k = \eta_k\,, \qquad t_k = \varepsilon/N \qquad \text{when} \quad \mathfrak{p}(\xi_k) = \mathfrak{q}(\eta_k) = 0.$$

We then have

$$\theta = \sum_{k=1}^{N} t_k\,x_k \otimes y_k\,, \qquad \mathfrak{p}(x_k) \leqslant 1, \quad \mathfrak{q}(y_k) \leqslant 1 \qquad \text{for each} \quad k,$$

and

$$\sum_k |\,t_k\,| \leqslant \rho + \varepsilon.$$

This proves that $\theta \in (\rho + \varepsilon)W$; as ε is arbitrarily small, it proves that $(\mathfrak{p} \otimes \mathfrak{q})(\theta) \leqslant \rho$.

Note that we have, in particular, for all $x \in E$, $y \in F$,

$$(\mathfrak{p} \otimes \mathfrak{q})(x \otimes y) \leqslant \mathfrak{p}(x)\,\mathfrak{q}(y).$$

In view of the Hahn–Banach theorem, there is $x' \in E'$ (resp. $y' \in F'$) such that

$$\langle x', x \rangle = \mathfrak{p}(x), \qquad |\langle x', x_1 \rangle| \leqslant \mathfrak{p}(x_1) \qquad \text{for all} \quad x_1 \in E$$

(resp. $\langle y', y \rangle = \mathfrak{q}(y)$, $|\langle y', y_1 \rangle| \leqslant \mathfrak{q}(y_1)$ for all $y_1 \in F$). Let then $\theta = x \otimes y$ be equal to $\sum_k x_k \otimes y_k$; we have

$$|\langle x' \otimes y', \theta \rangle| \leqslant \sum_{\kappa} |\langle x', x_k \rangle \langle y', y_k \rangle| \leqslant \sum_k \mathfrak{p}(x_k)\,\mathfrak{q}(y_k).$$

As this is true for all representations $\theta = \sum_k x_k \otimes y_k$, we see that

$$|\langle x' \otimes y', \theta \rangle| \leqslant (\mathfrak{p} \otimes \mathfrak{q})(x \otimes y).$$

But since, on the other hand, $\theta = x \otimes y$, we have

$$\langle x' \otimes y', \theta \rangle = \langle x', x \rangle \langle y', y \rangle = \mathfrak{p}(x)\,\mathfrak{q}(y),$$

whence

$$\mathfrak{p}(x)\,\mathfrak{q}(y) \leqslant (\mathfrak{p} \otimes \mathfrak{q})(x \otimes y).$$

By combining this with the estimate in the other direction (p. 436), we obtain (43.2).

PROPOSITION 43.2.   *The seminorm* $\mathfrak{p} \otimes \mathfrak{q}$ *is a norm if and only if both* $\mathfrak{p}$ *and* $\mathfrak{q}$ *are norms.*

Proof. If either $\mathfrak{p}$ or $\mathfrak{q}$ are not norms, this must also be true of $\mathfrak{p} \otimes \mathfrak{q}$, as follows immediately from (43.2). Observing that a seminormed space is normed if and only if it is Hausdorff, the converse follows from the following more general result:

PROPOSITION 43.3.   *Let* E, F *be two locally convex TVS; then* $E \otimes_\pi F$ *is Hausdorff if and only if both* E *and* F *are Hausdorff.*

Proof. If $E \otimes_\pi F$ is Hausdorff, the same must be true of E and F; this is obvious. Conversely, let us assume that both E and F are Hausdorff and let us show that, given any element $\theta \neq 0$ of $E \otimes F$, there is a continuous linear form θ' on $E \otimes_\pi F$ such that $\langle \theta', \theta \rangle \neq 0$. It suffices to write

$$\theta = \sum_j x_j \otimes y_j\,,$$

where the sum is finite and the sets $\{x_j\}, \{y_j\}$ are linearly independent. Since E and F are Hausdorff, we can find $x' \in E'$ and $y' \in F'$ such that $\langle x', x_1 \rangle = \langle y', y_1 \rangle = 1$ and $\langle x', x_j \rangle = \langle y', y_j \rangle = 0$ for $j > 1$. Consider then the linear form on $E \otimes F$,

$$\theta' : \left(\sum_k \xi_k \otimes \eta_k \right) \rightsquigarrow \sum_k \langle x', \xi_k \rangle \langle y', \eta_k \rangle.$$

It is clearly continuous for the π-topology, and $\langle \theta', \theta \rangle = 1$.   Q.E.D.

Going back to the proof of Proposition 43.2, it suffices to observe that if $(E, \mathfrak{p})$ and $(F, \mathfrak{q})$ are seminormed spaces, the topology π on $E \otimes F$ can be defined by the single seminorm $\mathfrak{p} \otimes \mathfrak{q}$. If $\mathfrak{p}$ and $\mathfrak{q}$ are norms, $E \otimes_\pi F$ must be Hausdorff, i.e., $\mathfrak{p} \otimes \mathfrak{q}$ must be a norm.

Definition 43.4.   If $(E, \mathfrak{p})$ *and* $(F, \mathfrak{q})$ *are normed spaces, the normed* space $(E \otimes F, \mathfrak{p} \otimes \mathfrak{q})$ *will be called the projective tensor product of* $(E, \mathfrak{p})$ *and* $(F, \mathfrak{q})$.

When E and F are arbitrary locally convex spaces, we obtain a basis of continuous seminorms for the π-topology on E, F by taking a family $(\mathfrak{p}_\alpha \otimes \mathfrak{q}_\beta)$ $(\alpha \in A, \beta \in B)$, where $\{\mathfrak{p}_\alpha\}$ (resp. $\{\mathfrak{q}_\beta\}$) is a basis of continuous seminorms in E (resp. F).

In analogy with the algebraic case (Theorem 39.1), the space $E \otimes_\pi F$ possesses a "universal" property:

PROPOSITION 43.4. *Let* E, F *be locally convex spaces. The* π-topology *on* $E \otimes F$ *is the only locally convex topology on* $E \otimes F$ *having the following property:*

For every locally convex space G, *the canonical isomorphism of the space of bilinear mappings of* $E \times F$ *into* G *onto the space of linear mappings of* $E \otimes F$ *into* G (Theorem 39.1(b)) *induces an isomorphism of the space of continuous bilinear mappings of* $E \times F$ *into* G, $B(E, F; G)$, *onto the space of continuous linear mappings of* $E \otimes F$ *into* G, $L(E \otimes F; G)$.

In the property above, the word isomorphism is used in the purely algebraic sense.

Proof. In the algebraic correspondence between bilinear mappings $E \times F \to G$ and linear mappings $E \otimes F \to G$, if we take $G = E \otimes F$, the canonical bilinear mapping $E \times F \to E \otimes F$ corresponds to the identity mapping of $E \otimes F$. If $\mathscr{T}$ is a locally convex topology on $E \otimes F$ having the universal property of the statement, we see that the canonical bilinear mapping of $E \times F$ into $E \otimes F$ must be continuous, therefore $\mathscr{T}$ is weaker than π (Definition 43.2). But on the other hand, since the canonical mapping $E \times F \to E \otimes_\pi F$ is continuous, so must be the identity mapping $E \otimes_{\mathscr{T}} F \to E \otimes_\pi F$, which means that $\mathscr{T}$ is finer than π.   Q.E.D.

COROLLARY. *The dual of* $E \otimes_\pi F$ *is canonically isomorphic to* $B(E, F)$, *the space of continuous bilinear forms on* $E \times F$.

It suffices to take $G = \mathbf{C}$ in Proposition 43.4.

Exercises

Exercise 43.1. Prove that the canonical isomorphism between $B(E, F; G)$ and $L(E \otimes_\pi F; G)$ transforms equicontinuous sets into equicontinuous sets and that the projective topology on $E \otimes F$ is the topology of uniform convergence on the equicontinuous subsets of $B(E, F)$ regarded as dual of $E \otimes F$.

PROPOSITION 43.5. *On* $E \otimes F$, *the topology* π *is finer than the topology* ε.

Indeed, the canonical bilinear mapping of $E \times F$ into $E \otimes_\varepsilon F$ is continuous.

PROPOSITION 43.6.   Let E_i, F_i $(i = 1, 2)$ be four locally convex Hausdorff spaces. Let u (resp. v) be a continuous linear map of E_1 into E_2 (resp. of F_1 into F_2). If $\mathscr{T}$ is either the topology π or the topology ε, the tensor product $u \otimes v$ of u and v is a continuous linear map of $E_1 \otimes_\mathscr{T} F_1$ into $E_2 \otimes_\mathscr{T} F_2$.

Proof.   First, suppose that $\mathscr{T} = \pi$. Noting that $(x, y) \rightsquigarrow (u(x), v(y))$ is a continuous linear map of $E_1 \times F_1$ into $E_2 \times F_2$ and composing this mapping with the canonical bilinear mapping of $E_2 \times F_2$ into $E_2 \otimes_\pi F_2$, we obtain a continuous bilinear mapping of $E_1 \times F_1$ into $E_2 \otimes_\pi F_2$, to which is associated, in view of Proposition 43.4, a continuous linear mapping of $E_1 \otimes_\pi F_1$ into $E_2 \otimes_\pi F_2$, which is nothing else but $u \otimes v$.

Next, suppose that $\mathscr{T} = \varepsilon$. Let us identify $E_i \otimes F_i$ with $B((E_i)'_\sigma, (F_i)'_\sigma)$ $(i = 1, 2)$. Then $u \otimes v$ is immediately seen to be identified with the mapping which assigns to every continuous bilinear form on $(E_1)'_\sigma \times (F_1)'_\sigma$,

$$(x'_1, y'_1) \rightsquigarrow \Phi(x'_1, y'_1),$$

the continuous bilinear form on $(E_2)'_\sigma \times (F_2)'_\sigma$,

(43.3) $$(x'_2, y'_2) \rightsquigarrow \Phi({}^t u(x'_2), {}^t v(y'_2)),$$

where ${}^t u$ (resp. ${}^t v$) is the transpose of u (resp. v). If A'_2 (resp. B'_2) is an equicontinuous subset of E'_2 (resp. F'_2), ${}^t u(A'_2)$ (resp. ${}^t v(B'_2)$) is an equicontinuous subset of E'_1 (resp. F'_1), so that, if the absolute value of Φ is $\leqslant 1$ on ${}^t u(A'_2) \times {}^t v(B'_2)$, the absolute value of (43.3) is $\leqslant 1$ on $A'_2 \times B'_2$.
                                                                Q.E.D.

COROLLARY.   Let E, F be two locally convex Hausdorff spaces; $E' \otimes F'$ is canonically isomorphic to a linear subspace of the dual of $E \otimes_\varepsilon F$ (and a fortiori, also of $E \otimes_\pi F$).

It suffices to take $E_1 = E$, $F_1 = F$, $E_2 = F_2 = \mathbf{C}$ in Proposition 43.6. Let E, F be two locally convex Hausdorff TVS.

Definition 43.5.   *We shall denote by* $E \widehat{\otimes}_\varepsilon F$ (resp. $E \widehat{\otimes}_\pi F$) *the completion of* $E \otimes_\varepsilon F$ (resp. $E \otimes_\pi F$).

Let E_i, F_i $(i = 1, 2)$, u and v be as in Proposition 43.6.

Definition 43.6.   *We shall denote by* $u \widehat{\otimes}_\varepsilon v$ (resp. $u \widehat{\otimes}_\pi v$) *the extension of* $u \otimes v$ *as a continuous linear map of* $E_1 \widehat{\otimes}_\varepsilon F_1$ *into* $E_2 \widehat{\otimes}_\varepsilon F_2$ (resp. of $E_1 \widehat{\otimes}_\pi F_1$ into $E_2 \widehat{\otimes}_\pi F_2$).

When E and F are both complete, we know (Proposition 42.3) that $\mathscr{B}_\varepsilon(E'_\sigma, F'_\sigma)$ is complete; in this case, $E \widehat{\otimes}_\varepsilon F$ can be canonically identified with the closure of $E \otimes F = B(E'_\sigma, F'_\sigma)$ in $\mathscr{B}_\varepsilon(E'_\sigma, F'_\sigma)$. In the general case, when it is not necessarily true that both E and F are complete, we may observe that the topologies $\hat{\sigma} = \sigma(E', \hat{E})$ and $\sigma(F', \hat{F})$ on E' and F', respectively, are finer than σ, whereas the equicontinuous subsets of E' and F' are the same, whether we regard these spaces as duals of E and F, or of $\hat{E}$ and $\hat{F}$, respectively. This means that $B(E'_\sigma, F'_\sigma) \subset B(E'_\hat{\sigma}, F'_\hat{\sigma})$ and $\mathscr{B}(E'_\sigma, F'_\sigma) \subset \mathscr{B}(E'_\hat{\sigma}, F'_\hat{\sigma})$ and that the ε-topology on the last one of these spaces of bilinear forms induces the ε-topology on all the others. Thus $E \widehat{\otimes}_\varepsilon F$ can be regarded as a subspace of $\mathscr{B}_\varepsilon(E'_\hat{\sigma}, F'_\hat{\sigma})$ and $E \widehat{\otimes}_\varepsilon F$, as the closure of $E \otimes F$, or of $\hat{E} \otimes \hat{F}$, in $\mathscr{B}_\varepsilon(E'_\hat{\sigma}, F'_\hat{\sigma})$. It is trivial that $E \otimes_\varepsilon F$ is dense in $\hat{E} \otimes_\varepsilon \hat{F}$.

PROPOSITION 43.7. *Let E_i, F_i $(i = 1, 2)$, u, and v be as in Proposition 43.6. Suppose that u (resp. v) is an isomorphism of E_1 into E_2 (resp. of F_1 into F_2).*

Then $u \widehat{\otimes}_\varepsilon v$ is an isomorphism of $E_1 \widehat{\otimes}_\varepsilon F_1$ into $E_2 \widehat{\otimes}_\varepsilon F_2$.

Proof. The extension by continuity to $\hat{E}$ of an isomorphism of E into F is an isomorphism of $\hat{E}$ into $\hat{F}$. It suffices therefore to show that $u \otimes v$ is an isomorphism of $E_1 \otimes_\varepsilon F_1$ into $E_2 \otimes_\varepsilon F_2$ (we know already that it is a continuous injection; cf. Exercise 39.3). If we identify a tensor $\theta \in E_1 \otimes F_1$ to a bilinear form $\Phi \in B((E_1)'_\sigma, (F_1)'_\sigma)$, $(u \otimes v)(\theta)$ will be identified to the form

$$(43.4) \qquad (x'_2, y'_2) \rightsquigarrow \Phi({}^tu(x'_2), {}^tv(y'_2)).$$

Let U_1, V_1 be arbitrary neighborhoods of zero in E_1 and F_1 respectively. Let us select U_2, V_2 neighborhoods of zero in E_2 and F_2 such that

$$u(U_1) \supset U_2 \cap u(E_1), \qquad v(V_1) \supset V_2 \cap v(F_1).$$

By taking the polars of all sides and observing that ${}^tu$ and ${}^tv$ are onto, we see that

$${}^tu^{-1}(U_1^0) \subset U_2^0 + \operatorname{Ker} {}^tu, \qquad {}^tv^{-1}(V_1^0) \subset V_2^0 + \operatorname{Ker} {}^tv,$$

that is to say

$$U_1^0 \subset {}^tu(U_2^0), \qquad V_1^0 \subset {}^tv(V_2^0).$$

Now, if the form (43.4) converges to zero uniformly on $U_2^0 \times V_2^0$, it is clear that Φ must converge to zero uniformly on $U_1^0 \times V_1^0$.   Q.E.D.

COROLLARY. *If E_1 (resp. F_1) is a linear subspace of E (resp. F), $E_1 \otimes_\varepsilon F_1$ (resp. $E_1 \widehat{\otimes}_\varepsilon F_1$) is canonically isomorphic to a linear subspace of $E \otimes_\varepsilon F$ (resp. $E \widehat{\otimes}_\varepsilon F$).*

In this corollary, isomorphic is meant in the sense of the TVS structure.

We switch now to the π-topology. Since the dual of a locally convex Hausdorff TVS E is canonically identifiable with the dual of the completion of E, we derive from the corollary of Proposition 43.4:

PROPOSITION 43.8. *The dual of $E \widehat{\otimes}_\pi F$ is canonically isomorphic to $B(E, F)$.*

The transpose of

$$u \widehat{\otimes}_\pi v : E_1 \widehat{\otimes}_\pi F_1 \to E_2 \widehat{\otimes}_\pi F_2$$

is the mapping

$$\Psi \rightsquigarrow \{(x_1, y_1) \rightsquigarrow \Psi(u(x_1), v(y_1))\}$$

of $B(E_2, F_2)$ into $B(E_1, F_1)$.

We may then prove the following statement:

PROPOSITION 43.9. *Let E_i, F_i $(i = 1, 2)$, u, and v be as in Proposition 43.6. Suppose that u (resp. v) is a homomorphism of E_1 (resp. F_1) onto a dense linear subspace of E_2 (resp. F_2).*

Then $u \widehat{\otimes}_\pi v$ is a homomorphism of $E_1 \widehat{\otimes}_\pi F_1$ onto a dense subspace of $E_2 \widehat{\otimes}_\pi F_2$ which is identical to $E_2 \widehat{\otimes}_\pi F_2$ when E_1 and F_1 are metrizable.

Proof. First, we make use of the fact that the image of a continuous linear map $u : E \to F$ is dense if and only if ${}^t u$ is one-to-one. If $u(E_1)$ (resp. $v(F_1)$) is dense in E_2 (resp. F_2), $\Psi \in B(E_2, F_2)$ cannot vanish on $u(E_1) \times v(F_1)$ without being identically equal to zero; thus ${}^t(u \widehat{\otimes}_\pi v)$ is one-to-one, therefore the image of $u \widehat{\otimes}_\pi v$ is dense.

Let, now, $\Phi \in B(E_1, F_1)$ have the property that $\Phi(x_1, y_1) = 0$ as soon as either $x_1 \in \mathrm{Ker}\, u$ or $y_1 \in \mathrm{Ker}\, v$. Let us define the following bilinear functional on $u(E_1) \times v(F_1)$:

$$\Psi(u(x_1), v(y_1)) = \Phi(x_1, y_1).$$

This is obviously a correct definition. Let us consider a neighborhood of zero U_1 (resp. V_1) in E_1 (resp. F_1) such that

$$|\Phi(U_1, V_1)| \leqslant 1.$$

Select, then, a neighborhood of zero U_2 (resp. V_2) in E_2 (resp. F_2) such that $U_2 \cap u(E_1) \subset u(U_1)$, $V_2 \cap v(F_1) \subset v(V_1)$. We have, for $u(x_1) \in U_2$ and $v(y_1) \in V_2$, $|\Psi(u(x_1), v(y_1))| \leqslant 1$. This shows that Ψ is continuous for the topology induced by $E_2 \times F_2$. It can be extended in a unique manner as a continuous bilinear functional on $E_2 \times F_2$. We denote this extension also by Ψ. We remark that, if Φ belongs to an equicontinuous subset of $B(E_1, F_1)$, H_1, the neighborhoods U_1 and V_1, and therefore also U_2 and V_2, can be chosen independently of Φ, depending only on H_1, and Ψ then belongs to

$$H_2 = \{X \in B(E_2, F_2) \mid |X(U_2, V_2)| \leqslant 1\},$$

which is obviously an equicontinuous subset of $B(E_2, F_2)$. These facts imply all we want. Indeed, they prove that the image of ${}^t(u \widehat{\otimes}_\pi v)$ is exactly equal to the set of $\Phi \in B(E_1, F_1)$ which vanish on

$$(\text{Ker } u) \times F_1 + E_1 \times (\text{Ker } v);$$

in particular, this image is weakly closed. Furthermore, any equicontinuous subset contained in it is the image of an equicontinuous subset of $B(E_2, F_2)$ under ${}^t(u \widehat{\otimes}_\pi v)$. It suffices, then, to apply Lemma 37.7 (Exercise 37.8).

If E_1 and F_1 are metrizable, this is also true of $E_1 \widehat{\otimes}_\pi F_1$; then $E_1 \widehat{\otimes}_\pi F_1$ is a Fréchet space; its image under $(u \widehat{\otimes}_\pi v)$ is isomorphic to $E_1 \widehat{\otimes}_\pi F_1 / \text{Ker}(u \widehat{\otimes}_\pi v)$, which is a Fréchet space, is therefore complete, i.e., is closed in $E_2 \widehat{\otimes}_\pi F_2$. Since it is also dense in this space, according to the first part, it must be equal to it.   Q.E.D.

Remark 43.1.   It is not difficult to find examples of homomorphisms onto, $u : E_1 \to E_2$, $v : F_1 \to F_2$, such that $u \widehat{\otimes}_\pi v$ is *not* onto.

Remark 43.2.   If u and v are isomorphisms into, it is not true in general that $u \widehat{\otimes}_\pi v$ will be an isomorphism into. In order that this be true, it is necessary and sufficient that every equicontinuous subset of $B(E_1, F_1)$ be the image, under ${}^t(u \widehat{\otimes}_\pi v)$, of an equicontinuous subset of $B(E_2, F_2)$. And it is sufficient that $u(E_1)$ and $v(F_1)$ have a topological supplementary in E_2 and F_2 respectively.

Remark 43.3.   If u and v are homomorphisms onto, it is not true, in general, that $u \widehat{\otimes}_\varepsilon v$ will be a homomorphism. Comparing with Remark 43.2, we see that the ε-completion and the π-completion of $E \otimes F$ behave in quite different manners with respect to isomorphisms into and homomorphisms. It is also clear, if we look at Propositions 43.7 and 43.9, that, in the cases where the two completions will be isomorphic, the extensions of tensor products of linear mappings $u \otimes v$ will have very convenient properties.

We close this section with a few words about the case where E and F are normed spaces (we denote by $\| \ \|$ the norm in both of them, as well as the norm on their duals, on the related spaces of continuous linear mappings, etc.). In particular, the space of continuous bilinear forms on $E' \times F'$, $B(E', F')$, carries a canonical norm, which is the maximum of the absolute value of a form on the product of the unit balls of E and F. This induces a norm on every one of its subspaces, in particular on $\mathscr{B}(E'_\sigma, F'_\sigma)$ and on $B(E'_\sigma, F'_\sigma) \cong E \otimes F$.

PROPOSITION 43.10.   *The ε-topology on the tensor product $E \otimes F$ of two normed spaces E and F is defined by the canonical norm on $B(E'_\sigma, F'_\sigma)$.*

The proof is straightforward. We shall denote by $\| \ \|_\varepsilon$ the canonical norm on $E \otimes F$ which defines the ε-topology. We have already the notion of π-norm: it is the tensor product of the norms of E and F (Definition 43.3, Proposition 43.2); we shall denote it by $\| \ \|_\pi$.

We keep assuming that E and F are normed spaces.

PROPOSITION 43.11.   *The canonical mapping of $E \times F$ into $E \widehat{\otimes}_\varepsilon F$ and into $E \otimes_\pi F$ has norm one.*

Proof.   By Proposition 43.1, we have

$$\| x \otimes y \|_\pi = \| x \| \| y \|.$$

If we identify $E \otimes F$ with $B(E'_\sigma, F'_\sigma)$, $x \otimes y$ is identified with the bilinear form

$$(x', y') \rightsquigarrow \langle x', x \rangle \langle y', y \rangle,$$

and

$$\| x \otimes y \|_\varepsilon = \sup_{\| x' \| = 1,} \sup_{\| y' \| = 1} |\langle x', x \rangle \langle y', y \rangle| = \| x \| \| y \|.$$

In the case of normed spaces, the universal property stated in Proposition 43.4 can be made more precise:

PROPOSITION 43.12.   *Let E, F be two normed spaces.*

(a) *Any norm on $E \otimes F$, such that the canonical bilinear mapping of $E \times F$ into $E \otimes F$ is continuous if norm $\leqslant 1$, is $\leqslant$ to the π-norm.*

(b) *For all normed spaces G, the canonical isomorphism of $B(E, F; G)$ onto $L(E \otimes_\pi F; G)$ is an isometry.*

(c) *Any norm on $E \otimes F$, such that the canonical linear mapping of $B(E, F)$ into $(E \otimes F)^*$, algebraic dual of $E \otimes F$, is an isometry of $B(E, F)$ onto $(E \otimes F)'$, is equal to the π-norm.*

Proof. Note that (b) $\Rightarrow$ (a) trivially: if $\mathfrak{N}$ is a norm on $E \otimes F$ such that the canonical mapping $\phi : E \times F \to E \otimes_{\mathfrak{N}} F$ has norm $\leqslant 1$, it suffices to apply (b) with $G = E \otimes_{\mathfrak{N}} F$; we know that the mapping $E \times F \to G$ corresponding in this case to ϕ is the identity mapping of $E \otimes F$.

Let us prove (b). Let $\tilde{u} \in B(E, F; G)$, $u \in L(E \otimes_{\pi} F; G)$ be canonically associated. For all $x \in E$, $y \in F$, we have

$$\| \tilde{u}(x, y)\| \leqslant \| u(x \otimes y)\| \leqslant \| u \| \, \| x \otimes y \|_{\pi} = \| u \| \, \| x \| \, \| y \|.$$

This proves that $\| \tilde{u} \| \leqslant \| u \|$. Let $\theta \in E \otimes F$ with $\| \theta \|_{\pi} = 1$. We may find a decomposition $\theta = \sum_j x_j \otimes y_j$ with $\sum_j \| x_j \| \, \| y_j \| \leqslant 1 + \varepsilon$. But then $\| u(\theta) \| = \| \sum_j \tilde{u}(x_j , y_j) \| \leqslant \| \tilde{u} \| (1 + \varepsilon)$. By taking $\varepsilon \to 0$, we conclude that the maximum of $\| u(\theta) \|$ on the unit sphere of $E \otimes_{\pi} F$ is $\leqslant \| \tilde{u} \|$, which proves what we wanted.

Finally, we prove (c). Let $\mathfrak{N}$ be a norm on $E \otimes F$ such that $B(E, F)$ and $(E \otimes_{\mathfrak{N}} F)'$ are canonically isometric. By (b) this is true when $\mathfrak{N} = \pi$. Therefore, both $E \otimes_{\pi} F$ and $E \otimes_{\mathfrak{N}} F$ are mapped isometrically (by the mapping: value at $\theta \in E \otimes F$) onto the same subspace of the dual of $B(E, F)$.   Q.E.D.

COROLLARY 1. *The canonical isomorphism of $B(E, F)$ onto $(E \otimes_{\pi} F)'$ is an isometry.*

COROLLARY 2. *For all $\theta \in E \otimes F$, we have $\| \theta \|_{\varepsilon} \leqslant \| \theta \|_{\pi}$.*

Proof of Corollary 2. Apply Proposition 43.12(a), as the canonical bilinear map $E \times F \to E \otimes_{\varepsilon} F$ is continuous and has norm one.

PROPOSITION 43.13. *Let E_i, F_i ($i = 1, 2$) be four normed spaces, and $u : E_1 \to E_2$, $v : F_1 \to F_2$ two continuous linear mappings. Then*

$$\| u \otimes v \|_{\pi} = \| u \otimes v \|_{\varepsilon} = \| u \| \, \| v \|.$$

Proof. For all $x \in E_1$, $y \in E_2$, we have (cf. Proposition 43.11)

$$\|(u \otimes v)(x \otimes y)\|_{\omega} = \| u(x) \otimes v(y) \|_{\omega} = \| u(x) \| \, \| v(y) \|,$$

where ω stands either for π or for ε. Take, now, $\| x \| = 1, \| y \| = 1$ such that $\| u(x) \|$ and $\| v(y) \|$ be arbitrarily close to $\| u \|$ and $\| v \|$, respectively. We obtain

$$\| u \| \, \| v \| \leqslant \| u \otimes v \|_{\omega}.$$

Let us prove the converse inequality: first, when $\omega = \pi$. Consider the bilinear map of $E_1 \times F_1$ into $E_2 \otimes_\pi F_2$,

$$(x, y) \rightsquigarrow u(x) \otimes v(y);$$

it has a norm $\leqslant \| u \| \| v \|$. By applying (b) in Proposition 43.12, we see that the associated linear map of $E_1 \otimes_\pi E_1$ into $E_2 \otimes_\pi F_2$, that is to say $u \otimes v$, must have norm $\leqslant \| u \| \| v \|$.

Finally we look at the case where $\omega = \varepsilon$. Let θ be an arbitrary element of $E_1 \otimes F_1$; its ε-norm is the supremum of its absolute value (when we regard θ as a bilinear form on $E_1' \times F_1'$) over the product of the unit balls of E_1' and F_1' (Proposition 43.10). We denote by sup the supremum over $x' \in E_2'$, $y' \in F_2'$, $\| x' \| = \| y' \| = 1$; we then have

$$\|(u \otimes v)(\theta)\|_\varepsilon = \sup |(u \otimes v)(\theta)(x' \otimes y')|$$
$$= \sup | \theta({}^t u(x'), {}^t v(y'))| \leqslant \| \theta \|_\varepsilon \| {}^t u \| \| {}^t v \| = \| \theta \|_\varepsilon \| u \| \| v \|$$

Q.E.D.

Exercises

43.2. Let E_i, F_i $(i = 1, 2)$, u, and v be as in Proposition 43.6. Prove that the kernel of $u \otimes_\pi v$ is equal to the closed linear subspace of $E_1 \otimes_\pi F_1$ spanned by the tensors of the form $x_1 \otimes y_1$ such that either $u(x_1) = 0$ or $v(x_1) = 0$ or both.

43.3. In relation with Remark 43.2, prove the following lemma (which can be viewed as a complement to Lemma 37.7, Exercise 37.8, and which is going to be used, later on):

LEMMA 43.1. *Let E, F be two Hausdorff locally convex spaces. A continuous linear map $u : E \to F$ is an isomorphism if and only if every equicontinuous subset of E' is the image under ${}^t u$ of an equicontinuous subset of F'.*

43.4. Let E_i, F_i $(i = 1, 2)$ be four normed spaces, and $u : E_1 \to E_2$, $v : F_1 \to F_2$ two isometries. Prove that $u \otimes v$ is an isometry of $E_1 \otimes_\pi F_1$ into $E_2 \otimes_\pi F_2$ if and only if every continuous bilinear form on $E_1 \times F_1$ is the image, under ${}^t(u \otimes_\pi v)$, of a continuous bilinear form on $E_2 \times F_2$ having the same norm.

43.5. Making use of Remark 43.2, prove the following statement: Let F be a Banach space, and E a closed linear subspace of F having a topological supplementary in its bidual E''. The canonical mapping $E \otimes_\pi E' \to F \otimes_\pi E'$ is an isomorphism (for the TVS structures) if and only if E has a topological supplementary in F.

43.6. By using Exercise 43.4, prove the following result: if E and F are normed spaces, the canonical linear mapping $E \otimes_\pi F \to E'' \otimes_\pi F$ is an isometry (into).

43.7. Prove that, if E and F are two Fréchet spaces, $E \otimes_\pi F$ is a barreled space (hint: use Exercise 33.4).

43.8. Let $\{E_\alpha\}$ be a family of locally convex spaces, and F a locally convex space. Prove the canonical isomorphism

$$\left(\prod_\alpha E_\alpha \right) \otimes_\pi F \cong \prod_\alpha (E_\alpha \otimes_\pi F).$$

44

Examples of Completion of Topological Tensor Products: Products ε

Example 44.1. The Space $\mathscr{C}^m(X; E)$ **of** $\mathscr{C}^m$ **Functions Valued in a Locally Convex Hausdorff Space** E $(0 \leqslant m \leqslant +\infty)$

In the discussion that follows, X is either a locally compact topological space, and then m can only be equal to zero, or else an open subset of $\mathbf{R}^n$, and then m can be any integer, or $+\infty$. The space $\mathscr{C}^m(X; E)$ carries its natural $\mathscr{C}^m$ topology (see Definition 40.2); when $m = 0$ and X is a locally compact space, that definition still holds: the topology of $\mathscr{C}^0(X; E)$ is the topology of uniform convergence on the compact subsets of X.

PROPOSITION 44.1. *If E is complete, so is $\mathscr{C}^m(X; E)$.*

Proof. As E is complete, a Cauchy filter $\mathscr{F}$ on $\mathscr{C}^m(X; E)$ converges pointwise to a function $f : X \to E$. As X is locally compact, the convergence of $\mathscr{F}$ is uniform on a neighborhood of every point (as it is uniform on compact subsets of X), hence the limit f is continuous. This proves the result when $m = 0$. Suppose now that X is an open subset of $\mathbf{R}^n$ and that $m > 0$. As a matter of fact, it suffices to show that $\operatorname{grad} f$ (grad $= gradient\ of$) exists, is a continuous function, and is the limit of $\operatorname{grad} \mathscr{F}$ (obvious notation); having done this, an obvious reasoning by induction on the order of differentiation easily completes the proof. We may even suppose that we are dealing with only one variable, so as to simplify the notation. Extension to $n > 1$ variables will be evident. We recall (Theorem 27.1) that, for all $e' \in E'$, $\phi \in \mathscr{C}^m(X; E)$,

$$(\partial/\partial x_1) \langle e', \phi(x) \rangle = \langle e', (\partial/\partial x_1) \phi(x) \rangle.$$

Now, by the preceding argument, we know that $(\partial/\partial x_1)\mathscr{F}$ converges, uniformly on the compact subsets of X, to a continuous function f_1.

446

A fortiori, $(\partial/\partial x_1) \langle e', \mathscr{F} \rangle$ converges to $\langle e', f_1 \rangle$. We conclude that the complex-valued continuous function $\langle e', f_1 \rangle$ is the derivative of the function $\langle e', f \rangle$. We then have

$$\left\langle e', \frac{f(x_1 + h) - f(x_1)}{h} - f_1(x_1) \right\rangle = \frac{1}{h} \int_{x_1}^{x_1+h} \langle e', f_1(t) - f_1(x_1) \rangle \, dt,$$

where $h \neq 0$. Let, then, U be a convex closed balanced neighborhood of 0 in E and take e' arbitrary in the polar U^0 of U. Because of the continuity of f_1, we may find $|h|$ so small that $f_1(t) - f_1(x_1) \in U$ for all t in the segment joining x_1 to $x_1 + h$; then the integrand, on the right-hand side, and, as a consequence, the left-hand side, have their absolute value $\leqslant 1$. This means that

$$h^{-1}\{f(x_1 + h) - f(x_1)\} \in U + f_1(x_1),$$

hence that f_1 is the first derivative of f.

COROLLARY 1.   *Suppose that X is countable at infinity and that E is a Fréchet space. Then $\mathscr{C}^m(X; E)$ is a Fréchet space.*

It is evident, on the definition of the topology of $\mathscr{C}^m(X; E)$, that it is metrizable whenever X is countable at infinity (i.e., a countable union of compact subsets) and E is metrizable.

COROLLARY 2.   *Suppose that X is compact and E a Banach space. Then $\mathscr{C}^0(X; E)$, equipped with the norm*

$$f \rightsquigarrow \sup_{x \in X} \|f(x)\| \qquad (\| \ \|: \text{norm in } E),$$

is a Banach space.

We denote by $\mathscr{C}^m(X) \otimes E$ the subspace of $\mathscr{C}^m(X; E)$ consisting of the functions whose image is contained in a finite dimensional subspace of E (cf. Proposition 40.2); $\mathscr{C}^m(X) \otimes E$ is a tensor product of $\mathscr{C}^m(X)$ and E.

Let $\phi \in \mathscr{C}^m(X) \otimes E$, $\mathbf{e}_1, \ldots, \mathbf{e}_d$ be linearly independent vectors of E such that the image of ϕ is contained in the linear subspace of E they span. Thus we may write, for all $x \in X$,

$$\phi(x) = \phi_1(x) \, \mathbf{e}_1 + \cdots + \phi_d(x) \, \mathbf{e}_d,$$

where $\phi_1, \ldots, \phi_d$ are complex-valued functions. These functions are m times continuously differentiable, as one sees immediately by writing

$$\phi_j(x) = \langle \mathbf{e}_j', \phi(x) \rangle$$

and then applying Theorem 27.1; here $\mathbf{e}'_j \in E'$ is such that

$$\langle \mathbf{e}'_j, \mathbf{e}_k \rangle = \delta_{jk} \qquad \text{(Kronecker symbol).}$$

We may also write

$$\underline{\phi} = \phi_1 \otimes \mathbf{e}_1 + \cdots + \phi_d \otimes \mathbf{e}_d.$$

PROPOSITION 44.2. *If X is a locally compact space, $\mathscr{C}^0_c(X) \otimes E$ is dense in $\mathscr{C}^0(X; E)$.*
  If X is an open subset of $\mathbf{R}^n$, $\mathscr{C}^\infty_c(X) \otimes E$ is dense in $\mathscr{C}^m(X; E)$.

Proof 1. *X is a locally compact space, $m = 0$.* Let $f \in \mathscr{C}^0(X; E)$, p be a continuous seminorm on E, K a compact subset of X, and ε a number $>$ 0. We may find a finite covering $U_1, ..., U_r$ of K, by relatively compact open subsets of X, such that, for each $j = 1, ..., r$, and each pair x, $y \in U_j$,

$$p(f(x) - f(y)) < \varepsilon.$$

It is a general property of compact sets (see, e.g., N. Bourbaki, "Topologie générale,") that we can find a continuous partition of unity subordinated to the above covering, i.e., r continuous functions in X, g_j $(1 \leqslant j \leqslant r)$ such that: (a) for each j, supp $g_j \subset U_j$; (b) $\sum_{j=1}^r g_j(x) = 1$ for all $x \in K$. In each set U_j we pick up a point x_j. We have, for $x \in K$, $\underline{f}(x) = \sum_{j=1}^r g_j(x)\underline{f}(x)$, hence

$$p(\underline{f}(x) - \sum_{j=1}^r g_j(x)\underline{f}(x_j)) \leqslant \sum_{j=1}^r g_j(x) \, p(\underline{f}(x) - \underline{f}(x_j)) \leqslant \varepsilon$$

as $p(\underline{f}(x) - \underline{f}(x_j)) \leqslant \varepsilon$ if $x \in \operatorname{supp} g_j$.

Proof 2. *X is an open subset of $\mathbf{R}^n$, m is arbitrary.* The proof consists of a few easy steps. First of all, we see that $\mathscr{C}^m_c(X; E)$ is dense in $\mathscr{C}^m(X; E)$: consider a sequence of complex-valued functions $g_\nu \in \mathscr{C}^\infty_c(X)$ $(\nu = 1, 2, ...)$ equal to one on increasingly large open sets Ω_ν whose union is equal to X; each $f \in \mathscr{C}^m(X; E)$ is the limit of the $g_\nu f$ (in $\mathscr{C}^m(X; E)$) as $\nu \to \infty$.

Let, now, $f \in \mathscr{C}^m_c(X; E)$ and $\rho_\varepsilon \in \mathscr{C}^\infty_c(\mathbf{R}^n)$ be the usual mollifiers (cf. p. 156): supp $\rho_\varepsilon \subset \{x \in \mathbf{R}^n; \mid x \mid \leqslant \varepsilon\}$; $\int \rho_\varepsilon(x) \, dx = 1$. We can define the integral (for fixed $x \in \mathbf{R}^n$)

$$\int \rho_\varepsilon(x - y) \underline{f}(y) \, dy$$

as the limit of the Riemann sums. The latter define a Cauchy filter which converges in the completion $\hat{E}$ of E. When x varies, we obtain a function of x with values in $\hat{E}$, which it is natural to denote by $\rho_\varepsilon * f$; this is an element of $\mathscr{C}_c^\infty(\mathbf{R}^n; \hat{E})$ which converges to f in $\mathscr{C}^m(\mathbf{R}^n; \hat{E})$ as $\varepsilon \to 0$. These statements are easy to check, by duplicating what is done in the scalar case.

The last step goes as follows. In view of the first part of the proof, f is the limit of functions $f_j \in \mathscr{C}_c^0(X) \otimes E$ in the $\mathscr{C}^0$ topology. By using a cutting off function, we may assume that the supports of the f_j lie in an arbitrary neighborhood U of supp f. But then, it is not difficult to see that, for each $\varepsilon > 0$, $\rho_\varepsilon * f$ is the limit of the $\rho_\varepsilon * f_j$ in $\mathscr{C}_c^\infty(\mathbf{R}^n; \hat{E})$. If U is a relatively compact open subset of X and ε is small enough, the $\rho_\varepsilon * f_j$ and $\rho_\varepsilon * f$ have all their support contained in U. Finally we see that the $\rho_\varepsilon * f_j$, which belong to $\mathscr{C}_c^\infty(U) \otimes E$ as soon as ε is sufficiently small, converge, as j varies, to $\rho_\varepsilon * f$ in $\mathscr{C}_c^\infty(X; \hat{E})$. As $\rho_\varepsilon * f \to f$ in $\mathscr{C}^m(X; \hat{E})$ when $\varepsilon \to 0$, the proof is complete.

THEOREM 44.1.   *If E is complete, $\mathscr{C}^m(X; E) \cong \mathscr{C}^m(X) \widehat{\otimes}_\varepsilon E$.*

The meaning of the isomorphism stated in Theorem 44.1 is the following: $\mathscr{C}^m(X; E)$ induces on its linear subspace $\mathscr{C}^m(X) \otimes E$ the topology ε; therefore, the natural injection of the latter into the former extends as an isomorphism of $\mathscr{C}^m(X) \widehat{\otimes}_\varepsilon E$ into the completion of $\mathscr{C}^m(X; E)$; but the latter is complete, by Proposition 44.1, and the isomorphism of $\mathscr{C}^m(X) \widehat{\otimes}_\varepsilon E$ into $\mathscr{C}^m(X; E)$ is onto, by Proposition 44.2.

Proof of Theorem 44.1.   As we have just said, it suffices to show that $\mathscr{C}^m(X; E)$ induces on $\mathscr{C}^m(X) \otimes E$ the topology ε.

We observe, first, that $\mathscr{C}^m(X; E)$ can be canonically injected in $L(E_\tau'; \mathscr{C}^m(X))$. Indeed, let $f \in \mathscr{C}^m(X; E)$ and consider the complex-valued function, defined in X,

(44.1) $x \rightsquigarrow \langle e', f(x) \rangle$,

where e' is an arbitrary element of E'. We know (cf. Theorem 27.1) that this function is $\mathscr{C}^m$. Now let p be an arbitrary n-tuple such that $|p| < m + 1$ and K is a compact subset of X. Then $(\partial/\partial x)^p f(x)$ stays in a compact subset $\mathscr{K}$ of E as x varies in K; but the closed convex balanced hull of $\mathscr{K}$, $\Gamma(\mathscr{K})$, is also compact since E is complete; it is, a fortiori, weakly compact. If e' belongs to the polar of $(1/\varepsilon)\,\Gamma(\mathscr{K})$, which is a neighborhood of zero in E_τ', we have

$$\sup_{x \in K} |(\partial/\partial x)^p \langle e', f(x) \rangle| = \sup_{x \in K} |\langle e', (\partial/\partial x)^p f(x) \rangle| \leqslant \varepsilon.$$

This shows that the mapping

(44.2) $e' \rightsquigarrow (x \rightsquigarrow \langle e', f(x) \rangle)$

is continuous from E'_τ into $\mathscr{C}^m(X)$.

The proof of Theorem 44.1 will be complete if we show that the topology $\mathscr{C}^m$ on $\mathscr{C}^m(X; E)$ is equal to the topology induced by $L_\varepsilon(E'_\tau ; \mathscr{C}^m(X))$. Let U be a closed convex balanced neighborhood of zero in E, U^0 its polar, K a compact subset of X, and $p \in \mathbf{N}^n$ such that $|p| < m + 1$. Then it is equivalent to say that $(\partial/\partial x)^p f(x) \in U$ for all $x \in K$, or to say that $|(\partial/\partial x)^p < e', f(x) >| \leqslant 1$ for all $x \in K$ and all $e' \in U^0$.

Exercises

44.1. Let E be a normed *space*, with norm $\| \ \|$. Let K be a compact set. Prove that the norm

$$f \rightsquigarrow \sup_{x \in K} \| f(x) \|$$

on $\mathscr{C}(K) \otimes E$ is equal to the ε-norm.

44.2. Let H, K be two compact sets. Prove that

$$\mathscr{C}(H \times K) \cong \mathscr{C}(H) \widehat{\otimes}_\varepsilon \mathscr{C}(K),$$

with the isomorphism to be understood as a Banach space isomorphism.

44.3. Let Ω be an open subset of $\mathbf{R}^n$, and m an integer $\geqslant 0$ or $+\infty$. Let E be a locally convex Hausdorff space. Prove that $\mathscr{C}^m(\Omega; E)$ is identical to the space of scalarly $\mathscr{C}^m$ functions valued in E, that is to say of functions f such that, for each $e' \in E'$, $x \rightsquigarrow \langle e', f(x) \rangle$ is a $\mathscr{C}^m$ complex-valued function in Ω.

44.4. Let $\widetilde{\mathscr{C}}^m_c(\Omega; E)$ be the space of functions f, defined in the open set $\Omega \subset \mathbf{R}^n$ and valued in the locally convex Hausdorff space E, such that, for every $e' \in E'$, the function $x \rightsquigarrow \langle e', f(x) \rangle$ belongs to $\mathscr{C}^m_c(\Omega)$. Prove that, if E is a normed space,

$$\widetilde{\mathscr{C}}^m_c(\Omega; E) = \mathscr{C}^m_c(\Omega; E).$$

44.5. With the notation of Exercise 44.4, prove that

$$\widetilde{\mathscr{C}}^\infty_c(\mathbf{R}^n; \mathscr{D}'(\mathbf{R}^n)) \neq \mathscr{C}^\infty_c(\mathbf{R}^n; \mathscr{D}'(\mathbf{R}^n)).$$

44.6. Let us denote by $\mathscr{S}(\mathbf{R}^n; E)$ the space of functions $f \in \mathscr{C}^\infty(\mathbf{R}^n; E)$ such that, for all pairs of polynomials P, Q in n variables, with complex coefficients, $P(x)Q(\partial/\partial x)f(x)$ remains in a bounded subset of E as x varies over $\mathbf{R}^n$. We equip $\mathscr{S}(\mathbf{R}^n; E)$ with its natural topology: the topology of uniform convergence of the functions $P(x) Q(\partial/\partial x) f$ over the whole of $\mathbf{R}^n$, for all possible P and Q. Prove that, if E is complete,

$$\mathscr{S}(\mathbf{R}^n; E) \cong \mathscr{S}(\mathbf{R}^n) \widehat{\otimes}_\varepsilon E.$$

44.7. Let l_∞ be the space of complex sequences converging to zero, with the norm induced by l^∞,

$$\sigma = (\sigma_n)_{(n=1,2,\dots)} \rightsquigarrow |\sigma|_\infty = \sup_n |\sigma_n|.$$

Let E be a Banach space, with norm $\| \quad \|$. Prove that $l_\infty \widehat{\otimes}_\varepsilon E$ is canonically isomorphic, as a Banach space, to the space of sequences in E which converge to zero, equipped with the norm

$$e = (e_n) \rightsquigarrow \sup_n \| e_n \|.$$

Example 44.2.
Summable Sequences in a Locally Convex Hausdorff Space

We shall need some results about the Banach space l^1 of complex sequences $\sigma = (\sigma_n)$ $(n = 0, 1, \ldots)$ such that

$$| \sigma |_1 = \sum_{n=0}^{\infty} | \sigma_n | < +\infty \qquad \text{(Chapter 11, Example IV).}$$

We shall say that a subset $\sum$ of l^1 is *equismall at infinity* if, to every $\varepsilon > 0$, there is an integer $n_\varepsilon \geqslant 0$ such that

$$\sum_{n \geqslant n_\varepsilon} | \sigma_n | < \varepsilon \qquad \text{for all} \quad \sigma \in \sum.$$

We recall that l^∞ "is" the dual of l^1 (Theorem 20.1). We shall make use of the following result:

THEOREM 44.2.   *The following properties of a sequence $S \subset l^1$ are equivalent:*

(a)   *S is weakly (i.e., for the topology $\sigma(l^1, l^\infty)$) convergent;*

(b)   *S is convergent (for the norm on l^1).*

The following properties of a subset K of l^1 are equivalent:

(a_1) *K is weakly compact;*

(b_1) *K is compact;*

(c_1) *K is bounded, closed, and equismall at infinity.*

Proof.   We begin by proving that a sequence S which converges weakly in l^1 is equismall at infinity. We may assume that S converges weakly to zero. We shall then reason by contradiction. Suppose that there is a sequence of integers n_k $(k = 1, 2, \ldots)$, strictly increasing, and a sequence of elements of S, $\{\sigma^{(k)}\}$, such that, for all k,

$$\sum_{n \geqslant n_k} | \sigma_n^{(k)} | \geqslant c.$$

For each k, we select an integer $n'_k > n_k$ such that

$$\sum_{n \geqslant n'_k} |\sigma_n^{(k)}| < c/4.$$

Next, we construct a sequence of integers $k_\nu \to +\infty$, in the following manner: $k_1 = 1$, and k_2 is an integer such that the following conditions are satisfied:

$$\text{(i)} \quad \sum_{n_1 \leqslant n \leqslant n'_1} |\sigma_n^{(k_2)}| < c/4; \qquad \text{(ii)} \quad n'_1 < n_{k_2};$$

and so on; k_ν is an integer such that

$$\text{(i)} \quad \sum_{n_1 \leqslant n \leqslant n'_{k_{\nu-1}}} |\sigma_n^{(k_\nu)}| < c/4; \qquad \text{(ii)} \quad n'_{k_{\nu-1}} < n_{k_\nu}.$$

These two conditions can be fulfilled, by induction on ν, since $n_k \to +\infty$ as $k \to +\infty$, and since the fact that the sequence of sequences S converges weakly to zero implies that, for each n separately, the $\sigma_n^{(k)}$ converge to zero.

For the sake of simplicity, we shall write ν instead of k_ν, hence $\sigma^{(\nu)}$ instead of $\sigma^{(k_\nu)}$, n_ν and n'_ν instead of n_{k_ν} and n'_{k_ν}. Observe that we have, for all $\nu > 1$,

$$\sum_{n_1 \leqslant n \leqslant n'_{\nu-1}} |\sigma_n^{(\nu)}| < c/4;$$

$$n'_{\nu-1} < n_\nu \quad \text{and} \quad \sum_{n_\nu \leqslant n < n'_\nu} |\sigma_n^{(\nu)}| > 3c/4;$$

finally,

$$\sum_{n'_\nu \leqslant n} |\sigma_n^{(\nu)}| < c/4.$$

Let us then define a sequence $\tau = (\tau_n)$ in the following way:

$$\tau_n = \overline{\sigma_n^{(\nu)}}/|\sigma_n^{(\nu)}| \quad \text{if} \quad n_\nu \leqslant n < n'_\nu \text{ and if } \sigma_n^{(\nu)} \neq 0,$$

$$\tau_n = 0 \quad \text{otherwise.}$$

A quick computation shows that, for all $\nu > 1$,

$$|\langle \tau, \sigma^{(\nu)} \rangle| \geqslant \sum_{n_\nu \leqslant n < n'_\nu} |\sigma_n^{(\nu)}| - c/2 > c/4;$$

this means that the $\sigma^{(\nu)}$ cannot converge weakly to zero. It proves our assertion.

Now we prove that (a) $\Rightarrow$ (b). Again, we may assume that S converges weakly to zero. In view of the first part, this implies that to every $\varepsilon > 0$ there is an integer $n_\varepsilon \geqslant 0$ such that, for all $\sigma \in S$,

$$\sum_{n \geqslant n_\varepsilon} |\sigma_n| < \varepsilon.$$

On the other hand, as σ varies over S, for each n, $\sigma_n \to 0$. It follows that there is a finite subset A of S such that, for $\sigma \in S$, $\sigma \notin A$,

$$\sum_{n < n_\varepsilon} |\sigma_n| < \varepsilon.$$

Finally, we see that, for $\sigma \in S$, $\sigma \notin A$, $|\sigma|_1 < 2\varepsilon$. This proves that S converges to 0 for the norm.

Let us now prove that $(a_1) \Rightarrow (b_1)$. We must show that every sequence in K, S, contains a subsequence which converges for the norm. But S contains a subsequence which converges weakly in K. Hence, it suffices to take into account the implication (a) $\Rightarrow$ (b).

Next we prove that, if K is compact, K is equismall at infinity. If K were not equismall at infinity, there would be a number $c > 0$, a sequence $\{\sigma^{(\nu)}\}$ in K, such that

$$\sum_{n \geqslant \nu} |\sigma_n^{(\nu)}| \geqslant c.$$

The sequence $\{\sigma^{(\nu)}\}$ cannot possibly contain a subsequence which converges to zero.

Finally, let us prove that $(c_1) \Rightarrow (b_1)$. Let S be a sequence contained in K. As K is equismall at infinity, we can select a sequence of integers $0 < n_1 < n_2 < \cdots < n_k < \cdots$ such that, for every k, and every $\sigma \in S$,

(44.3) $$\sum_{n \geqslant n_k} |\sigma_n| < 1/k.$$

Observe that since K is bounded there is a subsequence $S^{(1)}$ of S such that $\sigma_n^{(1)}$ converges to some complex number σ_n if $n < n_1$, as $\sigma^{(1)}$ ranges over $S^{(1)}$; then we may select a subsequence $S^{(2)}$ of $S^{(1)}$ such that $\sigma_n^{(2)}$ converges to a complex number σ_n if $n < n_2$, as $\sigma^{(2)}$ ranges over $S^{(2)}$; etc. Let us denote by σ the sequence (σ_n) thus defined. Now, in each sequence $S^{(k)}$ we select a sequence $\tau^{(k)}$ such that $|\tau_n^{(k)} - \sigma_n| < 1/k$ for all $n < n_k$. We see immediately that $\{\tau^{(k)}\}$ $(k = 1, 2, \ldots)$ is a Cauchy sequence in l^1 (for the norm; take into account (44.3)); therefore it converges in K, which is closed (of course, its limit is the sequence σ, which, thus, belongs to l^1).   Q.E.D.

We need the following consequence of Theorem 44.2:

LEMMA 44.1.   *The identity mapping of l^1 is the limit, for the topology of uniform convergence on the compact subsets of l^1, of a sequence of continuous linear mappings whose image is finite dimensional.*

Proof.   Let us denote by ϕ_n the multiplication mapping, in l^1, by the sequence $1_{(n)}$ whose terms of rank $\leqslant n$ are equal to one, whereas the other ones are all equal to zero: if $\sigma \in l^1$, $\phi_n(\sigma)$ is the sequence whose terms are all equal to zero if their rank is $> n$, and are equal to the terms of the same rank in σ otherwise; obviously, ϕ_n is a continuous linear map of l^1 into itself with finite dimensional image. It is also obvious that $\phi_n \to I$ uniformly on every subset of l^1 which is equismall at infinity.
                                                    Q.E.D.

Now let E be a locally convex Hausdorff TVS. We derive from Theorem 44.2:

LEMMA 44.2.   *Every linear map $u : E' \to l^1$, which is continuous when E' carries the weak topology $\sigma(E', E)$ and l^1, the topology $\sigma(l^1, l^\infty)$, transforms any equicontinuous subset of E' into a relatively compact subset of l^1.*

Proof.   If $A' \subset E'$ is equicontinuous, it is relatively weakly compact therefore $u(A')$ is relatively weakly compact, hence compact by Theorem 44.2.

Let us call *weakly continuous* any linear mapping $u : E' \to l^1$ which is continuous for the topologies $\sigma(E', E)$ and $\sigma(l^1, l^\infty)$, and any linear mapping $v : l^\infty \to E$ continuous for $\sigma(l^\infty, l^1)$ and $\sigma(E, E')$.

LEMMA 44.3.   *Let v be a linear mapping $l^\infty \to E$. The following properties are equivalent:*

(a)   *the image of the unit ball B of l^∞ under v is precompact;*

(b)   *v is weakly continuous;*

(c)   *v is the transpose of a mapping $u : E' \to l^1$ which is weakly continuous.*

Proof.   (a) implies that v is continuous, hence weakly continuous; (b) and (c) are equivalent, simply because the transpose of a continuous linear map is continuous. Let us show that (c) $\Rightarrow$ (a).

Let V be a closed convex balanced neighborhood of zero in E. The polar V^0 of V is weakly compact in E', hence $u(V^0)$ is a compact subset of l^1. This implies that we can find a finite subset of points of V^0, $y'_1, ..., y'_s$, such that, for every $y' \in V^0$, there is j, $1 \leqslant j \leqslant s$, such that:

$$(44.4) \qquad |\, {}^t v(y') - {}^t v(y'_j)|_1 \leqslant \tfrac{1}{4}.$$

On the other hand, the unit ball B of l^∞ is weakly relatively compact for $\sigma(l^\infty, l^1)$, in view of the Banach–Steinhaus theorem (Theorem 33.2). This implies that there is a finite family of points $x_1, ..., x_r$ in B such that, to every $x \in B$, there is i, $1 \leqslant i \leqslant r$, such that

$$(44.5) \qquad \sup_{1 \leqslant j \leqslant s} |\langle x - x_i, {}^t v(y'_j)\rangle| \leqslant \tfrac{1}{4}.$$

Now let $x \in B$, $y' \in V^0$ be arbitrary. Let us select an index i so as to have (44.5) and an index j so as to have (44.4). We have

$$|\langle x - x_i, {}^t v(y')\rangle| \leqslant |\langle x - x_i, {}^t v(y'_j)\rangle| + |\langle x - x_i, {}^t v(y') - {}^t v(y'_j)\rangle|.$$

But $x - x_i \in 2B$, hence, in view of (44.4),

$$|\langle x - x_i, {}^t v(y') - {}^t v(y'_j)\rangle| \leqslant \tfrac{1}{2},$$

and, therefore, by (44.5),

$$|\langle x - x_i, {}^t v(y')\rangle| \leqslant 1.$$

This proves that $v(x) - v(x_i) \in V = V^{00}$, hence $v(x) \in V + v(x_i)$. Thus we see that $v(B)$ is covered by the sets $V + v(x_i)$ $(1 \leqslant i \leqslant r)$. Proposition 6.9 implies that $v(B)$ is precompact. Q.E.D.

Of course, when E is complete, $v(B)$ is relatively compact (i.e., $\overline{v(B)}$ is compact).

Let us observe that, if u is weakly continuous, then it is a continuous mapping of E'_τ into l^1 (Proposition 42.2). We contend that u is the limit, in $L_\varepsilon(E'_\tau; l^1)$, of a sequence of continuous linear mappings with finite dimensional images. Indeed, with the notation of the proof of Lemma 44.1, it suffices to take the sequence of mappings $\phi_n \circ u$ $(n = 1, 2, ...)$. Indeed, if A' is an equicontinuous subset of E', $u(A')$ is relatively compact in l^1 by Lemma 44.2, hence ϕ_n converges to the identity of l^1, uniformly on $u(A')$ by Lemma 44.1; this proves that $\phi_n \circ u$ converges to u uniformly on A'. Q.E.D.

A consequence of what we have just seen is that, *when E is complete,*

$$(44.6) \qquad l^1 \mathbin{\widehat{\otimes}}_\varepsilon E \cong L_\varepsilon(E'_\tau; l^1).$$

As a next step, we show that there is a canonical correspondence between weakly continuous mappings $u : E' \to l^1$, or, equivalently, weakly continuous mappings $v : l^\infty \to E$, and *summable sequences* in E. We define the latter concept:

Definition 44.1.   A sequence $\{x_n\}$ in a TVS E is said to be summable if to every neighborhood of zero V in E there is an integer $n_V > 0$ such that, for all finite subsets J of integers $n \geqslant n_V$,

$$\sum_{n \in J} x_n \in V.$$

If $\{x_n\}$ is a summable sequence in a TVS E, the partial sums

$$\sum_{n \geqslant p} x_n \quad (p = 0, 1,...)$$

form a Cauchy sequence. Therefore they converge if E is complete; when they converge, their limit is denoted by

$$\sum_{n=0}^{\infty} x_n \, ,$$

The space of all sequences in E, i.e., of all mappings from the set $\mathbf{N}$ of nonnegative integers into E, induces a structure of linear space on the set of all summable sequences in E. We shall not put a topology on this set.

1. Let $v : l^\infty \to E$ be a weakly continuous mapping. Let us denote by e_n the sequence with all terms equal to zero, except the nth one, equal to one, and let us set $x_n = v(e_n)$. The sequence (e_n) $(n = 0, 1,...)$ is weakly summable in l^∞. Therefore, the sequence (x_n) is weakly summable in E. But on the other hand, all the partial finite sums

$$\sum_{n \in J} e_n \quad (J: \text{finite set of integers} \geqslant 0),$$

belong to the unit ball B of l^∞. By Lemma 44.3, $v(B)$ is a precompact subset A of E. On $\hat{A}$, closure of A in $\hat{E}$, which is compact, the weak topology $\sigma(\hat{E}, E')$ coincides with the topology of $\hat{E}$ as completion of E (equipped with its initial topology). We conclude that the sequence $\{x_n\}$ is summable in $\hat{E}$, hence in E.

Thus, with every weakly continuous linear mapping of l^∞ into E we have associated a summable sequence in E.

2. Conversely, let (x_n) be a summable sequence in E. Let us suppose now that E *is complete*. I contend that

$$v : (\tau_n) \rightsquigarrow \sum_{n=0}^{\infty} \tau_n x_n$$

is a weakly continuous linear mapping $l^\infty \to E$. We show first that v indeed maps l^∞ into E. Possibly by separating each τ_n into its real and imaginary parts, we may assume that every one of them is real; next, by dividing each τ_n by $\sup_n | \tau_n |$ (supposed to be $\neq 0$!), we may assume that $| \tau_n | \leqslant 1$ for all n. Let J be an arbitrary finite set of integers; let us denote by $\sum_J$ the (finite) family of finite sequences $\varepsilon = (\varepsilon_n)_{n \in J}$ such that $\varepsilon_n = \pm 1$ for every $n \in J$. If we embed any sequence $(\sigma_n)_{n \in J}$ in the Euclidean space with dimension equal to the number of elements of J, in the obvious canonical way, we see that, if $| \sigma_n | \leqslant 1$ for all $n \in J$, the sequence $(\sigma_n)_{n \in J}$ belongs to the convex hull of $\sum_J$: this is the same as saying that a hypercube is the convex hull of its vertices—which it is! In particular, the finite subsequence $(\tau_n)_{n \in J}$ belongs to the convex hull of $\sum_J$.

Now let p be a continuous seminorm on E, and ε a number > 0. Since the sequence (x_n) is summable, we may select an integer $n_\varepsilon > 0$ such that, for every finite set of integers $n \geqslant n_\varepsilon$, J, we have

$$p \left(\sum_{n \in J} x_n \right) \leqslant \varepsilon/2.$$

Let us take a finite sequence (ε_n) belonging to $\sum_J$. We have

$$p \left(\sum_{n \in J} \varepsilon_n x_n \right) \leqslant p \left(\sum_{n \in J, \varepsilon_n = +1} x_n \right) + p \left(\sum_{n \in J, \varepsilon_n = -1} x_n \right) \leqslant \varepsilon.$$

This immediately implies what we wanted.

Next, we must show that the mapping v is weakly continuous. It suffices to show that, for every $x' \in E'$, the sequence $\{\langle x', x_n \rangle\}$ ($n = 0$, 1,...) belongs to l^1. It is obvious that, to every $\varepsilon > 0$, there is $n_\varepsilon \geqslant 0$ such that, for every finite set J of integers $n \geqslant n_\varepsilon$,

$$\left| \sum_{n \in J} \langle x', x_n \rangle \right| \leqslant \varepsilon.$$

In taking for J any finite set of integers $n \geqslant n_\varepsilon$ such that $\mathrm{Re}\langle x', x_n \rangle$ is $\geqslant 0$ for all $n \in J$, then a set such that $\mathrm{Re}\langle x', x_n \rangle < 0$ for all $n \in J$, and doing this again with $\mathrm{Im}\langle x', x_n \rangle$, we reach the desired conclusion.

Finally, we observe that $v(e_n) = x_n$ for every n, which shows that the mapping $(x_n) \rightsquigarrow v$ is the inverse of the mapping $v \rightsquigarrow (x_n)$ introduced in 1. We may summarize:

THEOREM 44.3. *Let E be a complete locally convex Hausdorff space. Then:*

$$(x_n)_{n \in \mathbf{N}} \rightsquigarrow \left((\tau_n)_{n \in \mathbf{N}} \rightsquigarrow \sum_{n=0}^{\infty} \tau_n x_n \right)$$

is a one-to-one linear map of the space of summable sequences in E onto the space of weakly continuous linear mappings of l^∞ into E.

The latter space, canonically isomorphic to $L(E'_\tau ; l^1)$, when carrying the topology of uniform convergence on the equicontinuous subsets of E', is canonically isomorphic (for the TVS structure) to $l^1 \widehat{\otimes}_\varepsilon E$.

Thus $l^1 \widehat{\otimes}_\varepsilon E$ may be identified with the space of summable sequences in E.

Exercises

44.8. By making use of Theorem 44.2, prove that l^1 and l^∞ are not reflexive.

44.9. Let dx be the Lebesgue measure on $\mathbf{R}^n$, and L^p ($1 \leqslant p < +\infty$) the Banach space of (classes of) functions f such that $|f|^p$ is Lebesgue integrable. Prove that a subset A of L^p is compact if and only if it has the following three properties:

   (i) A is bounded in L^p (in the sense of the L^p norm);

   (ii) A is equismall at infinity, i.e., to every $\varepsilon > 0$ there is $\rho > 0$ such that, for *all* $f \in A$,

$$\int_{|x| > \rho} |f(x)|^p \, dx < \varepsilon;$$

   (iii)   to every $\varepsilon > 0$, there is $\eta > 0$ such that, for all $a \in \mathbf{R}^n$ such that $|a| \leqslant \eta$ and all $f \in A$,

$$\int |f(x - a) - f(x)|^p \, dx \leqslant \varepsilon.$$

45

Examples of Completion of Topological Tensor Products: Completed π-Product of Two Fréchet Spaces

We give the definition of an *absolutely summable* sequence in a locally convex Hausdorff space E.

Definition 45.1. *A sequence $\{z_n\}$ $(n = 0, 1,...)$ in E is said to be absolutely summable if, for every continuous seminorm p on E, the sequence of non-negative numbers $p(z_n)$ is summable.*

An absolutely summable sequence (z_n) is summable (Definition 44.1). If E is complete, the partial sums $\sum_{n \leqslant p} z_n$ converge, as $p \to +\infty$. Their limit, $\sum_{n=0}^{\infty} z_n$, is called an *absolutely convergent series*.

We state now the main theorem:

THEOREM 45.1. *Let E, F be two Fréchet spaces. Every element $\theta \in E \,\widehat{\otimes}_\pi\, F$ is the sum of an absolutely convergent series*

$$(45.1) \qquad \theta = \sum_{n=0}^{\infty} \lambda_n\, x_n \otimes y_n\,,$$

where (λ_n) is a sequence of complex numbers such that $\sum_{n=0}^{\infty} |\lambda_n| < 1$, and (x_n) (resp. (y_n)) is a sequence converging to zero in E (resp. F).

It is important, in various applications, to have a strengthened form of Theorem 45.1:

THEOREM 45.2. *Let E, F be two Fréchet spaces, and U (resp. V) a convex balanced open neighborhood of zero in E (resp. F).*

Let K_0 be a compact subset of the convex balanced hull of $U \otimes V$. There is a compact subset K_1 of the unit ball of l^1, a sequence $\{x_n\}$ (resp. $\{y_n\}$) contained in U (resp. V) and converging to zero in E (resp. F), such that, for every $\theta \in K_0$, (45.1) holds for some $(\lambda_n) \in K_1$.

Proof of Theorem 45.2. Let us denote by $\mathscr{X}_1$ the vector space of all complex-valued functions on the set $E \times F$ such that

$$\|f\|_{\mathfrak{p},\mathfrak{q}} = \sum_{(x,y)\in E\times F} |f(x,y)|\, \mathfrak{p}(x)\, \mathfrak{q}(y) < +\infty$$

for all continuous seminorms $\mathfrak{p}$ and $\mathfrak{q}$ on E and F, respectively. The space $\mathscr{X}_1$ contains the space of functions on $E \times F$ which have a finite support; as in the proof of Theorem 39.1, we denote the latter by $\mathscr{X}$. The seminorms $\|\ \|_{\mathfrak{p},\mathfrak{q}}$ define a locally convex topology on $\mathscr{X}_1$ which is certainly not Hausdorff, since all these seminorms vanish on the linear subspace

$$M_1 = \{f\in\mathscr{X}_1;\, x\neq 0 \text{ and } y\neq 0 \text{ implies } f(x,y)=0\}.$$

As a matter of fact, when E and F are Hausdorff, M_1 is exactly the intersection of the kernels of the seminorms $\|\ \|_{\mathfrak{p},\mathfrak{q}}$, and the Hausdorff space associated with $\mathscr{X}_1$ can be canonically identified to the space of complex functions on $T = (E - \{0\}) \times (F - \{0\})$, which we denote by Λ_1 (we denote by Λ the subspace consisting of the functions in T which have *finite* support). Let us denote by supp $\mathfrak{p}$ the complement in E of Ker $\mathfrak{p}$; similarly for $\mathfrak{q}$. If $f \in \Lambda_1$, we denote by supp f the set of points $(x, y) \in T$ such that $f(x, y) \neq 0$ (this is consistent with the usual definition of the support if we consider the discrete topology on $E - \{0\}, F - \{0\}$ and T). From the fact that $\|f\|_{\mathfrak{p},\mathfrak{q}} < +\infty$, we derive that supp f intersects (supp $\mathfrak{p}$) $\times$ (supp $\mathfrak{q}$) according to a *countable* subset. It is also evident that, to every $n = 1, 2,...$, there is a function f_n with finite support, such that $\|f - f_n\|_{\mathfrak{p},\mathfrak{q}} \leqslant 1/n$: Λ is dense in Λ_1.

Now, there is a canonical mapping of Λ *onto* $E \otimes F$, namely

$$(45.2) \qquad\qquad f \rightsquigarrow \sum_{(x,y)\in T} f(x,y)\, x \otimes y.$$

It is the definition of the π-topology that this mapping is a homomorphism of Λ (equipped with the topology defined by the seminorms $\|\ \|_{\mathfrak{p},\mathfrak{q}}$) onto $E \otimes_\pi F$.

We have not yet exploited the fact that E and F are Fréchet spaces. As both are metrizable, so is Λ_1: this is a trivial consequence of the fact that, if $\mathfrak{p}, \mathfrak{p}'$ (resp. $\mathfrak{q}, \mathfrak{q}'$) are two continuous seminorms on E (resp. F) and if $\mathfrak{p} \leqslant \mathfrak{p}', \mathfrak{q} \leqslant \mathfrak{q}'$, we have $\|f\|_{\mathfrak{p},\mathfrak{q}} \leqslant \|f\|_{\mathfrak{p}',\mathfrak{q}'}$ for all $f \in \Lambda_1$. Furthermore, the support of each $f \in \Lambda_1$ is countable: indeed, it is the union of a countable family of countable sets, its intersections with (supp $\mathfrak{p}_k$) $\times$ (supp $\mathfrak{q}_k$), where $(\mathfrak{p}_k)$ and $(\mathfrak{q}_k)$ are countable bases of continuous seminorms in E and F, respectively. Another consequence of the fact that

E and F are metrizable is that Λ_1 is complete. We leave the proof of this fact to the reader. Then let J_1 be the extension to Λ_1 of the canonical mapping of Λ onto $E \otimes F$; J_1 is a continuous linear mapping of Λ_1 into $E \,\widehat{\otimes}_\pi F$. As a matter of fact, it is a *homomorphism* of Λ_1 onto $E \,\widehat{\otimes}_\pi F$. In order to see this, it suffices to look at the usual diagram:

$$
\begin{array}{ccc}
\Lambda_1 & \xrightarrow{\;J_1\;} & E \,\widehat{\otimes}_\pi F \\[2pt]
{\scriptstyle \phi_1}\downarrow & \nearrow{\scriptstyle \tilde{J}_1} & \\[2pt]
\Lambda_1/\mathrm{Ker}\, J_1 & &
\end{array}
$$

Indeed, the image under ϕ_1 of Λ is a dense linear subspace of $\Lambda_1/\mathrm{Ker}\, J_1$ which is isomorphic, via $\tilde{J}_1$, to $E \otimes_\pi F$; the inverse of this mapping, defined on $E \otimes_\pi F$, can be extended to $E \,\widehat{\otimes}_\pi F$ by continuity, and this extension must be the inverse of $\tilde{J}_1$ (defined on the whole of $\Lambda_1/\mathrm{Ker}\, J_1$). Thus $\tilde{J}_1$ is an isomorphism onto (we have used the completeness of Λ_1 only to the extent that it implies the completeness of $\Lambda_1/\mathrm{Ker}\, J_1$). The expression of the homomorphism J_1 is given by (45.2): in the present situation, every $f \in \Lambda_1$ vanishes outside a countable subset of T, so that $\sum_{(x,y)\in T} f(x, y)\, x \otimes y$ is a series in $E \,\widehat{\otimes}_\pi F$, obviously absolutely convergent (see Definition 45.1).

We have thus obtained a representation of every element θ of $E \,\widehat{\otimes}_\pi F$ as an absolutely convergent series, closely resembling (45.2). We are going to show, now, that this series can be made to possess all the properties announced in Theorem 45.2. We begin by selecting two increasing sequences of continuous seminorms $(\mathfrak{p}_m)$, $(\mathfrak{q}_m)$ ($m = 0$, $1,...$) in E and F, respectively, such that every continuous seminorm $\mathfrak{p}$ (resp. $\mathfrak{q}$) on E (resp. F) $\leqslant$ some $\mathfrak{p}_n$ (resp. $\mathfrak{q}_n$). Furthermore, we start these two sequences by two seminorms $\mathfrak{p}_0$ and $\mathfrak{q}_0$ such that

$$ U = \{x \in E;\ \mathfrak{p}_0(x) < 1\}, \qquad V = \{y \in F;\ \mathfrak{q}_0(y) < 1\}. $$

It is then quite evident that the unit semiball

$$ \mathcal{O} = \{f \in \Lambda_1;\quad \|f\|_{\mathfrak{p}_0, \mathfrak{q}_0} < 1\} $$

is mapped by J_1 onto the convex balanced hull of $U \otimes V$, which we shall denote by W. We shall apply the following lemma of point-set topology:

LEMMA 45.1.   *Let $\mathscr{A}$, $\mathscr{B}$ be two complete metric spaces, u an open continuous mapping of $\mathscr{A}$ onto $\mathscr{B}$, and $\mathcal{O}$ an open subset of $\mathscr{A}$. Every compact* *subset K of $u(\mathcal{O})$ is the image $u(H)$ of a compact subset H of $\mathcal{O}$, which is the closure of a countable subset of $u^{-1}(K)$.*

Proof. Consider the family of all open subsets U of $\mathcal{O}$ whose closure is contained in $\mathcal{O}$; a finite number of open sets $u(U)$ cover K, which is the same as saying that there is one such set U with $u(U) \supset K$. Let us construct, by induction on $k = 1, 2,\ldots$, an increasing sequence of finite subsets of U, $A_1 \subset A_2 \subset \cdots \subset A_k \subset \cdots$, with the following properties:

(i) the A_k's are all contained in the preimage of K;

(ii) A_{k+1} is contained in the set A'_k of points of U lying at a distance $< 2^{-k}$ from A_k ;

(iii) $u(A'_k) \supset K$.

The possibility of finding A_1 is obvious. Suppose that A_k has been determined and call A''_k the set of points of U which lie at a distance $< 2^{-k-1}$ from A_k. We can find a subset B_k of $u^{-1}(K) \cap A'_k$ such that the set of points in U which lie at a distance $< 2^{-k-1}$ from B_k is mapped onto a subset of $u(U)$ containing $K - K \cap u(A''_k)$. We may then take $A_{k+1} = A_k \cup B_k$. Let us call A the union of the sets A_k. Whatever k is, every point of A lies at a distance $< 2^{-k-1}$ from A_k: this follows immediately from Property (ii). It implies immediately that A is precompact (cf. Proposition 6.9); let H be the closure of A in $\mathscr{A}$: H is compact and contained in $\bar{U}$, hence in $\mathcal{O}$. Given any $\varepsilon > 0$, there is k such that A'_k is contained in $\bar{U}$. If H'_k is the set of points of $\mathcal{O}$ at a distance $\leqslant 2^{-k}$ from H, we have $K \subset u(H'_k)$ since $A'_k \subset H'_k$; we derive immediately from this that $K \subset u(H)$. Since H is the closure of $A \subset u^{-1}(K)$ and since $u^{-1}(K)$ is closed, we have $u(H) = K$.   Q.E.D.

COROLLARY. *Let K be a compact subset of a complete metric space $\mathscr{A}$; K contains a subset which is everywhere dense and countable.*

Proof. It suffices to apply Lemma 45.1 with $\mathscr{A} = \mathscr{B}$ and u the identity mapping of $\mathscr{A}$.

Let us go back to the proof of Theorem 45.2. Let K be a compact subset of W, the convex balanced hull of $U \otimes V$. There is a compact subset H of $\mathcal{O} \subset \Lambda_1$ such that $J_1(H) \subset K$. This follows from Lemma 45.1; from its corollary it follows that there is a countable subset N of $T = (E - \{0\}) \times (F - \{0\})$ which contains the support of every $f \in H$. Indeed, there is a countable subset S of H which is everywhere dense in H; let N be the union of the supports of the $f \in S$. Since every $f \in H$ is the limit of a subsequence of the sequence S, f must be identically zero outside of N; and N is countable, as a countable union of countable sets. We order N in an arbitrary way: $N = \{(\xi_n, \eta_n)\}_{n=0,1,\ldots}$. For every $\theta \in K$, we have

$$\theta = \sum_{n=0}^{\infty} f(\xi_n, \eta_n)\, \xi_n \otimes \eta_n$$

for some $f \in H$. Needless to say, the series is absolutely convergent. Let us set $a_{m,n} = \mathrm{p}_m(\xi_n)$, $b_{m,n} = \mathrm{q}_m(\eta_n)$, $f_n = f(\xi_n, \eta_n)$. We shall construct two sequences of numbers > 0 (a_n), (b_n) $(n = 0, 1, \ldots)$ with the following properties:

(45.3)   For every m, $\lim_{n \to \infty}(a_{m,n}/a_n + b_{m,n}/b_n) = 0$;

(45.4)   the mapping $f \rightsquigarrow (f_n a_n b_n)_{(n=0,1,\ldots)}$ maps H into a compact subset K_1 of l^1.

Observe that, for each m, $f \rightsquigarrow (f_n a_{m,n} b_{m,n})_{(n=0,1,\ldots)}$ maps H into a compact subset of l^1 (by definition of the topology of Λ_1). But every compact subset of l^1 is equismall at infinity (Theorem 44.2), therefore, for each m, we may select an integer $N_m \geqslant 0$ such that

(45.5) $$\sum_{n \geqslant N_m} |f_n|\, a_{m,n} b_{m,n} < 8^{-m} \qquad \text{for all} \quad f \in H.$$

Since $H \subset \mathcal{O}$, we may even take $N_0 = 0$. Let us first choose $\varepsilon_n > 0$ such that, for all $f \in H$,

(45.6) $$\sum_{n \geqslant N_m} |f_n|(2^m a_{m,n} + 2^m b_{m,n} + \varepsilon_n)\varepsilon_n < 2^{-m}.$$

This is possible, since, for fixed $n \geqslant 0$, the set $\{|f_n|\}$ is bounded as f ranges over H. We then set, if $N_m \leqslant n < N_{m+1}$,

$$a_n = \varepsilon_n + 2^m a_{m,n}, \qquad b_n = \varepsilon_n + 2^m b_{m,n}.$$

Recalling that $a_{m,n}$ and $b_{m,n}$ are nondecreasing with m for fixed n, we see easily that (45.3) is satisfied. On the other hand, by combining (45.5) and (45.6), we obtain

$$\sum_{N_m \leqslant n \leqslant N_{m+1}} |f_n|\, a_n b_n \leqslant 2^{-(m-1)},$$

whence

$$\sum_{N_m \leqslant n} |f_n|\, a_n b_n \leqslant 2^{-(m-2)}.$$

This shows that (45.4) holds: indeed, by taking $m = 0$, hence $N_m = 0$, we see that the sequences $(f_n a_n b_n)$ remain bounded in l^1 when f varies over H; then, by taking $m = 1, 2, ...$, we see that these sequences form a subset of l^1 which is equismall at infinity, hence relatively compact by virtue of Theorem 44.2.

From there on, the proof of Theorem 45.2 is easy to complete. We set, for every $f \in H$ and every $n = 0, 1, ...,$

$$\tilde{f}_n = f_n a_n b_n , \qquad \tilde{\xi}_n = \xi_n/a_n , \qquad \tilde{\eta}_n = \eta_n/b_n .$$

This is possible since a_n , $b_n \geqslant \varepsilon_n > 0$. From (45.3) we derive immediately that the $\tilde{\xi}_n$ (resp. $\tilde{\eta}_n$) converge to zero in E (resp. F). From (45.4), we derive that $\{\tilde{f}_n\}_{n=0,1,...}$ varies in a compact subset $\tilde{H}$ of l^1 as f varies in H. Furthermore, note that there is a number $\kappa < 1$ such that, for all $f \in H$,

$$\sum_{n=0}^{+\infty} | \tilde{f}_n | \, \mathfrak{p}_0(\tilde{\xi}_n) \, \mathfrak{q}_0(\tilde{\eta}_n) = \sum_{n=0}^{+\infty} | f_n | \, \mathfrak{p}_0(\xi_n) \, \mathfrak{q}_0(\eta_n) < \kappa.$$

There is an integer $N \geqslant 0$ such that

$$\mathfrak{p}_0(\tilde{\xi}_n) < 1, \qquad \mathfrak{q}_0(\tilde{\eta}_n) < 1 \qquad \text{for all} \quad n \geqslant N,$$

$$\sum_{n \geqslant N} | \tilde{f}_n | < \frac{1 - \kappa}{2} \qquad \text{for all} \quad f \in H.$$

For $n \geqslant N$, we shall set $x_n = \tilde{\xi}_n , y_n = \tilde{\eta}_n , \lambda_n = \tilde{f}_n$. Let $\varepsilon > 0$ be a very small number. Let us select, for each n, two numbers ρ_n , σ_n such that

$$\mathfrak{p}_0(\tilde{\xi}_n) < \rho_n \leqslant \mathfrak{p}_0(\tilde{\xi}_n) + \varepsilon, \qquad \mathfrak{q}_0(\tilde{\eta}_n) < \sigma_n \leqslant \mathfrak{q}_0(\tilde{\eta}_n) + \varepsilon,$$

and let us set, for $n < N$,

$$x_n = \rho_n^{-1}\tilde{\xi}_n , \qquad y_n = \sigma_n^{-1}\tilde{\eta}_n , \qquad \lambda_n = \tilde{f}_n \rho_n \sigma_n .$$

It is then clear that the sequences (x_n) and (y_n) converge to zero and that they are contained in U and V, respectively. Furthermore, (λ_n) remains in a compact subset of l^1 when f ranges over H. Finally,

$$\sum_{n=0}^{\infty} | \lambda_n | \leqslant \sum_{n < N} | \tilde{f}_n | \, \rho_n \sigma_n + \tfrac{1}{2}(1 - \kappa)$$

$$\leqslant \sum_{n < N} | \tilde{f}_n | \, \mathfrak{p}_0(\tilde{\xi}_n) \, \mathfrak{q}_0(\tilde{\eta}_n) + \tfrac{1}{2}(1 - \kappa) + C\varepsilon \leqslant \tfrac{1}{2}(1 + \kappa) + C\varepsilon,$$

where C is a positive constant, depending on $\sup_n p_0(\tilde{\xi}_n)$, $\sup_n q_0(\tilde{\eta}_n)$, and on $\tilde{H}$. By taking ε sufficiently small, we see that (λ_n) remains inside a ball of radius < 1 (centered at 0) in l^1. The proof of Theorem 45.2 is complete.

COROLLARY 1. *Let E, F be two Banach spaces. Every element θ of the open unit ball of $E \hat{\otimes}_\pi F$ is equal to an absolutely convergent series*

$$\theta = \sum_{n=0}^{\infty} \lambda_n \, x_n \otimes y_n \,,$$

where (x_n) and (y_n) are sequences converging to zero in the open unit ball of E and F, respectively, and $\sum_{n=0}^{\infty} |\lambda_n| < 1$.

COROLLARY 2. *Let E and F be two Fréchet spaces. Every compact subset of $E \hat{\otimes}_\pi F$ is contained in the closed convex balanced hull of the tensor product of a compact subset of E with a compact subset of F.*

COROLLARY 3. *If E and F are Fréchet spaces and G a complete locally convex Hausdorff TVS, the canonical (algebraic) isomorphism of $B(E, F; G)$ onto $L(E \hat{\otimes}_\pi F; G)$ becomes a homeomorphism if the first space carries the topology of uniform convergence on the products of compact sets and the second one, the topology of compact convergence.*

Exercises

45.1. Prove that every compact subset of a Fréchet space E is contained in the closed convex balanced hull of a sequence converging to zero.

45.2. Let E and F be two Banach spaces, and G a complete locally convex Hausdorff space. Prove that the canonical algebraic isomorphism of $B(E, F; G)$ onto $L(E \hat{\otimes}_\pi F; G)$ is a homeomorphism when the first space carries the topology of uniform convergence on the products of bounded sets and the second space, the topology of bounded convergence.

45.3. Let E_j, F_j ($j = 1, 2$) be four Fréchet spaces, and $u_j : E_j \to F_j$ ($j = 1, 2$) two continuous linear mappings. Prove, by making use of Theorem 45.1, that, if both u_1 and u_2 are onto, then $u_1 \hat{\otimes}_\pi u_2$ is also onto.

45.4. Let H_z be the space of entire functions with respect to the variable z in $\mathbf{C}^n$, H'_ζ the space of analytic functionals in the variable $\zeta \in \mathbf{C}^n$, and $H_z(H'_\zeta)$ the space of entire functions of z with values in the space H'_ζ (all the spaces under consideration carry their natural topologies). Prove the following facts:

(i)   there is a canonical TVS isomorphism of $H_z \hat{\otimes}_\pi H'_\zeta$ onto $H_z(H'_\zeta)$;

(ii) every element $\theta(z, \zeta) \in H_z(H'_\zeta) \cong H_z \hat{\otimes}_\pi H'_\zeta$ is equal to the sum of an absolutely convergent series,

$$\theta(z, \zeta) = \sum_{p,q \in \mathbf{N}^n} \lambda_{p,q} \, z^p \otimes \delta_\zeta^{(q)},$$

where δ_ζ is the Dirac measure with respect to the variable ζ at the point $\zeta = 0$.

Give the expression of the numbers $\lambda_{p,q}$ in terms of $\theta(z, \zeta)$; prove that the function $\delta(z - \zeta)$, which assigns to each $z \in \mathbf{C}^n$ the Dirac measure (with respect to the variable ζ) at the point $\zeta = z$, belongs to $H_z(H'_\zeta)$, and compute the coefficient $\lambda_{p,q}$ when $\theta(z, \zeta) = \delta(z - \zeta)$.

45.5. Let $\mathscr{C}_x^\infty$ be the space of $\mathscr{C}^\infty$ functions with respect to the variable x in $\mathbf{R}^n$, $\mathscr{E}'_\xi$ the space of distributions with compact support in the variable $\xi \in \mathbf{R}^n$, and $\mathscr{C}_x^\infty(\mathscr{E}'_\xi)$ the space of $\mathscr{C}^\infty$ functions of $x \in \mathbf{R}^n$ with values in $\mathscr{E}'_\xi$ (all the spaces under consideration carry their usual topologies). Anticipating slightly what is to come, we admit the fact that we have, canonically,

$$\mathscr{C}_x^\infty(\mathscr{E}'_\xi) \cong \mathscr{C}_x^\infty \widehat{\otimes}_\pi \mathscr{E}'_\xi.$$

Let $\delta(x - y)$ be the function which assigns to every $x \in \mathbf{R}^n$ the Dirac measure, with respect to y, at the point $y = x$. Prove that $\delta(x - y)$ belongs to $\mathscr{C}_x^\infty(\mathscr{E}'_\xi)$. Prove that $\delta(x - y)$ is not equal to the sum of an absolutely convergent series, in $\mathscr{C}_x^\infty(\mathscr{E}'_\xi)$,

$$\delta(x - \xi) = \sum_{j=0}^\infty u_j(x) \otimes v_j(\xi),$$

with $\{u_j\} \subset \mathscr{C}^\infty$ and $\{v_j\} \subset \mathscr{E}'$ bounded sets (show that if this were true every function $u \in \mathscr{C}^\infty$ could be written in the form $u = \sum_{j=0}^\infty \lambda_j u_j$, which is not possible).

45.6. Let E be a Banach space, E' its strong dual, j the natural injection of $E \otimes_\pi E'$ into $L(E; E)$, and $\hat{j}$ the extension, to $E \widehat{\otimes}_\pi E'$, of the continuous linear map j. Prove (by applying Theorem 45.1) the equivalence of the following two facts:

(a)   the identity mapping $E \to E$ belongs to $\hat{j}(E \widehat{\otimes}_\pi E')$;

(b) $\dim E < +\infty$.

46

Examples of Completion of Topological Tensor Products: Completed π-Products with a Space L^1

In Chapter 44, we have studied the completed ε-product of a space E (locally convex Hausdorff and preferably complete) with the space l^1 of absolutely summable complex sequences. We have shown that, when E is complete, $l^1 \widehat{\otimes}_\varepsilon E$ is canonically isomorphic to the space of summable sequences in E. It is unnecessary to recall that l^1 is the space of integrable functions on the set $\mathbf{N}$ of nonnegative integers with respect to the measure dn whose mass at every point is $+1$. In the present chapter, we shall consider the space L^1 with respect to an arbitrary measure μ on a set X and study its π-product with a space E. We are going to show, among other properties, that, when E is a Banach space, $L^1 \widehat{\otimes}_\pi E$ is exactly "equal" to the space $L^1(E)$ of integrable functions with values in E. When $L^1 = l^1$, this means that $l^1 \widehat{\otimes}_\pi E$ can be identified with the space of *absolutely* summable sequences in E, and thus underlines the difference between completion of ε-products and π-products. One can then show that, if E is an infinite dimensional Banach space, the canonical mapping of $l^1 \widehat{\otimes}_\pi E$ into $l^1 \widehat{\otimes}_\varepsilon E$ is never onto. This implies immediately the theorem of Dvoretzky–Rogers: *in an infinite dimensional Banach space, there is at least one summable sequence which is not absolutely summable.* This stands in contrast with the case of a *nuclear* space E, whose theory we begin describing in the next chapter: these are the spaces such that $E \widehat{\otimes}_\pi F = E \widehat{\otimes}_\varepsilon F$ for all locally convex Hausdorff spaces F. It is obvious that, in such a space E, every summable sequence must be absolutely summable. Needless to say, no infinite dimensional Banach space is nuclear.

46.1. The Spaces $L^\alpha(E)$

Let X be a set, dx a positive measure on it, E a locally convex Hausdorff space, and α a number such that $1 \leqslant \alpha < +\infty$.

We denote by $\mathscr{F}^\alpha(E)$ the vector space of all functions $\mathbf{f} : X \to E$ such that, for every continuous seminorm p on E,

$$(46.1) \qquad \| \mathbf{f} \|_{L^\alpha, p} = \left(\int p(\mathbf{f}(x))^\alpha dx \right)^{1/\alpha},$$

where $\int$ denotes the upper integral; we provide $\mathscr{F}^\alpha(E)$ with the topology defined by the seminorms (46.1). Let us denote by $\sum \otimes E$ the linear subspace of $\mathscr{F}^\alpha(E)$ consisting of the *integrable step-functions* $\mathbf{s} : X \to E$. By this we mean the finite linear combinations of the form $\sum_j \sigma_j \mathbf{e}_j$, where the σ_j are complex integrable step-functions and the $\mathbf{e}_j$ are vectors belonging to E (of course, one may suppose that the σ_j are characteristic functions of integrable sets). We denote by $\mathscr{L}^\alpha(E)$ the closure in $\mathscr{F}^\alpha(E)$ of $\sum \otimes E$. As usual, we denote by $\mathscr{L}^\alpha$ the space $\mathscr{L}^\alpha(\mathbf{C})$. It is obvious, from the definition, that $\mathscr{L}^\alpha \otimes E$ is dense in $\mathscr{L}^\alpha(E)$. Generally, $\mathscr{L}^\alpha(E)$ is not Hausdorff and one denotes by $L^\alpha(E)$ the associated Hausdorff space; one sets $L^\alpha = L^\alpha(\mathbf{C})$. These notations are all right as long as one deals with a single measure dx; if more measures are introduced, the notation must be adapted so as not to create confusion.

One defines the space $\mathscr{F}^\infty(E)$ as the space of functions $\mathbf{f} : X \to E$ such that, for each continuous seminorm p on E, there is a number $M_p \geqslant 0$ and a set $N_p \subset X$ with measure zero such that

$$(46.2) \qquad p(\mathbf{f}(x)) \leqslant M_p \qquad \text{for all} \quad x \notin N_p.$$

One then defines the number $\| \mathbf{f} \|_{L^\infty, p}$ as the infimum of the numbers $M_p \geqslant 0$ such that (46.2) holds for *some* set N_p of measure zero. The topology of $\mathscr{F}^\infty(E)$ is then defined by the seminorms $\mathbf{f} \rightsquigarrow \| \mathbf{f} \|_{L^\infty, p}$; $\mathscr{L}^\infty(E)$ is the subspace of $\mathscr{F}^\infty(E)$ consisting of the functions $\mathbf{f}$ which are measurable. By $L^\infty(E)$ one denotes the associated Hausdorff space.

A straightforward generalization of the Fischer–Riesz theorem enables one to prove that, when E is a Fréchet space, the spaces $L^\alpha(E)(1 \leqslant \alpha \leqslant + \infty)$ are complete. But this is not so if E ceases to be metrizable—in general.

The canonical image of $\mathscr{L}^\alpha \otimes E$ into $L(E)$ is denoted by $L^\alpha \otimes E$; it is immediately seen to be a tensor product of L^α and E. For $\alpha \geqslant 1$ finite, $\mathscr{L}^\alpha \otimes E$ is dense in $\mathscr{L}^\alpha(E)$; as the canonical image of a dense subset, $L^\alpha \otimes E$ is dense in $L^\alpha(E)$.

We shall now focus our attention on the case of a Banach space E, with norm $\| \ \|$. In this case, the topology of $\mathscr{L}^\alpha(E)$ is defined by a single seminorm and $L^\alpha(E)$ is a normed space. Furthermore, $L^\alpha(E)$ is complete, hence a Banach space (theorem of Fisher–Riesz); its elements can be interpreted as classes of functions (belonging to $\mathscr{L}^\alpha(E)$) which are equal almost everywhere. It is immediately seen that the bilinear form

$$\langle \mathbf{f}, \mathbf{g} \rangle = \int \langle \mathbf{f}(x), \mathbf{g}(x) \rangle \, dx$$

is continuous on $L^\alpha(E) \times L^{\alpha'}(E')$, where E' is the Banach space dual of E and where $\alpha' = \alpha/(\alpha - 1)$. This defines a canonical mapping of $L^{\alpha'}(E')$ into $(L^\alpha(E))'$, which can be seen to be an isometry, but which in general is not onto.

If $f \in L^\alpha$ and $\mathbf{e} \in E$, we have

$$\left(\int \| f\mathbf{e} \|^\alpha \, dx \right)^{1/\alpha} = \left(\int |f|^\alpha \, dx \right)^{1/\alpha} \| \mathbf{e} \|.$$

This proves that the bilinear mapping $(f, \mathbf{e}) \rightsquigarrow f\mathbf{e}$ of $L^\alpha \times E$ into $L^\alpha(E)$ has norm one. But (Proposition 43.12) the π-norm on $L^\alpha \otimes E$ is the largest norm on this space such that the canonical mapping of $L^\alpha \times E$ into it has norm one. We conclude that the π-norm is larger than the norm induced by $L^\alpha(E)$. For $\alpha > 1$, the π-norm is strictly larger than the norm of $L^\alpha(E)$ (at least in general). If they were equal, it would mean (when α is finite) that $L^\alpha(E) = L^\alpha \hat{\otimes}_\pi E$. For instance, we would apply this to $E = $ the space L^α with respect to a positive measure dy on a set Y. In this case, $L^\alpha(E) = L^\alpha_{dx}(L^\alpha_{dy})$ can be canonically identified to the space $L^\alpha_{dx\,dy}$ with respect to the product measure $dx\,dy$ on the set $X \times Y$ (the identification extends to the norms!). Thus we would have $L^\alpha_{dx\,dy} \cong L^\alpha_{dx} \hat{\otimes}_\pi L^\alpha_{dy}$ (α: canonical isomorphism), which is generally not true, as we shall see in the case $\alpha = 2$ (Chapter 49).

However, it is true, and it is the main result of this chapter, that $L^1 \hat{\otimes}_\pi E = L^1(E)$.

46.2. The Theorem of Dunford–Pettis

As before, we consider a set X and a positive measure dx on it. By $\mathscr{B}^\infty$ we denote the space of complex-valued functions which are bounded on the whole of X, with its natural norm $f \rightsquigarrow \sup_{x \in X} |f(x)|$. The space $\mathscr{B}^\infty$ is obviously a Banach space; it is quite different from the space $\mathscr{F}^\infty$ introduced on p. 468, whose elements are not necessarily bounded functions. Let us denote by $\mathscr{N}$ the linear subspace of $\mathscr{B}^\infty$ consisting of the functions which vanish outside of a set of measure zero: $\mathscr{N}$ *is closed.* Indeed, if a sequence $\{f_k\}$ in $\mathscr{N}$ converges to f in $\mathscr{B}^\infty$, each f_k must vanish outside some set of measure zero, N_k, therefore $f = 0$ outside the union $\bigcup_k N_k$ which has also measure zero. We denote by

B^∞ the quotient $\mathscr{B}^\infty/\mathscr{N}$ equipped with its natural Banach space structure. Although $\mathscr{L}^\infty$ is not contained in $\mathscr{B}^\infty$, there is a natural isometry of L^∞ into B^∞: for given any function $f \in \mathscr{F}^\infty$ and any $\varepsilon > 0$, there is a function $g \in \mathscr{B}^\infty$ such that $f = g$ almost everywhere, and such that

$$\|f\|_{L^\infty} \leqslant \sup_{x \in X} |g(x)| \leqslant \|f\|_{L^\infty} + \varepsilon.$$

Of course, if f is measurable, so is g.

It will be handy to introduce two new terms in our vocabulary, so as to abbreviate the statements; the first one is widely used:

Definition 46.1. A subset A of a topological space is called separable if A contains a subset B which is countable and dense in A.

The second term is also commonly used: if we have a map f of a set E *onto* a set F, a *lift* of a subset V of F is a mapping g of V into E such that $f \circ g$ is the identity of V (if one prefers, g is a right inverse of f on V).

We may now state the following lemma, which will be used in the proof of the theorem of Dunford–Pettis:

LEMMA 46.1. *Every separable linear subspace V of B^∞ possesses a lift into $\mathscr{B}^\infty$ which is a linear isometry.*

The mapping of $\mathscr{B}^\infty$ onto B^∞ considered in this statement is, of course, the canonical homomorphism.

Proof. Let A be a countable and dense subspace of V, and W the set of all finite linear combinations of elements of A *with coefficients in the field $\mathbf{Q}$ of rational numbers*. Let τ be an arbitrary lift of W into $\mathscr{B}^\infty$ which is linear: such a lift τ is obtained by lifting in an arbitrary manner an algebraic basis of W and extending the lift by $\mathbf{Q}$-linearity. Let w be an element of W; two representatives of w in $\mathscr{B}^\infty$ differ only on a set of measure zero, therefore, for every $\varepsilon > 0$, there is a set of measure zero, N_ε, such that

$$|\tau(w)(x)| \leqslant \|w\|_{B^\infty} + \varepsilon \qquad \text{if} \quad x \notin N_\varepsilon.$$

If we denote by N_w the union of the sets $N_{1/n}$ as $n \to +\infty$, we see that

$$|\tau(w)(x)| \leqslant \|w\|_{B^\infty} \qquad \text{for} \quad x \notin N_w.$$

Let N be the union of the sets N_w as w ranges over W; since W is countable, the measure of N is equal to zero. Let us define a lift $\sigma : W \to \mathscr{B}^\infty$ by setting

$$\sigma(w)(x) = 0 \qquad \text{if} \quad x \in N,$$
$$\sigma(w)(x) = \tau(w)(x) \qquad \text{if} \quad x \notin N.$$

This is indeed a lift of W, hence $\| w \|_{B^\infty} \leqslant \sup_{x\in X} | \sigma(w)(x) |$; but $| \sigma(w)(x) | \leqslant \| w \|_{B^\infty}$ for all $x \in X$. We conclude that σ is an isometry, obviously linear (as τ was linear). By continuity, we can extend σ to the closure of W in B^∞, which contains V (taking advantage of the fact that $\mathscr{B}^\infty$ is complete!). The extension of a linear isometry is a linear isometry, and if σ was a right inverse of the canonical mapping of $\mathscr{B}^\infty$ onto B^∞ over W, the same is true of its extension—but, now, over $\bar{W} \supset V$.   Q.E.D.

Let E be a Banach space, and E' its dual with its Banach space structure; we shall denote by $\| \ \|$ the norms, both in E and in E'.

Definition 46.2.   A function $\mathbf{f} : X \to E'$ *is said to be scalarly measurable if, for every* $\mathbf{e} \in E$, *the complex-valued function*

$$x \rightsquigarrow \langle \mathbf{f}(x), \mathbf{e} \rangle$$

is measurable.

Here, measurable means dx-measurable in X. Every measurable function of X into E' is scalarly measurable, but the converse is not generally true.

Let us denote momentarily by π the canonical projection of $\mathscr{L}^\infty$ onto L^∞. As before, E is a Banach space.

PROPOSITION 46.1.   *Let* $\mathbf{g}$ *be a bounded function* $X \to E'$, *scalarly measurable. Then*

(46.3) $\mathbf{e} \rightsquigarrow \pi(x \rightsquigarrow (\langle \mathbf{g}(x), \mathbf{e} \rangle))$

is a continuous linear map of E *into* L^∞ *with norm* $\leqslant \sup_{x\in X} \| \mathbf{g}(x) \|$.

The theorem of Dunford–Pettis states that, under the assumption that E is separable, every continuous linear map of E into L^∞ is of the form (46.3); furthermore, the function $\mathbf{g}$ can be chosen so that the norm of the mapping (46.3) is exactly equal to the maximum of $\| \mathbf{g} \|$ on X.

THEOREM 46.1.   *Let* E *be a separable Banach space. To every continuous linear map of* E *into* L^∞, u, *there is a scalarly measurable bounded function* $\mathbf{g} : X \to E'$ *such that* u *is equal to* (46.3) *and such that the norm of* u *is equal to* $\sup_{x\in X} \| \mathbf{g}(x) \|$.

Proof.   As E is separable, $u(E)$ is a separable linear subspace of L^∞, therefore, by virtue of Lemma 46.1, there is a lift σ of $u(E)$ into $\mathscr{B}^\infty$ which is a linear isometry. For each $\mathbf{e} \in E$ and each $x \in X$, let us set $g_{\mathbf{e}}(x) = \sigma[u(\mathbf{e})](x)$; observe that the function $g_{\mathbf{e}}$, for fixed $\mathbf{e}$, belongs to $\mathscr{L}^\infty$ (indeed, it differs from any other representative of $u(\mathbf{e}) \in L^\infty$ only on a set of measure zero). Furthermore, we have

$$| g_{\mathbf{e}}(x)| \leqslant \sup_{x \in X} | \sigma[u(\mathbf{e})](x)| = \| u(\mathbf{e})\|_{L^\infty} \leqslant \| u \| \| \mathbf{e} \|.$$

This shows that, for fixed x, $\mathbf{e} \rightsquigarrow g_{\mathbf{e}}(x)$ is a continuous linear form on E, which we denote by $\mathbf{g}(x) \in E'$; of course, as $g_{\mathbf{e}} \in \mathscr{L}^\infty$ for all $\mathbf{e} \in E$, $x \rightsquigarrow \mathbf{g}(x)$ is a scalarly measurable function $X \to E'$. Furthermore,

$$\sup_{x \in X} \| \mathbf{g}(x)\| = \sup_{x \in X} (\sup_{\mathbf{e} \in E, \| \mathbf{e} \| = 1} | g_{\mathbf{e}}(x)|)$$

$$= \sup_{\| \mathbf{e} \| = 1} (\sup_{x \in X} | \sigma[u(\mathbf{e})](x)|) = \sup_{\| \mathbf{e} \| = 1} \| u(\mathbf{e})\|_{L^\infty} = \| u \|.$$

Finally, the mapping u is equal to (46.3) since, for all $\mathbf{e} \in E$ and $x \in X$,

$$\langle \mathbf{g}(x), \mathbf{e} \rangle = g_{\mathbf{e}}(x) = \sigma[u(\mathbf{e})](x). \qquad \text{Q.E.D.}$$

We give now, as corollaries, two equivalent statements of the theorem of Dunford–Pettis. We must, however, make a preliminary remark. Suppose that $\mathbf{g} : X \to E'$ is scalarly measurable and bounded. Then, for every $\mathbf{e} \in E$, the function $x \rightsquigarrow \langle \mathbf{g}(x), \mathbf{e} \rangle$ belongs to $\mathscr{L}^\infty$. We may compute the integral of its product with a function $f \in \mathscr{L}^1$. As we have

$$\left| \int \langle \mathbf{g}(x), \mathbf{e} \rangle f(x)\, dx \right| \leqslant \sup_{x \in X} |\langle \mathbf{g}(x), \mathbf{e} \rangle| \int | f(x)|\, dx$$

$$\leqslant (\sup_{x \in X} \| \mathbf{g}(x)\|) \| \mathbf{e} \| \int | f(x)|\, dx,$$

we see that

$$\mathbf{e} \rightsquigarrow \int \langle \mathbf{g}(x), \mathbf{e} \rangle f(x)\, dx$$

is a continuous linear form on E, which it is natural to denote by

$$(46.4) \qquad \int \mathbf{g}(x) f(x)\, dx.$$

The norm (in E') of (46.4) is $\leqslant (\sup_{x \in X} \| \mathbf{g}(x) \|) \| f \|_{L^1}$; it is also evident that (46.4) does not change value if we replace f by a function which is equal to f almost everywhere, so that

$$(46.5) \qquad f \rightsquigarrow \int \mathbf{g}(x) f(x)\, dx$$

might be viewed as a linear mapping of L^1 into E'; note that the norm of this mapping is $\leqslant \sup_{x \in X} \| \mathbf{g}(x) \|$.

We may now state the two equivalent versions of the theorem of Dunford–Pettis; we shall leave the proof of their equivalence to the student, as an (easy) exercise:

COROLLARY 1. *Let E be a separable Banach space. To every continuous linear map $v : L^1 \to E'$ there is a bounded, scalarly measurable function $\mathbf{g} : X \to E'$ such that v is given by (46.5) and such that the norm of v is equal to $\sup_{x \in X} \| \mathbf{g}(x) \|$.*

COROLLARY 2. *Let E be a separable B-space. To every continuous bilinear form Φ on $L^1 \times E$ there is a bounded, scalarly measurable function $\mathbf{g} : X \to E'$ such that, for all $f \in L^1$ and $\mathbf{e} \in E$,*

$$\Phi(f, \mathbf{e}) = \int \langle \mathbf{g}(x), \mathbf{e} \rangle f(x) \, dx,$$

and such that the norm of Φ is equal to $\sup_{x \in X} \| \mathbf{g}(x) \|$.

46.3. Application to $L^1 \widehat{\otimes}_\pi E$

Let λ be a continuous linear functional on $L^1(E)$ (see 46.1). The restriction of λ to $L^1 \otimes E$ defines a linear form on this vector space, hence (Theorem 39.1(b)) a bilinear form B_λ on $L^1 \times E$, which is immediately seen to be continuous; furthermore, $\| B_\lambda \| \leqslant \| \lambda \|$. Indeed, for all $f \in L^1$ and all $\mathbf{e} \in E$,

(46.6) $$B_\lambda(f, \mathbf{e}) = \lambda(f\mathbf{e}).$$

Observe that B_λ determines λ, since (46.6) determines the values of λ on $L^1 \otimes E$, which is dense in $L^1(E)$ ($p.$ 468).

THEOREM 46.2. *Let E be a Banach space. The canonical injection of $L^1 \otimes E$ into $L^1(E)$ can be extended as a linear isometry of $L^1 \widehat{\otimes}_\pi E$ onto $L^1(E)$.*

Proof 1. *E is separable.* We are going to show that the mapping considered before the statement of Theorem 46.2,

$$\lambda \rightsquigarrow B_\lambda : (f, \mathbf{e}) \rightsquigarrow \lambda(f\mathbf{e}),$$

is an isometry of the dual of $L^1(E)$ onto the Banach space $B(L^1, E)$ of continuous bilinear forms on $L^1 \times E$. This implies that the dual of $L^1 \otimes E$, dense linear subspace of $L^1(E)$ carrying the norm induced by $L^1(E)$, is equal to $B(L^1, E)$. By virtue of Part (c) of Proposition 43.12, we derive that the norm induced on $L^1 \otimes E$ by $L^1(E)$ is equal to the π-norm. This implies immediately the theorem when E is separable.

Let $B \in B(L^1, E)$. By virtue of Corollary 2 of Theorem 46.1, there is a bounded, scalarly measurable function $\mathbf{g} : X \to E'$ such that, for all $\mathbf{e} \in E$ and $f \in L^1$,

$$B(f, \mathbf{e}) = \int \langle \mathbf{g}(x), \mathbf{e} \rangle f(x) \, dx,$$

and such that

$$\| B \| = \sup_{x \in X} \| \mathbf{g}(x) \|.$$

Let us now consider $\mathbf{f} \in \mathscr{L}^1(E)$. To every $\varepsilon > 0$, there is an integrable step-function valued in E, $\mathbf{f}_\varepsilon$, such that

(46.7) $$\int \| \mathbf{f}(x) - \mathbf{f}_\varepsilon(x) \| \, dx < \varepsilon.$$

We have

$$\mathbf{f}_\varepsilon = \sum_{j=1}^{J_\varepsilon} \mathbf{f}_{\varepsilon,j} \chi_{\varepsilon,j}, \qquad J_\varepsilon < +\infty,$$

where the $\mathbf{f}_{\varepsilon,j}$ are elements of E and the $\chi_{\varepsilon,j}$ characteristic functions of integrable subsets of X. Note then that

$$\langle \mathbf{g}(x), \mathbf{f}_\varepsilon(x) \rangle = \sum_j \langle \mathbf{g}(x), \mathbf{f}_{\varepsilon,j} \rangle \chi_{\varepsilon,j}(x)$$

is a function of x which belongs to $\mathscr{L}^1$ since, for all j, the functions $\langle \mathbf{g}(x), \mathbf{f}_{\varepsilon,j} \rangle$ are bounded and measurable. From (46.7) we derive

(46.8) $$\int |\langle \mathbf{g}(x), \mathbf{f}(x) \rangle - \langle \mathbf{g}(x), \mathbf{f}_\varepsilon(x) \rangle| \, dx \leqslant \varepsilon \sup_{x \in X} \| \mathbf{g}(x) \|.$$

Here, the integral may be understood as the upper integral, or as the integral itself if we note that $\langle \mathbf{g}, \mathbf{f} \rangle$, being the limit, almost everywhere, of the measurable functions $\langle \mathbf{g}, \mathbf{f}_\varepsilon \rangle$ as $\varepsilon = 1/n, n = 1, 2, \ldots$, is measurable, and that $| \langle \mathbf{g}(x), \mathbf{f}(x) \rangle | \leqslant (\sup_{x \in X} \| \mathbf{g}(x) \|) \| \mathbf{f}(x) \|$. Either this or (46.8) shows that $\langle \mathbf{g}, f \rangle \in \mathscr{L}^1$. We set

$$\lambda(\mathbf{f}) = \int \langle \mathbf{g}(x), \mathbf{f}(x) \rangle \, dx.$$

We have

$$|\lambda(\mathbf{f})| \leqslant \|B\| \int \|\mathbf{f}(x)\| \, dx.$$

In particular, $\lambda(\mathbf{f}) = \lambda(\mathbf{f}_1)$ if $\mathbf{f} = \mathbf{f}_1$ a.e. Finally, we see that λ can be regarded as a continuous linear form on $L^1(E)$ with norm $\leqslant \|B\|$. But if $f \in L^1$ and $\mathbf{e} \in E$, $\lambda(f\mathbf{e}) = \int \langle \mathbf{g}(x), \mathbf{e} \rangle f(x) \, dx = B(f, \mathbf{e})$, so that $B = B_\lambda$. We have already seen (p. 473) that $\|B_\lambda\| \leqslant \|\lambda\|$, which proves that $\|\lambda\| = \|B\|$, hence our assertion: when E is separable, $\lambda \rightsquigarrow B_\lambda$ is an isometry of $(L^1(E))'$ onto $B(L^1, E)$.

Proof 2.   E arbitrary. If $\theta \in L^1 \otimes E$, there is a finite dimensional subspace M of E such that $\theta \in L^1 \otimes M$. Let us denote momentarily by $\| \ \|_{\pi_E}$ and $\| \ \|_{\pi_M}$ the respective norms in $L^1 \otimes_\pi E$ and $L^1 \otimes_\pi M$. Since the canonical mapping of $L^1 \times M$ into $L^1 \otimes M$ is continuous and has norm $\leqslant 1$ when the latter carries the norm $\| \ \|_{\pi_E}$, this norm is $\leqslant \| \ \|_{\pi_M}$ on $L^1 \otimes M$ by Proposition 43.12(a). But since M is separable, we have, by Part (1),

$$\|\theta\|_{\pi_M} = \int \|\theta(x)\| \, dx = \|\theta\|_{L^1(E)},$$

for we assume, needless to say, that M carries the norm induced by E. As the canonical mapping of $L^1 \times E$ into $L^1(E)$ has norm $\leqslant 1$, we have, again by Proposition 43.12(a),

$$\|\theta\|_{L^1(E)} \leqslant \|\theta\|_{\pi_E}.$$

Combining all these inequalities, we see that all the introduced norms of θ are equal. In particular,

$$\|\theta\|_{L^1(E)} = \|\theta\|_{\pi_E}. \qquad\qquad \text{Q.E.D.}$$

COROLLARY.   *Let E be a Banach space. Then*

$$\lambda \rightsquigarrow ((f, \mathbf{e}) \rightsquigarrow \lambda(f\mathbf{e}))$$

is an isometry of the strong dual of $L^1(E)$ onto $B(L^1, E)$.

Indeed, $B(L^1, E)$ is canonically isomorphic (as a normed space) with the dual of $L^1 \otimes_\pi E \cong L^1(E)$; this defines an isometry of the dual of $L^1(E)$ onto $B(L^1, E)$, which is immediately seen to be the one in the corollary. Note that this corollary extends to nonseparable Banach spaces the result stated and proved in Part (1) of the proof of Theorem 46.2.

Exercises

46.1. Let E be a Banach space, and F a linear subspace of E (equipped with the norm induced by E). Prove: (i) that the norm induced on $L^1 \otimes F$ by $L^1 \otimes_\pi E$ is equal to the π-norm on $L^1 \otimes F$; (ii) that every continuous bilinear form on $L^1 \times F$ can be extended as a continuous bilinear norm on $L^1 \times E$ having the same norm.

46.2. Let E and F be as in Exercise 46.1. Prove that every continuous linear map of F into L^∞ can be extended as a continuous linear map of E into L^∞ having the same norm.

46.3. Let X, Y be two sets, dx, dy two positive measures on X and Y, respectively, u a continuous linear map of L^1_{dx} into L^1_{dy}, and E a Banach space. Prove that there is a unique continuous linear map $\tilde{u} : L^1_{dx}(E) \to L^1_{dy}(E)$ such that $\tilde{u}(fe) = u(f) e$ for all $f \in L^1_{dx}$ and all $e \in E$. Prove also that $\| \tilde{u} \| = \| u \|$.

46.4. Let E be a locally convex Hausdorff space. Prove that the canonical injection of $L^1 \otimes E$ into $L^1(E)$ can be extended as an isomorphism (for the TVS structures) of $L^1 \widehat{\otimes}_\pi E$ onto $L^1(E)$.

46.5. Let X, Y, dx, and dy be as in Exercise 46.3. Prove that for all $\alpha \geqslant 1$ finite we have

$$L^\alpha_{dxdy} = L^\alpha_{dx}(L^\alpha_{dy}).$$

Prove that $L^1_{dxdy} \cong L^1_{dx} \widehat{\otimes}_\pi L^1_{dy}$.

46.6. Suppose that $X = Y = \mathbf{R}^n$ $(n \geqslant 1)$, $dx = dy =$ Lebesgue measure. Prove that

$$(f, g) \rightsquigarrow \int \exp(-2\pi i \langle x, y \rangle) f(x) \, g(y) \, dx \, dy$$

is a continuous bilinear form on $L^2_{dx} \times L^2_{dy}$ and that we have $L^2_{dxdy} \ncong L^2_{dx} \widehat{\otimes}_\pi L^2_{dy}$ (where $\cong$ would denote the canonical isomorphism).

46.7. Let dx be the Lebesgue measure on $X = \mathbf{R}^n$. Let E be a *finite* dimensional space equipped with a Hilbert norm. Prove that the π-norm on $L^2_{dx} \otimes E = L^2_{dx}(E)$ is equivalent to and *strictly* larger than the norm of $L^2(E)$.

47

Nuclear Mappings

We shall use systematically the concept of the space E_B introduced in Chapter 36: E is a locally convex Hausdorff space, B a convex balanced bounded subset of E, and E_B the subspace of E spanned by B, equipped with the norm $x \rightsquigarrow p_B(x) = \inf_{\rho > 0, x \in \rho B} \rho$ (see p. 370). We shall use the following definition:

Definition 47.1. A convex balanced bounded subset B of E is said to be infracomplete if the normed space E_B is a Banach space.

We have seen (Lemma 36.1 and corollary) that, if B itself is complete, in particular if B is compact, then B is infracomplete. But the converse is not necessarily true.

An important particular case is provided by the spaces $E'_{H'}$, where E' is the dual of a locally convex Hausdorff space E and H' is a weakly closed convex balanced equicontinuous subset of E' (hence H' is weakly compact). There is a natural interpretation of $E'_{H'}$ which should be kept in mind: let H'^0 be the polar of H' in E; H'^0 is a convex balanced closed neighborhood of zero in E (for the initial topology of E) which we denote by U, for simplicity. Let us then call E_U the space E equipped with the topology where a basis of neighborhoods of zero is formed by the multiples ρU, $\rho > 0$, of U. We see immediately that $E'_{H'}$ is nothing else but the space of linear forms on E which are continuous on E_U. In general, E_U is not Hausdorff; the associated Hausdorff space is obviously normed; its completion is a Banach space which we denote by $\hat{E}_U$. It should also be noted that every convex balanced closed neighborhood of zero of E is the polar of a weakly closed convex balanced equicontinuous subset of E': its own polar. Let p be the seminorm on E associated with U,

$$p(x) = \inf_{x \in \rho U, \rho > 0} \rho.$$

The Hausdorff space associated with E_U is nothing else but $E_U / \mathrm{Ker}\, p$, equipped with the quotient topology. The forms $x' \in E'_{U^0}$ (with the previous notation, $U^0 = H'$) are continuous on E for the topology defined by p, hence define continuous linear forms on $E_U/\text{Ker p}$, and by continuous extension, on $\hat{E}_U$. It is clear that E'_{U^0} can be identified as a Banach space (i.e., the identification extends to the norms) with the dual of $\hat{E}_U$. We have a canonical mapping of E into $\hat{E}_U$, the compose of the sequence

$$E \xrightarrow{h_U} E_U/\text{Ker p} \xrightarrow{j_U} \hat{E}_U,$$

where h_U is the canonical mapping and j_U the natural injection of a space into its completion. Note that, in general, h_U is not open, since $E_U/\text{Ker p}$ does not carry the quotient modulo Ker p of the topology of E but the quotient of the topology of E_U.

We consider two locally convex Hausdorff spaces E, F and the tensor product $E' \otimes F$ of the dual of E with F, regarded as a linear subspace of $L(E; F)$, space of continuous linear maps $E \to F$, namely the subspace of these maps whose image is finite dimensional. An element $\sum_j x'_j \otimes y_j$ (finite sum) of $E' \otimes F$ defines the mapping

$$x \rightsquigarrow \sum_j \langle x'_j, x \rangle y_j.$$

Let us consider momentarily the case where both E and F are Banach spaces. As usual, we denote by $L_b(E; F)$ the space $L(E; F)$ equipped with the topology of bounded convergence, i.e., with the topology defined by the operators norm,

$$\| u \| = \sup_{x \in E, \| x \| = 1} \| u(x) \|.$$

We observe that the canonical bilinear map of $E' \times F$ into $L_b(E; F)$,

$$(x', y) \rightsquigarrow (x \rightsquigarrow \langle x', x \rangle y),$$

is obviously continuous and has norm $\leqslant 1$. We therefore derive from Proposition 43.12(a), that the norm induced by $L_b(E; F)$ on $E' \otimes F$ is $\leqslant \| \quad \|_\pi$. The injection of $E' \otimes F$ into $L_b(E; F)$ can then be extended as a continuous linear map of $E' \hat{\otimes}_\pi F$ into $L_b(E; F)$, which we shall call canonical. It is not known if this canonical mapping of $E' \hat{\otimes}_\pi F$ into $L_b(E; F)$ is always injective, although it has been shown to be so in all the cases studied so far.

Definition 47.2. Let E, F be two Banach spaces. The image of $E' \hat{\otimes}_\pi F$ into $L(E; F)$ is denoted by $L^1(E; F)$. Its elements are called the nuclear mappings of E into F.

$L^1(E;F)$ is isomorphic, as a vector space, with $E' \otimes_\pi F/N$, where N is the kernel of the canonical mapping $E' \otimes_\pi F \to L_b(E;F)$; the above quotient is a Banach space (as the quotient of a B-space modulo a closed linear subspace). The norm on $L^1(E;F)$ transferred from $E' \otimes_\pi F/N$ is called the *trace-norm* (the motivation for this name will be seen later). Needless to say, the trace-norm restricted to $E' \otimes F$ is nothing else but the π-norm.

We shall now give the definition of a nuclear operator in the general case of two locally convex Hausdorff spaces E, F, not necessarily Banach spaces.

Let U be a convex balanced closed neighborhood of zero in E, and B a convex balanced infracomplete bounded subset of F. Let $u : \hat{E}_U \to F_B$ be a continuous linear map. We may define a map $\tilde{u} : E \to F$ by composing the sequence

$$E \xrightarrow{h_U} \hat{E}_U \xrightarrow{u} F_B \xrightarrow{i_B} F,$$

where h_U is the canonical mapping $E \to \hat{E}_U$ and i_B the natural injection. This correspondence $u \rightsquigarrow \tilde{u}$ yields an injection of $L(\hat{E}_U ; F_B)$ into $L(E;F)$: for if $\tilde{u} = 0$, it means that u vanishes on $h_U(E)$, which is dense in $\hat{E}_U$, hence $u = 0$. In the forthcoming, we consider $L(\hat{E}_U ; F_B)$ as a linear subspace of $L(E;F)$.

Definition 47.3. Let E, F be two locally convex Hausdorff spaces, $\mathfrak{U}$ the family of all convex balanced closed neighborhoods of zero in E, and $\mathfrak{B}$ the family of all convex balanced infracomplete bounded subsets of F.

The union, when U ranges over $\mathfrak{U}$ and B over $\mathfrak{B}$, of the subspaces

$$L^1(\hat{E}_U; F_B) \subset L(\hat{E}_U; F_B) \subset L(E;F),$$

is denoted by $L^1(E;F)$; its elements are called the nuclear mappings of E into F.

Suppose that E and F are Banach spaces. Then every $\hat{E}_U$ is canonically isomorphic (as a TVS, not as a Banach space) with E. On the other hand, $L^1(E;F_B)$ is canonically injected into $L^1(E;F)$ by composing the mappings with the injection $F_B \to F$. This shows that, when E and F are Banach spaces, Definition 47.3 coincides with Definition 47.2, as it should. It is not immediately evident that, in the general case (when E and F are not necessarily Banach spaces), $L^1(E;F)$ is a vector space. This will follow trivially from the next propositions, whose importance, however, goes beyond that consequence.

PROPOSITION 47.1. *Let E, F be two locally convex Hausdorff spaces, and u a nuclear mapping of E into F. Let G, H be two other locally convex*

Hausdorff spaces, and $g : G \to E, h : F \to H$ *two continuous linear mappings. Then* $h \circ u \circ g$ *is nuclear.*

Suppose that $E, F, G,$ *and* H *are Banach spaces. Then:*

$$\| h \circ u \circ g \|_{\mathrm{Tr}} \leqslant \| h \| \, \| u \|_{\mathrm{Tr}} \| g \|.$$

We have denoted by $\| \quad \|_{\mathrm{Tr}}$ the trace norm.

Proof. By hypothesis, we have a decomposition of u as a sequence

$$E \xrightarrow{\ h_U\ } \hat{E}_U \xrightarrow{\ \tilde{u}\ } F_B \xrightarrow{\ i_B\ } F,$$

where h_U and i_B are the canonical mappings and $\tilde{u}$ is nuclear. Let us denote by V the preimage of U in G under g, and by $\mathfrak{q}$ the associated seminorm, that is to say the seminorm

$$G \ni z \rightsquigarrow \| h_U(g(z)) \|,$$

where $\| \quad \|$ denotes the norm in $\hat{E}_U$. It is obvious that, in the commutative triangle

$$
\begin{array}{ccc}
G & \xrightarrow{\ g\ } E \xrightarrow{\ h_U\ } & \hat{E}_U \\
{\scriptstyle k_V} \downarrow & \nearrow {\scriptstyle \tilde{g}} & \\
G_V/\mathrm{Ker}\ \mathfrak{q} & &
\end{array}
$$

where k_V is the canonical mapping, $\tilde{g}$ is an isometry onto $h_U(g(G))$ and might therefore be extended as an isometry I of $\hat{G}_V$ onto the closure M of $h_U(g(G))$. On the other hand, $C = h(B)$ is convex balanced bounded in H. Furthermore, C is infracomplete: indeed, the restriction $h \mid F_B$ of h to F_B induces an isometry J of the Banach space $F_B/(F_B \cap \mathrm{Ker}\ h)$ onto H_C, as immediately seen; an isometric copy of a Banach space is a Banach space. Finally, we have decomposed $h \circ u \circ g$ into the sequence

$$G \xrightarrow{\ k_V\ } \hat{G}_V \xrightarrow{\ I\ } M \xrightarrow{\ \tilde{u}|M\ } F_B \xrightarrow{\ J\ } H_C \xrightarrow{\ j_C\ } H.$$

The notation is evident: k_V and j_C are the canonical mappings, I and J are the isometries defined above, $\tilde{u} \mid M$ is the restriction of $\tilde{u}$ to M. It remains to show that $J \circ (\tilde{u} \mid M) \circ I$ is nuclear. Thus we are reduced to the case where all the spaces involved are Banach spaces and where g and h are isometries onto.

We have the commutative diagram

$$(47.1) \qquad \begin{array}{ccc} E' \widehat{\otimes}_\pi F & \to L^1(E;F) & \to L(E;F) \\ {}^t g \otimes h \downarrow & \downarrow & \downarrow (*) \\ G' \widehat{\otimes}_\pi H & \to L^1(G;H) & \to L(G;H) \end{array}$$

where the horizontal arrows are the canonical mappings and (*) is the mapping $v \rightsquigarrow h \circ v \circ g$, which is an isometry onto. The vertical arrow at the center must obviously be an isometry onto; the image of $u \in L^1(E;F)$ under this isometry is of course $h \circ u \circ g$, which belongs to $L^1(G;H)$.

Still in the case where all the spaces are Banach spaces, we must prove the statement about the norms. For this we go back to Diagram (47.1), where we do not assume any more that g and h are isometries onto; then the vertical arrows are simply continuous linear mappings. We know (Proposition 43.13) that $\| {}^t g \otimes h \| \leqslant \| {}^t g \| \, \| h \|$; of course, $\| {}^t g \| = \| g \|$. If $\theta \in E' \widehat{\otimes}_\pi F$ defines a nuclear mapping u, then $h \circ u \circ g$ is defined by $({}^t g \otimes h)(\theta)$ and we have therefore

$$\| h \circ u \circ g \|_{\mathrm{Tr}} \leqslant \|({}^t g \otimes h)(\theta)\|_\pi \leqslant \| g \| \, \| h \| \, \| \theta \|_\pi .$$

By taking $\| \theta \|_\pi$ arbitrarily close to $\| u \|_{\mathrm{Tr}}$, we obtained the desired inequality.

The next result makes it easier to prove that certain operators are nuclear.

PROPOSITION 47.2. *Let E, F be two locally convex Hausdorff spaces, and $u : E \to F$ a continuous linear map. The following conditions are equivalent:*

(a) *u is nuclear;*

(b) *u is the compose of a sequence of continuous linear mappings*

$$E \xrightarrow{\;f\;} E_1 \xrightarrow{\;v\;} F_1 \xrightarrow{\;g\;} F,$$

where E_1 and F_1 are Banach spaces and v is nuclear;

(c) *there is an equicontinuous sequence $\{x'_k\}$ in E', a sequence $\{y_k\}$ contained in a convex balanced infracomplete bounded subset B of F, and a complex sequence $\{\lambda_k\}$ with $\sum_k |\lambda_k| < +\infty$ such that u is equal to the mapping*

$$(47.2) \qquad\qquad x \rightsquigarrow \sum_k \lambda_k \langle x'_k , x \rangle y_k .$$

If E and F are Banach spaces, u is nuclear if and only if there is a sequence in the closed unit ball of E', $\{x'_k\}$, a sequence in the closed unit ball of F, $\{y_k\}$, and a complex sequence $\{\lambda_k\}$ with $\sum_k |\lambda_k| < +\infty$ such that u is given by (47.2). Furthermore, the trace-norm $\|u\|_{\mathrm{Tr}}$ of u is equal to the infimum of the numbers $\sum_k |\lambda_k|$ over the set of all representations of u of the type (47.2).

Proof. Trivially, (a) $\Rightarrow$ (b); (b) $\Rightarrow$ (a) in view of Proposition 47.1. On the other hand, because of Theorem 45.1, (a) $\Rightarrow$ (c). Conversely, $\sum_k \lambda_k x'_k \otimes y_k$ converges absolutely in $E'_{H'} \widehat{\otimes}_\pi F_B$, where H' is the convex balanced weakly closed hull of the sequence $\{x'_k\}$: H' is an equicontinuous subset of E' and thus (c) $\Rightarrow$ (a). If E is a Banach space, the closed unit ball of E' is an equicontinuous subset of E', obviously convex balanced and weakly closed; if F is a Banach space, its closed unit ball is infracomplete. Finally, the statement about the trace-norm is a straightforward consequence of the definition.

COROLLARY 1. *Let E be barreled and F quasi-complete; u is nuclear if and only if u has a representation (47.2) with the sequence $\{x'_k\}$ bounded in E' and the sequence $\{y_k\}$ bounded in F.*

Proof. As E is barreled, a bounded sequence in E' is equicontinuous (Theorem 33.2). As F is quasi-complete, the closed convex balanced hull of a bounded subset, here the sequence $\{y_k\}$, is complete, a fortiori infracomplete (Lemma 36.1).

COROLLARY 2. *Let E, F be locally convex Hausdorff spaces; $L^1(E; F)$ is a linear subspace of $L(E; F)$.*

Proof. Let u_1, $u_2 : E \to F$ be nuclear. We have a diagram

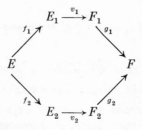

where all mappings are linear and continuous, where E_i, F_i ($i = 1, 2$) are B-spaces and where v_1, v_2 are nuclear; furthermore, the upper path gives u_1, the lower one u_2 (the diagram is not commutative!). Let G (resp. H) be the Banach space product $E_1 \times E_2$ (resp. $F_1 \times F_2$),

f the mapping $x \rightsquigarrow (f_1(x), f_2(x))$, and g the mapping $(y_1, y_2) \rightsquigarrow g_1(y_1) + g_2(y_2)$. The sequence

$$E \xrightarrow{\ f\ } G \xrightarrow{\ v\ } H \xrightarrow{\ g\ } F,$$

where $v = (v_1, v_2) : (x_1, x_2) \rightsquigarrow (v_1(x_1), v_2(x_2))$ consists of continuous linear mappings and constitutes a decomposition of $u = u_1 + u_2$. Then u is nuclear if v is nuclear. Suppose that v_i is defined by an element $\theta_i \in E_i' \widehat{\otimes}_\pi F_i \, (i = 1, 2)$. There is a canonical mapping of $(E_1' \otimes F_1) \times (E_2' \otimes F_2)$ into $(E_1' \times E_2') \otimes (F_1 \times F_2)$, $((\sum_j x_{1j}' \otimes y_{1j}), (\sum_k x_{2k}' \otimes y_{2k})) \rightsquigarrow \sum_{j,k} (x_{1j}', x_{2k}') \otimes (y_{1j}, y_{2k})$, which is immediately seen to be continuous when all the tensor products carry the projective topologies; this canonical mapping extends to the completions. Then it is easily seen that the image of (θ_1, θ_2) under this extension, an element $\theta \in (E_1' \times E_2') \widehat{\otimes}_\pi (F_1 \times F_2)$, defines v, which is therefore nuclear.

We introduce the following definition, familiar in Hilbert (or Banach) space theory:

Definition 47.4.   Let E, F be two locally convex Hausdorff spaces. A linear map $u : E \to F$ is called compact (or completely continuous) if there is a neighborhood U of 0 in E such that $u(U) \subset F$ is precompact.

PROPOSITION 47.3.   *Any nuclear map is compact.*

Proof. It suffices to go back to the proof of the implication (a) $\Rightarrow$ (c) in Proposition 47.2 and to observe that full use of Theorem 45.1 allows us to take the sequence $\{y_k\}$ converging to 0 in F_B, a fortiori in F. The closed convex hull Γ of this sequence is precompact (Proposition 7.11). Let U be the polar of the equicontinuous sequence $\{x_k'\}$: U is a neighborhood of zero in E, and if $x \in U$,

$$u(x) = \sum_k \lambda_k \langle x_k', x \rangle y_k \in \Gamma. \qquad \text{Q.E.D.}$$

We now study the transpose of a nuclear map.

PROPOSITION 47.4.   *Let E, F be two locally convex Hausdorff spaces, and u a nuclear map of E into F.*

Then ${}^t u : F' \to E'$ is nuclear when E' and F' carry their strong dual topology.

In the case of Banach spaces, some precision can be added to the preceding statement concerning the trace-norms:

PROPOSITION 47.5. *If E, F are Banach spaces and if $u \in L^1(E; F)$, then $^t u \in L^1(F'; E')$ and $\| \, ^t u \, \|_{\mathrm{Tr}} \leqslant \| u \|_{\mathrm{Tr}}$.*

Proof of Proposition 47.5. Let $u \in L(E; F)$ be defined by some element $\theta \in E' \widehat{\otimes}_\pi F$; suppose that θ is equal to an absolutely convergent series

$$\theta = \sum_{j=0}^{\infty} x'_j \otimes y_j \, .$$

Then $^t u$ is equal to the mapping

$$y' \rightsquigarrow \sum_{j=0}^{\infty} \langle y', y_j \rangle \, x'_j \, .$$

In other words, $^t u$ is defined by the element $(I \, \widehat{\otimes}_\pi \, i)(\theta) \in E' \, \widehat{\otimes}_\pi F''$, where I is the identity of E and i the canonical isometry of F into F'', $x \rightsquigarrow$ value at x. By Proposition 43.13, we know that

$$\| (I \, \widehat{\otimes}_\pi \, i)(\theta) \|_\pi \leqslant \| \, \theta \, \|_\pi \, .$$

By taking the infimum of both sides as θ varies over the set of representatives of u, we see that $\| \, ^t u \, \|_{\mathrm{Tr}} \leqslant \| u \|_{\mathrm{Tr}}$.

COROLLARY. *If E and F are Banach spaces and if F is reflexive, then u is nuclear if and only if $^t u$ is nuclear. Moreover, $\| u \|_{\mathrm{Tr}} = \| \, ^t u \, \|_{\mathrm{Tr}}$.*

If $^t u : F' \underset{^{t} t_u}{\rightarrow} E'$ is nuclear, then $^{tt} u : E'' \rightarrow F''$ is nuclear and so is $u : E \rightarrow E'' \overset{^t}{\rightarrow} F'' \cong F$, where the first arrow is the canonical isometry. Furthermore we have $\| u \|_{\mathrm{Tr}} \leqslant \| \, ^{tt} u \, \|_{\mathrm{Tr}} \leqslant \| \, ^t u \, \|_{\mathrm{Tr}}$ because of Proposition 47.5, which implies at once $\| u \|_{\mathrm{Tr}} = \| \, ^t u \, \|_{\mathrm{Tr}}$.

Remark 47.1. The conclusion of the previous corollary is valid in all the cases which are known, even when F is not reflexive.

Proof of Proposition 47.4. Here E and F are locally convex Hausdorff spaces, not necessarily Banach spaces. If $u : E \rightarrow F$ is nuclear, there exist two Banach spaces E_1, F_1, two continuous linear maps $f : E \rightarrow E_1$, $g : F \rightarrow F_1$, and a nuclear map $v : E_1 \rightarrow F_1$ such that $u = g \circ v \circ f$ (Proposition 47.2). Then $^t u = {}^t f \circ ({}^t v) \circ ({}^t g)$ is also nuclear, in view of Proposition 47.5.

We conclude these generalities about nuclear mappings by two results on extension and lifting of nuclear mappings:

PROPOSITION 47.6. *Let E, F, and G be three locally convex Hausdorff spaces, and j an isomorphism of E into F.*

(1) *Given any nuclear map $u : E \to G$ there is a nuclear map $v : F \to G$ such that $v \circ j = u$.*

(2) *Suppose that $j(E) \subset F$ is closed and let ϕ be the canonical map of F onto $F/j(E)$. Suppose moreover that every convex balanced compact subset of $F/j(E)$ is the image, under ϕ, of a convex balanced infracomplete bounded subset of F. Then, to every nuclear map $u : G \to F/j(E)$ there is a nuclear map $v : G \to F$ such that $\phi \circ v = u$.*

When E and F are Banach spaces and j is an isometry, for every $\varepsilon > 0$, v can be chosen (either in (1) or in (2)) *so as to have*

$$\| v \|_{\mathrm{Tr}} \leqslant \| u \|_{\mathrm{Tr}} + \varepsilon.$$

Proof of (1).   The map u is the canonical image of an element

$$\theta = \sum_k \lambda_k x_k' \otimes z_k ,$$

where the $\{x_k'\}$ form an equicontinuous sequence in E' and the sequence $\{z_k\}$ is a sequence contained in some infracomplete bounded subset of G; as usual the sequence (λ_k) belongs to l^1 (Proposition 47.2). By applying the Hahn–Banach theorem, we may lift the sequence $\{x_k'\} \subset E'$ into an equicontinuous sequence $\{y_k'\} \subset F'$; when E and F are Banach spaces (and j an isometry), the y_k' can be taken so as to have, for each k, the same norm as x_k' ; then v is defined by the element

$$\tilde{\theta} = \sum_k \lambda_k y_k' \otimes z_k .$$

When E and F are Banach spaces (and j is an isometry), we may take the x_k' (resp. the y_k' , resp. the z_k) in the unit ball of E' (resp. of F', resp. of G); then

$$\| v \|_{\mathrm{Tr}} \leqslant \sum_k | \lambda_k |,$$

and the right-hand side can be taken $\leqslant \| u \|_{\mathrm{Tr}} + \varepsilon$ (Proposition 47.2).
    The proof of (2) is similar and will be left to the student.

Remark 47.2.   If E and F are Fréchet spaces, every compact convex balanced subset of $F/j(E)$ is the canonical image of a compact convex balanced subset of F (Lemma 45.1). This is also true if F is the strong dual of a Fréchet space and if $j(E)$ is weakly closed in F.
    We conclude these generalities by a few words about the so-called *trace form*. This is a continuous linear functional on $E' \widehat{\otimes}_\pi E$, where E is a Banach space and E' the Banach space which is the strong dual

of E (with the dual norm). We observe that $(x, x') \rightsquigarrow \langle x', x \rangle$ is a continuous bilinear functional, with norm 1, on $E \times E'$; therefore, it corresponds to a continuous linear form, denoted by $\mathrm{Tr}(\cdot)$ and called trace form, on $E' \widehat{\otimes}_\pi E$; the norm of Tr is one. If we have

$$\theta = \sum_{k=0}^{\infty} x'_k \otimes x'_k, \qquad \text{with} \quad \sum_{k=0}^{\infty} \| x'_k \| \| x'_k \| < +\infty,$$

then we have

$$\mathrm{Tr}(\theta) = \sum_{k=0}^{\infty} \langle x'_k, x_k \rangle.$$

The motivation for the name trace form originates in the fact that, if $\theta \in E' \otimes E$, we may write

$$\theta = \sum_{i,j=1}^{r} \theta_{ij} \, e'_i \otimes e_j,$$

where $\langle e'_i, e_j \rangle = 1$ if $i = j$ and 0 otherwise, having then

$$\mathrm{Tr}(\theta) = \sum_{i=1}^{r} \theta_{ii},$$

which is the trace, in the usual sense, of the linear mapping of E into itself defined by θ.

Example. Nuclear Mappings of a Banach Space into a Space L^1

Let X be a set, and dx a positive measure on X; we assume σ-finiteness and we denote by L^1 the corresponding space of (classes of) integrable functions. Let E be a Banach space, and E' the dual Banach space. We have a canonical isometry

$$L^1(E') \cong E' \widehat{\otimes}_\pi L^1 \qquad \text{(Theorem 46.2)}.$$

By using the canonical mapping of $E' \widehat{\otimes}_\pi L^1$ onto the space of nuclear mappings of E into L^1, we see that every class of integrable functions $\mathbf{f} \in L^1(E')$ defines a nuclear map, namely

$$u_{\mathbf{f}} : \mathbf{e} \rightsquigarrow (x \rightsquigarrow \langle \mathbf{f}(x), \mathbf{e} \rangle),$$

and that $\mathbf{f} \rightsquigarrow u_{\mathbf{f}}$ is onto. Furthermore, this mapping is injective, for we have

$$\int |\langle \mathbf{f}(x), \mathbf{e} \rangle| \, dx = 0 \qquad \text{for all} \quad \mathbf{e} \in E,$$

if and only if the representatives of $\mathbf{f}$ vanish almost everywhere. But if the mapping $\mathbf{f} \to u_{\mathbf{f}}$ is injective, it is an isometry onto when we consider the trace-norm of nuclear operators; indeed, the canonical mapping of $E' \mathbin{\widehat{\otimes}_\pi} L^1$ onto $L^1(E; L^1)$ is then injective and the trace-norm is the quotient of the π-norm modulo the kernel of the canonical mapping. Thus

$$\| u_{\mathbf{f}} \|_{\mathrm{Tr}} = \int \| \mathbf{f}(x) \| \, dx.$$

The student may apply these considerations to the case where E itself is a space L^p $(1 \leqslant p < +\infty)$.

48

Nuclear Operators in Hilbert Spaces

Let E, F be two Hilbert spaces, u a continuous linear map of E into F, and J (resp. K) the canonical antilinear isometry of E onto its dual E' (resp. of F onto F'). We call *adjoint* of u and denote by u^* the compose of the sequence of mappings

$$F \xrightarrow{\ K\ } F' \xrightarrow{\ {}^t u\ } E' \xrightarrow{\ J^{-1}\ } E;$$

u^* is a continuous linear map of F into E; it has a norm equal to the one of u (Proposition 23.3). If we denote by $(\ |\ )$ and $\|\ \ \|$ the inner product and the norm, both in E and F, we have, for all $x \in E$, $y \in F$,

$$(u(x) \mid y) = (x \mid u^*(y)).$$

When $E = F$ and $u = u^*$, the operator u is said to be *self-adjoint*. A mapping $u \in L(E; E)$ is called *positive* if $(u(x) \mid x) \geqslant 0$ for all $x \in E$. A positive operator is self-adjoint: indeed, the bilinear form $(u(x) \mid y)$ is real when $x = y$ (see p. 113). An important and well-known property of positive operators is the one stated now:

LEMMA 48.1. *Let $u \in L(E; E)$ be positive. There is a unique positive map $v \in L(E; E)$ such that $v^2 = u$.*

Proof. By examining the coefficients of the Taylor expansion of the function

$$\mathbf{C}^n \ni z \rightsquigarrow (1 - z)^{1/2}$$

about $z = 0$, one sees that the Taylor series converges when $\mid z \mid = 1$. From this it follows immediately that the finite Taylor series, when z has been replaced by $w \in L(E; E)$ with $\| w \| \leqslant 1$, converge in $L_b(E; E)$, to a continuous linear map, which it is natural to denote by $(I - w)^{1/2}$. Observe that the latter commutes with any continuous linear map which commutes with w, as it is a limit (in the sense of the operators norm) of polynomials with respect to w. For the same reason, $(I - w)^{1/2}$ is self-adjoint whenever this is true of w. And of course $((I - w)^{1/2})^2 = I - w$. From this we derive that, if w is positive, so is $w(I - w) = ((I - w)^{1/2}) \, w((I - w)^{1/2})$. We see therefore that

$$\| (I - w)x \|^2 = ((I - w)x \mid x) - (w(I - w)x \mid x) \leqslant \|(I - w)x \| \| x \|,$$

which proves that, in the case where w is positive and $\| w \| \leqslant 1$, we have $\| I - w \| \leqslant 1$. But then we may consider

$$(I - (I - w))^{1/2},$$

which we shall denote precisely by $\sqrt{w}$. Now, if $\| u \|$ is arbitrary, we set $\sqrt{u} = \| u \|^{1/2} (\| u \|^{-1}u)^{1/2}$. This proves the existence of the mapping v (now denoted $\sqrt{u}$) of Lemma 48.1.

Suppose there was a second positive map $w \in L^2(E; E)$ such that $w^2 = u$. Then w commutes with u and therefore with $\sqrt{u}$, in view of the earlier considerations. This implies that we have

$$0 = w^2 - (\sqrt{u})^2 = (w - \sqrt{u})(w + \sqrt{u}).$$

We apply the right-hand side to a vector $x \in E$ and conclude that $w = \sqrt{u}$ on the image of $(w + \sqrt{u})$; the orthogonal of this image is the kernel of $(w + \sqrt{u})$ (as this operator is self-adjoint), which is the intersection of Ker w with Ker $\sqrt{u}$, as w and $\sqrt{u}$ are positive; and on this intersection, we have trivially $w = \sqrt{u}$.   Q.E.D.

COROLLARY. *Let u be a positive operator. If $x \in E$ satisfies the equation $(u(x) \mid x) = 0$, we have $u(x) = 0$.*

Proof. Let v be an operator such that $v^*v = u$; then $(u(x) \mid x) = \| v(x) \|^2$ and $v(x) = 0$ implies $u(x) = v^*(v(x)) = 0$.

The unique positive operator v of Lemma 48.1 is often denoted by $\sqrt{u}$ and called the *positive square root* of u.

Let us now consider an arbitrary continuous linear map of a Hilbert space E into another Hilbert space F; let u^* be the adjoint of u. Then $u^* u$ is a positive operator of E into E; let us denote by R its positive square root. We have $\| R(x) \| = \| u(x) \|$ for all $x \in E$ and therefore Ker $R = $ Ker u. Let us then define the following continuous linear map $U : E \to F$,

$$U(x) = u(x_1) \qquad \text{if} \quad x = R(x_1) \in \text{Im } R;$$

then U is extended by continuity to the closure of $\overline{\text{Im } R}$;

$$U(x) = 0 \qquad \text{if} \quad x \in \text{Ker } R.$$

We have, for all $x \in E$,

$$\| U(R(x)) \| = \| u(x) \| = \| R(x) \|.$$

This means that U is an isometry of Im R onto Im u. We have

$$u = U \circ R.$$

One refers often to R as the *absolute value* of u.

Our purpose is to study the operators $u : E \to F$ which are nuclear and to characterize them. But, because of the isometric properties of the operator U above, it is clear that u will be nuclear if and only if its absolute value R is nuclear. In other words, it suffices to study the nuclear operators which are positive mappings of a Hilbert space E into itself. But we know that nuclear mappings are compact: we may therefore restrict our attention to compact positive operators. These have simple and beautiful spectral properties, discovered at the beginning of this century by Fredholm and F. Riesz.

For the benefit of the student who does not have a treatise on Hilbert space theory within reach, we recall the statement and the proof of the main theorem on compact positive operators.

THEOREM 48.1. *Let E be a Hilbert space, and u a positive compact operator of E into itself. There is a sequence of positive numbers, decreasing and either finite or converging to zero,*

$$\lambda_1 > \lambda_2 > \cdots > \lambda_k > \cdots,$$

and a sequence of nonzero finite dimensional subspaces V_k of E ($k = 1, 2,...$) with the following properties:

   (1)   *the subspaces V_k are pairwise orthogonal;*
   (2)   *for each k and all $x \in V_k$, $u(x) = \lambda_k x$;*
   (3)   *the orthogonal of the subspace spanned by the union of the V_k is equal to the kernel of u.*

Proof. Let t be the supremum, on the unit sphere of E, of the non-negative function $(u(x) \mid x)$. We use the fact that the closed unit ball of E is weakly compact (Proposition 34.1); there is a weakly converging sequence $\{x_\nu\}, \| x_\nu \| = 1$, such that $(u(x_\nu) \mid x_\nu) \to t$. But as u itself is a compact operator, we may suppose that the sequence $\{u(x_\nu)\}$ converges in the sense of the operators norm. This implies at once that, if x is the (weak) limit of the x_ν, $(u(x) \mid x)$ is equal to the limit of the $(u(x_\nu) \mid x_\nu)$, i.e., to t. Thus we have

$$((tI - u)(x) \mid x) = 0.$$

But, because of our choice of t, $tI - u$ is positive. Now if v is positive and if $(v(x) \mid x) = 0$, we have $v(x) = 0$. Indeed, $(v(x) \mid x) = \| \sqrt{v}(x) \|^2$ and $\sqrt{v}(x) = 0$ implies $(\sqrt{v})^2(x) = 0$. Thus we have

$$u(x) = tx.$$

Let us denote by V_1 the linear subspace of elements $x \in E$ such that $u(x) = tx$. Let B_1 be the unit ball of V_1, i.e., the intersection of the unit ball of E with V_1. By hypothesis, $u(B_1)$ is precompact; but it is equal to tB_1, hence is closed and B_1 must be compact. Thus V_1 is finite dimensional. From now on we write λ_1 instead of t. If $V_1^\perp$ is the orthogonal of V_1, we have $u(V_1^\perp) \subset V_1^\perp$ as u is selfadjoint. Thus, by restriction, u defines a continuous linear map of $V_1^\perp$ into itself which is clearly compact and positive. We may repeat the above procedure with E replaced by $V_1^\perp$. The maximum λ_2 of the function $(u(x) \mid x)$ on the unit sphere of $V_1^\perp$ is $< \lambda_1$, for otherwise there would be an element x in this unit sphere such that $u(x) = \lambda_1 x$ and x would belong to V_1. We then build the sequences $\{\lambda_k\}$ and $\{V_k\}$ by induction on k, taking $V_k = \{x \in E;$ $u(x) = \lambda_k x\}$. If the procedure comes to a halt after a finite number of steps, say k steps, it means that the maximum of the function $(u(x) \mid x)$ on the orthogonal of $V_1 + \cdots + V_k$ is equal to zero; this orthogonal must then be the kernel of u (note that in this case the image of u is finite dimensional). If the procedure does not stop after a finite number of steps, the decreasing sequence $\{\lambda_k\}$ must converge to zero. Otherwise we would be able to find an orthonormal sequence of vectors x_ν such that $u(x_\nu) = \lambda_{k_\nu} x_\nu$ with $\lambda_{k_\nu} \geqslant c > 0$. But as u is compact, the sequence $\{u(x_\nu)\}$ should contain a converging subsequence, which is absurd as

$$\| u(x_\nu) - u(x_\mu) \|^2 = \lambda_{k_\nu}^2 + \lambda_{k_\mu}^2 \geqslant 2c.$$

Finally, if an element $x \in E$ is orthogonal to all the V_k, we must have, in view of their definition, $(u(x) \mid x) = 0$, hence $u(x) = 0$.   Q.E.D.

We recall the well-known terminology: the numbers λ_k are the *eigenvalues* of u; V_k is the *eigenspace* of u corresponding to the eigenvalue λ_k; dim V_k is sometimes called the *multiplicity* of the eigenvalue λ_k. The sum of the series

$$\sum_{k=1}^{\infty} \lambda_k \dim V_k$$

is called the *trace* of u and denoted by Tr u. For each k, let P_k be the orthogonal projection of E onto V_k (see p. 120). The finite sums

$$\sum_{k=1}^{K} \lambda_k P_k$$

converge to u for the operators norm as $K \to \infty$. We may write

$$(48.1) \qquad\qquad u = \sum_{k=1}^{\infty} \lambda_k P_k \, .$$

This representation of u is called the *spectral decomposition* of u. In the preceding notation, which makes use of infinite series, it is to be understood that, if the image of u is finite dimensional, the series in question are finite.

We may now state and prove the main theorem on the subject of nuclear operators in Hilbert spaces:

THEOREM 48.2. *Let E, F be two Hilbert spaces, $u : E \to F$ a continuous linear map, and R its absolute value.*

The following properties are equivalent:

(a) *u is nuclear;*
(b) *R is nuclear;*
(c) *R is compact and $\mathrm{Tr}\, R$ is finite.*

If u is nuclear, $\mathrm{Tr}\, R$ is equal to $\| u \|_{\mathrm{Tr}}$.

Proof. As we have already pointed out, we have $u = U \circ R$, where U is an isometry of $R(E)$ onto $u(E) \subset F$, and it suffices to prove the equivalence of (b) and (c) and the fact that, if R is nuclear, $\mathrm{Tr}\, R$ is equal to $\| R \|_{\mathrm{Tr}}$. In other words, we may as well suppose that $u \in L(E; E)$ and that u is positive.

Suppose that u is nuclear; then u is compact (Proposition 47.3) and has therefore a spectral decomposition

$$u = \sum_{k=1}^{\infty} \lambda_k P_k \, .$$

Let us consider the finite sums

$$u_K = \sum_{k=1}^{K} \lambda_k P_k;$$

they converge in norm to u. If we set $Q_K = P_1 + \cdots + P_K$, we see that we have $u_K = Q_K u$. Then (Proposition 47.1)

$$\| u_K \|_{\mathrm{Tr}} \leqslant \| Q_K \| \, \| u \|_{\mathrm{Tr}} = \| u \|_{\mathrm{Tr}} \, .$$

On the other hand, the trace form (see p. 485) is a continuous linear form on $E' \otimes_\pi E$ of norm one. Using the fact that, if a map $u : E \to E$ is defined by an element $\theta \in E' \otimes E$, we have $\| u \|_{\mathrm{Tr}} = \| \theta \|_\pi$, we have

$$| \mathrm{Tr}(u_K)| \leqslant \| u_K \|_{\mathrm{Tr}} \leqslant \| u \|_{\mathrm{Tr}} .$$

But as we have pointed out, when dealing with a linear map defined by an element $\theta \in E' \otimes E$, the trace form is equal to the trace in the usual sense, therefore

$$\sum_{k=1}^{K} \lambda_k \dim V_k \leqslant \| u \|_{\mathrm{Tr}} ,$$

where we have set $V_k = P_k(E)$. By taking $K \to + \infty$, we see that the trace of u is finite, which proves (c).

Conversely, suppose that u is compact and that $\mathrm{Tr}\, u$ is finite. By using the spectral decomposition of u, (48.1), we see that

$$\| u \|_{\mathrm{Tr}} \leqslant \sum_{k=1}^{\infty} \lambda_k \| P_k \|_{\mathrm{Tr}} ,$$

and the proof of Theorem 48.2 will be complete if we show that

$$\| P_k \|_{\mathrm{Tr}} \leqslant \dim V_k , \qquad V_k = P_k(E).$$

If we select an orthonormal basis $e_1 , ..., e_r$ in V_k, we can write

$$P_k = \sum_{i=1}^{r} e_i' \otimes e_i ,$$

where e_i' is the linear form $x \rightsquigarrow (x \mid e_i)$ on E. We have, therefore, in view of Proposition 47.2,

$$\| P_k \|_{\mathrm{Tr}} \leqslant \sum_{i=1}^{r} 1 = \mathrm{Tr}(P_K)] = \dim V_k . \qquad \text{Q.E.D.}$$

Theorem 48.2 provides a motivation of sorts for the name trace-norm. Observe that, if u is a positive nuclear map of a Hilbert space E into itself, its trace, as an operator, is equal to the trace of an element $\theta \in E' \otimes_\pi E$ representing it. As a matter of fact, one can show that such an element θ is unique, in other words, that the canonical mapping of $E' \otimes_\pi E$ onto $L^1(E; E)$ is injective; it follows then from the definition of the trace-norm that it is an isometry.

We leave the proof of the following corollary of Theorem 48.2 to the student:

COROLLARY. *A continuous linear map* $u : E \to F$ *of Hilbert spaces is nuclear if and only if there are two orthogonal sequences* $\{x_k\}$, $\{y_k\}$ *in E and F, respectively, and a sequence* $\{\lambda_k\}$ *in* l^1 *such that*

$$u : x \rightsquigarrow \sum_k \lambda_k(x \mid x_k)y_k .$$

We consider now the space $E \mathbin{\widehat{\otimes}_\varepsilon} F$ (E and F are Hilbert spaces). We recall that the topology ε on $E \otimes F$ is induced by $L_\varepsilon(E'_\tau ; F)$. The equicontinuous subsets of E' are identical to the bounded sets and, as E is reflexive, its weakly compact subsets are identical to its bounded subsets; in other words, $L_\varepsilon(E'_\tau ; F) = L_b(E'; F)$, the space of bounded linear operators $E' \to F$ equipped with the operators norm (E' carries its dual norm). As for $E \mathbin{\widehat{\otimes}_\varepsilon} F$, it is the closure, in the sense of the operators norm, of the continuous linear mappings whose image is finite dimensional.

THEOREM 48.3. *Let E and F be Hilbert spaces;* $E \mathbin{\widehat{\otimes}_\varepsilon} F$ *is identical to the space of compact operators of E' into F.*

Proof. $E \mathbin{\widehat{\otimes}_\varepsilon} F$ is contained in the set of compact operators; indeed every continuous linear map with finite dimensional image is obviously compact. On the other hand, we have the general result:

LEMMA 48.2. *Let E, F be two Banach spaces. The set of compact linear operators of E into F is closed in* $L_b(E; F)$.

Proof. Let $u : E \to F$ be the limit (for the operators norm) of a sequence of compact operators. Let ε be > 0 arbitrary; let us denote by B the closed unit ball of E, by B_1 the one of F. There is a compact operator $v : E \to F$ such that $(u - v)(B) \in \varepsilon B_1$. There is a finite set of points $x_1 ,..., x_r$ in B such that

$$v(B) \subset (v(x_1) + \varepsilon B_1) \cup \cdots \cup (v(x_r) + \varepsilon B_1)$$

whence

$$u(B) \subset (u(x_1) + 3\varepsilon B_1) \cup \cdots \cup (u(x_r) + 3\varepsilon B_1).$$

This proves the lemma and therefore that every element of $E \mathbin{\widehat{\otimes}_\varepsilon} F$ is a compact operator. In order to see that every compact operator $u : E' \to F$ belongs to $E \mathbin{\widehat{\otimes}_\varepsilon} F$, it suffices to observe that u is the limit, for the operators norm, of continuous linear mappings with a finite dimensional image. Indeed, we write $u = U \circ R$, with $R : E' \to E'$

compact positive and U, an isometry of $R(E')$ into F. We then use the spectral decomposition of R,

$$R = \sum_k \lambda_k P_k \quad \text{(cf. (48.1)).}$$

The sequence of mappings

$$u_K = \sum_{k=1}^{K} \lambda_k\, U \circ P_k \quad (K = 1, 2,...)$$

converges to u for the norm; but P_k maps E' onto a finite dimensional subspace, hence $U \circ P_k(E')$ is finite dimensional.   Q.E.D.

Let us take a look now at the space $B(E', F')$, the space of continuous bilinear forms on $E' \times F'$ equipped with its natural norm, the supremum of the absolute value on the product $B' \times B'_1$ of the unit ball of E' and the unit ball of F'. If $u : E' \to F$, we can associate with u the bilinear form on $E' \times F'$,

$$\tilde{u} : (x', y') \rightsquigarrow \langle y', u(x') \rangle,$$

which is obviously continuous. Conversely, let $\phi \in B(E', F')$; then $u_\phi : x' \rightsquigarrow (y' \rightsquigarrow \phi(x', y'))$ is a continuous linear map of E' into $F'' = F$; we see immediately that $\phi = \tilde{u}_\phi$. All this means that $B(E', F')$ is canonically isomorphic to $L(E'; F)$; it is evident that the isomorphism extends to the norms.

Thus we have the natural mappings

$$E \otimes_\pi F \to L^1(E'; F) \to E \otimes_\varepsilon F \to L(E'; F) \cong B(E', F').$$

The first space carries the π-norm, the second one the trace-norm, and the last two the operators norm. All the mappings are continuous; the last one is an isometry (into!). It will follow, from what we are going to say now, that the first mapping is an isometry (onto!). For this, we study the duals of the spaces above.

We know what the dual of the first one is: $B(E, F) \cong L(E; F')$. By transposing the mapping (with dense image) $E \otimes_\pi F \to E \otimes_\varepsilon F$, we obtain an injection of the dual of $E \otimes_\varepsilon F$ into $B(E, F)$. Its image is denoted (in general) by $J(E, F)$; the elements of $J(E, F)$ are called *integral forms* on $E \times F$ (see next chapter, Definition 49.1; this definition is hardly justified in the case of Hilbert spaces, as we shall soon see). At any event, the elements of $J(E, F)$ can be identified to certain continuous linear mappings of E into F'. What are they?

Before we can answer this question, we must recall the duality bracket between $E \widehat{\otimes}_\pi F$ and $L(E; F')$. Let θ be an element of $E \widehat{\otimes}_\pi F$, and u an element of $L(E; F')$. We may consider the extended tensor product

$$u \widehat{\otimes}_\pi I : E \widehat{\otimes}_\pi F \to F' \widehat{\otimes}_\pi F, \qquad I : \text{identity of } F.$$

Then $(u \widehat{\otimes}_\pi I)(\theta)$ is an element of $F' \widehat{\otimes}_\pi F$; we may consider its trace (see p. 486). We have

$$\langle u, \theta \rangle = \text{Tr}((u \widehat{\otimes}_\pi I)(\theta)).$$

If we use a representation of θ of the form

$$\theta = \sum_{k=0}^\infty \lambda_k \, x_k \otimes y_k$$

with $\sum_k |\lambda_k| < +\infty$, $\{x_k\}$ and $\{y_k\}$ being two sequences converging to zero in E and F, respectively, we have

(48.2) $$\langle u, \theta \rangle = \sum_{k=0}^\infty \lambda_k \langle u(x_k), y_k \rangle.$$

At this stage, we may show that we need not distinguish all the time between $E \widehat{\otimes}_\pi F$ and $L^1(E'; F)$. Since the trace-norm is the quotient of the π-norm modulo the kernel of the canonical mapping $E \widehat{\otimes}_\pi F \to L^1(F'; E)$, it suffices to show that this mapping is one-to-one. This is a consequence of the following result:

THEOREM 48.4. *If E and F are Hilbert spaces, the canonical mapping of $E \widehat{\otimes}_\pi F$ into $E \widehat{\otimes}_\varepsilon F$ is one-to-one.*

Proof. It suffices to show that the transpose of the mapping in question has a *weakly* dense image; since the mapping itself has trivially a dense image, its transpose is one-to-one. In other words, we must show that $J(E, F) \cong (E \widehat{\otimes}_\varepsilon F)'$ is weakly dense in $B(E, F) \cong (E \widehat{\otimes}_\pi F)'$; here, weakly has to be understood in the sense of the duality between $B(E, F)$ and $E \widehat{\otimes}_\pi F$. We identify $B(E, F)$ to $L(E; F')$ and we note that the set of $v : E \to F'$ with finite dimensional images (and which moreover are linear continuous!) belong clearly to $J(E, F)$. It will therefore suffice to show that, if $\theta \in E \widehat{\otimes}_\pi F$, to every $u \in L(E; F')$ there is such a v with the property that $|\langle u - v, \theta \rangle| < \varepsilon(\varepsilon > 0$ arbitrary). Suppose then that we have proved the following fact:

(48.3)   To every compact subset K of E and every neighborhood V of 0
   F', there is a continuous linear mapping $v : E \to F'$ with finite
   dimensional image such that

$$(u - v)(K) \subset V.$$

We shall combine (48.3) with Eq. (48.2): we shall take, as set K, the set $\{x_k\}_{k=0,1,\ldots} \cup \{0\}$ and, as set V, the multiple ρS^0 of the polar S^0 of the sequence $S = \{y_k\}$, with $\rho = \varepsilon/(\sum_k |\lambda_k|)$. We have then $|\langle u - v, \theta \rangle| \leqslant \varepsilon$. Now, Property (48.3) follows easily from the lemma:

LEMMA 48.3.   *Let H be a Hilbert space. Then, to every compact subset C of H and to every neighborhood of zero V in H, there is a continuous linear map $w : H \to H$ with finite dimensional image such that, for all $x \in C$,*

$$x - w(x) \in V.$$

Before proving the lemma, let us show how it enables us to complete the proof of Theorem 48.4. We apply it with $H = F'$ and $C = u(K)$, where K is an arbitrary compact subset of E. Let us choose w as in Lemma 48.3 and set $v = w \circ u$; the image of v is finite dimensional, and $(I' - w)(C) \subset V(I'$: identity of $F')$ is equivalent with $(u - v)(K) \subset V$, whence (48.3).

Proof of Lemma 48.3.   We may assume that V is a closed ball centered at the origin; then C can be covered by a finite number of balls of the form $x_0 + V$; let M be the (finite dimensional) linear subspace of H spanned by their centers. We have $C \subset M + V$; if w is the orthogonal projection of H onto M, the norm of $x - w(x)$ is necessarily $\leqslant$ radius of V, since there is a point $\tilde{x} \in M$ such that $x - \tilde{x} \in V$; thus $x - w(x) \in V$.

Let $u \in L(E; F') \cong B(E, F)$, and $\theta \in L^1(E'; F) \cong E \hat{\otimes}_\pi F$. The transpose of u, $^t u$, is a continuous linear operator of F into E'; the compose $\theta \circ {}^t u$ maps F into itself. It is a nuclear operator, by Proposition 47.1. As one sees immediately,

$$\langle u, \theta \rangle = \mathrm{Tr}(\theta \circ {}^t u).$$

We are going to show that the linear form $\theta \rightsquigarrow \langle u, \theta \rangle$ is continuous on $E \otimes F$ for the topology ε if and only if u is nuclear. This can be stated in the following way:

THEOREM 48.5.   *If E and F are Hilbert spaces, the dual of $E \hat{\otimes}_\varepsilon F$ is canonically isomorphic to $L^1(E; F')$.*

The isomorphism extends to the norms: $E \hat{\otimes}_\varepsilon F$ carries the operators norm (which is equal to the ε-norm); $L^1(E; F')$, the space of nuclear operators $E \to F'$, carries the trace-norm (which is equal to the π-norm). Theorem 48.5 is due to J. Dixmier and R. Schatten. If we use the canonical antilinear isometry of a Hilbert space onto its dual, we can give the following more striking statement of Theorem 48.5:

THEOREM 48.5'. *Let E and F be Hilbert spaces. The dual of the space of compact linear operators $E \to F$, equipped with the operators norm, is the space of nuclear operators $E \to F$, equipped with the trace-norm, and its bidual is the space $L_b(E; F)$ of all continuous linear operators $E \to F$.*

Proof of Theorem **48.5.** As we have said, we must show that, if $\theta \rightsquigarrow \langle u, \theta \rangle$ is continuous on $E \mathbin{\widehat{\otimes}_\varepsilon} F$, then u must be nuclear. Let us write

$$^t u = UR \qquad \text{(we omit the ``compose'' sign o),}$$

where R is the absolute value of $^t u$ and U is an isometry of $R(F)$ into E'. We shall take advantage of the following result of the spectral theory of linear operators in Hilbert spaces: *if R is a positive bounded operator* (of F into itself), *which is not compact, there exists a bounded linear operator G such that GR is the orthogonal projection P onto a closed subspace of infinite dimension.* We then choose

$$\theta = vGU^{-1} \qquad (: E' \to F),$$

where v is a continuous linear map of F into itself having a finite dimensional image. We have then

$$\langle u, \theta \rangle = \mathrm{Tr}(\theta^t u) = \mathrm{Tr}(vP).$$

Now, if v remains in the unit ball of $L_b(F; F)$, we have

$$\| \theta \| \leqslant \| v \| \| G \| \leqslant \| G \|,$$

so that the norm of θ (as an operator!) remains bounded. But the ε-norm is precisely the operators norm. On the other hand, if v is the orthogonal projection of F onto a linear subspace of $P(F)$ of dimension n, we have $vP = v$ and $\mathrm{Tr}(v) = n$. Since $\dim P(F) = +\infty$, we may take $n \to +\infty$. In this way, we see that $| \langle u, \theta \rangle |$ does not remain bounded, although θ remains bounded in the ε-topology. We have reached a contradiction. It means that R, and therefore $^t u$, must be compact. But then we may write (cf. (48.1))

$$R = \sum_{k=1}^{\infty} \lambda_k P_k \,,$$

where the P_k are orthogonal projections on (pairwise orthogonal) finite dimensional subspaces V_k of F. We choose now $\theta = vU^{-1}$, whence $\mathrm{Tr}(\theta^t u) = \mathrm{Tr}(vR)$, and $v = \sum_{k=1}^{N} P_k$, whence

$$\mathrm{Tr}(vR) = \sum_{k=1}^{N} \lambda_k \dim V_k \,.$$

On the other hand, $|\langle u, \theta \rangle| \leqslant \text{const } \|\theta\|$ ($\| \quad \|$: operators norm). Thus we have

$$\sum_{k=1}^{N} \lambda_k \dim V_k \leqslant \text{const } \|\theta\|.$$

By going to the limit $N \to \infty$, we see that $\text{Tr } R$ is finite and is at most equal to the norm of the linear form $\theta \rightsquigarrow \langle u, \theta \rangle$ on $E \otimes_\varepsilon F$. From Theorem 48.2, we derive that R, and therefore u, is nuclear. We have

$$\| u \|_{\text{Tr}} \leqslant \sup_{\|\theta\| \leqslant 1} |\langle u, \theta \rangle|.$$

But on the other hand, if u is nuclear, so is ${}^t u$ and $\theta \circ {}^t u$. We have (Proposition 47.1; here θ is any element of $L(E'; F)$)

$$|\langle u, \theta \rangle| = |\text{Tr}(\theta \circ {}^t u)| \leqslant \| \theta \circ {}^t u \|_{\text{Tr}} \leqslant \| \theta \| \, \| {}^t u \|_{\text{Tr}}$$

and the proof is complete if we apply the corollary of Proposition 47.5 (in relation to the last estimate, we recall that the trace-form is a continuous linear functional on $F' \otimes_\pi F \cong L^1(F; F)$ of norm one).

49

The Dual of $E \otimes_\varepsilon F$. Integral Mappings

We recall that $E \otimes_\varepsilon F$ is identifiable with (or is) the space $B_\varepsilon(E'_\sigma, F'_\sigma)$ of *continuous* bilinear forms on $E'_\sigma \times F'_\sigma$, equipped with the topology of uniform convergence on the products $A' \times B'$, A' (resp. B') equicontinuous subset of E' (resp. F'). The identity mapping

$$E \otimes_\pi F \to E \otimes_\varepsilon F$$

gives rise, by transposition, to a (continuous) injection of the dual of $E \otimes_\varepsilon F$ into the one of $E \otimes_\pi F$, $B(E, F)$, the space of continuous bilinear forms on $E \times F$ (cf. Corollary of Proposition 43.4).

Definition 49.1. *The canonical image of the dual of $E \otimes_\varepsilon F$ into $B(E, F)$ is denoted by $J(E, F)$; its elements are called the integral forms on $E \times F$.*

From now on, we identify $J(E, F)$ to the dual of $E \otimes_\varepsilon F$. Note that $E \otimes F = B(E'_\sigma, F'_\sigma)$ can be identified, as a vector space, with the dual of $E'_\sigma \otimes_\pi F'_\sigma$; the ε-topology is then the topology of uniform convergence on the sets $A' \otimes B'$, with A' and B' equicontinuous. A basis of neighborhoods in the ε-topology consists of the polars $(A' \otimes B')^0$. On the other hand, $E' \otimes F'$ can be trivially regarded as a linear subspace of $J(E, F)$. An equicontinuous subset of $J(E, F)$, the dual of $E \otimes_\varepsilon F$, is then a subset of a bipolar $(A' \otimes B')^{00}$. It is on this remark that we shall base the integral representation of integral forms, motivating the name of the latter. But prior to proving it, we need a few facts about Radon measures.

Let K be a compact subset of a locally convex Hausdorff TVS G. We denote by $\mathscr{C}(K)$ the Banach space of complex continuous functions on K, with the maximum norm $f \rightsquigarrow \sup_{x \in K} |f(x)|$, by $\mathscr{C}'(K)$ its dual, which by definition is the space of Radon measures on K. There is a natural injection of K into $\mathscr{C}'(K)$: to each $x \in K$ we assign the Dirac measure δ_x at x (we recall that $\langle \delta_x, f \rangle = f(x)$). The mapping $x \rightsquigarrow \delta_x$ is clearly continuous when $\mathscr{C}'(K)$ carries its weak dual topology; as it is also one-to-one, it is a homeomorphism of K onto its image $\delta(K) \in$

$\mathscr{C}'(K)$. The inverse mapping can be extended, by linearity, as a continuous linear map of the linear subspace of $\mathscr{C}'(K)$ spanned by the Dirac measures δ_x, $x \in K$, into G. Here $\mathscr{C}'(K)$ carries its weak dual topology. Then, by continuity, we can extend the inverse mapping as a continuous linear mapping of the closure of the subspace spanned by the Dirac measures into the completion $\hat{G}$ of G. But the finite linear combinations of the Dirac measures are weakly dense in $\mathscr{C}'(K)$, in view of the Hahn–Banach theorem: a continuous function f which is orthogonal to all Dirac measures must necessarily vanish identically. Thus we have obtained a continuous linear map of $\mathscr{C}'(K)$ (equipped with the weak dual topology) into $\hat{G}$. Consider now the weakly closed convex hull $\delta(\hat{K})$ of $\delta(K)$: it is a weakly closed subset of the closed unit ball of $\mathscr{C}'(K)$, which is weakly compact, as $\mathscr{C}(K)$ is barreled. Therefore $\delta(\hat{K})$ is compact. Its image in G is a compact convex set, which contains K. It contains therefore the closed convex hull of K in $\hat{G}$, $\hat{K}$, which is a compact set, and, in fact, is identical to $\hat{K}$, as is easy to check. On the other hand, $\delta(\hat{K})$ consists of positive Radon measures on K of total mass $\leqslant 1$; in fact, $\delta(\hat{K})$ is exactly the set of all such Radon measures. For if $\mu \in \mathscr{C}'(K)$ is $\geqslant 0$ and $\mu(K) \leqslant 1$, we have $|\mu(f)| \leqslant 1$ for all $f \in \mathscr{C}(K)$ which belong to the polar of $\delta(K)$. Indeed, such a function f satisfies $|f(x)| \leqslant 1$ for all $x \in K$. Thus μ belongs to the bipolar of $\delta(K)$ and, as μ is positive, it belongs to the weakly closed convex hull of $\delta(K)$, as is easily verified. At any event, we see that every point x of the closed convex hull $\Gamma(K)$ of K in G is the image of a positive Radon measure μ of total mass $\leqslant 1$ on K. Then if x' is an arbitrary element of the dual G' of G, we have

$$(49.1) \qquad \langle x', x \rangle = \int_K \langle x', y \rangle \, d\mu(y).$$

Indeed, this is true when $x = \sum_{j=1}^r m_j x_j$, $x_j \in K$, $\sum_{j=1}^r m_j \leqslant 1$, taking then $\mu = \sum m_j \delta_{x_j}$. It remains true by going to the limit, for an arbitrary $\mu \in \delta(\hat{K})$.

We may, now, state and prove the integral representation formula of integral forms. We recall that weakly closed equicontinuous subsets of a dual are weakly compact sets.

PROPOSITION 49.1.   *A bilinear form u on $E \times F$ is integral if and only if there is a weakly closed equicontinuous subset A' (resp. B') of E' (resp. F') and a positive Radon measure μ on the compact set $A' \times B'$ with total mass $\leqslant 1$, such that, for all $x \in E$, $y \in F$,*

$$(49.2) \qquad u(x, y) = \int_{A' \times B'} \langle x', x \rangle \langle y', y \rangle \, d\mu(x', y').$$

Proof. It is evident that (49.2) defines an integral form. If we regard the right-hand side as a linear form on $E \otimes F$ and if $\theta \in E \otimes F$ (regarded as a bilinear form on $E' \times F'$) is such that $| \theta(A', B') | \leqslant 1$, we have $| \langle u, \theta \rangle | \leqslant \int_{A' \times B'} d\mu \leqslant 1$.

Conversely, an integral form u belongs to the bipolar of a set $A' \otimes B'$, that is to say to the weakly closed convex balanced hull of $A' \otimes B'$. Here A' and B' are equicontinuous sets, but we may assume that they are weakly closed, which makes them weakly compact. Note that the canonical bilinear mapping of $E'_\sigma \times F'_\sigma$ into $E' \otimes F'$, equipped with the topology induced by $\sigma(J(E, F), E \otimes F)$, induces a homeomorphism of $A' \times B'$ onto $A' \otimes B'$; thus we may transfer every Radon measure from $A' \times B'$ onto $A' \otimes B'$ and vice versa. In view of the considerations preceding the statement of Proposition 49.1 and by applying Eq. (49.1), we know that, for some positive Radon measure μ on $A' \otimes B'$ of mass $\leqslant 1$ and for all $\theta \in E \otimes F$,

$$\langle u, \theta \rangle = \int_{A' \times B'} (x' \otimes y')(\theta) \, d\mu(x' \otimes y').$$

If we then take $\theta = x \otimes y$ and transfer μ onto $A' \times B'$, we obtain precisely (49.2).

Remark 49.1.   If both E and F carry their Mackey topology (see p. 369) there is identity, in E' and F', between weakly compact and weakly closed equicontinuous subsets—provided that they are convex balanced! This applies, in particular, to the case where E and F are barreled or metrizable, in particular normed.

Consider the integral form u given by (49.2) and let us call U and V the polars of A' and B' in E and F, respectively; U and V are closed convex balanced neighborhoods of zero. Let us denote by $\mathfrak{p}$ and $\mathfrak{q}$ the associated seminorms ($\mathfrak{p}(x) = \inf_{x \in \rho U} \rho, \ \rho > 0$) and by $E_\mathfrak{p}$ and $F_\mathfrak{q}$ the normed spaces $E/\mathrm{Ker}\,\mathfrak{p}$ and $F/\mathrm{Ker}\,\mathfrak{q}$ equipped with the quotient norms $\mathfrak{p}/\mathrm{Ker}\,\mathfrak{p}$ and $\mathfrak{q}/\mathrm{Ker}\,\mathfrak{q}$, respectively. The canonical map of E onto $E_\mathfrak{p}$ defines, by transposition, a continuous injection of $E'_\mathfrak{p}$ into E' whose image is the subspace E'_{U_0} spanned by the polar U^0 of U. The dual norm of $\mathfrak{p}$ is exactly equal to the gauge $\mathfrak{p}_{U^0}$ of U^0; as we have already done (cf. p. 478) we identify $E'_\mathfrak{p}$ and E'_{U^0}; $E'_\mathfrak{p}$ is the space of linear forms on E which are continuous with respect to the seminorm $\mathfrak{p}$. It is also obvious that E'_{U^0} is the dual of $\hat{E}_\mathfrak{p}$, completion of $E_\mathfrak{p}$. We note then that A' is an equicontinuous subset of $E'_\mathfrak{p}$, and B' an equicontinuous subset of $F'_\mathfrak{q}$; in fact each is contained in the respective closed unit ball of the Banach space $E'_\mathfrak{p}$ or $F'_\mathfrak{q}$. Going back to the form u, we derive from (49.2) that, for all $x \in E, \ y \in F$,

$$| u(x, y) | \leqslant \mathfrak{p}(x) \, \mathfrak{q}(y)$$

and, therefore, that u defines a continuous bilinear form $\tilde{u}$ on $E_p \times F_q$ and, by continuity, on $\hat{E}_p \times \hat{F}_q$. Returning then to Eq. (49.2) and regarding A' (resp. B') as an equicontinuous subset of E'_p (resp. F'_q), we see that the form $\tilde{u}$ is integral over $\hat{E}_p \times \hat{F}_q$. Conversely, given an integral form $\tilde{u}$ on $\hat{E}_p \times \hat{F}_q$ we can pull it back as an integral form on $E \times F$, just by setting

$$u(x, y) = \tilde{u}(\phi_p(x), \psi_q(y)),$$

where ϕ_p (resp. ψ_q) is the canonical mapping of E into $\hat{E}_p$ (resp. F into $\hat{F}_q$). This reduction of arbitrary integral forms to integral forms on products of Banach spaces will be used to a considerable extent. It follows from Eq. (49.2).

Definition 49.2. Let E, F be two locally convex Hausdorff spaces. A continuous linear map $u : E \to F$ is called integral if the associated bilinear form on $E \times F'$,

$$(x, y') \rightsquigarrow \langle y', u(x) \rangle,$$

is integral.

By F' we have denoted the strong dual of F. From what precedes, we know that there are continuous seminorms p and q' on E and F', respectively, such that $\langle y', u(x) \rangle$ is integral on $\hat{E}_p \times \hat{F}'_{q'}$. Let us denote by U (resp. V') the closed unit semiball of p (resp. q'). By definition of the strong dual topology, the polar V'^0 of V' in F is a closed convex balanced bounded subset of F. As we have

$$|\langle y', u(x) \rangle| \leqslant p(x)\, q'(y')$$

(for suitable choices of p and q'; see above), u maps U into V'^0. Thus u defines a continuous linear map of the normed space E_p into the normed space $F_{V'^0}$ (for the definition of the latter, see p. 370). As the dual of $F_{V'^0}$ can be identified, as a Banach space, with $\hat{F}'_{q'}$, we obtain a factorization of u into the sequence

$$E \xrightarrow{\ i\ } E_p \xrightarrow{\ \tilde{u}\ } F_{V'^0} \xrightarrow{\ j\ } F,$$

where i and j are the natural mappings and $\tilde{u}$ is integral. Here again, we are reduced to the case of normed spaces.

PROPOSITION 49.2. *Let E, F, G, and H be four locally convex Hausdorff spaces, $u : E \to F$ an integral map, and $f : G \to E$, $g : F \to H$ continuous linear maps. The compose $g \circ u \circ f$ is integral.*

Proof. It suffices to verify the following. If Φ is an integral form on $E \times F'$, the form on $G \times H'$, $(\xi, \eta') \rightsquigarrow \Phi(f(\xi), {}^tg(\eta'))$, is integral. But this follows at once from the fact that $f \otimes {}^tg$ is a continuous linear map of $G \otimes_\varepsilon H'$ into $E \otimes_\varepsilon F'$.

PROPOSITION 49.3.   *If a continuous linear map* $u : E \to F$ *is integral, its transpose* ${}^tu : F' \to E'$ *is also integral.*

Proof. The integral map u can be factorized as follows:

$$(49.3) \qquad\qquad E \xrightarrow{\ i\ } E_\mathrm{p} \xrightarrow{\ \tilde{u}\ } F_B \xrightarrow{\ j\ } F,$$

where p is a continuous seminorm on E, B is a closed convex balanced bounded subset of F, i and j are the natural mappings, and $\tilde{u}$ is integral (see above). The transpose of u is factorized by transposing the sequence (49.3); it suffices therefore to show that the transpose of $\tilde{u}$, ${}^t\tilde{u} : \hat{F}'_{q'} \to \hat{E}'_\mathrm{p}$ (q': gauge of the polar B^0 of B), is integral. In other words, it can be assumed that E and F are normed spaces. Let us denote by A' (resp. B'') the closed unit ball in E' (resp. F''). There is a positive Radon measure μ on $A' \times B''$ of total mass $\leqslant 1$ such that, for all $x \in E$, $y' \in F$,

$$(49.4) \qquad \langle {}^tu(y'), x \rangle = \langle y', u(x) \rangle = \int_{A' \times B''} \langle x', x \rangle \langle y'', y' \rangle \, d\mu(x', y'').$$

It is clear that the right-hand side, hence the left-hand side, can be extended from E to E'' and the form $(y', x'') \rightsquigarrow \langle {}^tu(y'), x'' \rangle$ is therefore integral.

PROPOSITION 49.4.   *Suppose that the canonical injections* $E \to E''$ *and* $F' \to F'''$ *are isomorphisms into (which is the case, e.g., when E and F' are barreled or metrizable). If* $u : E \to F$ *is a continuous linear map whose transpose is integral, u is integral.*

Proof. By Proposition 49.3, we know that the bitranspose ${}^{tt}u : E'' \to F''$ of u is integral. Let $i : E \to E''$ be the natural injection; viewed as a mapping of E into F'', u is equal to ${}^{tt}u \circ i$. By hypothesis i is continuous, therefore $u : E \to F''$ is integral. Thus we have an integral representation of the type (49.2), or rather (49.4),

$$\langle y''', u(x) \rangle = \int_{A' \times B''''} \langle x', x \rangle \langle y'''', y''' \rangle \, d\mu(x', y'''').$$

Taking the restriction of both sides to $E \times F'$, i.e., replacing $y''' \in E'''$

by $y' \in F'$, and taking into account the hypothesis that the natural injection $F' \to F'''$ is an isomorphism into, yields easily the result: for we may write

$$\langle y', u(x) \rangle = \int_{A' \times B''} \langle x', x \rangle \langle y'', y' \rangle \, d\tilde{\mu}(x', y''),$$

where B'' is the image of B'''' under the canonical mapping $F'''' \to F''$, transpose of $F' \to F'''$, and $\tilde{\mu}$ is the image of μ under that mapping. It is clear that B'' is a weakly closed equicontinuous subset of F''.

COROLLARY. *If E and F are normed spaces, $u : E \to F$ is integral if and only if ${}^t u : F' \to E'$ is integral.*

PROPOSITION 49.5.   *A nuclear map $u : E \to F$ is integral.*

Proof. Let us for instance use the representation (47.2) of u (Proposition 47.2). We see that the bilinear form on $E \times F'$ associated with u is equal to

$$\tilde{u}(x, y') = \sum_k \lambda_k \langle x_k', x \rangle \langle y', y_k \rangle.$$

We recall that $\{x_k'\}$ is an equicontinuous sequence S' in E', whereas the y_k are all contained in some bounded subset B of F. The polar B^0 of B is a neighborhood of zero in E', by definition of the strong dual topology. The polar of B^0 in E'' is therefore an equicontinuous subset B'' in E'', trivially weakly closed. If then $\theta \in E \otimes F'$ takes values on $S' \times B''$ which are bounded, in absolute value, by one, we have $| \langle \tilde{u}, \theta \rangle | \leqslant \sum_k | \lambda_k | < + \infty$, which implies immediately that u is integral.

Example 49.1.   Let X be a compact topological space, and dx a positive Radon measure on X of total mass $\leqslant 1$. The bilinear form

(49.5) $$(f, g) \rightsquigarrow \int_X f(x) g(x) \, dx$$

on $\mathscr{C}(X) \times \mathscr{C}(X)$ ($\mathscr{C}(X)$: space of continuous complex functions in X with the norm $f \rightsquigarrow \sup_{x \in X} | f(x) |$) is integral, as it can be written

$$(f, g) \rightsquigarrow \int \langle \delta_x, f \rangle \langle \delta_y, g \rangle \, d\mu(\delta_x, \delta_y),$$

where δ_x is the Dirac measure at x and μ is the Radon measure which is the image of dx via the mapping $x \rightsquigarrow (\delta_x, \delta_x)$ of X onto a weakly compact subset of the diagonal of $\mathscr{C}'(X) \times \mathscr{C}'(X)$. By $\mathscr{C}'(X)$ we denote the Banach space of Radon measures on X, the dual of $\mathscr{C}'(X)$. It can easily be checked that the form (49.5) is associated with the mapping $g \rightsquigarrow g(x)\, dx$ of $\mathscr{C}(X)$ into $\mathscr{C}'(X)$.

Remark 49.2.　Example 49.1 shows that there are integral mappings which are not nuclear. Indeed, the mapping $g \rightsquigarrow g\, dx$ above, of $\mathscr{C}(X)$ into $\mathscr{C}'(X)$, is integral. But, in general, it will not be nuclear; for then it would have to be compact. Take for instance X equal to the closed unit interval $[0, 1]$ of the real line, and dx equal to the induced Lebesgue measure; the image, under $g \rightsquigarrow g\, dx$, of the unit ball of $\mathscr{C}(X)$, is not a relatively compact subset of $\mathscr{C}'(X)$.

However, as a consequence of Theorem 48.3, every integral map of a Hilbert space into another one is nuclear:

PROPOSITION 49.6.　*Let E, F be two Hilbert spaces, and $u : E \to F$ an integral mapping. The operator u is nuclear.*

This is merely a restatement of Theorem 48.5.

We are now going to use Example 49.1 in order to obtain a useful factorization of any integral map. Let E, F be two locally convex Hausdorff TVS, and $u : E \to F$ an integral map. Consider the representation

$$\langle y', u(x) \rangle = \int_{A' \times B''} \langle x', x \rangle \langle y'', y' \rangle \, d\mu(x', y'').$$

We recall that A' and B'' are weakly closed equicontinuous subsets, for the dualities between E and E' and F' and F'', respectively. We choose, as compact topological space X, the product $A' \times B''$. We define two mappings $S : E \to \mathscr{C}(X)$ and $T : F' \to \mathscr{C}(X)$ in the following manner. For every $x \in E$ and every $y' \in F'$,

$$S(x) : (x', y'') \rightsquigarrow \langle x', x \rangle,$$

$$T(y') : (x', y'') \rightsquigarrow \langle y'', y' \rangle.$$

Then let $\tilde{u}$ be the mapping $f \rightsquigarrow f(x', y'')\, d\mu(x', y'')$ of $\mathscr{C}(X)$ into $\mathscr{C}'(X)$. It is immediately seen that u is the compose of the sequence

(49.6)
$$E \xrightarrow{\ S\ } \mathscr{C}(X) \xrightarrow{\ \tilde{u}\ } \mathscr{C}'(X) \xrightarrow{\ {}^tT\ } F''$$

(where we regard u as valued in F''); the central arrow $\tilde{u}$ is an integral mapping, as we have seen when looking at Example 49.1. From the factorization (49.6) of u, we derive the following important result:

PROPOSITION 49.7.   *Let $u : E \to F$ be an integral mapping. If the locally convex Hausdorff space F is complete, there is a Hilbert space H and two continuous linear mappings $\alpha : E \to H$, $\beta : H \to F$ such that $u = \beta \circ \alpha$.*

Proof.   In (49.6), we observe that $\tilde{u}$ can be factorized into

$$\mathscr{C}(X) \rightsquigarrow L^2(X, d\mu) \rightsquigarrow \mathscr{C}'(X),$$

where the first arrow is the natural mapping of the space of continuous functions into the space of classes of square integrable functions, and the second arrow is the injection $L^2(X, d\mu) \ni f \rightsquigarrow f \, d\mu \in \mathscr{C}'(X)$. By combining this with (49.6) we obtain a factorization of $u : E \to F''$ into a sequence

$$E \xrightarrow{\ \sigma\ } H_0 \xrightarrow{\ \tau\ } F'',$$

where H_0 is a Hilbert space. Let H_1 be the linear subspace of H_0 which is the image of E under σ; as $u(E) \subset F$, we have $\tau(H_1) \subset F$. We suppose that H_1 carries the pre-Hilbert structure induced by the structure of H_0. We have now a factorization of u into

$$E \xrightarrow{\ \sigma_1\ } H_1 \xrightarrow{\ \tau_1\ } F.$$

Since F is complete we may extend τ_1 by continuity to the completion H of H_1; we call this extension β; H is a Hilbert space. We call α the map σ_1 viewed as a map of E into H.   Q.E.D.

Remark 49.3.   There are linear mappings $u : E \to F$ which can be factorized into $E \xrightarrow{\alpha} H \xrightarrow{\beta} F$, with α, β continuous and H a Hilbert space, without u being integral. Trivial example: the identity mapping I of an infinite dimensional Hilbert space; if I were integral, it would be nuclear (Proposition 49.6), hence compact.

A simple and important consequence of Propositions 49.6 and 49.7 is the following one:

PROPOSITION 49.8.   *Let A, B, C, and D be four locally convex Hausdorff spaces. Suppose that B and D are complete. Then the compose*

(49.7) $A \xrightarrow{\ u\ } B \xrightarrow{\ v\ } C \xrightarrow{\ w\ } D$

of three integral mappings is nuclear.

Proof. By applying Proposition 49.7, we may factorize (49.7) into

$$A \xrightarrow{u_1} H_1 \xrightarrow{u_2} B \xrightarrow{v} C \xrightarrow{w_1} H_2 \xrightarrow{w_2} D,$$

where H_j is a Hilbert space, u_j and v_j are continuous linear mappings ($j = 1, 2$). As v is integral, the compose $w_1 \circ v \circ u_2$ must be integral (Proposition 49.2), hence nuclear (Proposition 49.6). But then the total compose, which is equal to $w \circ u \circ v$, must be nuclear (Proposition 47.1).

We shall soon make use of Proposition 49.8. Let us point out an easy consequence of it: let E be a Fréchet space, $u : E \to E$ a continuous linear mapping onto; then u cannot be integral. We leave the proof of this fact as an exercise to the student.

50

Nuclear Spaces

In Chapters 44, 45, and 46, we have seen a few examples of how completion of topological tensor products $E \otimes_\varepsilon F$ or $E \otimes_\pi F$ may yield new representations for known "functional" spaces. There is a number of reasons for the usefulness of such representations. An important one is that they make the extension of certain mappings automatic, showing that it is the extension of a tensor product of mappings of the type $u \widehat{\otimes}_\varepsilon v$ or $u \widehat{\otimes}_\pi v$. Also, these representations might bring to light certain interesting properties of the spaces under consideration, which could have gone unnoticed otherwise or which would have remained mysterious. A startling example of the latter is L. Schwartz's *kernels theorem*: it states, essentially, that every continuous linear map of the space $(\mathscr{C}_c^\infty)_x$ of test functions in some variable x, into the space $\mathscr{D}'_y$ of distributions in a second variable y, is given by a (unique) distribution $K_{x,y}$ in both variables x, y, according to the formula

$$\phi \rightsquigarrow \langle K_{x,y}, \phi(x) \rangle.$$

The reader will realize the peculiarity of this situation if he compares it to those of a more "classical" nature, such as, for instance, the one occurring in the L^2 theory. It is trivially false that every bounded operator of L_x^2 into L_y^2 can be represented as a kernel $K(x, y) \in L_{x,y}^2$, that is to say, can be written as

$$f \rightsquigarrow \int K(x, y) f(x) \, dx.$$

The identity mapping itself cannot be written in that way! We know indeed that the kernel (distribution) defining the identity is a Dirac kernel, $\delta(x - y)$—which is not a function.

What then are the reasons for such a striking difference between continuous linear operators into $\mathscr{D}'$ and the ones on L^2? This is the question which is at the origin of the theory of completed topological tensor products and nuclear spaces, due to A. Grothendieck. The answer to it is, very roughly speaking, that the spaces $\mathscr{D}'$ (as well as $\mathscr{C}^\infty$, $\mathscr{C}_c^\infty$, $\mathscr{E}'$, $\mathscr{S}$, etc.) are all *nuclear*, whereas no infinite dimensional Banach space is. A nuclear space E is a locally convex Hausdorff space such that, for any other space F, $E \otimes_\pi F = E \otimes_\varepsilon F$ (the sign $=$ stands for the canonical mapping, which is *onto* when E is nuclear). It is at once evident, on this definition, that nuclear spaces will be endowed with nice properties, in relation with extension of tensor products of two mappings, $u \otimes v$: for the ε-topology is well behaved when we deal with isomorphisms into, whereas the π-topology is well behaved when we deal with homomorphisms onto. But going back to the kernels theorem, it is readily seen that $\mathscr{D}'_{x,y}$ induces on $\mathscr{D}'_x \otimes \mathscr{D}'_y$ the ε-topology (or the π one), and that we have therefore

$$\mathscr{D}'_{x,y} \cong \mathscr{D}'_x \,\widehat{\otimes}\, \mathscr{D}'_y \,,$$

where we omit the indices π or ε to the symbol $\widehat{\otimes}$. In addition to this, it is easy to see that

$$\mathscr{D}'_x \,\widehat{\otimes}\, \mathscr{D}'_y \cong L_\varepsilon((\mathscr{C}_c^\infty)_x; \mathscr{D}'_y),$$

where $(\mathscr{C}_c^\infty)_x$ is supposed to carry the τ-topology. As $\mathscr{C}_c^\infty$ is a Montel space, its τ-topology is equal to its initial topology, and every equicontinuous subset (we are viewing $\mathscr{C}_c^\infty$ as the dual of $\mathscr{D}'$) is bounded ($=$ relatively compact). Finally, we see that

$$\mathscr{D}'_{x,y} \cong L_b((\mathscr{C}_c^\infty)_x; \mathscr{D}'_y).$$

We proceed now to define the nuclear spaces and give their basic properties.

Let E be a locally convex Hausdorff space, and $\mathfrak{p}$ a continuous seminorm on E. We recall that $\hat{E}_\mathfrak{p}$ is the completion of the normed space $E/\mathrm{Ker}\,\mathfrak{p}$ (the latter is a normed space if we put on it the quotient mod $\mathrm{Ker}\,\mathfrak{p}$ of the seminorm $\mathfrak{p}$); thus $\hat{E}_\mathfrak{p}$ is a Banach space.

Definition 50.1. The locally convex Hausdorff TVS E is said to be nuclear if to every continuous seminorm $\mathfrak{p}$ on E there is another continuous seminorm on E, $\mathfrak{q} \geqslant \mathfrak{p}$, such that the canonical mapping $\hat{E}_\mathfrak{q} \to \hat{E}_\mathfrak{p}$ is nuclear.

What the canonical mapping $\hat{E}_\mathfrak{q} \to \hat{E}_\mathfrak{p}$ is, should be easy to guess: since $\mathfrak{q} \geqslant \mathfrak{p}$, $\mathrm{Ker}\,\mathfrak{p} \supset \mathrm{Ker}\,\mathfrak{q}$, hence there is a canonical mapping of $E/\mathrm{Ker}\,\mathfrak{q}$ onto $E/\mathrm{Ker}\,\mathfrak{p}$; furthermore this mapping is continuous if we provide the first space with the norm $\mathfrak{q}/\mathrm{Ker}\,\mathfrak{q}$ and the second one with $\mathfrak{p}/\mathrm{Ker}\,\mathfrak{p}$.

We recall that $E'_\mathfrak{p}$, the dual of $\hat{E}_\mathfrak{p}$, can be identified with the subspace of E' consisting of the functionals x' which are continuous when E carries the topology defined by the single seminorm $\mathfrak{p}$; $E'_\mathfrak{p}$ carries its dual Banach space structure. We recall that this structure may be defined without reference to $\hat{E}_\mathfrak{p}$: indeed, $E'_\mathfrak{p}$ is the subspace of E' spanned by the polar $U^0_\mathfrak{p}$ of the closed unit semiball $U_\mathfrak{p}$ of $\mathfrak{p}$; its norm is the one associated with the weakly compact convex balanced set $U^0_\mathfrak{p}$ (cf. pp. 478 and 502).

The introduction of nuclear spaces is justified by the following theorem:

THEOREM 50.1. *The following properties of a locally convex Hausdorff TVS E are equivalent:*

(a) *E is nuclear;*

(b) *to every continuous seminorm $\mathfrak{p}$ on E there is another continuous seminorm $\mathfrak{q}$ on E, $\mathfrak{q} \geqslant \mathfrak{p}$, such that the canonical injection of $E'_\mathfrak{p}$ into $E'_\mathfrak{q}$ is nuclear;*

(c) *every continuous linear map of E into a Banach space is nuclear;*

(d) *every linear map of a Banach space into E', which transforms the unit ball into an equicontinuous set, is nuclear;*

(e) *for every Banach space F, the canonical map of $E \otimes_\pi F$ into $E \otimes_\varepsilon F$ is an isomorphism onto;*

(f) *for every locally convex Hausdorff TVS F, the canonical map of $E \otimes_\pi F$ into $E \otimes_\varepsilon F$ is an isomorphism onto.*

Proof. (a) $\Rightarrow$ (b) in view of Definition 50.1 and of the fact that the transpose of a nuclear map is nuclear.

Let us show that (b) $\Rightarrow$ (a). The canonical injection $j : E'_\mathfrak{p} \to E'_\mathfrak{q}$ is the transpose of the canonical map $i : \hat{E}_\mathfrak{q} \to \hat{E}_\mathfrak{p}$. From the corollary of Proposition 49.4 we know that i is integral. We can find two more continuous seminorms $\mathfrak{s} \geqslant \mathfrak{r} \geqslant \mathfrak{q}$ on E such that the canonical mappings

$$i' : \hat{E}_\mathfrak{r} \to \hat{E}_\mathfrak{q}, \qquad i'' : \hat{E}_\mathfrak{s} \to \hat{E}_\mathfrak{r}$$

are integral. In view of Proposition 49.8, the compose $i'' \circ i' \circ i$, which is the canonical mapping $\hat{E}_\mathfrak{s} \to \hat{E}_\mathfrak{p}$, is nuclear.

Let F be a Banach space, and $u : E \to F$ a continuous linear map. The preimage under u of the closed unit ball of F is a closed convex balanced neighborhood of zero U in E. Let $\mathfrak{p}$ be the gauge of U. If E is nuclear, there is $\mathfrak{q} \geqslant \mathfrak{p}$ such that $\hat{E}_\mathfrak{q} \to \hat{E}_\mathfrak{p}$ is nuclear. But we can factorize u into the sequence

$$E \xrightarrow{\; i \;} \hat{E}_\mathfrak{q} \xrightarrow{\; j \;} \hat{E}_\mathfrak{p} \xrightarrow{\; \hat{u} \;} F,$$

where i and j are the canonical mappings, and $\tilde{u}$ the extension by continuity of the mapping $E/\mathrm{Ker}\,\mathfrak{p} \to F$ defined by u ($E/\mathrm{Ker}\,\mathfrak{p}$ carries the norm $\mathfrak{p}/\mathrm{Ker}\,\mathfrak{p}$). As j is nuclear, so is u (Proposition 47.1). This shows that (a) $\Rightarrow$ (c). To see that (c) $\Rightarrow$ (a), we choose arbitrarily a continuous seminorm $\mathfrak{p}$ on E and we apply (c) with $F = \hat{E}_\mathfrak{p}$ and u the canonical mapping of E into $\hat{E}_\mathfrak{p}$. But by Definition 47.3, to say that u is nuclear is equivalent to saying that there is a continuous seminorm $\mathfrak{q}$ on E, $\mathfrak{q} \geqslant \mathfrak{p}$, such that the canonical map $\hat{E}_\mathfrak{q} \to \hat{E}_\mathfrak{p}$ is nuclear. This shows that (c) $\Rightarrow$ (a).

The equivalence of (b) and (d) is shown in pretty much the same fashion as the equivalence of (a) $\Leftrightarrow$ (c). We shall leave this part of the proof to the student. We shall concentrate on the proof of the fact that (e) and (f) are equivalent to the other properties. We begin by showing that (e) $\Leftrightarrow$ (f).

Trivially, (f) $\Rightarrow$ (e). Note that (f) (resp. (e)) means that, for all locally convex Hausdorff spaces (resp. Banach spaces) F, we have

$$(50.1) \qquad\qquad E \otimes_\pi F = E \otimes_\varepsilon F,$$

where the equality extends to the topologies. If we wish to prove (50.1) we may apply Lemma 43.1 (Exercise 43.3) and show that every equicontinuous subset of the dual $B(E, F)$ of $E \otimes_\pi F$ is an equicontinuous subset of the dual $J(E, F)$ of $E \otimes_\varepsilon F$. We recall that $B(E, F)$ is the space of all continuous bilinear forms on $E \times F$ and that $J(E, F)$ is the space of integral forms on $E \times F$ (see Chapter 49). To every equicontinuous subset Φ of $B(E, F)$ there are continuous seminorms $\mathfrak{p}$ and $\mathfrak{q}$ on E and F, respectively, such that Φ is an equicontinuous subset of $B(E_\mathfrak{p}, F_\mathfrak{q})$ ($E_\mathfrak{p}$ is the space E equipped with the single seminorm $\mathfrak{p}$; analog for $F_\mathfrak{q}$). But we may then go to the quotient spaces $E_\mathfrak{p} \to E_\mathfrak{p}/\mathrm{Ker}\,\mathfrak{p}$ and $F_\mathfrak{q} \to F_\mathfrak{q}/\mathrm{Ker}\,\mathfrak{q}$, and then to the completions of the quotient spaces. Thus Φ defines (canonically) an equicontinuous subset Φ_1 in $B(E_\mathfrak{p}, \hat{F}_\mathfrak{q})$ and one, Φ_2, in $B(\hat{E}_\mathfrak{p}, \hat{F}_\mathfrak{q})$. Of course, Φ_1 can also be regarded as an equicontinuous subset of $B(E, \hat{F}_\mathfrak{q})$. On the other hand, the equicontinuous subsets of $J(E, \hat{F}_\mathfrak{q})$ are identifiable to those of $J(E, \hat{F}_\mathfrak{q})$ and these, in turn, are equicontinuous subsets of $J(E, F)$ (this is only saying that the identity mapping is continuous from $E \otimes_\varepsilon F$ into $E \otimes_\varepsilon F_\mathfrak{q}$!). Suppose then that (e) has been proved. It implies that every set like Φ_1 is an equicontinuous subset of $J(E, \hat{F}_\mathfrak{q})$, hence of $J(E, F)$. This proves the equivalence of (e) and (f).

Let us suppose now that E is nuclear, F a Banach space, and let us show that (50.1) holds. Let Φ, $\mathfrak{p}$ be as above ($\mathfrak{q}$ is now the norm of F). Let $\mathfrak{p}_0$ be a continuous seminorm on E, $\mathfrak{p}_0 \geqslant \mathfrak{p}$, such that the canonical map $E_p' \to E_{p_0}'$ is nuclear (p_0 exists by (b)). For $B \in B(E, F)$, let us set

$$u_B : y \rightsquigarrow (x \rightsquigarrow B(x, y)).$$

When B ranges over Φ, we may regard the u_B as an equicontinuous subset of $L(E, E_p')$; if we compose the u_B, $B \in \Phi$, with the canonical injection $E_p' \to E_{p_0}'$, we obtain a bounded subset of $L^1(F; E_{p_0}')$ (equipped with the trace-norm). From this it follows immediately (cf. Proposition 49.5 and proof) that the bilinear forms on $F \times (E_{p_0}')'$ associated with the mappings u_B form an equicontinuous subset of $J(F, (E_{p_0}')')$, Φ''. But the natural map of $\hat{E}_{p_0}$ into its bidual, $(E_{p_0}')'$, is an isomorphism into, and therefore, by restriction to $\hat{E}_{p_0}$, Φ'' defines an equicontinuous subset Φ_0 of $J(\hat{E}_{p_0}, F)$. Let $i_0 : E \to \hat{E}_{p_0}$ be the canonical mapping, and set, for each $B_0 \in \Phi_0$, $B(x, y) = B_0(i_0(x), y)$, $x \in E$, $y \in F$. When B_0 ranges over Φ_0, B ranges over Φ, as immediately checked. But on the other hand, B ranges over an equicontinuous subset of $J(E, F)$ in view of the equicontinuity of Φ_0 and the continuity of i_0. Thus (50.1) holds when E is nuclear.

Conversely, let us suppose that (50.1) holds for all Banach spaces F and derive from this that E is nuclear. Then let p be an arbitrary continuous seminorm on E, and $u : E \to \hat{E}_p$ the canonical mapping. The bilinear form associated with u is trivially continuous on $E \times E_p'$, hence, since $B(E, E_p') = J(E, E_p')$ by hypothesis, it is integral; in other words, the mapping u is integral. But by the factorizations property of integral mappings (see p. 503), we know that there is a continuous seminorm q on E, which we can take $\geqslant p$, such that u decomposes into

$$E \xrightarrow{v} \hat{E}_q \xrightarrow{\tilde{u}} \hat{E}_p,$$

where v is the canonical mapping and $\tilde{u}$ (also canonical) is integral. By reasoning then as in the proof of the implication (b) $\Rightarrow$ (a), we conclude easily that E is nuclear.

Remark 50.1.   Inspection of the proof of Theorem 50.1 shows immediately that we have proved that Property (a), E nuclear, is equivalent with each one of the properties (b), (c), and (d) where the word *nuclear* is replaced by *integral* and that E is nuclear if and only if, to every continuous seminorm p on E there is a continuous seminorm $q \geqslant p$ on E such that the canonical mapping $\hat{E}_q \to \hat{E}_p$ is integral.

The basic properties of nuclear spaces are now easy to derive, either by direct derivation from Definition 50.1 or by application of Theorem 50.1. Let us begin with the so-called stability properties:

PROPOSITION 50.1.

(50.2)　A locally convex Hausdorff TVS E is nuclear if and only if its completion $\hat{E}$ is nuclear.

(50.3)　A linear subspace of a nuclear space is nuclear.

(50.4)　The quotient of a nuclear space modulo a closed linear subspace is nuclear.

(50.5)　A product of nuclear spaces is nuclear.

(50.6)　A countable topological direct sum of nuclear spaces is nuclear.

(50.7)　A Hausdorff projective limit of nuclear spaces is nuclear.

(50.8)　A countable inductive limit of nuclear spaces is nuclear.

(50.9)　If E and F are two nuclear spaces, $E \otimes F$ is nuclear.

Before giving the proof of Proposition 50.1, we recall the definitions of certain terms used in its statement. First of all (and this will be valid from now on), if E is nuclear, we write $E \otimes F$ instead of $E \otimes_\pi F$ or $E \otimes_\epsilon F$. Let $\{E_\alpha\}$ be a family of locally convex spaces, and E a vector space.

(i) Suppose that we are given, for each index α, a linear map $\phi_\alpha : E \to E_\alpha$. We then consider on E the least-fine topology, compatible with the linear structure of E, such that all the mappings ϕ_α be continuous. Equipped with it, E is called the *projective limit* of the spaces E_α with respect to the mappings ϕ_α. A basis of neighborhoods of zero in this topology is obtained as follows: in each E_α, we consider a basis of neighborhoods of zero $U_{\alpha,\beta}$ ($\beta \in B_\alpha$); let $V_{\alpha,\beta}$ be the preimage of $U_{\alpha,\beta}$ under ϕ_α; then, all the finite intersections of sets $V_{\alpha,\beta}$, when α and β vary in all possible ways, form a basis of neighborhoods of zero in the projective topology on E (this shows, in particular, that the said topology exists!); it is also clear that this topology is locally convex as soon as all the E_α are locally convex. It is Hausdorff if every one of the E_α is Hausdorff and if, for every $x \in E$, $x \neq 0$, there is at least one index α such that $\phi_\alpha(x) \neq 0$. If the latter condition is not satisfied, E_α cannot possibly be Hausdorff. Now a Hausdorff projective limit E of spaces E_α can be identified (topology included) to a linear subspace of the product space $\tilde{E} = \prod_\alpha E_\alpha$ via the mapping $x \rightsquigarrow (\phi_\alpha(x))_\alpha$.

(ii) Suppose now that we are given, for each index α, a linear map

$\phi_\alpha : E_\alpha \to E$, such that $E = \bigcup_\alpha \phi_\alpha(E_\alpha)$. Suppose that the E_α are all locally convex. We may then define on E the finest locally convex topology such that all the mappings ϕ_α be continuous. A *convex* subset U of E is a neighborhood of zero in this topology if, for every α, $U \cap \phi_\alpha(E_\alpha)$ is of the form $\phi_\alpha(U_\alpha)$, where U_α is a neighborhood of zero in E_α. When E is equipped with this topology, it is called the *inductive limit* of the spaces E_α. The student will readily perceive that *LF*-spaces are a special kind of inductive limit.

(iii)   A notion closely related to the preceding one, and as a matter of fact a particular case of it, is the notion of locally convex direct sum. In this case, we suppose that every mapping ϕ_α is injective and we replace the hypothesis that $E = \bigcup_\alpha \phi_\alpha(E_\alpha)$ by the hypothesis that E is the algebraic direct sum of the vector spaces E_α: every element x of E can be written in one and only one manner as a sum $x = \sum_\alpha \phi_\alpha(x_\alpha)$ in which all x_α are equal to zero except possibly a finite number of them. Then the *direct sum topology* on E is the finest locally convex topology such that all the mappings ϕ_α are continuous. We say, when E carries it, that it is the topological direct sum of the E_α (note that the latter are locally convex). A convex subset U of E is a neighborhood of zero if, for all α, $U \cap \phi_\alpha(E_\alpha) = \phi_\alpha(U_\alpha)$, where U_α is a neighborhood of zero in E_α. As we have said, this is a particular case of inductive limit, as can be seen in the following way. For each finite set A of indices α, let E_A be the direct sum of the E_α's, with its obvious topology (the one carried over from the product TVS $\prod_{\alpha \in A} E_\alpha$ canonically identified to E_A); let $\phi_A : E_A \to E$ be the linear map defined by

$$\sum_{\alpha \in A} x_\alpha \to \sum_{\alpha \in A} \phi_\alpha(x_\alpha).$$

Then the direct sum topology on E is nothing else but the inductive limit topology of the spaces E_A with respect to the mappings ϕ_A.

But conversely the inductive limit E of spaces E_α with respect to certain mappings ϕ_α may be regarded as a quotient, modulo a closed linear subspace M, of a direct sum of the E_α. As direct sum, we take the linear subspace $\tilde{E}_0$ of the product $\tilde{E}$ of the E_α consisting of those elements $(x_\alpha)_\alpha$ such that $x_\alpha = 0$ for all α except possibly a finite number of them. The injection of E_{α_0} into $\tilde{E}_0$ is the mapping $x_{\alpha_0} \to (\tilde{x}_\alpha)_\alpha$, where $\tilde{x}_\alpha = 0$ if $\alpha \neq \alpha_0$ and $\tilde{x}_{\alpha_0} = x_{\alpha_0}$. The linear subspace M of $\tilde{E}_0$ will then be the kernel of the linear map $\phi : \tilde{E}_0 \to E$ defined by $\phi((x_\alpha)_\alpha) = \sum \phi_\alpha(x_\alpha)$. We leave to the student the verification that ϕ is a homomorphism of $\tilde{E}_0$ onto E.

Proof of Proposition 50.1

Proof of (50.2). Evident by Theorem 50.1, since

$$E \mathbin{\widehat{\otimes}_\varepsilon} F \cong \hat{E} \mathbin{\widehat{\otimes}_\varepsilon} F, \qquad E \mathbin{\widehat{\otimes}_\pi} F \cong \hat{E} \mathbin{\widehat{\otimes}_\pi} F.$$

Proof of (50.3). Let E_1 be a linear subspace of a nuclear space E. Every continuous seminorm p_1 on E_1 is the restriction of a continuous seminorm p on E. By hypothesis, there are two continuous seminorms $r \geqslant q \geqslant p$ on E such that the canonical mappings

$$\hat{E}_r \to \hat{E}_q \to \hat{E}_p$$

are nuclear; we may say that the canonical mapping $E_r \to E_p$ is *polynuclear* in the sense that it is the compose of two (or more) nuclear mappings. Let r_1 be the restriction of r to E_1; there is a canonical mapping $\widehat{(E_1)}_{r_1} \to \hat{E}_r$ and therefore a canonical mapping

(50.10) $$\widehat{(E_1)}_{r_1} \to \hat{E}_p,$$

which is obviously polynuclear; but the image of (50.10) is contained in the closed linear subspace $\widehat{(E_1)}_{p_1}$. Assertion (50.3) will then follow from the lemma:

LEMMA 50.1. *If a linear mapping* $u : E \to F$ *is polynuclear* (i.e., the compose of at least two nuclear operators) *and if* $u(E)$ *is contained in a complete linear subspace* F_1 *of* F, *then the mapping* $u : E \to F_1$ *is also nuclear.*

Proof of Lemma 50.1. The mapping u can be factorized into

$$E \xrightarrow{\ v\ } G \xrightarrow{\ w\ } F$$

with v, w nuclear; we may suppose F complete, for if we regard w as taking its values in $\hat{F}$, w is still nuclear; in particular, w is integral and can be factorized into

$$G \xrightarrow{\ a\ } H \xrightarrow{\ b\ } F$$

with a, b continuous linear mappings and H a Hilbert space. We set $f = a \circ v$; f is nuclear and u is factorized into

$$E \xrightarrow{\ f\ } H \xrightarrow{\ b\ } F.$$

Let H_1 be the closure of $f(E)$ in H, and π the orthogonal projection of H onto H_1 ; let us set $g = \pi \circ f$; g is a nuclear operator and u can be factorized into

$$E \xrightarrow{\;g\;} H_1 \xrightarrow{\;b_1\;} F_1 \longrightarrow F,$$

where the last arrow is the natural injection. Indeed, $b(f(E)) \subset F_1$. As b_1 is continuous, $b_1 \circ g$ is nuclear.　　　　　　Q.E.D.

Proof of (50.4). Let E be a nuclear space, M a closed linear subspace of E, and $\phi : E \to E/M$ the canonical homomorphism. Let $\dot{\mathrm{p}}$ be a continuous seminorm on E/M; the seminorm on E, $\mathrm{p} = \dot{\mathrm{p}} \circ \phi$, is continuous. There exists a continuous seminorm $\mathrm{q} \geqslant \mathrm{p}$ on E such that the canonical mapping $\hat{E}_{\mathrm{q}} \to \hat{E}_{\mathrm{p}}$ is polynuclear. Let $\dot{\mathrm{q}}$ be the seminorm on E/M defined by $\mathrm{q} : \dot{\mathrm{q}}(\dot{x}) = \inf_{x \in \dot{x}+M} \mathrm{q}(x)$, and let us set $\mathrm{q}_1 = \dot{\mathrm{q}} \circ \phi$. The open unit ball U_1 of q_1 is equal to $U + M$, with U, the open unit ball of q. We derive from this that $\hat{E}_{\mathrm{q}_1}$ is a quotient space of $\hat{E}_{\mathrm{q}}$ and that the mapping $\hat{E}_{\mathrm{q}} \to \hat{E}_{\mathrm{p}}$ decomposes into

$$(50.11) \qquad\qquad \hat{E}_{\mathrm{q}} \to \hat{E}_{\mathrm{q}_1} \to \hat{E}_{\mathrm{p}} \,,$$

where the arrows denote the canonical mappings. But, as is easily seen, $\hat{E}_{\mathrm{q}_1} \cong (E/M)_{\dot{\mathrm{q}}}^{\hat{}}$ and $\hat{E}_{\mathrm{p}} \cong (E/M)_{\dot{\mathrm{p}}}^{\hat{}}$; therefore, it will suffice to show that the second arrow, in (50.11), is a nuclear map. This will follow from the next lemma:

LEMMA 50.2. *Let $u : E \to F$ be polynuclear and equal to the compose*

$$E \xrightarrow{\;\phi\;} E/N \xrightarrow{\;\tilde{u}\;} F,$$

where N is a closed linear subspace of E, $\phi : E \to E/N$ is the canonical homomorphism, and $\tilde{u}$ is continuous. If F is complete, $\tilde{u}$ is nuclear.

Proof. By hypothesis, u can be factorized into $E \xrightarrow{v} G \xrightarrow{w} F$ with v, w nuclear; we may suppose G complete after extending w to its completion (which is permitted, as F is complete) and regarding v as valued in $\hat{G}$. Then we decompose v into $E \xrightarrow{a} H \xrightarrow{b} G$, with a, b continuous and H a Hilbert space. Observe that $u(N) = 0$; then let H_1 be the closure of $a(N)$ in H. By going to the quotients, a defines a continuous linear map $\bar{a} : E/N \to H/H_1$; H/H_1 can be identified with the orthogonal of H_1 in H, on which b is defined; therefore let $\bar{b}$ be the restriction of b to H/H_1 . We have obtained a decomposition of $\tilde{u}$ into

$$E/N \xrightarrow{\;\bar{a}\;} H/H_1 \xrightarrow{\;\bar{b}\;} G \xrightarrow{\;w\;} F.$$

As w is nuclear, so is $\tilde{u}$.

Proof of (50.5). Let $E = \prod_\alpha E_\alpha$ be the product of a family of nuclear spaces E_α. Every continuous seminorm on E is at most equal to a seminorm of the form

$$\mathfrak{p}_A((x_\alpha)) = \sum_{\alpha \in A} \mathfrak{p}_\alpha(x_\alpha),$$

where A is a finite set of indices α, and where each $\mathfrak{p}_\alpha$ is a continuous seminorm on E_α. It is then clear that $\hat{E}_{\mathfrak{p}_A}$ is isomorphic to the finite product $\prod_{\alpha \in A} (\hat{E}_\alpha)_{\hat{\mathfrak{p}}}$. It suffices then to select, for each $\alpha \in A$, a continuous seminorm $\mathfrak{q}_\alpha \geqslant \mathfrak{p}_\alpha$ on E_α such that the canonical mapping $(\hat{E}_\alpha)_{\hat{\mathfrak{q}}_\alpha} \to (\hat{E}_\alpha)_{\hat{\mathfrak{p}}_\alpha}$ is nuclear, and set

$$\mathfrak{q}((x_\alpha)) = \sum_{\alpha \in A} \mathfrak{q}_\alpha(x_\alpha).$$

The canonical mapping of $\hat{E}_\mathfrak{q} \cong \prod_{\alpha \in A} (\hat{E}_\alpha)_{\hat{\mathfrak{q}}_\alpha}$ into $\hat{E}_{\mathfrak{p}_A}$ is then nuclear, as a product of nuclear mappings.

Proof of (50.6). Let E be the topological direct sum of a sequence of nuclear spaces E_k ($k = 0, 1, \ldots$). Let $\mathfrak{p}$ be a continuous seminorm on E, and $\mathfrak{p}_k$ its restriction to E_k (regarded as a linear subspace of E). By hypothesis, there is a continuous seminorm $\mathfrak{q}_k \geqslant \mathfrak{p}_k$ on E_k such that the canonical mapping $u_k : (E_k)_{\hat{\mathfrak{q}}_k} \to (E_k)_{\hat{\mathfrak{p}}_k}$ is nuclear. This means that there is a sequence $(x'_{k,n})$ in the unit ball of $(E_k)'_{\mathfrak{q}_k}$, a sequence $(y_{k,n})$ in the unit ball of $(E_k)_{\hat{\mathfrak{p}}_k}$, and a sequence $\{\lambda_{k;n}\}$ in l^1 such that u_k is given by

$$x_k \rightsquigarrow \sum_{n=0}^\infty \lambda_{k,n} \langle x'_{k,n}, x_k \rangle y_{k,n}.$$

Observe that the dual E'_k of E_k can be regarded as a linear subspace of E', via the mapping $x'_k \rightsquigarrow (x \rightsquigarrow \langle x'_k, x_k \rangle)$, where x_k is the kth component of x. Consider then the mapping

$$u : x = \sum_k x_k \rightsquigarrow \sum_{k=0}^\infty \sum_{n=0}^\infty \rho_k^{-1} \lambda_{k,n} \langle \rho_k x'_{k,n}, x_k \rangle y_{k,n}.$$

Here ρ_k is a number $\geqslant 2^k \sum_{n=0}^\infty |\lambda_{k,n}|$ and $y_{k,n}$ is regarded as an element of the direct sum of the Banach spaces $(E_k)_{\hat{\mathfrak{p}}_k}$; this direct sum is trivially isomorphic to $\hat{E}_\mathfrak{p}$. On the other hand, let us set, for $x = \sum_k x_k \in E$,

$$\mathfrak{q}(x) = \sup_k \rho_k \, \mathfrak{q}_k(x).$$

Obviously, $\mathfrak{q}$ is a continuous seminorm on E and the $\rho_k x'_{k,n}$ all belong to the closed unit ball of E'_q. Furthermore, the mapping u above induces the canonical mapping of $\hat{E}_q$ into $\hat{E}_p$, which is therefore nuclear.

Proof of (50.7). Follows from the combination of (50.3) and (50.5) since a Hausdorff projective limit is a linear subspace of a product (see preliminary remarks).

Proof of (50.8). Follows from the combination of (50.4) and (50.6) since an inductive limit is a quotient of a topological direct sum (see preliminary remarks).

Proof of (50.9). For all locally convex Hausdorff spaces G, we have

$$(E \otimes F) \otimes_\varepsilon G = E \otimes_\varepsilon (F \otimes_\varepsilon G) = E \otimes_\pi (F \otimes_\pi G) = (E \otimes F) \otimes_\pi G,$$

where $E \otimes F$ stands for $E \otimes_\varepsilon F = E \otimes_\pi F$. The associativity of topological tensor products π and ε follows straightforwardly from their definitions.

PROPOSITION 50.2.   *Let E be a nuclear space. Then:*

(50.12)   *every bounded subset of E is precompact;*

(50.13)   *every closed equicontinuous subset of the dual E' of E is a metrizable compact set* (for the strong dual topology);

(50.14)   *E is a linear subspace of a product of Hilbert spaces.*

Proof of (50.12). Let B be a bounded subset of E, p a continuous seminorm on E, and $q \geqslant p$ another one such that $\hat{E}_q \to \hat{E}_p$ be nuclear. The canonical mapping of E into $\hat{E}_p$ can be decomposed into

$$E \to \hat{E}_q \to \hat{E}_p;$$

as the arrows denote continuous linear mappings and the last one a compact mapping, the canonical image of B in $\hat{E}_p$ is precompact. As p is arbitrary, this implies immediately that B is precompact.

Proof of (50.13). Let A' be a closed equicontinuous subset of E'; we may assume that A' is convex and balanced. There is another such set $B' \supset A'$ with the property that the injection $E'_{A'} \to E'_{B'}$ is nuclear. The image of A' in $E'_{B'}$ is precompact; as it is closed and as $E'_{B'}$ is a Banach space (we may choose B' weakly compact), A', regarded as a subset of $E'_{B'}$, is compact and, of course, metrizable. Since the topology induced by E' on A' is weaker than the one induced by $E'_{B'}$, A' is a compact metrizable subset of E'.

Proof of (50.14). Let p be an arbitrary continuous seminorm on E, and $q \geqslant p$ another continuous seminorm such that the canonical map $\hat{E}_q \to \hat{E}_p$ is nuclear, hence integral. This implies that there is a Hilbert space H_p such that that canonical mapping has the factorization

$$\hat{E}_q \to H_p \to \hat{E}_p \, .$$

By introducing the canonical mapping of E into $\hat{E}_q$, we obtain a factorization

$$E \to H_p \to \hat{E}_p \, .$$

Let us denote by i_p the map corresponding to the first arrow and by $\mathscr{H}$ the product TVS $\prod_p H_p$. It is immediately verified that the mapping $x \rightsquigarrow (i_p(x))_p$ of E into $\mathscr{H}$ is an isomorphism into (for the TVS structure).

COROLLARY 1. *Let E be a quasi-complete nuclear space. Every closed bounded subset of E is compact.*

We recall that a TVS E is said to be quasi-complete if every closed bounded subset of E is complete. Banach spaces being complete are a fortiori quasi-complete; we have:

COROLLARY 2. *A normable space E is nuclear if and only if it is finite dimensional.*

We recall that E is nuclear if and only if $\hat{E}$ is nuclear.

COROLLARY 3. *A quasi-complete barreled space which is nuclear is a Montel space.*

We shall discuss some examples of nuclear spaces and of spaces which are neither nuclear nor normable in the next chapter. In particular, we shall see that the countability restriction in Properties (50.6) and (50.8) cannot be dropped (in general).

PROPOSITION 50.3. *Let E be a nuclear space. The identity mapping of E is the uniform limit over the compact subsets of E of continuous linear mappings of E into itself whose image is finite dimensional, i.e., E has the following approximation property:*

(A) *For every compact subset K and every neighborhood of zero U in E there exists a continuous linear map $u : E \to E$ with finite dimensional image, such that, for all $x \in K$, $u(x) - x \in U$.*

Proof. Suppose that U is the closed unit semiball of a continuous seminorm p on E and let us select another continuous seminorm

$q \geqslant \dot{p}$ such that the canonical mapping $\hat{E}_q \to \hat{E}_p$ is nuclear. We have the commutative diagram

$$
\begin{array}{ccc}
E & \xrightarrow{\;I\;} & E \\
\downarrow & & \downarrow \\
\hat{E}_q & \longrightarrow & \hat{E}_p
\end{array}
$$

where I is the identity mapping and the other arrows are the canonical ones. The lower horizontal one is of the form

$$(50.15) \qquad \hat{x} \rightsquigarrow \sum_{k=0}^{\infty} \lambda_k \langle x'_k , \hat{x} \rangle \hat{y}_k \,,$$

where the x'_k are bounded in E'_q , the $\hat{y}_k$ are bounded in $\hat{E}_p$, and $(\lambda_k) \in l^1$. The mapping (50.15) is the limit, in $L(\hat{E}_q \,;\, \hat{E}_p)$ for the operators norm, of mappings

$$\hat{x} \rightsquigarrow \sum_{k=0}^{n} \lambda_k \langle x'_k , \hat{x} \rangle \hat{y}_k \qquad (n = 0, 1,...),$$

which, in turn, are the limit of mappings

$$\hat{x} \rightsquigarrow \sum_{k=0}^{n} \lambda_k \langle x'_k , \hat{x} \rangle y_{j,k} \qquad (j = 0, 1,...),$$

where now the $y_{j,k}$ belong to the image of E in $\hat{E}_p$; this follows from the fact that the image in question is dense (by definition of $\hat{E}_p$ as the completion of $E/\mathrm{Ker}\ p$). We reach the conclusion that to every $\varepsilon > 0$ there are integers j and n sufficiently large that, for all $x \in E$ satisfying $q(x) \leqslant 1$,

$$x - \sum_{k=0}^{n} \lambda_k \langle x'_k , x \rangle y_{j,k} \in \varepsilon U.$$

If K is now an arbitrary bounded subset of E, we select $\varepsilon > 0$ such that $\varepsilon K \subset U_q = \{x \in E; q(x) \leqslant 1\}$.   Q.E.D.

Remark 50.2. The reader should not think that Property (A) is in any way a prerogative of nuclear spaces: it is not difficult to see that all Hilbert spaces have it (Lemma 48.3), that the spaces L^p $(1 \leqslant p \leqslant \infty)$ have it, and that the same is true of the space $\mathscr{C}(X)$ of continuous complex functions on a compact (or on a locally compact) space. As a matter of fact, no space is known which does not possess (A)!

Let E be a nuclear space; every weakly compact subset of E is bounded for the initial topology (this is true in general, in view of Mackey's theorem). But conversely, every bounded subset of E is precompact. It follows from this that, if E is nuclear and complete, the topology $\tau(E', E)$ on E is identical to the strong dual topology (which we have sometimes denoted by $b = b(E', E)$).

PROPOSITION 50.4. *Let E be a nuclear and complete space, and F a complete locally convex Hausdorff space. Then*

$$E \mathbin{\widehat{\otimes}} F \cong L_\varepsilon(E'_b; F) \cong L_\varepsilon(F'_\tau; E).$$

Proof. We know that $L_\varepsilon(E'_b; F)$ is complete (Proposition 42.3); it suffices to show that its linear subspace consisting of the mappings with finite dimensional image is dense. Note that $L_\varepsilon(E'_b; F) \cong L_\varepsilon(F'_\sigma; E_\sigma)$ and consider an element $v : F'_\sigma \to E_\sigma$ in this space. Let B' be an equicontinuous subset of F'; if we suppose that B' is weakly closed, which we may, B' is weakly compact, hence $v(B')$ is weakly compact in E (as v is continuous). But then $v(B')$ is closed and bounded, hence compact in E. We take into account Property (A) with $K = v(B')$ and U arbitrary; there is $u : E \to E$ with finite dimensional image such that, for all $y' \in B'$,

$$v(y') - u(v(y')) \in U.$$

The mapping $u \circ v : F'_\sigma \to E_\sigma$ has finite dimensional image and is continuous.

PROPOSITION 50.5. *Let E, F be two locally convex Hausdorff spaces. We make the following hypotheses:* (i) *E and F are complete;* (ii) *E is barreled;* (iii) *E' is nuclear and complete. Then $L_b(E; F)$ is complete, and we have*

$$E' \mathbin{\widehat{\otimes}} F \cong L_b(E; F).$$

Proof. We begin by showing that E' is semireflexive (i.e., equal to its bidual, E''', as a vector space). We apply Theorem 36.3: a closed and bounded subset A' of E', in the sense of $\sigma(E', E'')$, is of course closed and also, by Mackey's theorem, bounded in the sense of $b(E', E)$. As E' is nuclear and complete, A' is compact and, a fortiori, compact for $\sigma(E', E'')$, which proves our assertion.

As E' is semireflexive, E is semireflexive. Indeed, as E carries its Mackey topology $\tau(E, E')$, the natural injection $E \to E''$ is an isomorphism into; as E is complete, it is a closed subspace of E''. But E'' and $E = E''$

have the same dual, $E' = E'''$, hence they must be equal. Incidentally, we note (by Proposition 36.4) that E' is barreled, therefore E' is a Montel space. Then E, as the strong dual of a Montel space, is also a Montel space (Proposition 36.10).

At any event, as E' and F are both complete, we may apply Proposition 50.4. We obtain

$$E' \hat{\otimes} F = L_e(E; F).$$

But in the dual of a Montel space, there is identity between equicontinuous sets and strongly bounded sets (Theorem 33.2), whence Proposition 50.5.

Proposition 50.5 has applications to many situations occurring in distribution theory.

COROLLARY. *Under the hypotheses of Proposition 50.5, $L_b(E, F)$ is a nuclear space.*

It suffices to combine Proposition 50.5 with (50.9).

We close this chapter with some results about Fréchet spaces.

PROPOSITION 50.6. *A Fréchet space E is nuclear if and only if its strong dual is nuclear.*

Proof 1. Suppose that E is nuclear. Then (Corollary 3 of Proposition 50.2) E is a Montel space, therefore E is reflexive. Let $u : F \to E$ be a linear map of a Banach space F into E. Let us set $G = F'$; the bitranspose of u, ${}^{tt}u : G' \to E'' = E$, is weakly continuous (here weakly means for the topology $\sigma(G', G)$ on G and $\sigma(E, E')$ on E), which means that ${}^{tt}u \in L(G'_\sigma ; E_\sigma) = L(E'_\tau ; G)$ (Proposition 42.2). As E is nuclear, the latter space is equal to $E \hat{\otimes} G$ (Proposition 50.4). We now use the fundamental theorem on completed π-products of Fréchet spaces (Theorem 45.1): the mapping ${}^{tt}u$ is represented by an element $\theta \in E_A \hat{\otimes} G$, with A a closed convex balanced bounded subset of E. This implies immediately that ${}^{tt}u$ is nuclear. But then the restriction of ${}^{tt}u$ to the closed linear subspace $F \subset G' = F''$, restriction which is equal to u, is also nuclear. Thus Condition (c) in Theorem 50.1 is satisfied: we conclude that E' is nuclear.

Proof 2. Suppose now that E' is nuclear. As E' is complete (Corollary 2 of Theorem 32.2), E' is a Montel space (Corollary 3 of Proposition 50.2), hence E' is reflexive. On the other hand, E can be regarded as a closed linear subspace of its bidual E''; as E and E'' have the same dual, they must be equal. Thus E is semireflexive; but a barreled space which is semireflexive is reflexive. We are going to show that, given any Banach space F, the canonical mapping $E \mathbin{\widehat{\otimes}_\pi} F \to E \mathbin{\widehat{\otimes}_\varepsilon} F$ is an isomorphism.

First of all, that mapping is onto. Indeed, let $\theta \in E \mathbin{\widehat{\otimes}_\varepsilon} F$; θ defines a continuous linear map $E' \to F$. In view of the nuclearity of E' and of Theorem 50.1, Part (c), the mapping defined by θ is nuclear, hence defined by some element of $E'' \mathbin{\widehat{\otimes}_\pi} F$; but $E'' = E$.

Next we show that the mapping $E \mathbin{\widehat{\otimes}_\pi} F \to E \mathbin{\widehat{\otimes}_\varepsilon} F$ is one-to-one. The transpose of this mapping is the natural injection $J(E, F) \to B(E, F)$; it will suffice to show that its image is dense for the *weak* dual topology on $B(E, F) = (E \mathbin{\widehat{\otimes}_\pi} F)'$. We now apply Corollary 2 of Theorem 45.1: every element $\theta \in E \mathbin{\widehat{\otimes}_\pi} F$ belongs to the closed convex balanced hull $\Gamma(H \otimes K)$ of the tensor product $H \otimes K$ of a compact subset H of E and a compact subset K of F. It will therefore suffice to show that $E' \otimes F' \subset J(E, F)$ is dense in $B(E, F)$ for the topology of uniform convergence on products of two compact sets. Let us regard an element of $B(E, F)$ as a mapping $u : F \to E'$; the elements of $E' \otimes F'$ are then the mappings $v : F \to E'$ with finite dimensional image. As E' is nuclear, it has Property (A) in Proposition 50.3. We can approximate the identity mapping of E' uniformly over the compact subsets of E' by continuous linear mappings $w : E' \to E'$ with finite dimensional image. It is then clear that mappings of the form $w \circ u \in E' \otimes F'$ will converge to u uniformly on compact subsets of F. This proves the density of $E' \otimes F'$ in $B(E, F)$.

We have shown that the mapping $E \mathbin{\widehat{\otimes}_\pi} F \to E \mathbin{\widehat{\otimes}_\varepsilon} F$ is both onto and one-to-one; as it is continuous and as the two completed tensor products are Fréchet spaces, it follows from the open mapping theorem that this mapping is an isomorphism.

PROPOSITION 50.7. *Let E, F be two Fréchet spaces. If E is nuclear, we have the canonical isomorphisms*

$$E' \mathbin{\widehat{\otimes}} F' \cong B(E, F) \cong (E \mathbin{\widehat{\otimes}} F)'.$$

We shall give the proof only in the case where F is also nuclear; the general case requires a slightly lengthier treatment.

Proof of Proposition 50.7 (when both E and F are nuclear). Let $u : E \to F'$ be a continuous linear map; then the bilinear form on $E \times F$, $(x, y) \rightsquigarrow \langle u(x), y \rangle$, is separately continuous, hence continuous. This means that we have the vector space isomorphism

(50.16) $$B(E, F) \cong L(E; F').$$

In view of Proposition 50.5, where the hypotheses are obviously satisfied, we have

$$E' \mathbin{\widehat{\otimes}} F' \cong L(E; F').$$

Here the isomorphism extends to the topologies ($L(E; F)$ carries the topology of bounded convergence). It suffices therefore to show that (50.16) also extends to the topologies. The topology of bounded convergence in $L(E; F')$ is equivalent to the topology of uniform convergence on the products $A \times B \subset E \times F$, with A (resp. B) bounded in E (resp. F), in $B(E, F)$, or, if one prefers to use the duality between $B(E, F)$ and $E \otimes F$, to the topology of uniform convergence on the closed convex balanced hulls of tensor products $A \otimes B$ (with A, B as above). On the other hand, $B(E, F)$ carries the topology of uniform convergence on the bounded subsets of $E \mathbin{\widehat{\otimes}} F$. The result will therefore be proved if we show that every bounded subset of $E \mathbin{\widehat{\otimes}} F$ is contained in the closed convex balanced hull of a tensor product $A \otimes B$ of bounded subsets of E and F. This is true under the hypothesis that E is nuclear (F does not have to be nuclear); however, in the situation where both E and F are nuclear, we also know that $E \mathbin{\widehat{\otimes}} F$ is nuclear (Proposition 50.1, (50.9)). Therefore, the bounded subsets of the three nuclear Fréchet spaces E, F, and $E \mathbin{\widehat{\otimes}} F$ are relatively compact. It then suffices to apply Corollary 2 of Theorem 45.1.   Q.E.D.

By virtue of Propositions 50.4, 50.5, and 50.6 we see that when E and F are Fréchet spaces, E (and therefore E') being nuclear, we have

(50.17) $$E \mathbin{\widehat{\otimes}} F \cong L(E'; F),$$

(50.18) $$E' \mathbin{\widehat{\otimes}} F \cong L(E; F),$$

(50.19) $$E' \mathbin{\widehat{\otimes}} F' \cong (E \mathbin{\widehat{\otimes}} F)' \cong B(E, F),$$

where the duals carry the strong dual topology, the spaces of continuous linear maps carry the topology of uniform convergence on the bounded subsets, and $B(E, F)$ carries the topology of uniform convergence on the products of bounded sets.

51

Examples of Nuclear Spaces.
The Kernels Theorem

In this chapter, we shall prove that the most important spaces occurring in distribution theory, $\mathscr{C}_c^\infty$, $\mathscr{C}^\infty$, $\mathscr{S}$, $\mathscr{D}'$, $\mathscr{E}'$, and $\mathscr{S}'$, are nuclear; this is also true of the space of holomorphic functions $H(\Omega)$ in some open subset Ω of $\mathbf{C}^n$, of the space of polynomials in n variables, $\mathscr{P}_n$, and of its dual, the space of formal power series in n variables, $\mathscr{Q}_n$. To prove the nuclearity of the latter is most easy. Indeed we have:

THEOREM 51.1. *Let S be an arbitrary set; the product space* $\mathbf{C}^S$ *is nuclear.*

It suffices to apply Proposition 50.1, (50.5). Note that $\mathbf{C}^S$ is the space of all functions $S \to \mathbf{C}$ provided with the topology of pointwise convergence. Now we may identify $\mathscr{Q}_n$ to the space of sequences in n indices, or, equivalently, to the space of functions on $\mathbf{N}^n$, the set of n-tuples $p = (p_1, ..., p_n)$ consisting of n integers $\geqslant 0$. The identification is the obvious one: to $u \in \mathscr{Q}_n$ corresponds the function which assigns to every $p \in \mathbf{N}^n$ the pth coefficient u_p of u. By taking $S = \mathbf{N}^n$ in Theorem 51.1, we obtain:

COROLLARY 1. *The space* $\mathscr{Q}_n$ *of formal power series in n variables, equipped with the topology of simple convergence of the coefficients, is nuclear.*

COROLLARY 2. *The space* $\mathscr{P}_n$ *of polynomials in n variables, equipped with its LF topology, is nuclear.*

Indeed, $\mathscr{P}_n$ is the strong dual of $\mathscr{Q}_n$ and the latter is a nuclear Fréchet space. It suffices therefore to apply Proposition 50.6.

One can also prove Corollary 2 above in the following manner: $\mathscr{P}_n$ can be identified with the topological direct sum $\sum_{p \in \mathbf{N}^n} \mathbf{C}_p$, where each $\mathbf{C}_p$ is a copy of the complex plane $\mathbf{C}$. It then suffices to apply Proposition 50.1, (50.6). In relation with this, let us show that the countability restriction in (50.6) and (50.8) cannot be dropped:

THEOREM 51.2. *If the set S is not countable, the topological direct sum*

$$\mathbf{C}_S = \sum_{s \in S} \mathbf{C}_s \qquad (\mathbf{C}_s \cong \mathbf{C} \text{ for all } s)$$

is not nuclear.

Proof. The dual of $\mathbf{C}_S$ is easily seen to be isomorphic to the product space $\mathbf{C}^S = \prod_{s \in S} \mathbf{C}_s$. The closed equicontinuous subsets of the dual $\mathbf{C}^S$ of $\mathbf{C}_S$ are the compact subsets of $\mathbf{C}^S$. If $\mathbf{C}_S$ were nuclear, these compact subsets should be metrizable (Proposition 50.2). But a compact set such as $[0, 1]^S$ is metrizable (if and) only if S is countable.

Remark 51.1. Theorems 51.1 and 51.2 show that the dual of a nuclear space (e.g., $\mathbf{C}^S$ with S noncountable) is not necessarily nuclear. Thus Proposition 50.6 expresses a property of Fréchet spaces which, although not altogether characteristic of these, is not true for general spaces.

The nuclearity of the spaces of type $\mathscr{C}^\infty$ will follow from the fact that the space s of *rapidly decreasing sequences* is nuclear. We proceed to define and to study s.

The sequences which we are now going to consider will have n indices $p_1, ..., p_n$; but these indices will be positive, negative, or zero integers. We shall then use the notation $|p| = \sum_{j=1}^n |p_j|$; the set of the n-tuples $p = (p_1, ..., p_n)$, p_j : integers $\geqslant 0$ or < 0, will be denoted by $\mathbf{Z}^n$. This slight departure from our previous practice is due to the fact that we wish to relate rapidly decreasing sequences to Fourier series.

A complex sequence $\sigma = (\sigma_p)_{p \in \mathbf{Z}^n}$ is said to be rapidly decreasing if, for every constant $k \geqslant 0$, the quantity

$$(51.1) \qquad \sum_{p \in \mathbf{Z}^n} (1 + |p|)^k |\sigma_p|$$

is finite. The rapidly decreasing sequences form a vector space, which we denote by s and on which we put the topology defined by the seminorms (51.1) for $k = 0, 1, 2, ...$ It is easy to check that s is a Fréchet space.

A sequence $\tau = (\tau_p)$ is said to be *slowly growing* if there is a constant $k \geqslant 0$ such that the sequence $\{(1 + |p|)^{-k} \tau_p\}$ is bounded. We leave to the student the verification of the fact that the mapping

$$\tau = (\tau_p) \rightsquigarrow (\sigma = (\sigma_p) \rightsquigarrow \sum_{p \in \mathbf{Z}^n} \sigma_p \tau_p)$$

is an isomorphism (for the vector space structures) of the space of slowly growing sequences (which we shall denote by s') onto the dual of s.

For us, the importance of rapidly decreasing sequences stems from the fact that the Fourier coefficients of a periodic $\mathscr{C}^\infty$ function form such a sequence. Let us denote by $\mathbf{I}^n$ the hypercube $[0, 1]^n$ (regarded as a subset of $\mathbf{R}^n$). We denote then by $\mathscr{C}^\infty(\mathbf{R}^n)^\natural$ the vector space of periodic $\mathscr{C}^\infty$ functions in $\mathbf{R}^n$ with $\mathbf{I}^n$ as period. We provide $\mathscr{C}^\infty(\mathbf{R}^n)^\natural$ with the topology induced by $\mathscr{C}^\infty(\mathbf{R}^n)$ (i.e., with the topology of uniform convergence of the functions and all their derivatives); it turns it into a Fréchet space. Then, for every $u \in \mathscr{C}^\infty(\mathbf{R}^n)^\natural$ and every $p \in \mathbf{Z}^n$, we set

$$\hat{u}_p = \int_{\mathbf{I}^n} \exp(-2i\pi \langle p, x \rangle)\, u(x)\, dx,$$

where $\langle p, x \rangle = p_1 x_1 + \cdots + p_n x_n$. Let now $\alpha \in \mathbf{N}^n$ be arbitrary. We have, because of the smoothness and the periodicity of u,

$$(2i\pi p)^\alpha \hat{u}_p = \int_{\mathbf{I}^n} \exp(-2i\pi \langle p, x \rangle)\, (\partial/\partial x)^\alpha\, u(x)\, dx,$$

which implies immediately that the sequence $(\hat{u}_p)$ belongs to s. Conversely, let $\sigma = (\sigma_p)$ be a sequence belonging to s; the series

$$\sum_{p \in \mathbf{Z}^n} \sigma_p \exp(2i\pi \langle p, x \rangle)$$

converges in $\mathscr{C}^\infty(\mathbf{R}^n)$ to a periodic function u whose pth Fourier coefficient is equal to σ_p. Because of obvious continuity properties, we may state:

THEOREM 51.3. *The Fourier expansion $u \rightsquigarrow (\hat{u}_p)_{p \in \mathbf{Z}^n}$ is an isomorphism of the space of periodic $\mathscr{C}^\infty$ functions, $\mathscr{C}^\infty(\mathbf{R}^n)^\natural$, onto the space of rapidly decreasing sequences, s.*

In this statement, isomorphism is meant in the sense of TVS structures.

Stressing the analogy between the spaces s and $\mathscr{S}$ on one hand, their duals s' and $\mathscr{S}'$ on the other, is hardly needed. The Fourier expansion of periodic *distributions* in $\mathbf{R}^n$, $\mathscr{D}'(\mathbf{R}^n)^\natural$, is an isomorphism of this space onto $\mathscr{S}'$. Furthermore, the space $\mathscr{S}$ itself can be embedded isomorphically into the space $\mathscr{C}^\infty(\mathbf{R}^n)^\natural$. We shall now briefly describe such an embedding. Let us denote by $\mathscr{C}_0^\infty(\mathbf{R}^n)^\natural$ the subspace of $\mathscr{C}^\infty(\mathbf{R}^n)^\natural$ consisting of those functions which vanish *of infinite order* at the boundary of the hypercube $\mathbf{I}^n$. We put on $\mathscr{C}_0^\infty(\mathbf{R}^n)^\natural$ the topology induced by $\mathscr{C}^\infty(\mathbf{R}^n)^\natural$; it becomes a Fréchet space.

Let us denote by $h(t)$ the function defined for $0 < t < 1$ by

$$h(t) = \frac{1}{1-t} - \frac{1}{t}.$$

Then, for every $\phi \in \mathscr{S}(\mathbf{R}^n)$, we set

$$\tilde{\phi}(x) = \phi(h(x_1), ..., h(x_n)), \qquad 0 < x_j < 1, \quad 1 \leqslant j \leqslant n.$$

It is immediately seen that the function $\tilde{\phi}$ vanishes at the boundary of $\mathbf{I}^n$. We denote by $\phi^\natural$ the periodic function (with $\mathbf{I}^n$ as period) which is equal to $\tilde{\phi}$ in $\mathbf{I}^n$. We leave to the student, as an exercise, the proof of the following result:

THEOREM 51.4.   *The mapping $\phi \rightsquigarrow \phi^\natural$ is an isomorphism (for the TVS structures) of the space $\mathscr{S}(\mathbf{R}^n)$ of $\mathscr{C}^\infty$ functions in $\mathbf{R}^n$, rapidly decreasing at infinity, onto the space $\mathscr{C}_0^\infty(\mathbf{R}^n)^\natural$ of periodic $\mathscr{C}^\infty$ functions with period $\mathbf{I}^n$, which vanish of infinite order at the boundary of $\mathbf{I}^n$.*

We recall that $\mathbf{I}^n = [0, 1]^n \subset \mathbf{R}^n$.

The nuclearity of the main spaces occurring in distribution theory will follow from the next result:

THEOREM 51.5.   *The Fréchet space s of rapidly decreasing sequences is nuclear.*

Proof. Let $\mathfrak{p}$ be the seminorm (51.1); of course, $\mathfrak{p}$ is a norm. The completion $\hat{s}_\mathfrak{p}$ is the Banach space $s_{(k)}$ of complex sequences $\sigma = (\sigma_p)_{p \in \mathbf{Z}^n}$ such that (51.1) is finite. Its dual is the Banach space $s'_{(k)}$ of complex sequences $\tau = (\tau_p)$ such that

$$(51.2) \qquad \sup_p (1 + |p|)^{-k} |\tau_p|$$

is finite ($s'_{(k)}$ is equipped with the norm (51.2)). Let us denote by e_p the sequence whose pth term is equal to one while all the others are equal to zero. We then consider the following element of $s'_{(k+n+1)} \otimes s_{(k)}$:

$$\theta = \sum_{p \in \mathbf{Z}^n} \lambda_p x'_p \otimes y_p,$$

where

$$\lambda_p = (1 + |p|)^{-n-1}, \qquad \text{thus the sequence } (\lambda_p) \text{ is summable};$$

$$x'_p = (1 + |p|)^{k+n+1} e_p \quad \text{(belongs to the unit ball of } s'_{(k+n+1)}\text{)};$$

$$y_p = (1 + |p|)^{-k} e_p \quad \text{(belongs to the unit ball of } s_{(k)}\text{)}.$$

Now let $\sigma = (\sigma_p)$ be an arbitrary element of $s_{(k+n+1)}$; we have

$$\theta(\sigma) = \sum_{p \in \mathbf{Z}^n} \lambda_p (1 + |p|)^{n+1} \sigma_p e_p = \sum_{p \subset \mathbf{Z}^n} \sigma_p e_p = \sigma.$$

This means that θ is the natural injection of $s_{(k+n+1)}$ into $s_{(k)}$; we have thus proved that this injection is nuclear. We may then interpret $s_{(k+n+1)}$ as the completion $\hat{s}_q$ of the space s with respect to a norm q which is (51.2) where k has been replaced by $k + n + 1$. We see that the nuclearity of s is thus established.

COROLLARY.   *The following spaces are nuclear:*

$\mathscr{S}(\mathbf{R}^n)$, $\mathscr{S}'(\mathbf{R}^n)$, $\mathscr{C}_c^\infty(K)$ (K: *compact subset of* $\mathbf{R}^n$);

$\mathscr{C}_c^\infty(\Omega)$, $\mathscr{C}^\infty(\Omega)$, $\mathscr{D}'(\Omega)$, $\mathscr{E}'(\Omega)$ (Ω: *open subset of* $\mathbf{R}^n$);

$H(\Omega)$, $H'(\Omega)$ (Ω: *open subset of* $\mathbf{C}^n$).

We recall that $H(\Omega)$ is the space of holomorphic functions in Ω and $H'(\Omega)$, its dual, is the space of analytic functionals in Ω.

Proof of Corollary. $\mathscr{S}(\mathbf{R}^n)$ is isomorphic to a subspace of s, as we see by combining Theorems 51.3 and 51.4, hence is nuclear (by (50.3), Proposition 50.1). So is its dual, $\mathscr{S}'(\mathbf{R}^n)$, by Proposition 50.6, and its linear subspace $\mathscr{C}_c^\infty(K)$ (again by (50.3)). But then $\mathscr{C}_c^\infty(\Omega)$, as a countable inductive limit of spaces $\mathscr{C}_c^\infty(K)$, is nuclear, by virtue of (50.8). The dual of the nuclear Fréchet space $\mathscr{C}_c^\infty(K)$ is nuclear; as $\mathscr{D}'(\Omega)$ is a projective limit of such duals, as is seen at once, we derive from (50.7) that $\mathscr{D}'(\Omega)$ is nuclear. On the other hand, consider, for each compact set $K \subset \Omega \subset \mathbf{R}^n$, a function $\phi_K \in \mathscr{C}_c^\infty(K)$, in such a way that, to every point $x \in \Omega$, there is a set K such that ϕ_K is different from zero at every point of some neighborhood of x. Then $\mathscr{C}^\infty(\Omega)$ is the projective limit of the spaces $\mathscr{C}_c^\infty(K)$ with respect to the mappings $f \rightsquigarrow \phi_K f$; it follows then from (50.7) that $\mathscr{C}^\infty(\Omega)$ is nuclear and so is $\mathscr{E}'(\Omega)$, by Proposition 50.6. Finally, $H(\Omega)$ is nuclear as it is a subspace of $\mathscr{C}^\infty(\Omega)$ (where we identify Ω to a subspace of $\mathbf{R}^{2n}$) and $H'(\Omega)$ is nuclear as the dual of a nuclear Fréchet space (Proposition 50.6).

THEOREM 51.6.   *We have the following canonical isomorphisms:*

(51.3) $\mathscr{S}(\mathbf{R}^m) \mathbin{\hat{\otimes}} \mathscr{S}(\mathbf{R}^n) \cong \mathscr{S}(\mathbf{R}^{m+n})$;

(51.4) $\mathscr{C}^\infty(X) \mathbin{\hat{\otimes}} \mathscr{C}^\infty(Y) \cong \mathscr{C}^\infty(X \times Y)$ ($X \subset \mathbf{R}^m$, $Y \subset \mathbf{R}^n$ *open sets*);

(51.5) $\mathscr{C}_c^\infty(K) \mathbin{\hat{\otimes}} \mathscr{C}_c^\infty(L) \cong \mathscr{C}_c^\infty(K \times L)$ ($K \subset \mathbf{R}^m$, $L \subset \mathbf{R}^n$ *compact sets*);

(51.6) $H(X) \mathbin{\hat{\otimes}} H(Y) \cong H(X \times Y)$ ($X \subset \mathbf{C}^m$, $Y \subset \mathbf{C}^n$ *open sets*).

Proof.   We shall prove only (51.3); the proof is the same in all the other cases. It follows from Theorem 39.2 that $\mathscr{S}(\mathbf{R}^m) \otimes \mathscr{S}(\mathbf{R}^n)$ is dense in $\mathscr{S}(\mathbf{R}^m \times \mathbf{R}^n)$; it suffices to show that the latter induces on the former the topology $\pi = \varepsilon$. It induces a weaker topology, since the bilinear mapping $(u, v) \rightsquigarrow u \otimes v$ of $\mathscr{S}(\mathbf{R}^m) \times \mathscr{S}(\mathbf{R}^n)$ into $\mathscr{S}(\mathbf{R}^m \times \mathbf{R}^n)$ is continuous (it is separately continuous!). On the other hand, if $\{f_\nu\}$ is a sequence converging to zero in $\mathscr{S}(\mathbf{R}^{m+n})$, it converges to zero uniformly on the equicontinuous subsets of $\mathscr{S}'(\mathbf{R}^{m+n})$, by a general property of locally convex spaces. In particular, sets of the form $A' \otimes B'$, with $A' \subset \mathscr{S}'(\mathbf{R}^m)$ and $B' \subset \mathscr{S}'(\mathbf{R}^n)$ equicontinuous, are equicontinuous in $\mathscr{S}'(\mathbf{R}^{m+n})$. Thus $\mathscr{S}(\mathbf{R}^{m+n})$ induces on $\mathscr{S}(\mathbf{R}^m) \otimes \mathscr{S}(\mathbf{R}^n)$ a topology which is finer than the ε one.

COROLLARY.   *We have, with the notation of Theorem* 51.6,

$$(51.7) \quad \mathscr{S}'(\mathbf{R}^m) \hat{\otimes} \mathscr{S}'(\mathbf{R}^n) \cong L(\mathscr{S}(\mathbf{R}^m); \mathscr{S}'(\mathbf{R}^n)) \cong \mathscr{S}'(\mathbf{R}^{m+n});$$

$$(51.8) \quad \mathscr{E}'(X) \hat{\otimes} \mathscr{E}'(Y) \cong L(\mathscr{C}^\infty(X); \mathscr{E}'(Y)) \cong \mathscr{E}'(X \times Y);$$

$$(51.9) \quad H'(X) \hat{\otimes} H'(Y) \cong L(H(X); H'(Y)) \cong H'(X \times Y).$$

Proof. It suffices to apply Proposition 50.5 and combine Theorem 51.6 with (50.19).

The isomorphisms (51.7), (51.8), and (51.9) can be regarded as variants of the *kernels theorem,* due to L. Schwartz, which we proceed now to state and prove:

THEOREM 51.7.   *We have the canonical isomorphisms:*

$$(51.10) \quad \mathscr{D}'(X \times Y) \cong \mathscr{D}'(X) \hat{\otimes} \mathscr{D}'(Y) \cong L(\mathscr{C}_c^\infty(Y); \mathscr{D}'(X))$$

$$(X \subset \mathbf{R}^m, \, Y \subset \mathbf{R}^n \text{ open sets}).$$

Proof. The second isomorphism is a straightforward application of Proposition 50.5. The conditions there are satisfied if we take $E = \mathscr{C}_c^\infty(Y)$ and $F = \mathscr{D}'(X)$: indeed, E, F, and F' are complete (Theorem 13.1; Corollary 3 of Theorem 32.2); E is barreled; E' is nuclear. Therefore $E' \hat{\otimes} F \cong L_b(E; F)$.

It remains to show that $\mathscr{D}'(X \times Y)$ induces on its dense (Proposition 40.4) linear subspace $\mathscr{D}'(X) \otimes \mathscr{D}'(Y)$ the topology $\varepsilon = \pi$. By a now standard argument, we see that it is enough to show that every compact subset of $\mathscr{C}_c^\infty(X \times Y)$ is contained in the closed convex balanced hull of the tensor product of a compact subset of $\mathscr{C}_c^\infty(X)$ with a compact subset of $\mathscr{C}_c^\infty(Y)$. But a compact (i.e., closed and bounded) subset of $\mathscr{C}_c^\infty(X \times Y)$ is contained (and is compact) in a subspace $\mathscr{C}_c^\infty(K \times L)$ with K (resp. L) a compact subset of X (resp. Y). Our assertion follows then if we combine the isomorphism (51.5) with Corollary 2 of Theorem 45.1.

Q.E.D.

The isomorphism (51.10) calls for some comment. To every kernel-distribution $K(x, y)$ on $X \times Y$ we may associate a continuous linear mapping K of $\mathscr{C}_c^\infty(Y)$ into $\mathscr{D}'(X)$ in the following manner: if $v \in \mathscr{C}_c^\infty(Y)$, then Kv is the distribution on X,

$$\mathscr{C}_c^\infty(X) \ni u \rightsquigarrow \langle K(x, y), u(x)\, v(y) \rangle.$$

It is traditional to write

$$(Kv)(x) = \int K(x, y)\, v(y)\, dy.$$

Theorem 51.7 states that the correspondence $K(x, y) \leftrightarrow K$ is an isomorphism.

Note that the transpose of the mapping K is given by

$$\mathscr{C}_c^\infty(X) \ni u \rightsquigarrow ({}^t Ku)(y) = \int K(x, y)\, u(x)\, dx.$$

Concerning kernels, the following terminology is commonly used:

(1) The kernel $K(x, y)$ or its associated map K are said to be *semiregular in x* if K maps $\mathscr{C}_c^\infty(Y)$ into $\mathscr{C}^\infty(X)$ (then K is a continuous linear map $\mathscr{C}_c^\infty(Y) \to \mathscr{C}^\infty(X)$). The kernels which are semiregular in x are the elements of the space

$$\mathscr{C}^\infty(X) \,\widehat{\otimes}\, \mathscr{D}'(Y) \cong L(\mathscr{C}_c^\infty(Y); \mathscr{C}^\infty(X)).$$

By virtue of Theorem 44.1, we have

$$\mathscr{C}^\infty(X) \,\widehat{\otimes}\, \mathscr{D}'(Y) \cong \mathscr{C}^\infty(X; \mathscr{D}'(Y)).$$

Thus, the kernels semiregular in x can also be identified with the $\mathscr{C}^\infty$ functions of x valued in the space of distributions with respect to y.

(2) $K(x, y)$ is said to be *semiregular in y* if its associated mapping K can be extended as a continuous linear map of $\mathscr{E}'(Y)$ into $\mathscr{D}'(X)$. Then we see that the transpose of K continuously maps $\mathscr{C}_c^\infty(X)$ into $\mathscr{C}^\infty(Y)$. In other words, we are considering the same property of kernels as in (1) but with x and y (as well as K and ${}^t K$) exchanged. In particular, the kernels semiregular in y are the elements of

$$\mathscr{D}'(X) \,\widehat{\otimes}\, \mathscr{C}^\infty(Y) \cong L(\mathscr{C}_c^\infty(X); \mathscr{C}^\infty(Y)) \cong \mathscr{C}^\infty(Y; \mathscr{D}'(X)).$$

(3) $K(x, y)$ is called a *regularizing kernel* if the associated mapping K can be extended as a continuous linear map of $\mathscr{E}'(Y)$ into $\mathscr{C}^\infty(X)$, in other words if

$$K(x, y) \in \mathscr{C}^\infty(X) \otimes \mathscr{C}^\infty(Y) \cong \mathscr{C}^\infty(X \times Y).$$

The student should not think that a kernel $K(x, y)$ which is semi-regular both in x and y is regularizing! A look at the identity mapping of $\mathscr{C}_c^\infty$ into $\mathscr{D}'$ easily clarifies this question. Indeed, let us take $X = Y = \Omega$, an open subset of $\mathbf{R}^n$. The kernel distribution in $\Omega \times \Omega$, associated with the natural injection of $\mathscr{C}_c^\infty(\Omega)$ into $\mathscr{D}'(\Omega)$, is the "Dirac measure" on the diagonal of $\Omega \times \Omega$, which is always denoted by $\delta(x - y)$. As a distribution with respect to (x, y), it is defined by the formula

$$\langle \delta(x - y), \phi(x, y) \rangle = \int \phi(x, x) \, dx, \qquad \phi \in \mathscr{C}_c^\infty(\Omega \times \Omega).$$

But $\delta(x - y)$ can also be viewed as a distribution in x depending on the "parameter" y: it is then the Dirac measure $\delta_y(x)$ in Ω_x at the point $y \in \Omega$. Of course, $\delta(x - y)$ is symmetric in x and y. Clearly, the kernel $\delta(x - y)$ is semiregular in both x and y: the natural map $\mathscr{C}_c^\infty(\Omega) \to \mathscr{D}'(\Omega)$ is a continuous linear map of $\mathscr{C}_c^\infty(\Omega)$ into $\mathscr{C}^\infty(\Omega)'$. In other words,

$$\delta(x - y) \in \{\mathscr{C}^\infty(\Omega_x) \,\hat\otimes\, \mathscr{D}'(\Omega_y)\} \cap \{\mathscr{D}'(\Omega_x) \,\hat\otimes\, \mathscr{C}^\infty(\Omega_y)\}.$$

But $\delta(x - y)$ is obviously not regularizing, as it is not a $\mathscr{C}^\infty$ function in $\Omega \times \Omega$.

Finally, let E be a locally convex space, Hausdorff, and complete. By virtue of Theorem 44.1 and of the nuclearity of $\mathscr{C}^\infty(X)$, we have

$$\mathscr{C}^\infty(X; E) \cong \mathscr{C}^\infty(X) \,\hat\otimes\, E.$$

Similarly (cf. Exercise 44.6),

$$\mathscr{S}(\mathbf{R}^n; E) \cong \mathscr{S}(\mathbf{R}^n) \,\hat\otimes\, E.$$

On the other hand, it is natural to define a distribution $\mathbf{T}$ in the open set $X \subset \mathbf{R}^n$ with values in the space E, as a continuous linear map of $\mathscr{C}_c^\infty(X)$ into E. This is indeed the definition when $E = \mathbf{C}$. When E is finite dimensional, it corresponds to the natural idea of what should be a vector-valued distribution (its components with respect to a basis should be complex-valued distributions). In other words, the space of E-valued distributions in X will be, by definition,

$$\mathscr{D}'(X; E) = L(\mathscr{C}_c^\infty(X); E).$$

The latter space, equipped with the topology of bounded convergence (i.e., the topology of uniform convergence on the equicontinuous subsets of $\mathscr{C}_c^\infty(X)$ regarded as the dual of $\mathscr{D}'(X)$; cf. Proposition 50.4), is identical to $\mathscr{D}'(X) \widehat{\otimes} E$, so that finally

$$\mathscr{D}'(X; E) \cong \mathscr{D}'(X) \widehat{\otimes} E.$$

This can also be done for tempered distributions:

$$\mathscr{S}'(\mathbf{R}^n; E) = L(\mathscr{S}(\mathbf{R}^n); E) \cong \mathscr{S}'(\mathbf{R}^n) \widehat{\otimes} E.$$

Note that if F' is the strong dual of a Fréchet space F, we have, by (50.19),

$$\mathscr{S}'(\mathbf{R}^n; F') \cong \mathscr{S}(\mathbf{R}^n; F)'.$$

Let $\mathscr{F}$ be the Fourier transformation $\mathscr{S}' \to \mathscr{S}'$ and I the identity mapping of E into itself. Then (Proposition 43.7) $\mathscr{F} \otimes I$ is an isomorphism of $\mathscr{S}' \widehat{\otimes} E$ into itself; as its image is both dense and complete, it is onto. This defines the Fourier transformation of E-valued tempered distributions.

52

Applications

In this chapter, we shall present some applications of kernels and topological tensor products theory to linear partial differential equations. Kernels play a role in this connection as inverses, or as approximations of inverses, of differential operators. We shall deal with a differential operator D defined in some open subset Ω of $\mathbf{R}^n$; we recall that, with our definition of differential operators, the coefficients of these are $\mathscr{C}^\infty$ functions. We shall have to deal with the product $\Omega \times \Omega$; the variable in this product will be (x, y); the *diagonal* in $\Omega \times \Omega$ is the set of points (x, y) such that $x = y$. The operator D acts on distributions in Ω; if D acts on distributions on $\Omega \times \Omega$, we must indicate clearly in what variable D operates: for instance, if D acts in the variable x, we shall write D_x rather than D. Note that the operator

$$D_x : \mathscr{D}'(\Omega_x \times \Omega_y) \to \mathscr{D}'(\Omega_x \times \Omega_y)$$

is nothing else but the extended tensor product $D_x \widehat{\otimes} I_y$, with I_y the identity mapping of $\mathscr{D}'(\Omega_y)$, taking into account the canonical isomorphism $\mathscr{D}'(\Omega_x \times \Omega_y) \cong \mathscr{D}'(\Omega_x) \widehat{\otimes} \mathscr{D}'(\Omega_y)$. We now introduce some of the terminology of kernels and differential operators:

Definition 52.1. *A kernel* $K(x, y) \in \mathscr{D}'(\Omega_x \times \Omega_y)$ *is called a fundamental kernel* (resp. *a parametrix*) *of the differential operator* D_x *if*

$$D_x K(x, y) - \delta(x - y) = 0 \qquad (\text{resp. } belongs\ to\ \mathscr{C}^\infty(\Omega_x \times \Omega_y)).$$

We have denoted by $\delta(x - y)$ the "Dirac measure" on the diagonal of $\Omega_x \times \Omega_y$ (see p. 533).

Definition 52.2. *The differential operator* D *is said to be hypoelliptic in* Ω *if, for every open subset* Ω' *of* Ω *and every distribution* u *in* Ω, *the fact that* Du *is a* $\mathscr{C}^\infty$ *function in* Ω' *implies that* u *is a* $\mathscr{C}^\infty$ *function in* Ω'.

Hypoelliptic differential operators form an important class of differential operators, to which belong the elliptic and the parabolic operators (but not the hyperbolic ones). In the forthcoming statements concerning hypoelliptic differential operators, we make use of the terms *"regular kernel"* for a kernel $K(x, y)$ semiregular both in x and y (see p. 532) and *"very regular kernel"* for a kernel $K(x, y)$ which is regular and which, moreover, is a $\mathscr{C}^\infty$ function of (x, y) in the complement of the diagonal.

The following result is due to L. Schwartz:

THEOREM 52.1.   *Suppose that the transpose ^tD of D has a parametrix which is very regular. Then D is hypoelliptic. Furthermore, the topologies induced on the kernel of D,*

$$\mathscr{N}_D = \{u \in \mathscr{D}'(\Omega); \, Du = 0\} \subset \mathscr{C}^\infty(\Omega),$$

by $\mathscr{D}'(\Omega)$ and by $\mathscr{C}^\infty(\Omega)$ are equal and turn $\mathscr{N}_D$ into a Fréchet space.

In the proof which follows, we suppose that ^tD acts in the variable y : if $K(x, y)$ is the parametrix of ^tD, we have $^tD_y K(x, y) - \delta(x - y) \in \mathscr{C}^\infty(\Omega \times \Omega)$.

Proof of Theorem 52.1.   Let $u \in \mathscr{D}'(\Omega)$ be such that Du is a $\mathscr{C}^\infty$ function in a neighborhood ω of a point $a \in \Omega$. Let $g \in \mathscr{C}_c^\infty(\omega)$ be equal to one in some neighborhood ω' of a and $\rho \in \mathscr{C}_c^\infty(\mathbf{R}^n)$ be equal to one in some neighborhood of 0; we suppose furthermore that $\rho(x) = 0$ for $|x| > \varepsilon$, $\varepsilon > 0$ to be chosen later. We consider then

$$w(x) = \int \rho(x - y) \, K(x, y) \, D_y[g(y) \, u(y)] \, dy.$$

It should be underlined that this makes sense only because we have assumed that $K(x, y)$ and therefore $\rho(x - y) \, K(x, y)$ is semiregular in y. For then, the mapping it defines can be extended as a continuous linear mapping of $\mathscr{E}'(\Omega_y)$ (to which $D_y[g(y) \, u(y)]$ belongs) into $\mathscr{D}'(\Omega_x)$. By integration by parts, we see that

$$w(x) = \int {}^tD_y[\rho(x - y) \, K(x, y)] \, g(y) \, u(y) \, dy.$$

But, since $K(x, y)$ is $\mathscr{C}^\infty$ for $x \neq y$, and $\rho \equiv 1$ near 0,

$$K_1(x, y) = {}^tD_y[\rho(x - y) \, K(x, y)] - \rho(x - y) \, {}^tD_y K(x, y)$$

is a $\mathscr{C}^\infty$ function of (x, y). Now, since $K(x, y)$ is a parametrix of ^tD_y and since $\rho(x - y) \, \delta(x - y) = \delta(x - y)$ (as $\rho(0) = 1$),

$$K_2(x, y) = \rho(x - y) \, {}^tD_y K(x - y) - \delta(x - y)$$

is also a $\mathscr{C}^\infty$ function of (x, y). We reach the conclusion that

$$w(x) - g(x)\, u(x) \qquad \text{is a } \mathscr{C}^\infty \text{ function in } \Omega.$$

As $g \equiv 1$ near a, it remains to show that w is $\mathscr{C}^\infty$ near a for a suitable choice of ε. We observe that

$$D_y[g(y)\, u(y)] - g(y)\, D_y\, u(y) = 0$$

in the neighborhood ω' of a. Let us then choose $\varepsilon > 0$ so small that the support of $\rho(x - y)$, as a function of y, is contained in ω' when $|x - a| < \varepsilon$. Then, for these x's,

$$w(x) = \int \rho(x - y)\, K(x, y)\, g(y)\, D_y\, u(y)\, dy.$$

But it is our hypothesis that $g\, Du \in \mathscr{C}_c^\infty(\Omega)$ and that $\rho(x - y)\, K(x, y)$ is semiregular in x, therefore maps $\mathscr{C}_c^\infty(\Omega_y)$ into $\mathscr{C}^\infty(\Omega_x)$. This proves that D is hypoelliptic.

Consider now a filter of elements u of $\mathscr{N}_D$ converging to 0 in $\mathscr{D}'(\Omega)$. Using the above notation, we may write

$$w(x) - g(x)u(x) = \int [K_1(x, y) + K_2(x, y)]\, g(y)u(y)\, dy.$$

As $K_j(x, y) \in \mathscr{C}^\infty(\Omega \times \Omega), j = 1, 2$, and as gu converges to 0 in $\mathscr{E}'(\Omega)$, we see that $w - gu$ converges to 0 in $\mathscr{C}^\infty(\Omega)$. But now, if $\varepsilon > 0$ is small enough, $w(x) = 0$ for $|x - a| < \varepsilon$, hence $gu = u$ in

$$\omega'' = \{x;\, |x - a| < \varepsilon\},$$

and converges to 0 in $\mathscr{C}^\infty(\omega'')$. As a is arbitrary, we conclude that u converges to 0 in $\mathscr{C}^\infty(\Omega)$.   Q.E.D.

Let $K(x, y) \in \mathscr{D}'(\Omega \times \Omega)$, and K be the map $\mathscr{C}_c^\infty(\Omega) \to \mathscr{D}'(\Omega)$ defined by $K(x, y)$. We say that $K(x, y)$ is a *two-sided fundamental kernel* of the differential operator D if, for all $\phi \in \mathscr{C}_c^\infty(\Omega)$,

$$K\, D\phi = D\, K\phi = \phi.$$

We see that $K(x, y)$ is a two-sided fundamental kernel of D if and only if

$$D_x\, K(x, y) = \delta(x - y), \qquad {}^t D_y\, K(x, y) = \delta(x - y).$$

THEOREM 52.2. *If D is a hypoelliptic differential operator in Ω, every point of Ω has an open neighborhood in which ${}^t D$ has a fundamental kernel. If ${}^t D$ is also hypoelliptic, every point of Ω has an open neighborhood where D has a two-sided fundamental kernel, which is very regular.*

Proof. We shall introduce the spaces H^s (see Definition 31.4) and their norm $\| \ \|_s$ (p. 330). Let K be an arbitrary compact subset of Ω, and s a real number. We denote by $\Phi^s(K)$ the space $\mathscr{C}_c^\infty(K)$ equipped with the topology defined by the seminorms

$$\phi \rightsquigarrow \| D\phi \|_t + \| \phi \|_s$$

as t ranges over the set of real numbers or, equivalently, over the set of positive integers; $\Phi^s(K)$ is a metrizable space. Its completion, $\hat{\Phi}^s(K)$, can be identified with a linear subspace of H^s, hence of $\mathscr{D}'(\Omega)$. Note however that $u \rightsquigarrow Du$ is a continuous linear map of $\Phi^s(K)$ into $\mathscr{C}^\infty(\Omega)$, hence all the distributions belonging to $\hat{\Phi}^s(K)$ must be $\mathscr{C}^\infty$ functions; as they have obviously their support in K, we have $\hat{\Phi}^s(K) = \Phi^s(K) = \mathscr{C}_c^\infty(K)$. As the identity mapping $\mathscr{C}_c^\infty(K) \to \Phi^s(K)$ is continuous, it is an isomorphism, by virtue of the open mapping theorem. We reach the conclusion that to every real number r, there is a real number t and a constant $C(r, s) > 0$ such that, for all $\phi \in \mathscr{C}_c^\infty(K)$,

$$(52.1) \qquad \| \phi \|_r \leqslant C(r, s) \, (\| D\phi \|_t + \| \phi \|_s).$$

At this stage, we choose $r = 1$, $s = 0$ and we make use of the following fact, whose verification will be left to the student. To every $\varepsilon > 0$, there is $\eta > 0$ such that, if diam(supp ϕ) $< \eta$,

$$\| \phi \|_0 \leqslant \varepsilon \| \phi \|_1 .$$

We apply Estimate (52.1) to the closure K of an arbitrary relatively compact open neighborhood Ω' of an arbitrary point x^0 of Ω. It is clear that there is another open neighborhood $U \subset \Omega'$ of x^0 such that we have, for all $\phi \in \mathscr{C}_c^\infty(U)$,

$$\| \phi \|_1 \leqslant \text{const} \| D\phi \|_t .$$

Finally, by enlarging t if necessary, we see that there is a constant $C' > 0$ such that

$$(52.2) \qquad \| \phi \|_{-t} \leqslant C' \| D\phi \|_t \qquad \text{for all} \quad \phi \in \mathscr{C}_c^\infty(U).$$

If also ^tD is hypoelliptic, we may further enlarge t and C' and possibly shrink U so as to also have

$$(52.3) \qquad \| \phi \|_{-t} \leqslant C' \| {}^tD\phi \|_t \qquad \text{for all} \quad \phi \in \mathscr{C}_c^\infty(U).$$

Let M be the closure, in H^t, of the set of distributions of the form $D\phi$, $\phi \in \mathscr{C}_c^\infty(U)$, and let p_M be the orthogonal projection of H^t onto M.

Estimate (52.2) means that mapping $D\phi \leadsto \phi$ can be extended as a continuous linear map of M (equipped with the H^l norm) into H^{-l}. It can be further extended as a continuous linear map $G : H^l \to H^{-l}$ by setting it equal to zero on the orthogonal of M. The transpose ${}^t G$ of G is a continuous linear map of H^l into H^{-l} such that, for all $f \in H^l$, we have in U

$$ {}^t D({}^t Gf) = f. $$

As $\mathscr{C}_c^\infty(U)$ is continuously embedded in H^l, ${}^t G$ induces a continuous linear map of $\mathscr{C}_c^\infty(U)$ into H^{-l}, a fortiori into $\mathscr{D}'(U)$; the kernel associated with it is a fundamental kernel (in U) of ${}^t D$. This proves the first part of the statement.

If ${}^t D$ is also hypoelliptic, we derive from (52.3) that there is a continuous linear operator $\Gamma : H^l \to H^{-l}$ such that, for all $f \in H^l$, $D(\Gamma f) = f$ in U. Let us then set

$$ E = G p_M + \Gamma(I - p_M), $$

where I is the identity mapping of H^l; E is a bounded linear operator $H^l \to H^{-l}$. We have $p_M D\phi = D\phi$ for all $\phi \in \mathscr{C}_c^\infty(U)$, hence $E(D\phi) = G(D\phi) = \phi$. On the other hand, in U,

$$ D(E\phi) = D\, G(p_M \phi) + (I - p_M)\phi. $$

Let $\{\phi_k\}$ be a sequence in $\mathscr{C}_c^\infty(U)$ such that $D\phi_k \to p_M \phi$ in H^l; as G is continuous, $\phi_k = GD\phi_k$ converges to $G(p_M \phi)$ in H^{-l} and $D\phi_k$ converges to $D\, G(p_M \phi)$ in $\mathscr{D}'$; but $D\phi_k$ converges also to $p_M \phi$, which must therefore be equal to $D\, G(p_M \phi)$. Finally we have, in U, $D(E\phi) = \phi$. This proves that the kernel $K(x, y)$ associated with the continuous linear operator $\phi \leadsto E\phi \,|\, U$ (restriction of $E\phi$ to U) is a two-sided fundamental kernel of D. The fact that K is very regular follows from the following lemma:

LEMMA 52.1. *Let D and ${}^t D$ be hypoelliptic in Ω. Every two-sided fundamental kernel $K(x, y)$ of D in Ω is very regular.*

Proof. We have, for every $\phi \in \mathscr{C}_c^\infty(\Omega)$,

$$ D_x \int K(x, y)\,\phi(y)\,dy = \phi(x) $$

and, as D is hypoelliptic, we must have $\int K(x, y)\,\phi(y)\,dy \in \mathscr{C}^\infty(\Omega_x)$. Thus $K : \phi(y) \leadsto \int K(x, y)\,\phi(y)\,dy$ maps $\mathscr{C}_c^\infty(\Omega)$ into $\mathscr{C}^\infty(\Omega)$ and, in this sense, its graph is closed since K is continuous when taking its values in $\mathscr{D}'(\Omega)$. We conclude that $K : \mathscr{C}_c^\infty(\Omega) \to \mathscr{C}^\infty(\Omega)$ is continuous. This shows that $K(x, y)$ is semiregular in x. By interchanging x and y, D and ${}^t D$, we see that $K(x, y)$ is also semiregular in y. Thus $K(x, y)$

is regular. It remains to prove that $K(x, y)$ is a $\mathscr{C}^\infty$ function of (x, y) in the complement of the diagonal of $\Omega \times \Omega$.

Let U and V be two open subsets of Ω such that $U \cap V = \varnothing$; in what follows, we denote by $K(x, y)$ the restriction to $U \times V$ of what has been denoted until now by $K(x, y)$. We have just seen that

$$K(x, y) \in \mathscr{C}^\infty(V_y; \mathscr{D}'(U_x)) \cong \mathscr{D}'(U_x) \,\widehat{\otimes}\, \mathscr{C}^\infty(V_y).$$

For every n-tuple q, we have

$$D_x[(\partial/\partial y)^q K(x, y)] = 0 \quad \text{in} \quad U \times V.$$

Let H be a compact neighborhood of an arbitrary point y^0 in V. The image of H under the mapping $y \rightsquigarrow (\partial/\partial y)^q K(x, y)$, which is continuous from V into $\mathscr{D}'(U)$, is a compact subset $\mathscr{H}$ of $\mathscr{D}'(U)$. Let U' be any relatively compact open subset of U, $\mathscr{H} \mid U'$ the set of restrictions to U' of the elements of $\mathscr{H}$. One can show (cf. Theorem 34.3, p. 359) that $\mathscr{H} \mid U'$ is contained in some space $H^s_{\text{loc}}(U')$ (Chap. 31–11) and compact there. But $\mathscr{H} \mid U'$ is also contained in $\mathscr{N}_D(U')$, the space of solutions in U' of the homogeneous equation $Du = 0$. Since D is hypoelliptic, $\mathscr{N}_D(U') \subset \mathscr{C}^\infty(U')$ and $\mathscr{N}_D(U')$ is closed in both Fréchet spaces $\mathscr{C}^\infty(U')$ and $H^s_{\text{loc}}(U')$. They necessarily induce the same topology on $\mathscr{N}_D(U')$. Therefore, $\mathscr{C}^\infty(U')$ and $\mathscr{D}'(U')$ induce the same topology on $\mathscr{H} \mid U'$, which implies at once that $\mathscr{C}^\infty(U)$ and $\mathscr{D}'(U)$ induce the same topology on $\mathscr{H}$. Hence, $y \rightsquigarrow (\partial/\partial y)^q K(x, y)$ is a continuous map of H into $\mathscr{C}^\infty(U)$; as y^0 is arbitrary, it is a continuous map of V into $\mathscr{C}^\infty(U)$. As q is arbitrary, we reach the conclusion that $K(x, y)$ is a $\mathscr{C}^\infty$ function of (x, y) in $U \times V$.

COROLLARY 1.   *The following conditions are equivalent:*

(a)   *D and ${}^tD$ are hypoelliptic;*

(b)   *every point of Ω has an open neighborhood where D has a two-sided fundamental kernel (which is very regular).*

If they are satisfied, the topologies induced on $\mathscr{N}_D$ by $\mathscr{D}'(\Omega)$ and by $\mathscr{C}^\infty(\Omega)$ are equal and turn $\mathscr{N}_D$ into a Fréchet space.

It suffices to combine Theorems 52.1 and 52.2.

It is well known that, for harmonic or for holomorphic functions, the uniform convergence of functions on compact sets and the uniform convergence, still on compact sets, of the functions and all their derivatives, is one and the same thing. Granting that the Laplace and the Cauchy–Riemann operator are hypoelliptic, which they are, Corollary 1 of Theorem 52.2 strengthens and generalizes this convergence property.

COROLLARY 2. *Same hypotheses as in Theorem 52.2. Let K be a compact subset of Ω. The linear space $\mathcal{N}_D(K)$ of (distributions) solutions of the homogeneous equation $Du = 0$ which have their support in K is finite dimensional.*

Proof. Let E be a Banach space of distributions in Ω, e.g., $E = L^2(\Omega)$, which contains $\mathscr{C}_c^\infty(K)$ and therefore induces on it a topology finer than the one induced by $\mathscr{D}'(\Omega)$. From Corollary 1 of Theorem 52.2 we derive that E induces on $\mathcal{N}_D(K) \subset \mathscr{C}_c^\infty(K)$ the same topology as $\mathscr{D}'(\Omega)$. But the unit ball of $\mathcal{N}_D(K)$ for the norm induced by E is bounded in $\mathscr{D}'(\Omega)$, hence precompact, whence the corollary.

The previous results will be used to prove the following theorem, due to B. Malgrange, asserting, under suitable conditions, the existence of a *global* two-sided fundamental kernel:

THEOREM 52.3. *Let D be a differential operator in Ω, and ${}^tD$ its transpose. Suppose that both D and ${}^tD$ are hypoelliptic and map $\mathscr{C}^\infty(\Omega)$ onto itself. Then D has a two-sided fundamental kernel in Ω, which is very regular.*

Proof. Let F be a Fréchet space. We begin by proving that

$$(52.4) \qquad (D \,\widehat{\otimes}\, I)(\mathscr{C}^\infty(\Omega) \,\widehat{\otimes}\, F) = \mathscr{C}^\infty(\Omega) \,\widehat{\otimes}\, F,$$

$$(52.5) \qquad (D \,\widehat{\otimes}\, I)(\mathscr{D}'(\Omega) \,\widehat{\otimes}\, F) = \mathscr{D}'(\Omega) \,\widehat{\otimes}\, F,$$

where I is the identity mapping of F. (52.4) is a trivial consequence of Proposition 43.9 and of the fact that $D\mathscr{C}^\infty(\Omega) = \mathscr{C}^\infty(\Omega)$; hypoellipticity is irrelevant in this connection (cf. Theorem 52.5).

We proceed to prove (52.5). As D and ${}^tD$ are both hypoelliptic, we may apply Theorem 52.2. We see that there is an open covering $\{U_i\}$ of Ω such that D has a two-sided, very regular, fundamental kernel in every U_i. We may assume the covering $\{U_i\}$ to be locally finite and consisting of relatively compact open sets. Let then $\{V_i\}$ be another locally finite open covering of Ω such that $\bar{V}_i \subset U_i$ for every i; and let $\{g_i\}$ be a partition of unity in $\mathscr{C}_c^\infty(\Omega)$ subordinated to the covering $\{V_i\}$. For each i, let $h_i \in \mathscr{C}_c^\infty(U_i)$ be equal to one on V_i. Multiplication by h_i, $S \rightsquigarrow h_iS$, in $\mathscr{D}'(\Omega)$, gives rise to the extended tensor product

$$h_i \,\widehat{\otimes}\, I : \mathscr{D}'(\Omega) \,\widehat{\otimes}\, F \to \mathscr{E}'(U_i) \,\widehat{\otimes}\, F.$$

We shall write h_iS rather than $(h_i \,\widehat{\otimes}\, I)S$ for $S \in \mathscr{D}'(\Omega) \,\widehat{\otimes}\, F$. On the other hand, if K_i is a two-sided (very regular) fundamental kernel of D in U_i, we may consider the mapping

$$K_i \,\widehat{\otimes}\, I : \mathscr{E}'(U_i) \,\widehat{\otimes}\, F \to \mathscr{D}'(U_i) \,\widehat{\otimes}\, F$$

which satisfies, for all $\mu \in \mathscr{E}'(U_i) \,\hat{\otimes}\, F$,

$$(D \,\hat{\otimes}\, I)(K_i \,\hat{\otimes}\, I)\mu = \mu.$$

Then let $\mathbf{S}$ be an arbitrary element of $\mathscr{D}'(\Omega) \,\hat{\otimes}\, F$, and set

$$\mathbf{T}_i = (K_i \,\hat{\otimes}\, I)(h_i \mathbf{S}).$$

Observe that the restriction of $\mathbf{T}_i - \mathbf{T}_j$ to $V_{ij} = V_i \cap V_j$ belongs to $\mathscr{D}'(V_{ij}) \,\hat{\otimes}\, F$ and satisfies there

$$(D \,\hat{\otimes}\, I)(\mathbf{T}_i - \mathbf{T}_j) = (h_i - h_j)\mathbf{S} = 0$$

since $h_i = h_j = 1$ in V_{ij}. Let us now use the nuclearity of $\mathscr{D}'$ and recall that

$$\mathscr{D}'(V_{ij}) \,\hat{\otimes}\, F \cong L_e(F'_\tau; \mathscr{D}'(V_{ij})) .$$

But we have just seen that $\mathbf{T}_i - \mathbf{T}_j$, viewed as a mapping of F' into $\mathscr{D}'(V_{ij})$, takes its values in $\mathscr{N}_D(V_{ij})$, the space of distributions in V_{ij} solutions of the equation $Du = 0$. By Corollary 1 of Theorem 52.2, we know that the topology induced on $\mathscr{N}_D(V_{ij})$ by $\mathscr{D}'(V_{ij})$ and by $\mathscr{C}^\infty(V_{ij})$ are the same. Hence

(52.6) $$\mathbf{T}_i - \mathbf{T}_j \in L_e(F'_\tau; \mathscr{C}^\infty(V_{ij})) \cong \mathscr{C}^\infty(V_{ij}) \,\hat{\otimes}\, F.$$

We set

$$\mathbf{T} = \sum_j g_j \mathbf{T}_j .$$

We have, in V_i,

$$\mathbf{S} - (D \,\hat{\otimes}\, I)\mathbf{T} = (D \,\hat{\otimes}\, I)\mathbf{T}_i - (D \,\hat{\otimes}\, I)\mathbf{T} = \sum_j g_j(\mathbf{T}_i - \mathbf{T}_j),$$

and each term $g_j(\mathbf{T}_i - \mathbf{T}_j)$ belongs to $\mathscr{C}^\infty(\Omega) \,\hat{\otimes}\, F$, by virtue of (52.6). As the V_i's cover Ω, we conclude that

$$\mathbf{S} - (D \,\hat{\otimes}\, I)\mathbf{T} \in \mathscr{C}^\infty(\Omega) \,\hat{\otimes}\, F.$$

By (52.4) we derive the existence of an element $\phi \in \mathscr{C}^\infty(\Omega) \,\hat{\otimes}\, F$ such that $(D \,\hat{\otimes}\, I)\phi = \mathbf{S} - (D \,\hat{\otimes}\, I)\mathbf{T}$, hence

$$(D \,\hat{\otimes}\, I)(\mathbf{T} + \phi) = \mathbf{S},$$

and this proves (52.5).

We begin by applying (52.5) with $F = \mathscr{C}^\infty(\Omega)$. We know that $\delta(x-y) \in \mathscr{D}'(\Omega_x) \widehat{\otimes} \mathscr{C}^\infty(\Omega_y)$. Therefore, by (52.5), there is a kernel $K_1(x, y)$ in the same completed tensor product, such that $(D_x \widehat{\otimes} I_y) K_1(x, y) = \delta(x - y)$, i.e., $D_x K_1 = I_{x,y}$. (We denote by I_x, I_y, and $I_{x,y}$ the identity mappings in the spaces of distributions in the variables x, y, and (x, y).) Let us then set

$$L = I_{x,y} - (I_x \widehat{\otimes} {}^tD_y)K_1 \in \mathscr{D}'(\Omega_x) \widehat{\otimes} \mathscr{D}'(\Omega_y).$$

We have

$$(D_x \widehat{\otimes} I_y)L = (D_x \widehat{\otimes} I_y)I_{x,y} - (I_x \widehat{\otimes} {}^tD_y)(D_x \widehat{\otimes} I_y)K_1$$

$$= (D_x \widehat{\otimes} I_y - I_x \widehat{\otimes} {}^tD_y)I_{x,y} = 0$$

by definition of ${}^tD$. Thus we see that L defines a linear map of $\mathscr{C}_c^\infty(\Omega_y)$ into $\mathscr{N}_D(\Omega_x) = \{u \in \mathscr{D}'(\Omega_x); Du = 0\}$, which is continuous when the latter carries the topology induced by $\mathscr{D}'(\Omega_x)$. This topology turns $\mathscr{N}_D(\Omega_x)$ into a Fréchet space (Corollary 1 of Theorem 52.2). Therefore, in view of (52.5) applied with $F = \mathscr{N}_D(\Omega_x)$ and ${}^tD$ instead of D, there exists $K_2 \in \mathscr{N}_D(\Omega_x) \widehat{\otimes} \mathscr{D}'(\Omega_y)$, hence satisfying $(D_x \widehat{\otimes} I_y)K_2 = 0$, such, furthermore, that $(I_x \widehat{\otimes} {}^tD_y)K_2 = L$. Then, if we set $K = K_1 + K_2$, we have

$$(I_x \widehat{\otimes} {}^tD_y)K = I_{x,y}, \qquad (D_x \widehat{\otimes} I_y)K = I_{x,y}.$$

This means precisely that K is a two-sided fundamental kernel of D; Lemma 52.1 implies, then, that K is very regular.   Q.E.D.

Let E, F be two Hausdorff TVS, and $u : E \to F$ a homomorphism onto. We say that u has a *continuous right inverse* if there is a continuous linear map $v : F \to E$ such that $u \circ v =$ identity of F. We recall that a linear subspace M of E has a *topological supplementary* if there is another linear subspace N of E such that the mapping $(x_1, x_2) \leadsto x_1 + x_2$ of $M \times N$ into E is an isomorphism onto (for the TVS structures: $M \times N$ carries the product structure, M and N carry their induced one). Then it is easy to see that *u has a continuous right inverse if and only if* Ker *u has a topological supplementary*. Indeed, if v is a continuous right inverse of u, $v(F)$ is a topological supplementary of Ker u. Conversely, if M is a topological supplementary of Ker u, the restriction of u to M is an isomorphism of M onto F whose inverse is a continuous right inverse of u.

The next theorem, due to A. Grothendieck, shows that an important class of differential operators ("most" elliptic operators with analytic coefficients), although they map $\mathscr{C}^\infty$ onto itself, have no continuous right inverse. From what we have said above, this implies that their

kernels have no topological supplementary. As the class includes all
elliptic operators with constant coefficients, in particular the Laplace
operator and the Cauchy–Riemann operator, we see that the space of
harmonic functions in an open subset Ω of $\mathbf{R}^n$ or the space of holomorphic ones in an open subset Ω of $\mathbf{C}^n$ have no topological supplementary in $\mathscr{C}^\infty(\Omega)$.

THEOREM 52.4. *Let D be a differential operator in an open set $\Omega \subset \mathbf{R}^n$
having the following properties:*

(I)   *For every open subset Ω' of Ω and every distribution S in Ω,
      $DS = 0$ in Ω' implies that S is an analytic[†] function in Ω'.*

(II) *D maps $\mathscr{C}^\infty(\Omega)$ onto itself.*

(III) *To every open set $\Omega' \subset \Omega$ there is another open subset $\Omega'' \subset \Omega'$
      such that ${}^tDT = 0$ for some $T \in \mathscr{D}'(\Omega'')$, $T \neq 0$.*

Under these conditions, D has no continuous right inverse in $\mathscr{C}^\infty(\Omega)$.

Proof. We reason by contradiction. Suppose that D had a continuous
right inverse G in $\mathscr{C}^\infty(\Omega)$. To every compact subset K of Ω there is
another compact subset K' of Ω, an integer $m \geqslant 0$, and a constant
$C > 0$ such that, for all $\phi \in \mathscr{C}^\infty(\Omega)$,

$$\sup_{x \in K} | G\phi(x)| \leqslant C \sup_{x \in K'} \sum_{|p| \leqslant m} |(\partial/\partial x)^p \phi(x)|.$$

Suppose then that ϕ vanishes in a neighborhood of K': $G\phi$ must vanish
in K. Let us choose K with a nonempty interior $\mathcal{O}$. Let Ω_0 be an open
ball, contained in the intersection of the complement of $K \cup K'$ with
some connected component Ω_1 of Ω which intersects $\mathcal{O}$. For every
$\phi \in \mathscr{C}_c^\infty(\Omega_0)$, we have $G\phi = 0$ in K, in particular in $\mathcal{O}$. On the other
hand, $D(G\phi) = 0$ in the complement of supp ϕ, hence $G\phi$ is analytic
in this complement (in view of (I)). We derive from this that $G\phi = 0$
in every connected component of $\Omega_1 - $ supp ϕ which intersects $\mathcal{O}$, i.e.,
in every nonrelatively compact connected component of $\Omega_1 - $ supp ϕ
(relatively compact with respect to Ω_1). This implies immediately
that $\Omega_1 \cap ($supp $G\phi)$ is a compact subset of Ω_0. As $D(G\phi) = \phi$ we see
that $D\mathscr{C}_c^\infty(\Omega_0) = \mathscr{C}_c^\infty(\Omega_0)$, hence that ${}^tD : \mathscr{D}'(\Omega_0) \to \mathscr{D}'(\Omega_0)$ is one-to-
one. But exactly the same reasoning applies to any open ball contained
in Ω_0 so that we may suppose (by (III)) that Ω_0 itself is contained in
some open set Ω'' where the homogeneous equation ${}^tDT = 0$ has a

---

† *Analytic* is meant here in the real sense, i.e., the Taylor expansion about each point
converges in some neighborhood of that point.

solution $T \in \mathscr{D}'(\Omega'')$ whose restriction to Ω_0 is nonzero. We have thus reached a contradiction.

Remark 52.1. Elliptic operators with analytic coefficients are such that (I) and (III) hold. If D is such an operator, its transpose ${}^tD$ is also elliptic (and has analytic coefficients). Then let K be an arbitrary compact subset of Ω, and $\hat{K}$ the union of K with all the connected components of $\Omega - K$ which are relatively compact; $\hat{K}$ is obviously a compact subset of Ω. If ${}^tD$ has Property (I) of Theorem 52.4, for every $\mu \in \mathscr{E}'(\Omega)$, supp ${}^tD\mu \subset K \Rightarrow$ supp $\mu \subset \hat{K}$. Thus Ω is D-convex (Definition 38.1). To show that Property (II) holds, it suffices therefore (by Theorem 38.2) to show that D is semiglobally solvable (Definition 38.2). All operators with constant coefficients are semiglobally solvable (Exercise 38.1). We see thus that all the elliptic operators with constant coefficients satisfy Conditions (I), (II), and (III) in Theorem 52.4 (regardless of what the open set Ω is).

Now let E be a nuclear Fréchet space, and $u : E \to E$ a homomorphism of E onto itself. Suppose that u has a continuous right inverse v. Let us identify v to an element of $E' \mathbin{\widehat{\otimes}} E$ (by (50.18)), say θ_v. Let $\theta_I \in E' \mathbin{\widehat{\otimes}} E$ be the element corresponding to the identity mapping of E. To say that v is a continuous inverse of u is equivalent with saying that

$$(I' \mathbin{\widehat{\otimes}} u)(\theta_v) = \theta_I .$$

We have denoted by I' the identity mapping of E'. Thus we see that, if u does *not* have a continuous right inverse, $I' \mathbin{\widehat{\otimes}} u$ does not map $E' \mathbin{\widehat{\otimes}} E$ onto itself—although both I' and u are mappings onto. In particular, Theorem 52.4 provides us with examples of tensor products of surjections whose extension to the completion of the tensor product (of $\mathscr{C}^\infty(\Omega)$ with $\mathscr{E}'(\Omega)$) is not surjective. It shows that the metrizability restriction, in Proposition 43.9, cannot be lightly brushed aside.

Let us go on considering a differential operator D on an open subset Ω of $\mathbf{R}^n$. If E is a complete locally convex Hausdorff space, we may make D operate on functions valued in E, in fact on distributions valued in E if the latter are defined as the elements of $\mathscr{D}'(\Omega) \mathbin{\widehat{\otimes}} E$. Then if $\mathbf{T} \in \mathscr{D}'(\Omega) \mathbin{\widehat{\otimes}} E \cong L(\mathscr{C}_c^\infty(\Omega); E)$ (Proposition 50.5), we define $D_x\mathbf{T}$ as the value at $\mathbf{T}$ of the mapping $D_x \mathbin{\widehat{\otimes}} I$, where I is the identity of E. Note that if $\phi \in \mathscr{C}_c^\infty(\Omega)$, we have

$$D_x \mathbf{T}(\phi) = \mathbf{T}({}^tD\phi).$$

If $\mathbf{f}$ is a function valued in E, sufficiently smooth, say $\mathbf{f} \in \mathscr{C}^\infty(\Omega; E)$, we may define Df directly. Needless to say, the two definitions agree, as

$$\mathscr{C}^{\infty}(\Omega; E) \simeq \mathscr{C}^{\infty}(\Omega) \,\widehat{\otimes}\, E \simeq L(\mathscr{E}'(\Omega); E)$$

is canonically identifiable to a subspace of $\mathscr{D}'(\Omega) \,\widehat{\otimes}\, E$. The next result is a straightforward consequence of Proposition 43.9:

THEOREM 52.5. *Suppose that* $D\mathscr{C}^{\infty}(\Omega) = \mathscr{C}^{\infty}(\Omega)$. *Then, whatever the Fréchet space E is,* $D\mathscr{C}^{\infty}(\Omega; E) = \mathscr{C}^{\infty}(\Omega; E)$.

We consider now a differential operator in Ω of the form

$$P(x, y, \partial/\partial x) = \sum_{|p| \leqslant m} a_p(x, y)\,(\partial/\partial x)^p,$$

where y is the variable point of some open subset Y of a Euclidean space $\mathbf{R}^d$ and where the coefficients $a_p(x, y)$ are $\mathscr{C}^{\infty}$ functions of (x, y) in $\Omega \times Y$.

THEOREM 52.6. *Suppose that to every function* $f \in \mathscr{C}^{\infty}(\Omega)$ *there is a function* $u \in \mathscr{C}^{\infty}(\Omega \times Y)$ *such that* $P(x, y, \partial/\partial x)\, u(x, y) = f(x)$. *Let then $F(Y)$ be a Fréchet space of distributions in Y with the property that* $\psi g \in F(Y)$ *for all* $\psi \in \mathscr{C}^{\infty}(Y), g \in F(Y)$. *Then*

$$P(x, y, \partial/\partial x)\, \mathscr{C}^{\infty}(\Omega; F(Y)) = \mathscr{C}^{\infty}(\Omega; F(Y)).$$

Proof. We take advantage of the fact that

$$\mathscr{C}^{\infty}(\Omega; F(Y)) \simeq \mathscr{C}^{\infty}(\Omega) \,\widehat{\otimes}\, F(Y).$$

And we apply Theorem 45.1: every element f of $\mathscr{C}^{\infty}(\Omega; F(Y))$ can be written as a series

$$f(x, y) = \sum_k \lambda_k f_k(x)\, g_k(y),$$

with $\{\lambda_k\} \in l^1$, $f_k \to 0$ in $\mathscr{C}^{\infty}(\Omega)$, $g_k \to 0$ in $F(Y)$. Let us denote by G the linear subspace of $\mathscr{C}^{\infty}(\Omega \times Y)$ consisting of the functions $u(x, y)$ such that $P(x, y, \partial/\partial x)\, u(x, y) \in \mathscr{C}^{\infty}(\Omega)$, i.e., is independent of y; obviously G is a closed linear subspace of $\mathscr{C}^{\infty}(\Omega \times Y)$, i.e., G is a Fréchet space for the induced topology. By hypothesis, the restriction of $P(x, y, \partial/\partial x)$ is a continuous linear map of G onto $\mathscr{C}^{\infty}(\Omega)$, hence it is a homomorphism onto. If we apply Lemma 45.1, we see that there is a sequence $\{u_k(x, y)\}$ relatively compact (or even converging!) in $\mathscr{C}^{\infty}(\Omega \times Y)$ such that, for each k, $P(x, y, \partial/\partial x)\, u_k(x, y) = f_k(x)$. We then set

$$u(x, y) = \sum_k \lambda_k u_k(x, y)\, g_k(y).$$

We claim that the series converges in $\mathscr{C}^\infty(\Omega) \widehat{\otimes} F(Y)$ and that $P(x, y, \partial/\partial x) u(x, y) = f(x, y)$ (this is a trivial consequence of that). Consider then the bilinear map $(\psi, g) \rightsquigarrow \psi g$ of $\mathscr{C}^\infty(Y) \times F(Y)$ into $\mathscr{D}'(Y)$; it is separately continuous and its image, by hypothesis, is contained in the Fréchet space $F(Y)$ (which is continuously embedded in $\mathscr{D}'(Y)$). Therefore, this bilinear map is separately continuous, hence continuous, when we regard it as a map valued in $F(Y)$ (simply remark that the graphs of the linear mappings $\psi \rightsquigarrow \psi g$ and $g \rightsquigarrow \psi g$ are closed!). In particular, when g remains in a bounded subset of $F(Y)$, a fortiori in a compact one, the mappings $\psi \rightsquigarrow \psi g$ form an equicontinuous subset of $L(\mathscr{C}^\infty(Y); F(Y))$. Given g in $F(Y)$, we may then consider the mapping $I \widehat{\otimes} g$ of $\mathscr{C}^\infty(\Omega \times Y)$ into $\mathscr{C}^\infty(\Omega) \widehat{\otimes} F(Y)$ (I: identity of $\mathscr{C}^\infty(\Omega)$). When g remains in a bounded subset of $F(Y)$, these mappings form an equicontinuous set. We derive immediately from this that, as k varies, $u_k(x, y) g_k(y)$ remains in a bounded subset of $F(Y)$. As $\sum_k |\lambda_k| < +\infty$, the series defining $u(x, y)$ converges absolutely in $F(Y)$, whence the result.

Theorem 52.6 can be applied to the case of $F(Y) = \mathscr{C}^h(Y)$ ($0 \leqslant h \leqslant +\infty$) or $F(Y) = L^p_{\mathrm{loc}}(Y)$ ($1 \leqslant p \leqslant +\infty$). Note that we have

$$\mathscr{C}^\infty(\Omega) \widehat{\otimes} \mathscr{C}^h(Y) \cong \mathscr{C}^\infty(\Omega; \mathscr{C}^h(Y)) \cong \mathscr{C}^h(Y; \mathscr{C}^\infty(\Omega)).$$

The theorem says that, in this case, if we know how to solve in $\mathscr{C}^\infty_{x,y}$ the equation

$$P(x, y, \partial/\partial x) u = f$$

for right-hand sides f independent of y, then we know also how to solve it in $\mathscr{C}^{\infty,h}_{x,y}$ (obvious notation) for arbitrary right-hand sides in this space.

Appendix:
The Borel Graph Theorem

Recently, L. Schwartz has proved the following result:

THEOREM A.1. *Let E, F be two locally convex Hausdorff TVS, u a linear map of E into F. If E is the inductive limit of an arbitrary family of Banach spaces, if F is a Souslin space, and if the graph of u is a Borel set in $E \times F$, then u is continuous.*

Theorem A.1 implies, as we are going to show, that the closed graph theorem is valid for linear mappings defined and valued in most spaces encountered in Analysis and, in particular, in distribution theory. The original proof of Schwartz is based on the theory of Radon measures on arbitrary topological spaces; we shall present here a proof due to A. Martineau, which is an adaptation of reasonings in the book [0] of S. Banach. Martineau has also succeeded in weakening the condition that F be a Souslin space; but then the graph has to be closed and not merely a Borel set (this restriction is hardly an inconvenience!). A few words about this extension will be found at the end of the Appendix; bibliographical references will also be found there. Independently and by different methods, D. Raïkov has also proved the closed graph theorem for a large class of spaces, including most spaces of Analysis, in particular $\mathscr{D}'$.

We begin by recalling the definitions of the various terms used in the statement of Theorem A.1. The meaning of an *inductive limit of locally convex spaces* has been given on pp. 514-515. We recall what a *Borel set* in a topological space X is (X does not have to carry any algebraic structure). A collection $\mathscr{L}$ of subsets of X is called a σ-algebra if $X\backslash A$ belongs to $\mathscr{L}$ whenever $A \in \mathscr{L}$, and if countable intersections of sets which belong to $\mathscr{L}$ also belong to $\mathscr{L}$ (this is then also true of countable unions). Given any collection $\mathfrak{S}$ of subsets of X, there is a smallest σ-algebra containing $\mathfrak{S}$; it is evident. If we apply this to the collection $\mathfrak{S}$ of all the closed subsets of X, we obtain the σ-algebra of Borel sets in X.

A topological space P is called *Polish* if there is a metric on P which defines the topology of P, such that P, equipped with this metric, is a complete metric space and, moreover, if there is a countable subset of P which is everywhere dense in P (the latter property is expressed by saying that P is *separable*).

Definition A.1. *A Hausdorff topological space S is called a Souslin space if there is a Polish space P and a continuous mapping of P onto S.*

We shall now state a certain number of "stability properties" of Polish and Souslin spaces, most of them without a proof (the proofs are all easy).

PROPOSITION A.1. *The following spaces are Polish*:

(a) *closed subsets of a Polish space;*

(b) *products and disjoint unions of countable families of Polish spaces;*

(c) *open subsets of a Polish space;*

(d) *locally compact spaces which are metrizable and countable at infinity;*

(e) *countable intersections of Polish subspaces of a Hausdorff topological space;*

(f) *the set of nonrational numbers with the topology induced by the real line $\mathbf{R}$.*

Let us, for instance, prove (c). Let E be a Polish space, U an open subset of E, and d a metric on E defining the topology of E. The product space $\mathbf{R} \times E$ is Polish by (b), and so is the subset V of $\mathbf{R} \times E$ consisting of the pairs (t, x) such that $t \cdot d(x, E \setminus U) = 1$; indeed, V is closed. The second coordinate projection induces a homeomorphism of V onto U, whence (c).

PROPOSITION A.2. *A subspace Q of a Polish space P is Polish if and only if Q is the intersection of a sequence of open subsets of P.*

The sufficiency is obvious by Proposition A.1; let us prove its necessity. Let d be a metric on Q, inducing the topology of Q and for which Q is a complete metric space. Let $\tilde{Q}$ be the closure of Q in P, and Q_n be the set of $x \in \tilde{Q}$ such that there is an open neighborhood U of x in P so that the diameter of $Q \cap U$ is $\leqslant 1/n$; Q_n is open in $\tilde{Q}$. Let $x^0 \in Q_n$ for all n; the filter of neighborhoods of x^0 in P obviously induces a Cauchy filter (for the metric d) on Q. This Cauchy filter has a limit which can only be x^0; thus, $x^0 \in Q$, i.e., $Q = \bigcap_{n=0}^{+\infty} Q_n$. For each n, let U_n be an open subset of P such that $Q_n = \bar{Q} \cap U_n$. On the other hand, let δ be a metric on P defining the topology of P, and set

$$V_m = \{x \in P; \delta(x, \bar{Q}) < 1/m\}, \qquad m = 1, 2, \dots .$$

It is clear that $Q = \bigcap_{m,n} U_n \cap V_m$.

Let us denote by $\mathbf{I}$ the closed unit interval $[0, 1]$ in the real line, and by $\mathbf{I}^N$ the cube which is the product of a countable infinity of copies of $\mathbf{I}$. Let X be a separable metric space; the metric in X will be denoted by d. We may and shall assume that the diameter of X is $\leqslant 1$; this can be obtained by replacing d with $x \rightsquigarrow \sup(d(x), 1)$ if necessary. Then let (x_n) $(n = 0, 1, \dots)$ be a sequence in X which is everywhere dense. The following mapping

$$x \rightsquigarrow (d(x, x_n))_{n=0,1,\dots}$$

is a homeomorphism of X onto a subset of $\mathbf{I}^N$, as the student may readily check. Combining this fact with Proposition A.2, we obtain:

PROPOSITION A.3. *A topological space X is Polish if and only if X is homeomorphic to the intersection of a sequence of open subsets of the cube $\mathbf{I}^N$.*

We switch now to Souslin spaces; they are obviously separable. They have the following stability properties:

PROPOSITION A.4. *The following are Souslin spaces:*

(a) *closed or open subspaces of a Souslin space;*

(b) *countable products and disjoint unions of Souslin spaces;*

(c) *countable intersections or countable unions of Souslin subspaces of a Hausdorff topological space;*

(d) *continuous images of Souslin spaces.*

Proposition A.4 follows easily from Proposition A.1. We shall limit ourselves to proving (c). Let $A_n(n = 0, 1, \dots)$ be a sequence of Souslin subspaces of a Hausdorff topological space X, A their disjoint union, which is a Souslin space in view of (b), ϕ the canonical mapping of A onto the union $\bigcup_n A_n$. As ϕ is continuous, this union is a Souslin space in view of (d). Let A' be the intersection of the A_n's, f the canonical map of X onto the diagonal of X^N (N equals the set of integers $\geqslant 0$). The image of A' under f is the intersection of this diagonal with $\prod_n A_n$; as X is Hausdorff, the diagonal is closed in X^N and f is a homeomorphism of X onto it. Thus, the restriction of f to A' is a homeomorphism of A' onto a closed subset of $\prod_n A_n$. It suffices then to apply (a) and (b).

PROPOSITION A.5.   *Borel subsets of a Souslin space are Souslin spaces.*

Proof. Let $\mathscr{L}$ be the family of subsets of a Hausdorff topological space X which are Souslin spaces and whose complements are also Souslin spaces. In view of Proposition A.4(c), $\mathscr{L}$ is a σ-algebra. If X is a Souslin space, $\mathscr{L}$ contains the closed sets, in view of Proposition A.4(a), hence, the Borel sets.

PROPOSITION A.6.   *Let S be a Souslin space. For each $n = 0, 1,...$ there is a countable set C_n and two mappings $p_n : C_{n+1} \to C_n$ and $\phi_n : C_n \to \mathfrak{P}(S)$, the set of all subsets of S, with the following properties:*

(A.1) p_n *is onto;*

(A.2) *for every $c \in C_n$,*

$$\phi_n(c) = \bigcup_{c' \in C_{n+1}, \, p_n(c')=c} \phi_{n+1}(c');$$

(A.3) *for each n, the sets $\phi_n(c)$, as c runs over C_n, are pairwise disjoint and their union is equal to S;*

(A.4) *for any sequence $\{c_n\}$ $(n = 0, 1,...)$, such that $c_n \in C_n$ and $p_n(c_{n+1}) = c_n$, the sequence of sets $\{\phi_n(c_n)\}$ is the basis of a convergent filter on S.*

Proof. Let P be a Polish space and f a continuous map of P onto S; we equip P with a metric d which turns it into a complete metric space. As P is separable, for arbitrary $n = 0, 1,...$, there is a covering of P consisting of a countable family of closed balls of radius $1/(n+1)$; we select such a covering and we order it in an arbitrary fashion. This yields a sequence $\mathscr{U}_n$ whose elements we denote by B_n^k $(k = 0, 1,...)$. Let us set first by induction on k,

$$\tilde{B}_n^{k+1} = f\left(B_n^{k+1}\right) \cap \left(S \setminus \bigcup_{j=0}^{k} \tilde{B}_n^j\right), \qquad \tilde{B}_n^0 = f\left(B_n^0\right).$$

It is clear that the union of the sets $\tilde{B}_n^k$, as k varies, is equal to S, and that they are pairwise disjoint. We denote by C_0 the collection of the sets $\tilde{B}_0^k$, $k = 0, 1,...$; then we denote by C_1 the collection of the sets $\tilde{B}_0^k \cap \tilde{B}_1^l$ which are not empty. By induction, we denote by C_m the collection of sets $\tilde{B}_m^l \cap c$, as $l = 0, 1,...$, and c varies over C_{m-1}, provided that $\tilde{B}_m^l \cap c \neq \varnothing$. Condition (A.3) is obviously satisfied. The mapping p_n is defined as follows: if $c' \in C_{n+1}$, $p_n(c')$ is the unique element of C_n which contains c'; then (A.2) is obviously satisfied (ϕ_n is the mapping which assigns to the element c of C_n the subset c of S). As for (A.4), it follows from the fact that for each $n = 0, 1,...$ there is a set A_n in P such that $f(A_n) = c_n$ and diameter of $A_n \leq 1/(n + 1)$, such that $A_{n+1} \subset A_n$. The A_n's form the basis of a Cauchy filter on P, which converges as P is complete, whence (A.4).

We say that a subset M of a topological space X is *meager* if M is contained in the union of a countable family $A_n (n = 1, 2,...)$ of closed sets none of which has an interior point. We recall that X is called a *Baire space* if no meager subset of X has interior points.

Definition A.2. *Let X be a topological space and Y a subset of X. We call Baire closure of Y in X the set of points $x \in X$ having the following property*:

(A.5) *every neighborhood of x contains a subset of Y which is not meager (in X).*

We shall denote by $b(Y)$ the Baire closure of Y; clearly $b(Y)$ is closed in X.

PROPOSITION A.7. *Let X be a Baire space and suppose that $Y \subset X$ is nonmeager. Then the interior $o(Y)$ of $b(Y)$ is nonempty.*

Proof. We must show that there is an open subset Ω of X such that, for any open subset Ω' of Ω, $\Omega' \cap Y$ is nonmeager. Let $\mathcal{M}$ be a maximal family of open sets $\mathcal{O}_i$, pairwise disjoint, such that $Y \cap \mathcal{O}_i$ be meager for every i. Let $\mathcal{O}$ be the union of the $\mathcal{O}_i$'s; then $\mathcal{O} \cap Y$ is meager. This is not because $\mathcal{O}$ is a countable union of sets whose intersection with Y is meager, for the set of indices i needs not be countable; it is because every connected component of $\mathcal{O}$, necessarily contained in some $\mathcal{O}_i$, intersects Y according to a meager set. The complement F of $\mathcal{O}$ is a nonmeager closed set, for Y is nonmeager; hence, its interior F^0 is nonempty. By definition of $\mathcal{M}$, no open subset of F^0 intersects Y according to a meager set, therefore $F^0 \subset o(Y)$. Q.E.D.

The subset of Y which lies outside $b(Y)$ is meager: $b(Y)$ is the smallest closed set containing the whole of Y except possibly a meager subset of Y; $o(Y)$ is the largest open set with the property that every open subset of it contains some nonmeager part of Y. Obviously $b(Y) \setminus o(Y)$ is meager; hence, $\overline{o(Y)} = b(Y)$.

PROPOSITION A.8. *Let $\{Y_n\}$ $(n = 0, 1,...)$ be a sequence of nonmeager subsets of X, Y their union. Then $b(Y) \setminus \bigcup_n o(Y_n)$ is a meager set.*

Proof. Every open subset of $o(Y)$ contains a nonmeager subset of Y, hence a nonmeager subset of some Y_n, and intersects some $o(Y_n)$, which means that $\bigcup_n o(Y_n)$ is a dense open subset of $o(Y)$. Its complement with respect to $o(Y)$ must be meager, as well as its complement with respect to $b(Y)$, since $b(Y) \setminus o(Y)$ is meager.

We come now to the main step in the proof of Theorem A.1.

THEOREM A.2. *Let X be a Hausdorff topological space, S a Souslin subspace of X. Then S and $o(S)$, therefore also S and $b(S)$, differ only by a meager set.*

Proof. We shall make use of a sequence of triples (C_n, p_n, ϕ_n) $(n = 0, 1,...)$ with the properties listed in Proposition A.6. For each $c \in C_n$, the set

$$M_n(c) = b[\phi_n(c)] \setminus \bigcup_{c' \in C_{n+1}, p_n(c')=c} b[\phi_{n+1}(c')]$$

is meager, by virtue of Proposition A.8 and of (A.2). Therefore,

$$M = \bigcup_{\substack{n=0,1,... \\ c \in C_n}} M_n(c)$$

is also meager. Thus, it will suffice to show that $o(S) \cap (X \setminus M) \subset S$. Let $x \in o(S) \cap (X \setminus M)$. As $S = \bigcup_{c \in C_0} \phi_0(c)$, there is $c_0 \in C_0$ such that $x \in b(\phi_0(c_0))$. Since $x \in o(S)$ but $x \notin M_1(c_0)$, there is $c_1 \in C_1$, $p_1(c_1) = c_0$, such that $x \in b[\phi_1(c_1)]$, etc. We find a sequence $\{c_n\}$ like the one in (A.4). We have, for each $n = 0, 1,..., x \in b(\phi_n(c_n)) \subset \overline{\phi_n(c_n)}$. As the sequence of sets $\phi_n(c_n)$ converges to a unique point of S, this must also be true of $\{\overline{\phi_n(c_n)}\}$, and this point must necessarily be x which therefore belongs to S.   Q.E.D.

THEOREM A.3. *Let E be a Baire Hausdorff TVS, S a convex balanced subset of E. If S is nonmeager and is a Souslin space, S is a neighborhood of O in E.*

Proof. As S is nonmeager, $o(S) \neq \varnothing$ (Proposition A.7); let $a \in S \cap o(S)$. The operation $S \rightsquigarrow o(S)$ is invariant under translation and scalar multiplication, therefore $o(2S) \supset o(S - a) = o(S) - a$. But $o(S)$ is an open set containing a, hence $o(S) - a$ is a neighborhood of O, and this is also true of $o(2S)$, hence of $o(S) = \frac{1}{2} o(2S)$. The proof will be complete if we show that $o(S) \subset 2S$. Let $x \in o(S)$. As $o(S)$ is a neighborhood of O, $o(S) \cap [x - o(S)]$ is an nonempty open set which differs from $S \cap (x - S)$ only by a meager set; hence, $S \cap (x - S) \neq \varnothing$, which means that there is $y \in S$ such that $x - y \in S$, therefore, $x \in 2S$.   Q.E.D.

PROOF OF THEOREM A.1. Suppose that E is the inductive limit of an arbitrary family $\{E_\alpha\}$ $(\alpha \in A)$ of Banach spaces, with respect to linear mappings $\phi_\alpha : E_\alpha \to E$; to say that $u : E \to F$ is continuous is equivalent to saying that, for each α, $u_\alpha = u \circ \phi_\alpha : E_\alpha \to F$ is continuous. The graph of u_α is the preimage of the graph of u under the mapping $(x_\alpha, y) \rightsquigarrow (\phi_\alpha(x_\alpha), y)$ of $E_\alpha \times F$ into $E \times F$; but the preimage of a Borel set is easily seen to be a Borel set. This shows that we may assume E to be a Banach space. Of course, it suffices to show that u is sequentially continuous and it is enough to consider the restriction of u to the smallest closed linear subspace of E containing a given, but arbitrary, sequence of E. In other words, we may assume that E is a *separable* Banach space.

Let G denote the graph of u, V an arbitrary closed convex balanced neighborhood of O in F, W the intersection of G with $E \times V$. Of course, W is a Borel set; on the other hand, $E \times F$ is a Souslin space, as are both E and F (E is a Polish space!). From Proposition A.5 we derive that W is also a Souslin space and, therefore, that this is also true of its image under the first coordinate projection, which is nothing else but $U = u^{-1}(V)$. Now, the convex and balanced subset U of E is nonmeager, since $E = \bigcup_{n=1}^\infty nU$. By Theorem A.3, U is a neighborhood of O in E.

Q.E.D.

COROLLARY 1. *Let* E, F *be locally convex Hausdorff spaces,* E *the inductive limit of a collection of Banach spaces,* F *a Souslin space,* $v : F \to E$ *a continuous linear map. If* v *is surjective,* v *is open.*

Proof. Let $\bar{v} : F/\mathrm{Ker}\, v \to E$ be the associated injective map; $\bar{v}$ is continuous, hence the graph of $\bar{v}^{-1}$ is closed. But $F/\mathrm{Ker}\, v$ is a Souslin space [Proposition A.4,(d)], hence $\bar{v}^{-1}$ is continuous by Theorem A.1.

Q.E.D.

COROLLARY 2. *Let* $\mathscr{T}_1$ *and* $\mathscr{T}_2$ *be two locally convex Hausdorff topologies on the same vector space* E, *both turning* E *into a Souslin space inductive limit of Banach spaces, such that the infimum of* $\mathscr{T}_1$ *and* $\mathscr{T}_2$ *is Hausdorff. Then* $\mathscr{T}_1 = \mathscr{T}_2$.

COROLLARY 3. *If a TVS* E, *which is both a Souslin space and the inductive limit of Banach spaces, is the algebraic direct sum of two closed subspaces, it is their topological direct sum.*

We show next, very quickly, that the most important spaces of distribution theory, in particular $\mathscr{D}'$, are Souslin spaces.

Of course, a separable Fréchet space is a Souslin space, in fact, a Polish space. But we also have:

PROPOSITION A.9. *The weak dual of a separable Fréchet space and the strong dual of a separable Fréchet-Montel space are Souslin spaces.*

Proof. Let E be a separable Fréchet space, $\{U_n\}$ ($n = 1, 2, ...$) be a basis of neighborhoods of O in E, U_n^0 the polar of U_n. The dual E' of E is the union of the (weakly compact convex balanced) sets U_n^0, $n = 1, 2, ...$. It suffices to show [Proposition A.4(c)] that under out hypotheses the U_n^0 are Souslin spaces. For the weak topology, this follows from Exercise 32.9. For the strong topology, when E is a Fréchet-Montel space, it follows from the result about the weak topology and from Proposition 34.6.

Proposition A.9 and the remark that precedes it tell us that the following are Souslin spaces (Ω is an open subset of $\mathbf{R}^n$, K a compact subset of $\mathbf{R}^n$): $L^p(\Omega)$, $L^p_{\mathrm{loc}}(\Omega)(1 \leqslant p < +\infty)$, $H^s_{\mathrm{loc}}(\Omega)(s \in \mathbf{R})$, $\mathscr{C}^\infty(\Omega)$, $\mathscr{S}(\mathbf{R}^n)$, $\mathscr{C}^\infty_c(K)$, $L^p_c(K)$, $H^s_c(K)$ (also for $1 \leqslant p < +\infty$, $s \in \mathbf{R}$); $\mathscr{E}'(\Omega)$, $\mathscr{S}'(\mathbf{R}^n)$, $(\mathscr{C}^\infty_c(K))'$. We now use the fact that countable inductive limits of Souslin spaces are Souslin spaces [Proposition A.4(c)]. We see thus that $L^p_c(\Omega)$ ($1 \leqslant p$ finite), $H^s_c(\Omega)$, and $\mathscr{C}^\infty_c(\Omega)$ are Souslin spaces.

Next we use the fact that projective limits of countable families of Souslin spaces are Souslin spaces; this follows from Proposition A.4 and from the fact that a projective limit of a collection of spaces E_α is a closed subspace of the product of the E_α's.

Since $\mathscr{C}^\infty_c(\Omega)$ is the inductive limit of the $\mathscr{C}^\infty_c(K)$ as K runs over an increasing sequence of compact subsets of Ω whose interiors fill Ω, its dual, $\mathscr{D}'(\Omega)$, is the projective limit of the duals $(\mathscr{C}^\infty_c(K))'$, which are Souslin spaces, therefore it is a Souslin space.

One can prove that if E is a countable inductive limit of separable Fréchet spaces and if F is a countable union of images, under continuous linear mappings, of separable Fréchet spaces, then $L_c(E; F)$ is a Souslin space.

We indicate now what is the improvement of Theorem A.1 obtained by A. Martineau. A topological space X is said to be a $K_{\sigma\delta}$ space if X is a countable intersection of countable unions of compact sets. A Hausdorff space X is said to be *K-analytic* if there is a $K_{\sigma\delta}$ space Y and a continuous mapping f of Y onto X. Martineau's theorem states that Theorem A.1 remains valid if we substitute the hypothesis "*F is a Souslin space*" by the hypothesis "*F is a K-analytic space*", provided that we also replace the hypothesis "*the graph of u is a Borel set*" by the one: "*the graph of u is closed.*"

Every Souslin space is K-analytic, as is easily derived from Proposition A.3. Every compact space is K-analytic, which implies that there are nonseparable K-analytic spaces. Thus, Martineau's theorem is a true generalization. For instance, every weak dual of a Fréchet space is K-analytic, although it is not necessarily separable (hence, not a Souslin space). The same remark applies to reflexive Fréchet spaces.

BIBLIOGRAPHY FOR APPENDIX

Bourbaki, N., "Topologie Générale," 2ème éd., Chapter 9 (Utilisation des nombres réels en topologie générale), Section 6. Hermann, Paris, 1958.

Martineau, A., Sur le théorème du graphe fermé. *Compt. Rend. Acad. Sci.* **263**, 870 (1966).

Raïkov, D., Two-sided closed graph theorem for topological vector spaces. *Sibirsk. Mat. Ž.* **7**, 353 (1966).

Schwartz, L., Sur le théorème du graphe fermé. *Compt. Rend. Acad. Sci.* **263**, 602 (1966).

GENERAL BIBLIOGRAPHY

(For Further Reading)

On the subject of topological vector spaces:

(0) Banach, S., "Théorie des Opérations Linéaires." Warsaw, 1932.

(1) Bourbaki, N., "Espaces Vectoriels Topologiques," 2 vols. Hermann, Paris, 1953 and 1955.

(2) Dieudonné, J., and Schwartz, L., La dualité dans les espaces ($\mathscr{F}$) et ($\mathscr{LF}$). *Ann. Inst. Fourier Grenoble* I (1949), 61–101.

(3) Horvath, J., "Topological Vector Spaces." Addison-Wesley, Reading, Massachusetts, 1966.

(4) Köthe, G., "Topologische Lineare Raume," Vol. I. Springer, Berlin, 1960.

(4') Yosida, K., "Functional Analysis." Academic Press, New York, and Springer, Berlin, 1965.

On the subject of normed spaces:

(5) Day, M. M., "Normed Linear Spaces," Ergebnisse der Mathematik.... Berlin, 1962 (2nd print.).

(6) Dunford, N., and Schwartz, J. T., "Linear Operators," Parts I and II, Wiley (Interscience), New York, 1958 and 1964.

On the subject of topological vector spaces and distribution theory:

(7) Edwards, R. E., "Functional Analysis." Holt, New York, 1965.

(8) Gelfand, I. M., and Silov, G., "Generalized Functions." Academic Press, New York, 1964.

(9) Schwartz, L., "Théorie des Distributions," Parts I and II. Hermann, Paris, 1957 and 1965 (3rd print.).

On the subject of distribution theory and partial differential equations:

(10) Friedman, A., "Generalized Functions and Partial Differential Equations." Prentice-Hall, Englewood Cliffs, New Jersey, 1963.

(11) Hörmander, L., "Linear Partial Differential Operators." Springer, Berlin, 1963.

(12) Treves, F., "Linear Partial Differential Operators with Constant Coefficients." Gordon and Breach, New York, 1966.

On the subject of topological tensor products and kernels:

(13) Grothendieck, A., Produits tensoriels topologiques et espaces nucléaires, *Mem. Amer. Math. Soc.* **16** (1955).

(14) Schwartz, L., Produits tensoriels topologiques. Séminaire, Paris, 1953–4.
   Schwartz, L., Distributions à valeurs vectorielles. Vols. I and II. *Ann. Inst. Fourier Grenoble* **VII** (1957), 1–141, **VIII** (1959), 1–207.

(Introductory reading)

(15) Goffmann, C., and Pedrick, G., "A First Course in Functional Analysis." Prentice-Hall, Englewood Cliffs, New Jersey, 1965.

(16) Robertson, A. P., and Robertson, W. J., "Topological Vector Spaces." Cambridge University Press, London and New York, 1964.

Index of Notation

1. Basic notations

$\mathbf{R}$, 9; $\quad \mathbf{Q}$, 9; $\quad \mathbf{C}$, 14; $\quad \mathbf{N}$, 29; $\quad \mathbf{N}^n$, 29; $\quad \mathbf{R}^n$, 54; $\quad \mathbf{R}^n_+$, 238; $\quad \mathbf{I}^{\mathbf{N}}$, $X^{\mathbf{N}}$, 551

$x = (x_1, \ldots, x_n)$, 85; $\quad \partial/\partial x_j$, 85; $\quad |x| = (x_1^2 + \cdots + x_n^2)^{1/2}$, 92; $|\zeta| = (|\zeta_1|^2 + \cdots + |\zeta_n|^2)^{1/2}$, 60; $\quad dx = dx_1 \ldots dx_n$, 99; $\quad \partial/\partial z_j$, 90; $\partial/\partial \bar{z}_j$, 231; $\quad \langle x, \xi \rangle$, 267

$p = (p_1, \ldots, p_n)$, 85; $\quad |p| = p_1 + \cdots + p_n$, 85; $\quad p! = p_1! \ldots p_n!$, 90; $\binom{p}{q}$, 249; $\quad q \leqslant p$, 249; $\quad (\partial/\partial x)^p$, 85; $\quad (\partial/\partial z)^p$, 90; $\quad X^p$, 91; $\quad P(x, \partial/\partial x)$, 247; ${}^t P(x, \partial/\partial x)$, 248; $\quad D$, ${}^t D$, 391

$\bar{A}$, 8; $\quad \mathring{A}$, 8; $\quad \mathscr{I} \geqslant \mathscr{I}'$, 10; $\quad \complement A$, complement of A, 158; $\quad X \setminus A$, complement of A with respect to X, 549; $\quad \text{meas}(K)$, 212; $\quad b(Y)$, $o(Y)$, 553; $I_A(\eta)$, 307

2. Abstract functional analysis

E/M, 15; $\quad \text{Im} f$, 16; $\quad \text{Ker} f$, 16; $\quad \mathscr{L}(E; F)$, 17; $\quad \mathscr{F}(E; F)$, 17; $\quad F^E$, 339; E^*, 17; $\quad E'$, 17, 35; $\quad \hat{E}$, 38; $\quad \bar{E}'$, 116; $\quad \langle x', x \rangle$ or $\langle x, x' \rangle$, 195; $\quad M^\perp$, 18; TVS, 20; $\quad LC$-space, LCS, 86; $\quad L(E; F)$, 35; $\quad \dim E$, 78; $\quad A^0$, 195; A^{00}, 362; $\quad \oplus$, 81; $\quad \overline{\{0\}}$, 32; $\quad U_p$, 61; $\quad \mathring{U}_p$, 61; $\quad \dot{p}$, 65; $\quad d(x, y)$, 70; $\quad \| \ \|$, 95; $\| \ \|_E$, 203; $\quad {}^t u$, 187, 199, 240; $\quad u^*$ (algebraic transpose), 17; $\quad u^*$ (adjoint), 252, 488; $\quad \sqrt{u}$, 489

$\mathfrak{S}$, 196, 335; $\quad E'_{\mathfrak{S}}$, 199; $\quad E'_\sigma$, E'_γ, 197; $\quad E'_c$, E'_b, 198; $\quad \sigma(E', E)$, 197; $\sigma(E, E')$, $\sigma(F, G)$, 362; $\quad \gamma(E', E)$, 197; $\quad c(E', E)$, $b(E', E)$, 198; $\quad \tau(E, E')$, 369; $\quad L_{\mathfrak{S}}(E; F)$, 336; $\quad L_\sigma(E; F)$, $L_\gamma(E; F)$, $L_c(E; F)$, $L_b(E; F)$, 337; $\quad E''$, 372; E''_b, 372

E_B, 370; $\quad E'_p$, 381; $\quad E'_{H'}$, 477; $\quad E_{U'}$, 477; $\quad \hat{E}_U$, 477; $\quad E_p$, $\hat{E}_p$, 510

$\mathscr{B}(E, F; G)$, $B(E, F; G)$, $\mathscr{B}(E, F)$, $B(E, F)$, 427; $\quad \mathscr{B}_\varepsilon(E'_b, F'_b; G)$, 428

$L_\varepsilon(E'_\sigma ; F'_\sigma)$, $L_\varepsilon(E'_\tau ; F)$, 428; Φ_x, 420

$E \otimes F$, 406; $u \otimes v$, 406; $f \otimes g$, 407; $S \otimes T$, 417; $\mathfrak{p} \otimes \mathfrak{q}$, 435; $E \otimes_\varepsilon F$, $E \otimes_\pi F$, 434; $E \widehat{\otimes}_\varepsilon F$, $E \widehat{\otimes}_\pi F$, 439; $u \widehat{\otimes}_\varepsilon v$, $u \widehat{\otimes}_\pi v$, 439; $\| \ \|_\varepsilon$, $\| \ \|_\pi$, 443; $\| \ \|_{\mathrm{Tr}}$, 480; $\mathrm{Tr}(\cdot)$, 486; Tr, 491; $L^1(E_U ; F_B)$, $L^1(E; F)$, 479; $J(E, F)$, 500

3. Functions and distributions

$| \zeta |_p$, $| \zeta |_\infty$, 59; $| \ |_{m,K}$, 86; ρ_ε, 156; $\mathrm{supp}\, f$, 103; $\mathrm{supp}\, T$, 255; δ_x, 217; $\mathrm{Re}\,\mu$, $\mathrm{Im}\,\mu$, 217; μ^+, μ^-, 220; $| \mu |$, 221; 1_K, 222; $\dot{f}$, $\dot{\phi}$, $\dot{T}$, 290; $f * g$, 278; $f * \mu$, 282; $T * \phi$, $\phi * T$, 287; $S * T$, 293; $\tau_a f$, $\tau_a T$, 296; $\check{\phi}$, $\check{f}$, 268, 305; $\mathscr{F}$, 238, 268; $\overline{\mathscr{F}}$, 268; $\mathscr{L}$, 277; $\hat{T}$, $\hat{T}(\xi)$, $\hat{T}(\zeta)$, 307, 318; $\triangleright$, 262; Y, $Y(x)$, 276; $\varDelta$, 324, 331; $\langle P, u \rangle$, 227; $\hat{\mu}$, 237;

4. Main spaces of functions

$\mathbf{C}[[X]]$, 25; $\mathbf{C}[X]$, 129; $\mathscr{P}^m_n$, 129, 227; $\mathscr{P}_n$, $\mathscr{Q}_n$, 227; $\mathscr{C}^k$, $\mathscr{C}^k(\Omega)$, 86; $\mathscr{C}^\infty$, $\mathscr{C}^\infty(\Omega)$, 86; $\mathscr{C}^\infty_c(K)$, 94; $\mathscr{C}^k(\overline{\Omega})$, 98; $\mathscr{C}^k_c(K)$, 131; $\mathscr{C}^k_c(\Omega)$, $\mathscr{C}^\infty_c(\Omega)$, 132; $\mathscr{C}^k(Y; E)$, $\mathscr{C}^k_c(Y; E)$, 413; $\mathscr{C}^\infty(\mathbf{R}^n)^\natural$, 528; $\mathscr{C}_\infty(\mathbf{R}^n)$, 110

l^p, 101; $\mathscr{F}^p$, $\mathscr{L}^p$, 101; $\mathscr{L}^p(X)$, $\mathscr{L}^p(X, dx)$, 104; L^p, 102; $L^p(\mathbf{R}^n)$, $L^p(X)$, $L^p(X, dx)$, 104; $L^p(\Omega)$, 210; $\mathscr{F}^\alpha(E)$, $\mathscr{L}^\alpha(E)$, $L^\alpha(E)$, 468; L^2, 123; $l^2(S)$, 122; l^∞, 101; $\mathscr{L}^\infty$, L^∞, 105; l_∞, 110; $\mathscr{B}^\infty$, 469; B^∞, 470; $L^p_c(\Omega)$, 132

s, 527; $\mathscr{S}$, $\mathscr{S}(\mathbf{R}^n)$, 4, 92, 267; $\mathscr{O}_M$, 275; $H(\Omega)$, 4, 89, 231; $H(\mathbf{C}^n)$, 152; $\mathrm{Exp}(K)$, $\overline{\mathrm{Exp}}(K)$, 232; Exp, 238

5. Main spaces of distributions

$\mathscr{D}'(\Omega)$, 226; $\mathscr{E}'(\Omega)$, 255; $\mathscr{D}'^m(\Omega)$, $\mathscr{D}'^F(\Omega)$, 258; $\mathscr{E}'_0$, $\mathscr{E}'^m_0$, 265; $\mathscr{S}'$, $\mathscr{S}'(\mathbf{R}^n)$, 271; $\mathscr{O}'_C$, 315; H', $H'(\mathbf{C}^n)$, 238, 266

$H^{p,m}(\Omega)$, 323; $H^{p,m}_0(\Omega)$, 324; $H^{p,-m}(\Omega)$, 324; H^s, $H^s(\mathbf{R}^n)$, 329; $H^{p,m}_{\mathrm{loc}}(\Omega)$, $H^{p,m}_c(\Omega)$, H^s_c, H^s_{loc}, $H^s_c(\Omega)$, $H^s_{\mathrm{loc}}(\Omega)$, 332

Norms and inner products in the Sobolev spaces

$\| \ \|_{p,m}$, $(,)_m$, 323; $\| \ \|_{p,-m}$, 326; $\| \ \|_s$, $(,)_s$, 330; operator U_s, 330

Subject Index